Agro-Waste Derived Biopolymers and Biocomposites

Scrivener Publishing
100 Cummings Center, Suite 541J
Beverly, MA 01915-6106

Publishers at Scrivener
Martin Scrivener (martin@scrivenerpublishing.com)
Phillip Carmical (pcarmical@scrivenerpublishing.com)

Agro-Waste Derived Biopolymers and Biocomposites

Innovations and Sustainability in Food Packaging

Edited by

Santosh Kumar

Dept. of Food Engineering and Technology, Central Institute of Technology Kokrajhar, Assam, India

Avik Mukherjee

Dept. of Food Engineering and Technology, Central Institute of Technology Kokrajhar, Assam, India

and

Vimal Katiyar

Dept. of Chemical Engineering, Indian Institute of Technology, Indian Institute of Technology Guwahati, Guwahati, Assam, India

Scrivener Publishing

WILEY

This edition first published 2024 by John Wiley & Sons, Inc., 111 River Street, Hoboken, NJ 07030, USA
and Scrivener Publishing LLC, 100 Cummings Center, Suite 541J, Beverly, MA 01915, USA

For more information about Scrivener publications please visit www.scrivenerpublishing.com.

Wiley Global Headquarters
111 River Street, Hoboken, NJ 07030, USA

For details of our global editorial offices, customer services, and more information about Wiley products visit us at www.wiley.com.

Library of Congress Cataloging-in-Publication Data

ISBN 978-1-394-17455-3

Cover image: Pixabay.Com
Cover design by Russell Richardson

Set in size of 11pt and Minion Pro by Manila Typesetting Company, Makati, Philippines

Printed in the USA

10 9 8 7 6 5 4 3 2 1

Contents

Preface xv

1 **Promising Agro-Wastes for Food Packaging** 1
L. Susmita Devi, Avik Mukherjee and Santosh Kumar
1.1 Introduction 2
1.2 Current Global Status of Agro-Wastes 4
1.3 Types of Agro-Wastes 5
1.3.1 Agro-Industrial Waste 5
1.3.2 Crop Residues 6
1.3.3 Animal Waste 7
1.3.4 Aquatic Waste 8
1.4 Extraction of Biopolymers from Agro-Wastes 8
1.4.1 Chemical Treatment 9
1.4.1.1 Acid Treatment 9
1.4.1.2 Alkali Treatment 10
1.4.1.3 Organic Solvent Treatment 10
1.4.2 Biological Treatment 11
1.4.3 Mechanical Processing 12
1.4.4 Thermochemical Processing 13
1.5 Extraction of Bioactive Compounds from Agro-Wastes 13
1.6 Conclusion and Future Perspectives 21
References 21

2 **Natural Fiber–Based Composite for Food Packaging** 33
Rishika Bora, Hemanta Chutia, Monika, Murchana Changmai, Charu Lata Mahanta, Vimal Katiyar and Tabli Ghosh
2.1 Introduction 34
2.2 Fiber Types 36
2.2.1 Natural Fibers 36
2.2.1.1 Plant-Based Fibers 37
2.2.1.2 Animal-Based Fibers 38

2.2.2 Man-Made Fibers 39
2.3 Plant Fiber–Based Composite for Food Packaging 39
2.3.1 Wood-Based Composite for Food Packaging 40
2.3.2 Stem/Bast-Based Composite for Food Packaging 40
2.3.2.1 Kenaf Fiber 40
2.3.2.2 Jute Fiber 41
2.3.2.3 Ramie Fiber 41
2.3.2.4 Hemp Fiber 41
2.3.3 Leaf-Based Composite for Food Packaging 42
2.3.3.1 Palm Fiber 42
2.3.3.2 Sisal Fiber 42
2.3.3.3 Pineapple Leaf Fiber 43
2.3.3.4 Banana Fiber 43
2.3.4 Seed/Fruit-Based Composite for Food Packaging 44
2.3.5 Grass-Based Composite for Food Packaging 44
2.3.5.1 Bamboo Fiber 44
2.3.5.2 Rice Husk and Wheat Straw 45
2.4 Animal Fiber–Based Composite for Food Packaging 46
2.4.1 Silk-Based Composite for Food Packaging 46
2.4.2 Wool-Based Composite for Food Packaging 47
2.4.3 Other Animal Fiber-Based Composite 50
2.5 Nanomaterials from Natural Fiber 51
2.6 Natural Fiber–Based Composite for Circular Economy 57
2.7 Conclusion and Future Perspective 59
Acknowledgment 59
References 59

3 Corncob Waste for Food Packaging 69
Antony Catherine Flora Louis and Sivakumar Venkatachalam
3.1 Introduction 70
3.2 Isolation of Cellulose from Corncob 72
3.2.1 Pretreatment of Corncob Waste Residues 73
3.2.1.1 Acidic Hydrolysis 73
3.2.1.2 Alkaline Treatment 74
3.2.2 Bleaching Process 75
3.2.3 Extraction of Nanocellulose 76
3.3 Isolation of Hemicellulose from Corncob 77
3.4 Microbial Biosynthesis of Polyhydroxy Butyrate (PHB) from Corncobs 78
3.5 Biopolymers-Based Food Packaging Reinforced with Corncob Fibers 80

3.6 Hybrid Nanocomposite of Corncob for Food Packaging 83
3.7 Conclusion and Future Perspectives 86
References 86

4 Coir Fibers for Sustainable Food Packaging 91
Sweety Kalita, Manashi Das Purkayastha, Avik Mukherjee and Santosh Kumar
4.1 Introduction 92
4.2 Coir Fibers as Reinforcement Material for Synthetic Polymers 93
4.2.1 Polyethylene-Based Composites Reinforced with Coir Fibers 93
4.2.2 Polypropylene-Based Composites Reinforced with Coir Fibers 95
4.2.3 Polyester-Based Composites Reinforced with Coir Fibers 96
4.3 Coir Fibers as Reinforcement Material in Biopolymers 98
4.3.1 Composites of Coir and Polylactic Acid (PLA) 98
4.3.2 Composites of Coir and Protein 99
4.3.3 Composites of Coir with Starch 101
4.3.4 Hybrid Composites of Coir 102
4.4 Biodegradable Package/Container from Coconut Coir 103
4.5 Conclusion and Future Perspective 107
References 108

5 Sugarcane Bagasse for Sustainable Food Packaging 115
Sweety Kalita, Avik Mukherjee and Santosh Kumar
5.1 Introduction 115
5.2 Chemical Composition and Characteristics of Sugarcane Bagasse (SB) 117
5.3 Cellulosic and Hemicellulosic Fractions of Sugarcane Bagasse 118
5.4 Pretreatment Approaches for SB 121
5.4.1 Physical Pretreatments 122
5.4.1.1 Mechanical Pretreatment 122
5.4.1.2 Microwave Pretreatment (MWP) 123
5.4.1.3 Ultrasound Pretreatment (USP) 124
5.4.2 Chemical Pretreatments 125
5.4.2.1 Acidic Pretreatment (AP) 125
5.4.2.2 Alkaline Treatment 125
5.4.2.3 Ionic Liquids Pretreatment 126
5.4.3 Physiochemical Pretreatment 127
5.4.3.1 Organosolv Pretreatments 127

5.4.3.2 Steam Explosion Pretreatment 128
5.4.3.3 Hot Water Pretreatments 129
5.4.4 Biological Treatment 129
5.5 Sugarcane Bagasse in Biopolymer Matrix as Reinforcement Filler 130
5.6 Food Containers and Trays Made From SB 135
5.7 Conclusion and Future Perspective 139
References 139

6 Husk and Straw of Cereals Grains for Sustainable Food Packaging 153
Pooja Baiju, Maria Jose Mukkadan and Preetha Radhakrishnan
Abbreviations 154
6.1 Introduction 155
6.2 Extraction and Purification of Cellulose from Husk and Straw 156
6.2.1 Pretreatment Methods 157
6.2.2 Purification Methods 158
6.2.2.1 Alkali Treatment 159
6.2.2.2 Bleaching 159
6.2.3 Extraction Methods 160
6.3 Cellulose Nanocrystals 160
6.3.1 Modifications and Functionalization of CNC 161
6.3.2 Applications of CNC in Packaging Films 163
6.4 Use of Cellulose and Its Derivatives in Food Packaging 164
6.4.1 Cellulose Ethers 164
6.4.1.1 Ethyl Cellulose (EC) and Methyl Cellulose (MC) 164
6.4.1.2 Carboxymethyl Cellulose (CMC) 166
6.4.1.3 Hydroxyethyl Cellulose (HEC) 167
6.4.1.4 Hydroxypropyl Cellulose (HPC) 167
6.4.2 Cellulose Esters 168
6.4.2.1 Cellulose Acetate (CA) 168
6.4.2.2 Cellulose Nitrate (CN) 168
6.4.2.3 Cellulose Sulfate (CS) 169
6.5 Paper-Based Package from Straw and Husk 170
6.6 Tableware and Food Containers from Straw and Husk 171
6.6.1 Compostable and Biodegradable Tableware and Containers 171
6.6.2 Wheat Straw Plastic 174
6.7 Conclusion and Future Perspective 174
References 175

7 **Sericulture Waste for Edible Films and Coating of Fruits and Vegetables** 187
Sweety Kalita, Avik Mukherjee and Santosh Kumar
7.1 Introduction 187
7.2 Sericulture Wastes 189
7.3 Extraction and Purification of Silk Protein/Fibroin 190
7.4 Silk Protein–Based Active Food Packaging 192
7.4.1 Silk Protein/Fibroin–Based Active Films 192
7.4.2 Silk Protein/Fibroin–Based Edible Coating 196
7.5 Toxicological and Food Allergy Assessment of Silk Protein/Fibroin 197
7.6 Conclusion and Future Perspective 200
References 200

8 **Functional Agents from Agro-Waste for Active and Intelligent Food Packaging** 205
Lokesh Kumar, Ram Kumar Deshmukh and Kirtiraj K. Gaikwad
8.1 Introduction 205
8.2 Functional Agents in Active and Intelligent Packaging 208
8.2.1 Polyphenolic Compounds from Agro-Waste 209
8.2.1.1 Polyphenolic Compounds from Apple Peel 209
8.2.1.2 Polyphenolic Compounds from Citrus Fruit 210
8.2.1.3 Polyphenolic Compounds from Potato Peel 211
8.2.1.4 Polyphenolic Compounds from Pineapple Peel 212
8.2.1.5 Polyphenolic Compounds from Mango Kernel 212
8.2.1.6 Polyphenolic Compounds from Grape 213
8.2.1.7 Polyphenolic Compounds from Pomegranate Peel 213
8.2.1.8 Polyphenolic Compounds from Banana Peel 214
8.2.1.9 Polyphenolic Compounds from Corncob 214
8.2.1.10 Polyphenolic Compounds from Wheat Straw 215
8.2.2 Antioxidants from Agro-Waste 216
8.2.2.1 Antioxidant Compounds from Apple Peel 216
8.2.2.2 Antioxidant Compounds from Citrus Fruit 217
8.2.2.3 Antioxidant Compounds from Potato Peel 217
8.2.2.4 Antioxidant Compounds from Pineapple Peel 218
8.2.2.5 Antioxidant Compounds from Mango Kernel 218
8.2.2.6 Antioxidant Compounds from Grape Pomace 218
8.2.2.7 Antioxidant Compounds from Pomegranate Peel 219
8.2.2.8 Antioxidant Compounds from Banana Peel 219

8.2.2.9 Antioxidant Compounds from Corncob 219
8.2.2.10 Antioxidant Compounds from Wheat Straw 220
8.2.3 Antimicrobials Compounds from Agro-Waste 220
8.2.3.1 Antimicrobials Compounds from Pomegranate Peel 221
8.2.3.2 Antimicrobials Compounds from Grape Pomace 221
8.2.3.3 Antimicrobials Compounds from Mango Kernel 222
8.2.3.4 Antimicrobials Compounds from Citrus Fruit 222
8.2.3.5 Antimicrobials Compounds from Banana Peel 223
8.2.3.6 Antimicrobials Compounds from Pineapple Peel 223
8.2.4 Biobased Indicators 223
8.3 Active and Intelligent Agents in Biopolymer-Based Food Packaging 225
8.3.1 Oxygen and Carbon Dioxide Indicators 225
8.3.2 Moisture or Humidity Indicator 228
8.3.3 pH Indicators 228
8.3.4 Temperature Indicator 229
8.3.5 Specific Chemical Indicator 229
8.4 Conclusion and Perspective 230
References 231

9 Starch from Agro-Waste for Food Packaging Applications 241
Shashank T. Mhaske, Jyoti Darsan Mohanty and Pavan Y. Borse
9.1 Introduction 242
9.2 Starch from Agro-Waste 243
9.2.1 Tuber Wastes 244
9.2.2 Seed Wastes 245
9.3 Modifications in Starch for Food Packaging 247
9.3.1 Chemical Modification 248
9.3.1.1 Acetylation 248
9.3.1.2 Acid Chloride Modification 249
9.3.1.3 Octenyl Succinic Anhydride (OSA) Modification 250
9.3.1.4 Hydropropylation 250
9.3.1.5 Oxidation of Starch 251
9.3.1.6 Cross-Linking of Starch 252

9.3.2 Physical Modification of Starch to Thermoplastic Starch (TPS) 252
9.4 Starch-Based Composite, Nanocomposite, and Hybrid Films 253
9.4.1 Starch-Based Blends 253
9.4.2 Starch-Based Composite and Nanocomposite 255
9.5 Food Packaging Applications 256
9.6 Conclusion and Perspectives 259
References 259

10 Chitosan from Agro-Waste for Food Packaging Applications 267
Mekala Pavani, Sushil Kumar Singh and Poonam Singha
List of Abbreviations 268
10.1 Introduction 268
10.2 Sources of Chitosan 269
10.2.1 Agro-Waste 269
10.2.2 Sources Other Than Agro-Waste 272
10.2.2.1 Terrestrial Insects 272
10.2.2.2 Microbial Sources 273
10.2.2.3 Marine Sources 273
10.3 Chitosan Extraction 274
10.4 Chitosan and Its Functional Properties 275
10.4.1 Antimicrobial Activity 276
10.4.2 Antioxidant Properties 277
10.4.3 Film-Forming Ability 277
10.4.4 Solubility 278
10.5 Chitosan-Based Composites and Nanocomposites 278
10.5.1 Coating Formulations 278
10.5.2 Packaging Films 279
10.6 Food Packaging Applications 281
10.6.1 Fish and Meat Products 281
10.6.2 Fruits and Vegetables 283
10.7 Conclusion and Future Perspectives 285
References 285

11 Biodegradable Synthetic Poly(Lactic Acid) (PLA) for Food Packaging Application 295
Shikha Sharma, Tabli Ghosh, Neha Mulchandani and Vimal Katiyar
11.1 Introduction 296
11.2 Synthesis of PLA 300
11.3 Properties of PLA 302

11.3.1 Composites of PLA 302
11.3.2 Stereocomplex of PLA 304
11.3.2.1 Stereocomplex PLA 305
11.3.2.2 Additives to Enhance Stereocomplexation of PLA 306
11.4 Food Packaging Applications of PLA 308
11.5 Conclusion and Future Prospects 314
References 314

12 Pectin from Agro-Waste Residues for Food Packaging 319
Kona Mondal and Vimal Katiyar
12.1 Introduction 320
12.2 Structure and Classification of Pectin 322
12.2.1 Structure of Pectin 322
12.2.2 Classifications of Pectin 324
12.3 Agro-Waste as Sources of Pectin 325
12.4 Techniques for Pectin Extraction from Agro-Waste 327
12.4.1 Hydrothermal Extraction 328
12.4.2 Ultrasound-Assisted Extraction 329
12.4.3 Hydrodynamic Cavitation 331
12.4.4 Microwave-Assisted Extraction 332
12.4.5 Subcritical Water Extraction 333
12.4.6 Enzyme-Assisted Extraction 333
12.5 Food Packaging Applications of Pectin-Based Films and Coatings 334
12.5.1 Pectin-Based Active Films 335
12.5.1.1 Film Fabrication by Casting 335
12.5.1.2 Film Fabrication by Spraying 336
12.5.1.3 Film Fabrication by Extrusion 336
12.5.2 Pectin-Based Active Coatings 337
12.5.2.1 Preservation of Fruits and Vegetables 337
12.5.2.2 Preservation of Meat, Poultry, and Aquatic Animal Products 338
12.6 Conclusion and Perspective 341
References 341

13 Cellulosic Nanomaterials and Its Derivatives from Agro-Waste for Food Packaging Applications 349
Rahul Ranjan, Rohit Rai, Vikash Kumar, Smruti B. Bhatt and Prodyut Dhar
13.1 Introduction 350

13.2 Cellulose Structure and Its Nano-Derivatives 353
13.2.1 Hemicellulose 355
13.2.2 Xylan 356
13.2.3 Mannans (Mannoglycans) 358
13.2.4 Xyloglucan 358
13.2.5 β-Glucan 359
13.2.6 Holocellulose 359
13.2.7 Bacterial Cellulose 360
13.3 Agro-Waste as Source of Cellulose 361
13.4 Extraction of Cellulose from Agro-Waste 362
13.5 Cellulose-Derived Biopolymers 366
13.5.1 Carboxymethyl Cellulose (CMC) 366
13.5.2 Cellulose Acetate (CA) 369
13.5.3 Cellulose Nanofiber (CNF) 371
13.6 Food Packaging Applications of Cellulose and Its Derivatives 373
13.7 Conclusion and Perspective 376
References 377

14 Biodegradability of Biopolymers 389
Anamika Nayak and Debjani Dutta
14.1 Introduction 389
14.2 Biodegradability of Traditional Food Packaging 390
14.3 Biodegradability of Biopolymers 392
14.3.1 Cellulose-Based Biopolymers 392
14.3.2 Starch-Based Biopolymers 392
14.3.3 Protein-Based Biopolymers 393
14.3.4 Polylactide (PLA) 394
14.3.5 Polyhydroxybutyrate (PHB) 395
14.3.6 Polyhydroxybutyrate-Co-Valerate (PHBV) 396
14.4 Mechanisms and Pathways of Biopolymer Degradation 396
14.4.1 Abiotic Degradation 397
14.4.2 Biodeterioration and Depolymerization 399
14.4.3 Biofragmentation 401
14.4.4 Assimilation and Mineralization 403
14.5 Biodegradation of Biopolymer-Based Food Packaging 404
14.5.1 Microbial Degradation 404
14.5.2 Enzymatic Degradation 405
14.6 Conclusion and Perspective 406
References 407

15 Migration Concerns of Biopolymer-Based Food Packaging 421
Vikramsingh Thakur and Bhabani K. Satapathy
15.1 Introduction 422
15.2 Migration Concerns of Biopolymer-Based Nanocomposite 423
15.2.1 Starch-Based Nanocomposites 424
15.2.2 PLA-Based Nanocomposites 426
15.2.3 Polyhydroxyalkanoates (PHA)-Based Nanocomposites 430
15.3 Migration of Oligomers From Biopolymers in Contact with Food 432
15.4 Migration of Nanomaterials 436
15.4.1 Silver Nanoparticle–Incorporated Biopolymer Composite Films 437
15.4.2 Clay-Based Biopolymer Films 438
15.4.3 Other Nanomaterials as Fillers in Biopolymer Films 439
15.5 Biopolymer-Based Nanocomposite Films and Coatings 441
15.6 Effect of Polymer and Migrant Geometry on Migration 442
15.7 Diffusion Modeling of Migration in Food Packaging 443
15.8 Conclusion and Future Perspectives 445
References 446

Index 451

Preface

The concept of this book formed sometime in early 2022, in light of tremendous global concerns associated with the indiscriminate use of synthetic plastics, and their impacts on the environment and human health. Given these concerns, agro- and food waste management and the valorization to obtain bioplastics for innovative food packaging applications have been of increased importance in the pursuit of a sustainable circular economy. The decision to develop a book on this emerging area of innovations and applications was compelled by the need for a sustainable future.

This book serves as a complete, systematic, comprehensive account of the contemporary developments in the area of novel and environment-friendly valorization of agro- and food wastes into value-added products like biodegradable polymer and active functional agents for food packaging applications. Also, the book describes the hurdles and challenges in the commercialization of these novel biopolymer-based materials, including their composites, their applications, safety, and legal ramifications. It is a unique resource material for educators and researchers, students, entrepreneurs, professionals in agriculture, food manufacturing, handling, and packaging industries, as well as other interested readers.

This book consists of fifteen chapters covering different aspects of agro- and food waste utilization, the development of biodegradable polymers, and their composites for sustainable food packaging applications.

The first thirteen chapters detail the processing of various agro- and food wastes of plant and animal origin to synthesize different biopolymers, such as starch, cellulose, chitosan, silk proteins, pectin, etc., and their applications for the fabrication of sustainable food packaging materials and composites that are attractive alternatives to synthetic plastic packaging. These chapters also summarize the effectiveness of these biopolymers and their composites in developing active films and edible coatings for shelf-life extension and preservation of perishable foods.

Chapter 14 discusses biodegradability, including analyses of various biodegradation reactions, such as depolymerization, mineralization, biochemical, and abiotic degradation both in soil and aquatic environments. On the other hand, the last chapter of the book addresses the concerns associated with the possible migration of components or additives from these biodegradable packaging into packaged food items.

It has been a great pleasure and a privilege to collaborate with the contributing authors affiliated with different academic institutes and universities of international repute in India and abroad. This book will benefit readers with interests in any one or more of the entire spectra of food, agriculture, waste management, agro- and food waste management, and valorization. The book will also have tremendous potential to appeal to industries, businesses, and government bodies. We could not have achieved success in our efforts without the support, cooperation, and understanding of our peers, friends, and families, which stems from our unwavering dedication and commitment to this book project.

Santosh Kumar
Avik Mukherjee
Vimal Katiyar
March 2024

1

Promising Agro-Wastes for Food Packaging

L. Susmita Devi, Avik Mukherjee and Santosh Kumar*

Department of Food Engineering and Technology, Central Institute of Technology Kokrajhar, Kokrajhar, BTR, Assam, India

Abstract

The agricultural sector produces a substantial amount of waste, typically left unused, and these wastes are a growing risk to human health, food security, and the environment. Transforming these waste materials into valuable products holds the potential to be advantageous for both the environment and significant cost savings. Agro- and food wastes originating from various sources have the potential to serve as a primary resource for creating value-added products, extracting food-related compounds, and producing biopolymers for sustainable food packaging. The increasing consumer demand has spurred the development of new eco-friendly food packaging, leading the packaging industry to focus on creating sustainable, renewable, biodegradable, and environment-friendly packaging materials from agro-based resources. Moreover, these agro-based materials hold immense potential for replacing petrochemical-based plastics in the packaging sector due to their abundant availability and biodegradable nature. Various agro-wastes, and their potential for extraction of different natural biopolymers, and bioactive compounds are discussed in detail in this chapter. In addition, the applications of agro-waste–based biopolymers and bioactive compounds in biodegradable sustainable food packaging systems have also been discussed. The chapter also delves into future considerations, particularly in the realm of enhancing the effectiveness and efficiency of agro-waste packaging systems, while simultaneously addressing the inherent production challenges. This article holds particular significance within the context of the biorefinery concept, which is dedicated to the conversion of agricultural waste into a diverse range of valuable products.

**Corresponding author*: s.kumar@cit.ac.in

Santosh Kumar, Avik Mukherjee and Vimal Katiyar (eds.) Agro-Waste Derived Biopolymers and Biocomposites: Innovations and Sustainability in Food Packaging, (1–32) © 2024 Scrivener Publishing LLC

Keywords: Solid waste, sustainable packaging, biopolymers, organic compounds, eco-friendly

1.1 Introduction

Rapid urbanization, population growth, and industrialization generate a vast amount of agro-waste every year [1]. Agro-wastes are the residue left after processing and producing different agro-products, such as fruits, vegetables, muscle foods, and dairy products that are liquids, slurries, or solids in nature [2]. These wastes are generally left untreated or unutilized and are either burned on-site, disposed of by dumping, or unplanned landfilling, leading to climate change by increasing greenhouse gas emissions causing a threat to the environment and human health [3, 4]. The lack of efficient waste management techniques creates a huge challenge for mankind to deal with such massive food and agro-wastes. Most of the food wastes are generated during production, harvesting, transportation, industrial processing, and consumption [5]. Agro-wastes have the potential to be used as raw materials for value-added products, renewable energy production, extraction of food compounds, and packaging material manufacturing [6]. Agro-industry byproducts are also composed of numerous bioactive compounds such as carbohydrates, proteins, and lipids that can be extracted and used to produce biodegradable food packaging [7]. Agro- and food wastes are rich in nutrients and organic compounds that can be processed to recover valuable products including biopolymers [8]. The biopolymers are directly extracted either by chemical synthesis or by genetically modified organisms from agro-industrial wastes and can be converted into biodegradable packaging materials. However, the treatment methods or downstream processing of such waste are usually complex, expensive, and time-consuming and cause secondary pollutants due to their heterogeneous nature [3]. On the other hand, waste generated from the food industries is extremely nutritious in nature, making substrates for microbial fermentation and enzyme production [9]. The oil industries also produce a large number of residues known as oil cakes that contain high concentrations of fats, oils, greases, suspended solids, and dissolved solids, contributing to air, water, and solid waste pollution [8]. Such environmental problems can be reduced or minimized if these wastes are utilized for the extraction of biopolymers and other valuable products.

Food packaging serves as a vital part of the food system that protects food and food products from degradation and damage, maintains food safety and hygiene, enhances shelf life, and reduces food waste and food losses [10]. A wide variety of synthetic plastic materials are being used in food packaging that raise environmental pollution concerns [11]. The food packaging sector consumes the largest amounts of synthetic plastics globally, and it is the primary contributor to synthetic plastic waste [12]. Waste generated from food packaging accounts for about 66% of total packaging waste, and thus, there is a growing prerequisite for sustainable and biodegradable packaging materials that have the desired physical, mechanical, and barrier properties for food packaging applications. Biopolymers derived from bioresources, i.e., agro- and food wastes have shown potential for use as food packaging (Figure 1.1). Extraction and purification of the biopolymers from agro-wastes to produce a food packaging material can replace petrochemical-based synthetic plastics in the food packaging industry [1]. The growing market for food delivery services is expected to boost packaging material consumption in the future, and a sustainable approach to materials development requires agro- and food waste–based materials to fill the gap between industrial supply and consumer demand [13]. The major benefits of agro-waste valorization include nontoxicity, improving ecosystems, encouraging circular economy, increasing employment, and biocompatibility with other materials [14, 15]. This chapter presents a summary of the potential uses of agro-wastes for the extraction of biopolymers for technological applications. Furthermore, the chapter summarizes existing information concerning various waste types of

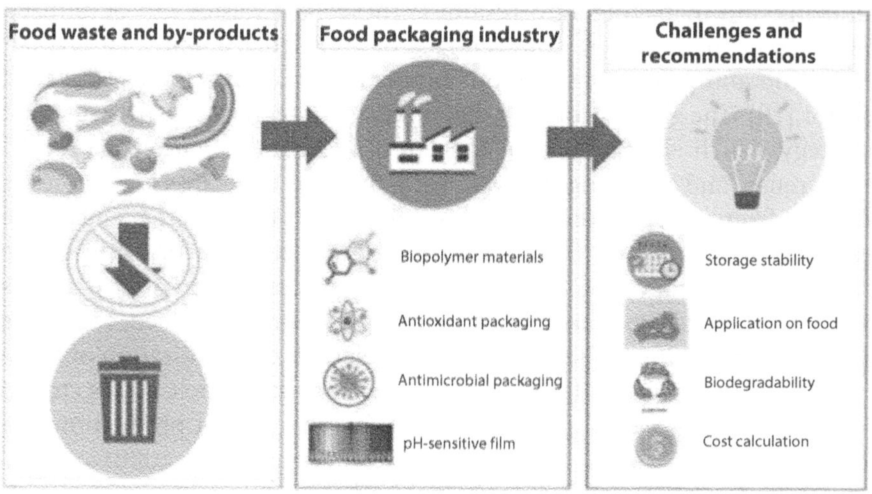

Figure 1.1 Utilization of agro- and food wastes in sustainable food packaging [16].

agro-wastes and their treatment for extracting valuable compounds for applications in the sustainable food packaging sector.

1.2 Current Global Status of Agro-Wastes

Solid waste generation increases with the growing population, urbanization, industrialization, living standards, disposable incomes, and consumption of goods and services [17]. In the last century, China, India, and Africa have experienced significant increases in agro-waste generation alongside rapid population growth and economic development. In India, agro-waste remains the predominant form of solid waste, with annual production ranging from 350 to 990 million tonnes (Mt). India produces more than 130 Mt of paddy straw each year, of which half is used as animal feed and the other half is dumped, making it the second-largest producer of agricultural waste in the world after China [18]. Approximately 34% of Asia's biomass residues are burned in the field, and open-field burning of crop residues contributes significantly to greenhouse gas emissions in countries like China and India [19]. Managing these solid wastes is a worldwide problem because waste collection, transportation, segregation, treatment, and disposal are complex processes and have detrimental effects on the environment. Additionally, inappropriate solid waste management results in several environmental pollutions and health problems including waterborne diseases, respiratory illnesses, soil pollution, and groundwater contamination [20]. Thus, it is important to identify efficient ways of utilizing and managing agro-residues. These agro-wastes can be utilized in different food sectors, however, lack of knowledge about their characterization and processing methods hinders their applications in the food sector [19]. The global economy is planning an integral waste management strategy, which is based on the concept of biorefinery and on the reduction, reuse, and recycling of waste with the goal of recovering resources derived from wastes as renewable resources [21]. Agriculture being one of the largest sectors with the highest waste production becomes an essential part of the bioeconomy, which promotes the conversion of agro-wastes into value-added products such as bioenergy, bioproducts, feed, and foods [22]. The increasing global population accelerates over utilization of natural resources and generates a large amount of waste, which is expected to double by 2050 to 4 billion tonnes at the current rate of production and consumption [23]. Agro-residues obtained after harvesting grains are used in different applications such as animal feed, raw material for animal bedding, loose-fill insulation, fuels, composites, bioplastics, biorefineries, and as raw material in the paper and pulp industry, as an absorbent of uranium residues, and as a substrate for solid state for microbial growth [1, 24].

The global population has increased from 3.7 billion to 8.0 billion during 1970–2022, and it is predicted to reach 8.5 billion, 9.7 billion, and 11.2 billion by 2030, 2050, and 2100, respectively. As a result, ensuring food security in the future would be a formidable challenge, and fulfilling the high demand of a growing population, livestock, and crop production has to increase significantly, which will simultaneously increase the generation of agricultural waste [25]. Currently, the global economy relies on biorefineries and the concept of reusing, recycling, and reducing waste with the aim of recovering valuable materials from waste [21]. There have been various studies from all around the world that have confirmed the importance of food waste reduction and its valorization. In recent years, various published literature discussed on utilization of agro-waste for various applications in food industries [26]. In addition, promoting the conversion of waste into value-added products such as food, feed, and bio-based products can also contribute to the development of new green markets and jobs [27]. Surendra *et al.* (2022) studied the bioconversion of agro-waste into fuel, feed, fertilizer, and biobased products [28]. Pomegranate wastes such as peel, seed, and pomace generated from the pomegranate juice and jam industry have high antimicrobial, antimutagenic, and free radicals scavenging properties that increased interest in this byproduct for the development of active food packaging films [29, 30]. Bagasse, a byproduct obtained from the sugar industry during juice extraction from sugarcane, also has several applications including the production of ethanol, paper, hydrogen, and bio-oil [31]. John *et al.* (2023) studied the feasibility of utilizing agricultural biomass such as wheat bran and rice bran to produce bioethanol by enzymatic hydrolysis (EH) followed by saccharification [32]. The waste generated after the processing of cashew nut is reportedly used in a variety of products, including adhesive resins, bioethanol, biodiesel, surface coating agents, dyes, insecticides, larvicides, anti-termite, and rubber [25, 33, 34].

1.3 Types of Agro-Wastes

1.3.1 Agro-Industrial Waste

Agro-industrial residues are wastes generated by the food and agriculture industries. The efficient transformation of these wastes has become a key objective in recent years, in which biodiesel production is getting a lot of attention. It is also possible to obtain new materials, chemicals, and valuable products from agro-industrial wastes [35, 36]. Agro-industrial wastes can be categorized into three main groups, namely, compostable wastes, non-compostable wastes, and hazardous wastes. Compostable wastes are recyclable agricultural residues generated during the processing of agricultural crops into various food products, for example, molasses, bagasse, peels, hulls, husks, seeds, slurry, and cakes. These wastes contain different compositions of cellulose, hemicellulose, lignin, moisture, ash, carbon, nitrogen, and many other compounds [8]. Non-compostable wastes include those produced from farm construction, farm machinery, and livestock protection that are usually nonrecyclable agro-industrial waste, whereas hazardous agro-industrial wastes such as acids, fertilizers, chemically contaminated water, foods, and other materials pose very serious problems if they are not managed properly [37]. Processing of industrial wastes or compostable waste can be economically beneficial, as they can be utilized as substrates for obtaining biopolymers having a wide array of functional properties including antimicrobial and antioxidant properties, biodegradability, and nontoxicity. One of the most effective options is to use these residues as a source of fibrous biomass, which can be used economically to produce paper and bioplastics at a far lower cost that can solve disposal issues and environmental problems [1]. Wastes like pumpkin seeds and peels, grapefruit peels, sunflower heads, sisals waste, pomegranate peels, eggplant peels, and orange peels are found to be used in developing biopolymers such as alginate and pectin [38, 39]. For example, garlic and onion peels contain flavonoids that have antioxidant, antimicrobial, and UV-resistant properties. Agro-industrial waste can also be utilized as low-cost raw materials for the microbial synthesis of valuable primary and secondary metabolites [40, 41].

1.3.2 Crop Residues

Harvest waste, commonly referred to as crop residue, encompasses both field residues left in agricultural fields after crop harvesting and process residues remaining after the crop has been transformed into a usable resource [42]. Crop residue includes leaf, stalk, root, stem, seed pod, and straw waste. Some residues are either used as animal feed or left on the ground to avoid erosion, but these approaches are less efficient and may encourage pest infestations [1]. As per the Ministry of New and Renewable Energy (MNRE), India produces an average of 500 Mt of crop residue annually. The MNRE report also indicates that a significant portion of this crop residue is effectively utilized as animal feed and as a fuel source for both household and industrial applications. Nonetheless, there remains an excess of 140 Mt, with 92 Mt being burned each year [42, 43]. Crop residues constitute a valuable component of agricultural waste, possessing an organic composition that can be harnessed for the betterment of society [42]. These crop wastes are also the most widely available and inexpensive organic waste, and they can be easily converted into a variety of high-value goods [44]. Such wastes are primary sources of polysaccharides that can be used as raw materials to produce a variety of valuable products such as biopolymers, ethanol, and biochemicals [45]. Different studies have shown that rice stubble residues can be converted into carboxymethyl cellulose, which is used as flexible packaging material. Chollakup *et al.* (2020) developed rice straw–based paper for antibacterial food packaging applications by coating them with pomelo peel extract [46]. Rice hulls and their nanofibers were used as reinforcement fillers in films for improving mechanical, optical, and barrier properties, demonstrating the potentiality of milling waste in biopolymer-based packaging [47]. Safont *et al.* (2018) used rice husk and almond shells, lignocellulosic wastes, in poly(3-hydroxybutyrate)/fiber composites as high-performance fillers for improving film properties and sustainable way to valorize crop residues [48].

1.3.3 Animal Waste

Animal wastes are produced during meat processing, animal husbandry (such as wash water, urine, dung, and waste feed), poultry waste (such as bedding material, feathers, droppings, and spilled feed), slaughterhouse waste (flesh, blood, hides, hair, and bones), etc. [22, 49]. India's livestock population was approximately 536.76 Mt, as reported in the 20th Livestock Census in 2019. This population generates around 3 Mt of livestock waste each year [50, 51]. Animal wastes represent substantial environmental challenges that, if not managed appropriately, can result in contamination of both ground and surface water sources [52]. Inappropriate discarding of such waste impacts air and water quality and also upsurges the risk of pathogenic microbes for human health [53]. Moreover, it has the potential to serve as a valuable source for biomass-based conversion processes, particularly in the production of biopolymers, bioenergy, and biofertilizers [54]. Wastes and byproducts from slaughterhouses are known for their rich content of proteins [55]. The extraction of proteins from byproducts primarily involves pretreatment, extraction, and downstream processing [56, 57]. Animal-based proteins are highly nutrient-dense and offer a range of beneficial properties such as film-forming, water- and oil-holding capacity, gelling, and emulsifying properties [58]. Proteins of animal sources like keratin, collagen, myofibrillar, and gelatin proteins, are widely used in biodegradable food packaging films [59].

1.3.4 Aquatic Waste

The fishery and aquaculture play an important role in food nutrition and security. Fish and seafood processing generate a large amount of solid waste from activities like beheading, de-shelling, degutting, removing fins and scales, and filleting [60, 61]. In general, fish byproducts are viscera, muscle tissues, carcasses, heads, fins, skin, scales, and bones, which normally weigh between 50 and 75% of the fresh weight of the fish depending on the species [62]. Fish industry byproducts are also used in animal feed to provide low-value nutrients; however, most are landfilled or incinerated, posing environmental and economic hazards [63, 64]. As fish byproducts contain high microbial load and endogenous enzymes, they are susceptible to rapid degradation if not properly processed or stored posing significant environmental and food-technology challenges. There are two types of aquatic waste: those that can easily degrade with a lot of enzymes, like viscera and blood, and those that are more stable (bones, heads, and skin) [64]. The timely collection and treatment of such wastes are critical to their quality, which makes them suitable raw materials for high-value compounds, like enzymes, lipids, gelatin, collagen, hydrolysates, and bioactive peptides [65]. For example, collagen is mainly obtained from fish skin, bones, fins, and scales, as well as jellyfish and starfish connective tissue. Fish skin contains about 70 to 80% collagen by dry weight [66, 67]. On the other hand, gelatin is a denatured protein formed during the process of partial hydrolysis of collagen and subsequent thermal treatment, which is composed mainly of proteins and polypeptides with different molecular weights [68]. Zavareze *et al.* (2014) developed a biodegradable film by extracting myofibrillar and protein isolate residue of Whitemouth croaker (*Micropogonias furnieri*) that achieves high tensile strength and low water vapor permeability at the lowest protein concentration; however, water vapor permeability was higher for films with higher protein concentrations [69]. Similarly, Nie *et al.* (2017) developed an edible film from silver carp myofibrillar proteins and tannins using alkaline and heat treatment methods [70]. The most valuable products obtained from marine waste are fishmeal and fish oil.

1.4 Extraction of Biopolymers from Agro-Wastes

Most of the agricultural wastes are lignocellulosic, which contains high amounts of polysaccharides such as cellulose, hemicellulose, lignin, and other nutrients such as proteins, lipids, and polyphenols [40]. Many lignocellulosic biomasses including agricultural wastes are difficult to degrade due to their composition and chemical structure, and to overcome this complication, a pretreatment process is used [71, 72]. During this process, agro-waste complex molecular structures are broken down into simpler monomers such as cellulose, hemicellulose, and lignin. A major goal of agricultural waste pretreatment is to reduce the energy required during transformation and cost-cutting and produce sugars directly from biomass [73, 74]. It is challenging to utilize agro-waste effectively for certain purposes due to its extremely heterogeneous nature. The biopolymers are extracted from agro-wastes commonly through chemical, biological, mechanical, and thermomechanical treatments, which are briefly discussed below.

1.4.1 Chemical Treatment

1.4.1.1 Acid Treatment

Acid treatment is an extensively used pretreatment process applied to the manufacturing of value-added compounds from lignocellulosic biomass [75]. In the pretreatment process, acids such as HCl, H_2SO_4, phosphoric acid, HNO_3, ethanoic acid, and formic acid are often utilized as reagents for the degradation of cellulose and hemicellulose and promote the release of fermentable sugar. Sulfuric acid treatment has typically proven to be more efficient than other acid treatments; however, strong acids can also cause significant loss of fermentable sugar due to an excessive breakdown of the complex substrate [76]. From an economic viewpoint, strong acids tend to negatively impact the anaerobic digestion (AD) process due to corrosion and the requirement of additional chemicals for neutralization [77]. Therefore, to reduce such risks of toxicity, corrosion, and handling, diluted acids (<4% w/w) are mostly used coupled with higher temperatures treatment (>100°C) in acid pretreatment [76]. According to Ali and Tindyala (2015), pretreatment of rice husk with H_2SO_4 significantly enhances the thermal conversion of rice husk, and the results showed that the degradation of treated husk initiates occurred at 185°C, whereas decomposition of untreated rice husk occurred nearly at 250°C [78]. In a recent study, Panchami and Gunasekaran (2017) obtained pectin from fruit waste such as citrus peel, apple pomace, mango peel, and banana peel by treating with 0.1 N HCl and boiling for 30 min, and the acid treatment resulted in higher pectin yields of 25.5, 12.5, 8.8, and 2.8%, respectively [79]. Tiwari *et al.* (2017) also extracted pectin from the peel of an orange upon citric acid treatment for 30 min at pH 1 and 65°C and finally washed with ethanol resulting in a pectin yield of 52.9% [80].

1.4.1.2 Alkali Treatment

Alkaline pretreatment entails using alkaline reagents like NaOH, hydrazine, anhydrous ammonia, KOH, or $Ca(OH)_2$ to liquefy lignin and certain hemicellulose to reduce cellulose crystallinity and to increase the conversion of biomass [81]. Additionally, alkali treatment increases active surface area, causes swelling, and facilitates the substrate's higher accessibility for anaerobic microbes [77]. The alkali pretreatment method is applied at room temperature or at a little higher temperature (around 40°C) with a long response time that usually takes many hours to days to complete [82]. Alkali treatment is more desirable than acid pretreatments because alkali treatment prevents pH reduction, which improves the conditions for the AD of the waste [77]. They are economical and less hazardous to the environment; however, they need a longer reaction time and neutralization of the pretreated slurry [81]. Ramesh and Radhakrishnan (2019) reported that potato peel powder was treated with 2% NaOH at 50°C for 180 min resulting in high production of cellulose with a yield of 39.8%. The obtained cellulose was subsequently incorporated into polyvinyl alcohol (PVA) films, leading to improved antibacterial characteristics and notably increased free radical scavenging activity, which is advantageous for food packaging applications [83]. As a result of the alkali treatment, Verma and Goh (2021) stated that the agro-waste fibers improved in roughness, thereby improving the adhesion of the fiber to the matrix and cellulose content on the surface of the fiber, which also enhanced the interfacial strength and bonding between the natural fiber and the matrix. By improving fiber interlocking with matrices, polymer-based green composites achieve higher mechanical properties [84].

1.4.1.3 Organic Solvent Treatment

The organic solvent treatment, also known as the organosolv treatment method entails the pretreatment of lignin and the production of other potentially valuable coproducts (such as acetone, butanol, ethanol, and biogas) [85]. Organosolv treatment uses organic solvents such as CH_3OH, C_2H_5OH, ethylene glycol, acetone, and tetrahydrofurfuryl alcohol in order to breakdown of the internal bonds of hemicelluloses and lignin to decompose lignocellulosic material and ensure efficient cellulose conversion [86]. Frequently organic acids like oxalic, acetylsalicylic, and salicylic acids are also utilized as catalysts to speed up the reaction process [87]. Organosolv pretreatments are very promising because of their inherent benefits, including the separation of high-purity cellulose with only minor degradation, the isolation of high-quality lignin, and the higher efficiency of hemicellulose fractionation when compared to other conventional treatments [88]. In a study, Salapa *et al.* (2017) treated wheat straw with ethanol at 180°C for 40 min that yielded maximum cellulose (89%) and ethanol (67%) [85]. Similarly, Zhao *et al.* (2009) reported that corn stover treatment with 60% (v/v) ethanol led to a delignification of about 81.7% and a yield of 83.2% total sugar. Such pretreatment is particularly useful for lignocellulosic biomass with a high lignin content because it breaks internal linkages between lignin and hemicellulose [89].

1.4.2 Biological Treatment

Biological treatment comprises practices of microorganisms, such as bacteria or fungi, to break down the lignin content in the agro-waste and promote the conversion of lignocellulose material into a useful product. This pretreatment technique is environmentally friendly, emits no harmful substances, uses less energy, and creates no or less inhibitor [90]. Biological treatment is an inexpensive, time-consuming, and slow process that requires longer residence time [91]. The most effective biological treatment method is one that can break down the majority of undesired substances while retaining a high proportion of valuable components in a short period of time [92]. The availability of nutrients like sugars, minerals, and proteins within the agro-waste provides favorable conditions for the abundant growth of microorganisms. These microorganisms have the potential to use the waste as ingredients or a substrate for their growth using fermentation techniques, leading to the creation of various valuable compounds. Utilizing this waste would not just assist in recycling, but it would also lower production expenses and significantly decrease pollution levels [8]. Cellulase, lignolytic enzymes, and industrial enzymes can all be used to digest the lignocellulosic components of agro-wastes. Even though the enzymatic response rate for lignocelluloses is quite slow and requires a large sample size and close monitoring of growth conditions of the microorganism to prevent contamination, biological treatment is known to be useful for biomass containing lignocelluloses because microbial enzymes are effective at decomposing lignocellulosic substances [77]. Bacteria and fungi produce cellulases that break down and hydrolyze lignocellulosic biomass to increase sugar yields and improve the efficiency of microbial fuel cells. The bacteria that generate cellulases include *Bacteroides*, *Erwinia*, *Clostridium*, *Bacillus*, *Cellulomonas*, *Termomonospora*, *Acetovibrio*, *Ruminococcus*, *Microbispora*, and *Streptomyces* [93, 94]. Gadgey and Bahekar (2017) used enzymatic and fermentation methods for the extraction of chitin, in which enzymes, such as proteases, are used to deproteinize crustacean shells. These enzymes are commonly gained from *Lactobacillus* spp., *Bacillus subtiltis*, *Serratia marcescens* FS-3, and *Pseudomonas aeruginosa* K-187 [95]. In the biological process, Nielsen *et al.* (2017) used microbial strains such as *Bacillus*, *Cupriavidus*, *Comamonass*, *Pseudomonas*, *Hydrogenophaga*, *Acinetobacter*, *Azotobacter*, *Burkholderia*, *Haloferax*, and *Azohydromonas* species to produce poly(3-hydroxybutyric acid) (PHB) and polyhydroxyalkanoates (PHA) from agro-waste [96]. PHB, which is a homopolymer of 3-hydroxybutyrate, is widely used; however, PHA is used less in industrial settings because of its high production costs [97].

1.4.3 Mechanical Processing

In general, lignocellulosic material is pretreated using grinding, chipping, and by using several milling techniques such as colloid milling, vibratory milling, hammer milling, two-roll milling, and other procedures before being treated by AD. In a prior study, meadow grass was processed with a checker plate, sandpaper, and mesh grating plate resulting in a high production of methane by 25 to 27% compared to that achieved without mechanical treatment [98, 99]. Mechanical processing was employed for particle size reduction of lignocellulosic material of particle sizes in the range of 0.2 to 2 mm and 10 to 30 mm by milling and chipping pretreatments, respectively [100, 101]. In another study, wheat straw subjected to roll milling, extrusion, pelletization, and hammer milling, maximum methane production per day (DMPmax), biochemical methane potential (BMP), and EH yield as the amount of released glucan/initial glucan content were used to analyze the impacts of the pretreatments on AD at 37°C and on EH with Cellic CTec2 at 50°C. Maximum BMP of 287 Nml CH_4 gVS1, up 21% from untreated wheat straw was produced in roll milling, although extrusion produced the highest rate of methane (52 Nml CH_4 gVS1 day^{-1}). According to EH results and specific surface analysis of 0.25- to 1-mm particles, the mechanical pretreatment possibly changes other physicochemical characteristics like crystallinity or melting of the outer wax layer of wheat straw, which in turn influences EH processes and AD [102]. Pretreatment of herbaceous materials requires less energy than pretreatment of woody materials. The energy required to reduce the size of one metric tonne of pine chips, maize stover, switch grass, and poplar was found to be 85.4, 11.0, 27.6, and 118.5 kWh, respectively [103]. Furthermore, increasing the surface area during pretreatment improves the EH of lignocellulosic material [104]. It was also shown that reducing the size of lignocellulosic materials during milling pretreatment increased the buffering capacity of AD, which is beneficial in reducing anaerobic digestion acidification [105].

1.4.4 Thermochemical Processing

This method involves hot water and steam explosion pretreatment, in which biomass is pretreated at higher pressure at a higher temperature (140°C to 220°C) without any chemical intervention [106, 107]. Pressurized water penetrates the biomass, hydrolyzes hemicellulose, increases the surface area, hydrates cellulose, and removes the lignin fraction [108]. This pretreatment attracted attention due to its low-cost and nonchemical requirement [109]. Moreover, this method can increase methane production from lignocellulosic materials and has been extensively used for AD [110]. The hydrothermal pretreatment of sunflower stalks for AD was performed, and the pretreatment increased methane yield by 88.7% [111]. In another study, sunflower stalk was processed for AD at 160°C, 180°C, 200°C, and 220°C by Lee and colleagues (2020), in which the methane yield increased significantly (213.87 to 289.47 ml/g VS) in comparison to the control (129 ml/g VS) [112]. Pretreatment with a steam explosion at 160°C to 250°C and 5- to 50-bar pressure for a small period using saturated steam, followed by a rapid decrease in pressure that destroys the lignocellulosic components. The lignin is partially changed during the steam explosion in addition to the hydrolysis of hemicellulose, which increases the degradability of the lignocellulose material [77, 108]. Zhou and team (2016) prepared rice straw for 2 min at 200°C and found a 51% increase in methane output during AD. Steam explosion pretreatment of olive mill solid waste (OMSW) at 200°C and 1.57 MPa for 5 min affected the production of a solid fraction (SF) and a liquid fraction (LF) during the AD [113]. BMP studies of both steam explosion–pretreated and –untreated OMSW showed that the LF had the highest methane production (589 ± 42 mL CH_4/g VS) [114]. Another investigation claimed that pretreatment with steam explosion might efficiently increase the generation of methane from water primrose, in spite of the plant's woody structure and higher lignin content [115].

1.5 Extraction of Bioactive Compounds from Agro-Wastes

Bioactive compounds are secondary molecules produced naturally in small amounts in plants and animals and possess certain functional properties like antioxidant, antimicrobial, anticarcinogenic, and anti-inflammatory [116]. Bioactive compounds are grouped into various classes including nitrogen-containing compounds, terpenoids, organosulfur compounds, alkaloids, and phenolics [117]. Bioactive compounds are being used to improve food quality (nutritional, sensory, and technological aspects), to manufacture functional foods such as nutraceuticals, and can be used to develop smart and active food packaging films and coatings [118–121].

The extraction process stands as the pivotal step in acquiring bioactive compounds from agro-waste, and the selection of optimal extraction methods plays a crucial role in determining both the specific types and quantities of bioactive compounds [122]. The extraction of bioactive compounds is driven by several critical factors, including extraction technique, properties of plant components, solvent used, and extraction condition (Figure 1.2) [123, 124]. Sample preparation constitutes another critical factor influencing the type and quantity of bioactive compounds extracted. For example, Ferri *et al.* (2020) applied a solvent-based extraction technique on grape pomace for the extraction of bioactive compounds, in which different solvent ratios (1:5 and 1:10) were mixed with the grind pomace and incubated at specific temperatures (50°C and 70°C) in a closed system. After incubation, the liquid extracts were separated from solid residues and stored at −20°C. Solvent types included 50% aqueous-ethanol, 50% aqueous-acetonitrile, and 75% aqueous-acetone. Results revealed that the highest amount of phenolics was obtained using 75% acetone followed by 50% aqueous-acetonitrile and 50% aqueous-ethanol [125]. Another study by Sharif *et al.* (2023) extracted bioactive phytochemicals from

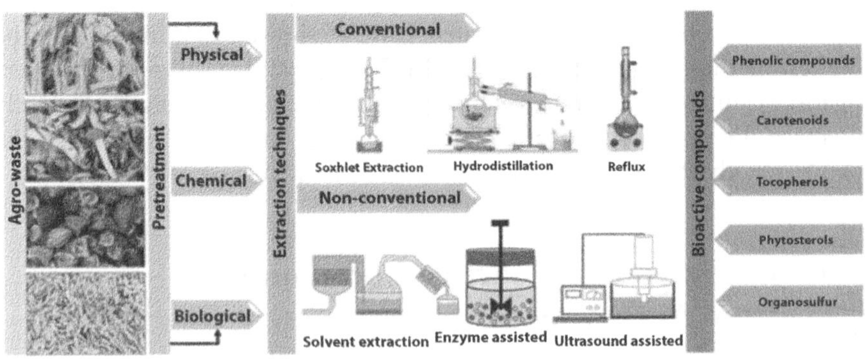

Figure 1.2 Bioactive compounds extraction from agro-waste employing various extraction techniques.

mango peel using a pretreatment combination of ultrasound irradiation and enzyme treatment. The mango peel was treated with enzymes under optimized conditions, followed by ultrasonic pretreatment. The enzymes were then deactivated by heating at 95°C for 5 min, and the treated peel was subjected to extraction using an ethanol-water solvent mixture (80:20 v/v). After incubation at 45°C in an orbital shaker at 120 rpm for 24 hours and centrifugation at 4500 rpm for 10 min, the extracts were filtered and concentrated using a rotary vacuum evaporator. The findings of the study indicated that the mango peel extracts contained a total phenolic content (TPC) of 33.56 ± 1.04 mg GAE/g of fresh weight. Moreover, the mango peel extracts treated under the optimized conditions exhibited a noteworthy Trolox equivalent antioxidant capacity (TEAC) value of 215.42 ± 1.21 mM TE/g [126].

Zhang *et al.* (2019) extracted anthocyanin from the black plum peel and effectively used it in chitosan and TiO_2 films, where incorporation produces strong barrier characteristics against water vapor and UV light as well as improved mechanical strength. The developed films resulted in higher free radical scavenging, antioxidant, antimicrobial, and ethylene scavenging activities having potential applications as pH-sensitive color-changing food packaging film [127]. Anthocyanins extracted from blueberry juice residue have also been used in producing smart films using cassava starch that can monitor the quality of orange juice, chicken pieces, and corn oil [128]. Betacyanin extracted from dragon fruit shells has been used in developing smart packaging film based on glucomannan-PVA for sensing the freshness of fish. Betacyanin encourages a visible color change from purple to yellowish because of fish deterioration leading to increased levels of TVB-N [129]. In addition, protein hydrolysates obtained from agro-industrial byproducts have also been incorporated as active agents in food packaging films. For example, active films developed from protein hydrolysates and alginate extracted from cotton seed byproducts had excellent antioxidant, light barrier, and antimicrobial properties [130]. Other bioactive compounds like curcumins, tannins, and carotenoids can be extracted from agro-wastes for application as active agents in the development of smart biodegradable food packaging films and coating [131]. There is substantial use of bioactive components in developing active and smart packaging due to their variety, potential interactions, and diverse functions; however, each component must be evaluated properly to ensure maximum effectiveness. Many studies reported the extraction of bioactive compounds from agro-waste and their uses in various food packaging systems (Table 1.1).

Table 1.1 Extraction of bioactive compound from agro-wastes.

Agro-waste	Bioactive compound	Extraction methods	Application	Remark	Reference
Blackberry and blueberry pomace	Anthocyanin	Microwave-assisted extraction	pH indicator film	The developed films had high total phenol content, excellent antioxidant activity, and showed intelligent character, i.e., change in color upon exposure to solutions of different pH	[132]
Blueberry agro-waste	Anthocyanin	Frozen at −40 ± 2°C prior to lyophilization followed by milling and sieving	pH indicator film	Blueberry powder can be used as a pH indicator for sensing deterioration in food	[133]
Blueberry pomace	Anthocyanin	Freeze-dried followed by milling and sieving	Biodegradable films	The developed films had UV light protection properties because of aromatic compounds and thus reduced food deterioration	[134]

(Continued)

Table 1.1 Extraction of bioactive compound from agro-wastes. (*Continued*)

Agro-waste	Bioactive compound	Extraction methods	Application	Remark	Reference
Jamun fruit skin	Anthocyanin	Solvent extraction	pH indicator	The developed indicator may be used to monitor the meat quality during refrigerated storage	[135]
Red pitaya peel	Betalains	Solvent extraction	Active and intelligent packaging films	The incorporation of the extract in the developed film improved UV light, water vapor barrier properties, and mechanical, antibacterial, and antioxidant activities. The film was also able to monitor shrimp freshness through visible color changes	[136]
Pomegranate flesh and peel	Polyphenols	Acid-base extraction	Active and intelligent films	The developed films showed a desirable tensile strength, water vapor, and UV light barrier ability of the film and exhibited a pH-sensitive property	[137]

(*Continued*)

Table 1.1 Extraction of bioactive compound from agro-wastes. (*Continued*)

Agro-waste	Bioactive compound	Extraction methods	Application	Remark	Reference
Cashew nut testa	Tannins	Aqueous extraction	Active packaging film	The fabricated films had high antimicrobial and antioxidant (91%) properties and thermal stability (290°C) and thus could be used for shelf-life extension of perishable foods	[138]
Rambutan peel	Polyphenols	Solvent extraction	Whey protein isolate film	The developed films showed high antioxidant ability and moderate antimicrobial properties and may be used for food packaging application	[139]
Jaboticaba peel	Anthocyanins	Microwave-assisted extraction	Carrageenan-based films	The film showed good antimicrobial activities and can extend the postharvest-life of food	[140]

(*Continued*)

Table 1.1 Extraction of bioactive compound from agro-wastes. (*Continued*)

Agro-waste	Bioactive compound	Extraction methods	Application	Remark	Reference
Apple peel	Polyphenols	Ultrasound extraction	Active food packaging film	The developed film improved the film's opacity, swelling level, and solubility due to the addition of polyphenols	[141]
Pineapple peel	Polyphenols	Solid-liquid extraction	Edible film	The developed films effectively preserved red beef meat, significantly inhibited microbial growth, and delayed lipid oxidation in meat during storage	[142]
Black plum peel	Anthocyanin	Acid-base extraction	Multifunctional food packaging films	The developed film significantly enhanced UV light and water vapor barrier properties, tensile strength, and antioxidant, antimicrobial, and pH-sensitive abilities due to the abundance of phenolics and anthocyanins content	[127]

(*Continued*)

Table 1.1 Extraction of bioactive compound from agro-wastes. (*Continued*)

Agro-waste	Bioactive compound	Extraction methods	Application	Remark	Reference
Hass avocado peel	Flavonoids and chlorophylls	Solvents extraction	Active films	The developed film had compatible and improved tensile strength and UV light barrier characteristics and effectively inhibited *Rhizopus stolonifera* and *Aspergillus niger* growth	[143]
Potato peel	Caffeic, chlorogenic, and neochlorogenic acids	Solvents extraction	Active edible film	The addition of extract in the developed film increased phenolic content and DPPH scavenging activity	[144]

1.6 Conclusion and Future Perspectives

Over the past few years, agricultural wastes have gained significant attention due to the increasing demand for a range of biobased products and environmental concerns. This prompted a shift toward utilizing agro-waste as a source of nutrients, bioactive compounds, and biopolymers. In this chapter, types of agro-waste, their processing, extraction methods, and uses have been discussed. Agro-waste can be recycled to produce various value-added products to ensure a safe and green environment. Agro-waste contains a variety of contaminants, including microorganisms, heavy metals, and pesticides, which may pose health risks to humans and animals. For agricultural sustainability and the security of human food and health, it is imperative to develop solutions for the safe utilization and valorization of agro-waste in order to create value-added products, farmer livelihoods, and job opportunities for youth. The agro-waste generated during the growing and first processing of agriculture can provide a resource for improving the nutritional value of food and improving food security. Future research efforts should also focus on the advancement of innovative biotechnological approaches to effectively valorize agricultural waste. The development of suitable green technologies is essential for the advancement of bioproducts. With effective policies and their proper execution, the economic and market resistance problems can be overcome over time. Managing agro-waste and valorizing it are essential for sustainable development. Considering the benefits of reusing agro-waste in the manufacturing process means resolving important problems today, i.e., environmental pollution and economic sustainability.

References

1. Varghese, S.A. *et al.*, Renovation of agro-waste for sustainable food packaging: A review. *Polymers*, 15, 3, 648, 2023.
2. Sharma, V. *et al.*, Agro-industrial food waste as a low-cost substrate for sustainable production of industrial enzymes: A critical review. *Catalysts*, 12, 11, 1373, 2022.
3. Thomas, C. *et al.*, Docetaxel-rechallenge in castration-resistant prostate cancer: Defining clinical factors for successful treatment response and improvement in overall survival. *Int. Urol. Nephrol.*, 50, 10, 1821–1827, 2018.
4. Dey, T. *et al.*, Valorization of agro-waste into value added products for sustainable development. *Bioresour. Technol. Rep.*, 16, 100834, 2021.

5. Singh, A. *et al.*, Production of bioethanol from food waste: Status and perspectives. *Bioresour. Technol.*, 360, 127651, 2022.
6. Valdés, A. *et al.*, Natural pectin polysaccharides as edible coatings. *Coatings*, 5, 4, 865–886, 2015.
7. Nemes, S.A., Szabo, K., Vodnar, D.C., Applicability of agro-industrial by-products in intelligent food packaging. *Coatings*, 10, 6, 550, 2020.
8. Sadh, P.K., Duhan, S., Duhan, J.S., Agro-industrial wastes and their utilization using solid state fermentation: A review. *Bioresour. Bioprocess.*, 5, 1, 1, 2018.
9. Salim, A.A. *et al.*, Production of enzymes by a newly isolated Bacillus sp. TMF-1 in solid state fermentation on agricultural by-products: The evaluation of substrate pretreatment methods. *Bioresour. Technol.*, 228, 193–200, 2017.
10. Trajkovska Petkoska, A. *et al.*, Edible packaging: Sustainable solutions and novel trends in food packaging. *Food Res. Int.*, 140, 109981, 2021.
11. Abdul Khalil, H.P.S. *et al.*, A review on nanocellulosic fibres as new material for sustainable packaging: Process and applications. *Renew. Sustain. Energy Rev.*, 64, 823–836, 2016.
12. Ncube, L.K. *et al.*, Environmental impact of food packaging materials: A review of contemporary development from conventional plastics to polylactic acid based materials. *Materials*, 13, 21, 4994, 2020.
13. Ma, J. *et al.*, From agricultural cellulosic waste to food delivery packaging: A mini-review. *Chin. Chem. Lett.*, 34, 2, 107407, 2023.
14. Maharana, T., Mohanty, B., Negi, Y.S., Melt–solid polycondensation of lactic acid and its biodegradability. *Prog. Polym. Sci.*, 34, 1, 99–124, 2009.
15. Oyeoka, H.C. *et al.*, Packaging and degradability properties of polyvinyl alcohol/gelatin nanocomposite films filled water hyacinth cellulose nano-crystals. *J. Bioresour. Bioprod.*, 6, 2, 168–185, 2021.
16. Zainal Arifin, M.A. *et al.*, Utilization of food waste and by-products in the fabrication of active and intelligent packaging for seafood and meat products. *Foods*, 12, 456, 2023, DOI: 10.3390/foods12030456.
17. Bayram, B. *et al.*, Valorization and application of fruit and vegetable wastes and by-products for food packaging materials. *Molecules*, 26, 13, 4031, 2021.
18. Koul, B., Yakoob, M., Shah, M.P., Agricultural waste management strategies for environmental sustainability. *Environ. Res.*, 206, 112285, 2022.
19. Rithuparna, R., Jittin, V., Bahurudeen, A., Influence of different processing methods on the recycling potential of agro-waste ashes for sustainable cement production: A review. *J. Cleaner Prod.*, 316, 128242, 2021.
20. Mozhiarasi, V. and Natarajan, T.S., Slaughterhouse and poultry wastes: Management practices, feedstocks for renewable energy production, and recovery of value added products. *Biomass Convers. Biorefin.*, 2022. https://doi.org/10.1007/s13399-022-02352-0

21. Flora, B.-R. *et al.*, Agro-industrial waste revalorization: The growing biorefinery, in: *Biomass for Bioenergy*, A. Abd El-Fatah (Ed.), IntechOpen, Rijeka, 2019, Ch. 5.
22. Duque-Acevedo, M. *et al.*, Agricultural waste: Review of the evolution, approaches and perspectives on alternative uses. *Glob. Ecol. Conserv.*, 22, e00902, 2020.
23. Zulkifli, A.A. *et al.*, Assessment of municipal solid waste generation in Universiti Putra Malaysia and its potential for green energy production. *Sustainability*, 11, 14, 3909, 2019.
24. Canizares, D., Angers, P., Ratti, C., Flax and wheat straw waxes: Material characterization, process development, and industrial applications. *Biomass Convers. Biorefin.*, 10, 2, 555–565, 2020.
25. Khalid Zafeer, M. and Subrahmanya Bhat, K., Valorisation of agro-waste cashew nut husk (Testa) for different value-added products. *Sustain. Chem. Clim. Action*, 2, 100014, 2023.
26. He, J., Kawasaki, S., Achal, V., The utilization of agricultural waste as agro-cement in concrete: A review. *Sustainability*, 12, 17, 6971, 2020.
27. Scarlat, N. *et al.*, The role of biomass and bioenergy in a future bioeconomy: Policies and facts. *Environ. Dev.*, 15, 3–34, 2015.
28. Surendra, K.C., Angelidaki, I., Khanal, S.K., Bioconversion of waste-to-resources (BWR-2021): Valorization of industrial and agro-wastes to fuel, feed, fertilizer, and biobased products. *Bioresour. Technol.*, 347, 126739, 2022.
29. Hanani, Z.A.N., Yee, F.C., Nor-Khaizura, M.A.R., Effect of pomegranate (Punica granatum L.) peel powder on the antioxidant and antimicrobial properties of fish gelatin films as active packaging. *Food Hydrocoll.*, 89, 253–259, 2019.
30. Nur Hanani, Z.A. *et al.*, Effect of different fruit peels on the functional properties of gelatin/polyethylene bilayer films for active packaging. *Food Packag. Shelf Life*, 18, 201–211, 2018.
31. Gopinath, A. *et al.*, A circular framework for the valorisation of sugar industry wastes: Review on the industrial symbiosis between sugar, construction and energy industries. *J. Cleaner Prod.*, 203, 89–108, 2018.
32. Ashwini John, J., Samuel, M.S., Selvarajan, E., Immobilized cellulase on Fe3O4/GO/CS nanocomposite as a magnetically recyclable catalyst for biofuel application. *Fuel*, 333, 126364, 2023.
33. Sanjeeva, S.K. *et al.*, Distilled technical cashew nut shell liquid (DT-CNSL) as an effective biofuel and additive to stabilize triglyceride biofuels in diesel. *Renewable Energy*, 71, 81–88, 2014.
34. Silva, J.S. *et al.*, Cashew apple bagasse as new feedstock for the hydrogen production using dark fermentation process. *J. Biotechnol.*, 286, 71–78, 2018.
35. Madeira, J.V. *et al.*, Chapter 18 - Agro-industrial residues and microbial enzymes: An overview on the eco-friendly bioconversion into

high value-added products, in: *Biotechnology of Microbial Enzymes*, G. Brahmachari (Ed.), pp. 475–511, Academic Press, London, UK, 2017.

36. Lopes, F.C. and Ligabue-Braun, R., Agro-industrial residues: Eco-friendly and inexpensive substrates for microbial pigments production. *Front. Sustain. Food Syst.*, 5, 2021. https://doi.org/10.3389/fsufs.2021.589414
37. Cosmas, C.O. and Stephen, N.O., Agro-industrial waste management: The circular and bioeconomic perspective, in: *Agricultural Waste - New Insights*, A. Fiaz and S. Muhammad (Eds.), p. Ch. 9, IntechOpen, Rijeka, 2023.
38. Kazemi, M. *et al.*, Ultrasonic and heating extraction of pistachio by-product pectin: Physicochemical, structural characterization and functional measurement. *J. Food Meas. Charact.*, 14, 2, 679–693, 2020.
39. Lalnunthari, C., Devi, L.M., Badwaik, L.S., Extraction of protein and pectin from pumpkin industry by-products and their utilization for developing edible film. *J. Food Sci. Technol.*, 57, 5, 1807–1816, 2020.
40. Ravindran, R. *et al.*, A review on bioconversion of agro-industrial wastes to industrially important enzymes. *Bioengineering*, 5, 4, 93, 2018.
41. Gao, Z. *et al.*, Waste cooking oil used as carbon source for microbial lipid production: Promoter or inhibitor. *Environ. Res.*, 203, 111881, 2022.
42. Bhuvaneshwari, S., Hettiarachchi, H., Meegoda, J.N., Crop residue burning in India: Policy challenges and potential solutions. *Int. J. Environ. Res. Public Health*, 16, 5, 832, 2019.
43. Meena, H.N., S., S.K., Meena, M.S., Narayan, R., Sen, B., Crop residue: Waste or wealth? *Tech. Bull. 2022*, pp. 1–30, 2022.
44. Nath, P.C. *et al.*, Biogeneration of valuable nanomaterials from agro-wastes: A comprehensive review. *Agronomy*, 13, 2, 561, 2023.
45. Singh, P. *et al.*, *Waste management and resource recycling in the developing world*, Elsevier, Amsterdam, Netherlands, 2022.
46. Chollakup, R. *et al.*, Development of rice straw paper coated with pomelo peel extract for bio-based and antibacterial packaging. *Key Eng. Mater.*, 847, 141–146, 2020.
47. Nascimento, P. *et al.*, Nanocellulose produced from rice hulls and its effect on the properties of biodegradable starch films. *Mater. Res.*, 19, 167–174, 2016.
48. Sánchez-Safont, E.L. *et al.*, Biocomposites of different lignocellulosic wastes for sustainable food packaging applications. *Compos. Part B: Eng.*, 145, 215–225, 2018.
49. Jayathilakan, K. *et al.*, Utilization of byproducts and waste materials from meat, poultry and fish processing industries: A review. *J. Food Sci. Technol.*, 49, 3, 278–293, 2012.
50. Kumari, M. and Dhawal, K.K., Utilization patterns of livestock waste: Waste is wealth?! *Vigyan Varta*, 7, 10, 63, 2020.
51. *Annual report 2021-22*, pp. 77–94, A.H.a.D.G.o.I. Department of Animal Husbandry and Dairying Ministry of Fisheries, New Delhi, India, 2022.

52. Matsagar, B.M. and Wu, K.C.W., Chapter 1 - Agricultural waste-derived biochar for environmental management, in: *Biochar in Agriculture for Achieving Sustainable Development Goals*, D.C.W. Tsang and Y.S. Ok (Eds.), pp. 3–13, Academic Press, London, UK, 2022.
53. Tolera, S.T. and Alemu, F.K., Potential of abattoir waste for bioenergy as sustainable management, Eastern Ethiopia, 2019. *J. Energy*, 2020, 6761328, 2020.
54. Balaman, Ş.Y., Chapter 1 - Introduction to biomass—Resources, production, harvesting, collection, and storage, in: *Decision-Making for Biomass-Based Production Chains*, Ş.Y. Balaman (Ed.), pp. 1–23, Academic Press, London, UK, 2019.
55. Álvarez, C. *et al.*, Chapter 2 - Classification and target compounds, in: *Food Waste Recovery*, Second Edition, C.M. Galanakis (Ed.), pp. 21–49, Academic Press, San Diego, 2021.
56. Lynch, S.A. *et al.*, Optimization of protein recovery from bovine lung by pH shift process using response surface methodology. *J. Sci. Food Agric.*, 98, 5, 1951–1960, 2018.
57. Drummond, L., Álvarez, C., Mullen, A.M., Chapter 4 - Proteins recovery from meat processing coproducts, in: *Sustainable Meat Production and Processing*, C.M. Galanakis (Ed.), pp. 69–83, Academic Press, London, UK, 2019.
58. Pérez-Andrés, J.M. *et al.*, Effect of cold plasma on the techno-functional properties of animal protein food ingredients. *Innovative Food Sci. Emerg. Technol.*, 58, 102205, 2019.
59. Khodaei, D., Álvarez, C., Mullen, A.M., Biodegradable packaging materials from animal processing co-products and wastes: An overview. *Polymers*, 13, 15, 2561, 2021.
60. Mishra, P.K. *et al.*, Synthesis of biodegradable films using gamma irradiation from fish waste. *Waste Biomass Valorization*, 12, 5, 2247–2257, 2021.
61. Hua, K. *et al.*, The future of aquatic protein: Implications for protein sources in aquaculture diets. *One Earth*, 1, 3, 316–329, 2019.
62. Vázquez, J.A. *et al.*, Production of valuable compounds and bioactive metabolites from by-products of fish discards using chemical processing, enzymatic hydrolysis, and bacterial fermentation. *Mar. Drugs*, 17, 3, 139, 2019.
63. Araújo, C.S. *et al.*, Optimizing process parameters to obtain a bioplastic using proteins from fish byproducts through the response surface methodology. *Food Packag. Shelf Life*, 16, 23–30, 2018.
64. Al Khawli, F. *et al.*, Innovative green technologies of intensification for valorization of seafood and their by-products. *Mar. Drugs*, 17, 12, 689, 2019.
65. Ideia, P. *et al.*, Fish processing industry residues: A review of valuable products extraction and characterization methods. *Waste Biomass Valorization*, 11, 7, 3223–3246, 2020.

66. Blanco, M. *et al.*, Collagen extraction optimization from the skin of the small-spotted catshark (S. canicula) by response surface methodology. *Mar. Drugs*, 17, 1, 40, 2019.
67. Pal, G.K. and Suresh, P.V., Sustainable valorisation of seafood by-products: Recovery of collagen and development of collagen-based novel functional food ingredients. *Innovative Food Sci. Emerg. Technol.*, 37, 201–215, 2016.
68. Bourbon, A.I. *et al.*, Protein-based nanostructures for food applications. *Gels*, 5, 1, 9, 2019.
69. Zavareze, E.d.R. *et al.*, Mechanical, barrier and morphological properties of biodegradable films based on muscle and waste proteins from the white-mouth croaker (Micropogonias furnieri). *J. Food Process. Preserv.*, 38, 4, 1973–1981, 2014.
70. Nie, X. *et al.*, Phenolics-protein interaction involved in silver carp myofibrilliar protein films with hydrolysable and condensed tannins. *LWT - Food Sci. Technol.*, 81, 258–264, 2017.
71. Awogbemi, O. and Kallon, D.V.V., Pretreatment techniques for agricultural waste. *Case Stud. Chem. Environ. Eng.*, 6, 100229, 2022.
72. Arif Hakimi Saadon, S.Z., Osman, N.B., Yusup, S., Chapter 5 – Pretreatment of fiber-based biomass material for lignin extraction, in: *Value-Chain of Biofuels*, S. Yusup and N.A. Rashidi (Eds.), pp. 105–135, Elsevier, Amsterdam, Netherlands, 2022.
73. Areepak, C. *et al.*, Improvement of lignocellulosic pretreatment efficiency by combined chemo - mechanical pretreatment for energy consumption reduction and biofuel production. *Renewable Energy*, 182, 1094–1102, 2022.
74. Zhang, R.-Y. *et al.*, Cellulose fibers extracted from sesame hull using subcritical water as a pretreatment. *Arabian J. Chem.*, 14, 6, 103178, 2021.
75. Baruah, J. *et al.*, Recent trends in the pretreatment of lignocellulosic biomass for value-added products. *Front. Energy Res.*, 6, 141, 2018.
76. Rezania, S. *et al.*, Different pretreatment technologies of lignocellulosic biomass for bioethanol production: An overview. *Energy*, 199, 117457, 2020.
77. Paudel, S.R. *et al.*, Pretreatment of agricultural biomass for anaerobic digestion: Current state and challenges. *Bioresour. Technol.*, 245, 1194–1205, 2017.
78. Ali, M. and Tindyala, M.A., Thermoanalytical studies on acid-treated rice husk and production of some silicon based ceramics from carbonised rice husk. *J. Asian Ceram. Soc.*, 3, 3, 311–316, 2015.
79. Panchami, P.S. and Gunasekaran, S., Extraction and characterization of pectin from fruit waste. *Int. J. Curr. Microbiol. Appl. Sci. (IJCMAS)*, 6, 8, 943–948, 2017.
80. Tiwari, A.K. *et al.*, Extraction and characterization of pectin from orange peels. *Int. J. Biotechnol. Biochem.*, 13, 1, 39–47, 2017.
81. Lorenci Woiciechowski, A. *et al.*, Lignocellulosic biomass: Acid and alkaline pretreatments and their effects on biomass recalcitrance – Conventional processing and recent advances. *Bioresour. Technol.*, 304, 122848, 2020.

82. Amin, F.R. *et al.*, Pretreatment methods of lignocellulosic biomass for anaerobic digestion. *AMB Express*, 7, 1, 72, 2017.
83. Ramesh, S. and Radhakrishnan, P., Cellulose nanoparticles from agro-industrial waste for the development of active packaging. *Appl. Surf. Sci.*, 484, 1274–1281, 2019.
84. Verma, D. and Goh, K.L., Effect of mercerization/alkali surface treatment of natural fibres and their utilization in polymer composites: Mechanical and morphological studies. *J. Compos. Sci.*, 5, 7, 175, 2021.
85. Salapa, I. *et al.*, Organosolv pretreatment of wheat straw for efficient ethanol production using various solvents. *Biomass Bioenergy*, 100, 10–16, 2017.
86. Mankar, A.R. *et al.*, Pretreatment of lignocellulosic biomass: A review on recent advances. *Bioresour. Technol.*, 334, 125235, 2021.
87. Kumari, D. and Singh, R., Pretreatment of lignocellulosic wastes for biofuel production: A critical review. *Renew. Sustain. Energy Rev.*, 90, 877–891, 2018.
88. Zhang, K., Pei, Z., Wang, D., Organic solvent pretreatment of lignocellulosic biomass for biofuels and biochemicals: A review. *Bioresour. Technol.*, 199, 21–33, 2016.
89. Zhao, X., Cheng, K., Liu, D., Organosolv pretreatment of lignocellulosic biomass for enzymatic hydrolysis. *Appl. Microbiol. Biotechnol.*, 82, 5, 815–827, 2009.
90. Sindhu, R., Binod, P., Pandey, A., Biological pretreatment of lignocellulosic biomass – An overview. *Bioresour. Technol.*, 199, 76–82, 2016.
91. Farkas, C. *et al.*, Microbial saccharification of wheat bran for bioethanol fermentation. *J. Cleaner Prod.*, 240, 118269, 2019.
92. Azelee, N.I.W. *et al.*, Biological treatment of agro-industrial waste, in: *Valorisation of Agro-industrial Residues–Volume I: Biological Approaches*, pp. 59–79, 2020.
93. Wagner, A.O. *et al.*, Biological pretreatment strategies for second-generation lignocellulosic resources to enhance biogas production. *Energies*, 11, 7, 1797, 2018.
94. Galbe, M. and Wallberg, O., Pretreatment for biorefineries: A review of common methods for efficient utilisation of lignocellulosic materials. *Biotechnol. Biofuels*, 12, 1, 294, 2019.
95. Gadgey, K.K. and Bahekar, A., Studies on extraction methods of chitin from crab shell and investigation of its mechanical properties. *Int. J. Mech. Eng. Technol.*, 8, 2, 220–231, 2017.
96. Nielsen, C. *et al.*, Food waste conversion to microbial polyhydroxyal-kanoates. *Microb. Biotechnol.*, 10, 6, 1338–1352, 2017.
97. Johnston, B. *et al.*, The microbial production of polyhydroxyalkanoates from waste polystyrene fragments attained using oxidative degradation. *Polymers*, 10, 9, 957, 2018.

98. Tsapekos, P. *et al.*, Improving methane production from digested manure biofibers by mechanical and thermal alkaline pretreatment. *Bioresour. Technol.*, 216, 545–552, 2016.
99. Tsapekos, P., Kougias, P.G., Angelidaki, I., Mechanical pretreatment for increased biogas production from lignocellulosic biomass; predicting the methane yield from structural plant components. *Waste Manage.*, 78, 903–910, 2018.
100. Menardo, S., Cacciatore, V., Balsari, P., Batch and continuous biogas production arising from feed varying in rice straw volumes following pre-treatment with extrusion. *Bioresour. Technol.*, 180, 154–161, 2015.
101. Patil, P.N. *et al.*, Intensification of biogas production using pretreatment based on hydrodynamic cavitation. *Ultrason. Sonochem.*, 30, 79–86, 2016.
102. Victorin, M., Davidsson, Å., Wallberg, O., Characterization of mechanically pretreated wheat straw for biogas production. *Bioenergy Res.*, 13, 3, 833–844, 2020.
103. Neshat, S.A. *et al.*, Anaerobic co-digestion of animal manures and lignocellulosic residues as a potent approach for sustainable biogas production. *Renew. Sustain. Energy Rev.*, 79, 308–322, 2017.
104. Sankaran, R. *et al.*, Recent advances in the pretreatment of microalgal and lignocellulosic biomass: A comprehensive review. *Bioresour. Technol.*, 298, 122476, 2020.
105. Mustafa, A.M. *et al.*, Combinations of fungal and milling pretreatments for enhancing rice straw biogas production during solid-state anaerobic digestion. *Bioresour. Technol.*, 224, 174–182, 2017.
106. Shi, Z. *et al.*, Genome-centric metatranscriptomics analysis reveals the role of hydrochar in anaerobic digestion of waste activated sludge. *Environ. Sci. Technol.*, 55, 12, 8351–8361, 2021.
107. Ren, S. *et al.*, Hydrochar-facilitated anaerobic digestion: Evidence for direct interspecies electron transfer mediated through surface oxygen-containing functional groups. *Environ. Sci. Technol.*, 54, 9, 5755–5766, 2020.
108. Zheng, Y. *et al.*, Pretreatment of lignocellulosic biomass for enhanced biogas production. *Prog. Energy Combust. Sci.*, 42, 35–53, 2014.
109. Usman, M. *et al.*, Hydrochar promoted anaerobic digestion of hydrothermal liquefaction wastewater: Focusing on the organic degradation and microbial community. *Chem. Eng. J.*, 399, 125766, 2020.
110. Shi, Z. *et al.*, Combined microbial transcript and metabolic analysis reveals the different roles of hydrochar and biochar in promoting anaerobic digestion of waste activated sludge. *Water Res.*, 205, 117679, 2021.
111. Hesami, S.M. *et al.*, Enhanced biogas production from sunflower stalks using hydrothermal and organosolv pretreatment. *Ind. Crops Prod.*, 76, 449–455, 2015.
112. Lee, J. *et al.*, Interactions between substrate characteristics and microbial communities on biogas production yield and rate. *Bioresour. Technol.*, 303, 122934, 2020.

113. Zhou, J. *et al.*, Effect of steam explosion pretreatment on the anaerobic digestion of rice straw. *RSC Adv.*, 6, 91, 88417–88425, 2016.
114. Rincón, B. *et al.*, Influence of a steam-explosion pre-treatment on the methane yield and kinetics of anaerobic digestion of two-phase olive mil solid waste or alperujo. *Process Saf. Environ. Prot.*, 102, 361–369, 2016.
115. Bhatia, P. *et al.*, Effect of steam explosion pre-treatment on methane generation from Ludwigia grandiflora. *Biomass Bioenergy*, 142, 105771, 2020.
116. Rodríguez García, S.L. and Raghavan, V., Green extraction techniques from fruit and vegetable waste to obtain bioactive compounds—A review. *Crit. Rev. Food Sci. Nutr.*, 62, 23, 6446–6466, 2022.
117. Altemimi, A. *et al.*, Phytochemicals: Extraction, isolation, and identification of bioactive compounds from plant extracts. *Plants*, 6, 4, 42, 2017.
118. Birch, C.S. and Bonwick, G.A., Ensuring the future of functional foods. *Int. J. Food Sci. Technol.*, 54, 5, 1467–1485, 2019.
119. Daliu, P., Santini, A., Novellino, E., A decade of nutraceutical patents: Where are we now in 2018? *Expert Opin. Ther. Pat.*, 28, 12, 875–882, 2018.
120. Reque, P.M. and Brandelli, A., Encapsulation of probiotics and nutraceuticals: Applications in functional food industry. *Trends Food Sci. Technol.*, 114, 1–10, 2021.
121. Oliveira Filho, J.G.d. *et al.*, The potential of anthocyanins in smart, active, and bioactive eco-friendly polymer-based films: A review. *Food Res. Int.*, 142, 110202, 2021.
122. Sagar, N.A. *et al.*, Fruit and vegetable waste: Bioactive compounds, their extraction, and possible utilization. *Compr. Rev. Food Sci. Food Saf.*, 17, 3, 512–531, 2018.
123. Jha, A.K. and Sit, N., Extraction of bioactive compounds from plant materials using combination of various novel methods: A review. *Trends Food Sci. Technol.*, 119, 579–591, 2022.
124. Naumovski, N. *et al.*, Bioactive compounds in agricultural and food production waste, in: *Utilisation of Bioactive Compounds from Agricultural and Food Waste*, pp. 1–26, 2017.
125. Ferri, M. *et al.*, From winery waste to bioactive compounds and new polymeric biocomposites: A contribution to the circular economy concept. *J. Adv. Res.*, 24, 1–11, 2020.
126. Sharif, T. *et al.*, Recovery of high-value bioactive phytochemicals from agro-waste of mango (Mangifera indica L.) using enzyme-assisted ultrasound pretreated extraction. *Biomass Convers. Biorefin.*, 13, 8, 6591–6599, 2023.
127. Zhang, X. *et al.*, Development of multifunctional food packaging films based on chitosan, TiO2 nanoparticles and anthocyanin-rich black plum peel extract. *Food Hydrocoll.*, 94, 80–92, 2019.

128. Luchese, C.L. *et al.*, Evaluation of blueberry residue incorporated cassava starch film as pH indicator in different simulants and foodstuffs. *Food Hydrocoll.*, 82, 209–218, 2018.
129. Ardiyansyah, *et al.*, The potency of betacyanins extract from a peel of dragon fruits as a source of colourimetric indicator to develop intelligent packaging for fish freshness monitoring. *IOP Conf. Ser.: Earth Environ. Sci.*, 207, 012038, IOP Publishing, 2018.
130. Oliveira Filho, J.G.d. *et al.*, Active food packaging: Alginate films with cottonseed protein hydrolysates. *Food Hydrocoll.*, 92, 267–275, 2019.
131. Alizadeh-Sani, M. *et al.*, pH-sensitive (halochromic) smart packaging films based on natural food colorants for the monitoring of food quality and safety. *Trends Food Sci. Technol.*, 105, 93–144, 2020.
132. Kurek, M. *et al.*, Development and evaluation of a novel antioxidant and pH indicator film based on chitosan and food waste sources of antioxidants. *Food Hydrocoll.*, 84, 238–246, 2018.
133. Luchese, C.L. *et al.*, Effect of blueberry agro-industrial waste addition to corn starch-based films for the production of a pH-indicator film. *Int. J. Biol. Macromol.*, 104, 11–18, 2017.
134. Luchese, C.L. *et al.*, Development and characterization of cassava starch films incorporated with blueberry pomace. *Int. J. Biol. Macromol.*, 106, 834–839, 2018.
135. Talukder, S. *et al.*, Jamun fruit (Syzgium cumini) skin extract based indicator for monitoring chicken patties quality during storage. *J. Food Sci. Technol.*, 57, 2, 537–548, 2020.
136. Qin, Y. *et al.*, Development of active and intelligent packaging by incorporating betalains from red pitaya (Hylocereus polyrhizus) peel into starch/polyvinyl alcohol films. *Food Hydrocoll.*, 100, 105410, 2020.
137. Liu, Y. *et al.*, Comparison of the structural, physical and functional properties of κ-carrageenan films incorporated with pomegranate flesh and peel extracts. *Int. J. Biol. Macromol.*, 147, 1076–1088, 2020.
138. Lee, J.J.L. *et al.*, Interfacial assembly of a cashew nut (Anacardium occidentale) testa extract onto a cellulose-based film from sugarcane bagasse to produce an active packaging film with pH-triggered release mechanism. *Food Bioprocess Technol.*, 13, 3, 501–510, 2020.
139. Sukatta, U. *et al.*, Rambutan (Nephelium lappaceum) peel extract: Antimicrobial and antioxidant activities and its application as a bioactive compound in whey protein isolate film. *Songklanakarin J. Sci. Technol.*, 43, 1, 37–44, 2021.
140. Avila, L.B. *et al.*, Carrageenan-based films incorporated with jaboticaba peel extract: An innovative material for active food packaging. *Molecules*, 25, 23, 5563, 2020.
141. Riaz, A. *et al.*, Preparation and characterization of chitosan-based antimicrobial active food packaging film incorporated with apple peel polyphenols. *Int. J. Biol. Macromol.*, 114, 547–555, 2018.

142. Lourenço, S.C. *et al.*, Application of edible alginate films with pineapple peel active compounds on beef meat preservation. *Antioxidants*, 9, 8, 667, 2020.
143. Vargas-Torrico, M.F., von Borries-Medrano, E., Aguilar-Méndez, M.A., Development of gelatin/carboxymethylcellulose active films containing Hass avocado peel extract and their application as a packaging for the preservation of berries. *Int. J. Biol. Macromol.*, 206, 1012–1025, 2022.
144. Gebrechristos, H.Y. *et al.*, Potato peel extracts as an antimicrobial and potential antioxidant in active edible film. *Food Sci. Nutr.*, 8, 12, 6338–6345, 2020.

2

Natural Fiber–Based Composite for Food Packaging

Rishika Bora[1], Hemanta Chutia[1], Monika[2], Murchana Changmai[3], Charu Lata Mahanta[1], Vimal Katiyar[4] and Tabli Ghosh[1]*

[1]Department of Food Engineering and Technology, School of Engineering, Tezpur University, Tezpur, Assam, India
[2]Department of Forestry and Wood Technology, Linnaeus University, Växjö, Sweden
[3]Department of Chemical Engineering, Birla Institute of Technology and Science Pilani, Dubai, United Arab Emirates
[4]Department of Chemical Engineering, Indian Institute of Technology Guwahati, Guwahati, Assam, India

Abstract

The recent progress in the use of renewable biobased feedstock for developing food packaging has received tremendous attention as a green alternative to fossil-derived synthetic materials. Among available biobased feedstock, natural fibers possess several noteworthy properties including renewability, low cost, lightweight, availability, and environmentally sustainable and tuneable surface properties. Fabrication of natural fiber–based biocomposites is carried out using several plants and animal fibers for improved properties of food packaging. The existing strategies for natural fiber–based composite modifications for improved properties include the use of plasticizers, antimicrobial agents, and other functional agents. The current chapter focuses on using natural fibers and their derivatives to develop composite for targeted packaging applications. Additionally, the innovative development of packaging materials based on natural fiber has received considerable attention from researchers, as the innovative packaging system can hygienically pack food products, extend their shelf life, and transport food products safely without affecting the environment.

**Corresponding author*: tablighosh1@gmail.com

Santosh Kumar, Avik Mukherjee and Vimal Katiyar (eds.) Agro-Waste Derived Biopolymers and Biocomposites: Innovations and Sustainability in Food Packaging, (33–68) © 2024 Scrivener Publishing LLC

Keywords: Natural fibers, plant fibers, animal fibers, biocomposite, food packaging, nanofibers

2.1 Introduction

Food packaging is an integral part of the safe delivery of food products to end consumers and the packaging of food provides various benefits including food preservation, protection from external agents, and promotion of packaged food products [1–3]. As shown in Figure 2.1, the important characteristic attributes of sustainable packaging materials include nontoxicity, biodegradability, thermal properties, barrier properties, mechanical properties, optical properties, and chemical properties. The inclusion of food packaging should have the function of protecting packaged products from various deteriorations/damages with respect to physical damage, chemical (natural or artificial) damage, and biological damage. However, the most important function of food packaging involves the retention of quality and acts as a safeguard for maintaining the safety attributes of the food products [4, 5]. Various commonly used packaging materials for food products are glass, plastics, paper and paperboard, and metals, among which plastic has gained wide usability due to low cost, low weight, ease of processing, versatility, etc. [6]. However, this fossil-derived packaging has

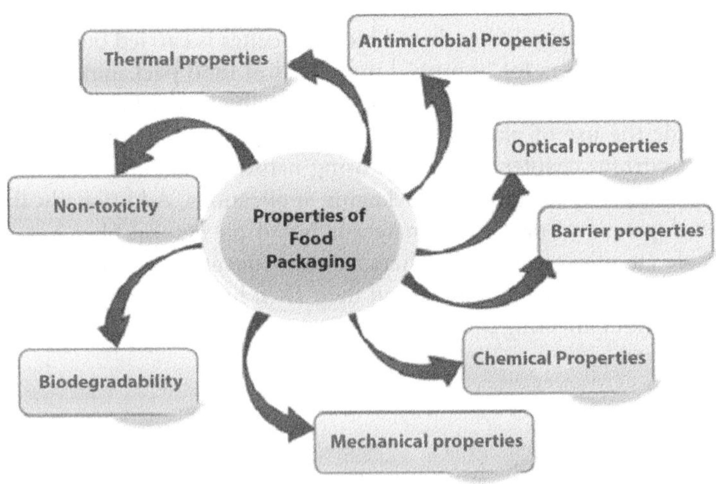

Figure 2.1 Properties of food packaging required for improved shelf life of food products.

created several problems including waste generation, pollution to the environment, nonbiodegradability, and being developed from nonrenewable sources [7]. In this context, several biobased plastics have been developed to replace fossil-derived synthetic plastics [7]. The available sustainable materials for food packaging include polysaccharides, proteins, lipids, and biodegradable polymers such as polylactic acid (PLA), polycaprolactone, poly(butylene succinate), and polyhydroxyalkanoates, etc. [7]. However, biobased and biodegradable polymers may have various limitations such as high cost, low processability, and poor functional properties. For example, starch has a limited application in food packaging due to poor mechanical properties and is hydrophilic in nature [5, 8]. However, the fabrication of starch-based nanocomposites by reinforcing natural fibers and their nanostructures can deliver tunable properties [8]. Furthermore, PLA has poor melt strength and melt elasticity, which may reduce the processing and application [9], which can be improved by developing biocomposites of PLA. Based on this, the fabrication of biocomposites using natural fibers as filler materials can solve the associated problems such as improved packaging properties (mechanical, barrier, and thermal properties).

Among available filler/reinforcement in developing polymer composite for food packaging applications, agricultural waste–based natural fibers are recognized as potential candidates to use as filler materials [7, 10]. The agricultural wastes are produced during various agricultural activities such as cultivation, aquaculture, and livestock production [11]. Polymer composites are defined as the composites of polymers as matrix materials and reinforcements, where the reinforcements/fillers can be of continuous and noncontinuous types in order to improve the properties of polymer materials [7]. Natural fiber–reinforced polymer composites can be used for various applications such as automobiles, the electronic industry, aerospace, packaging applications, biofuel production, and bioactive compound production [12]. Natural fibers (plant fibers and animal fibers) possess several noteworthy properties such as renewable nature, low cost, abundance, and sustainability. The plant fibers are stem/bast fibers, leaf fibers, wood fibers, stalk fibers, reed/grass fibers, seed/fruit fibers, etc.; on the other hand, the animal fibers include silk fibers, wool, and hair [13]. These plant and animal fibers are used for developing polymer composites for food packaging applications; however, the associated problems with natural fiber–reinforced polymer composites include interfacial adhesion issues and dispersibility issues in the polymer matrix. The existing strategies for improving the properties of natural fiber–based composite include the use of plasticizers, antimicrobial agents, and functionalizing agents.

Additionally, the natural fibers are also processed to fabricate nanostructured materials, which act as potential nanofillers in developing biocomposite-based packaging. Physical, chemical, biological, and oxidation methods are used for developing nanofibers from plant- and animal-based fibers. The different forms of nanostructured materials from cellulose are developed for the fabrication of nanocomposites for food packaging applications [14]. Cellulose nanomaterials are considered potential nanostructured materials for the fabrication of nanocomposite-based packaging materials [14]. The various available nanocellulosic materials are nanocrystals, nanofibers, nanoparticles, etc. [14]. Additionally, the animal fiber *viz* silk-derived silk nanodisc incorporated chitosan-based nanocomposites can be used in edible coating applications for the improved shelf life of fruits [15]. Considering the above discussion, the present chapter mainly demonstrates the development and use of natural fiber–based composite for targeted packaging applications.

2.2 Fiber Types

Fibers can be natural or man-made and have a significantly higher length compared to the width. Natural fibers are cheaper compared to synthetic fibers and can be obtained in higher amounts [16]. The fibers are generally used in improving the properties of other materials such as fiber-based composites. Carbon fibers are also used to develop polymer composites; however, the brittle nature and high cost of carbon fibers reduce their applications [16]. Aramid fibers are synthetic and are also very expensive, due to which there has been an upsurge in the use of natural fibers. Based on this, a discussion on the different types of fibers has been made for targeted composite formulation for food packaging applications.

2.2.1 Natural Fibers

Natural fibers are derived from several natural resources including plant-based resources, animal-based resources, and minerals. Some of the common natural fibers are jute, silk, wool, cotton, etc. The attractive properties of natural fibers include low density, low cost, high specific modulus values, nontoxicity, and recyclability. In Figure 2.2, the classification of natural fibers has been represented which include plant-based fibers (wood, leaf, stem, seed, and grass), animal fibers (silk, hair, and wool), and minerals (asbestos).

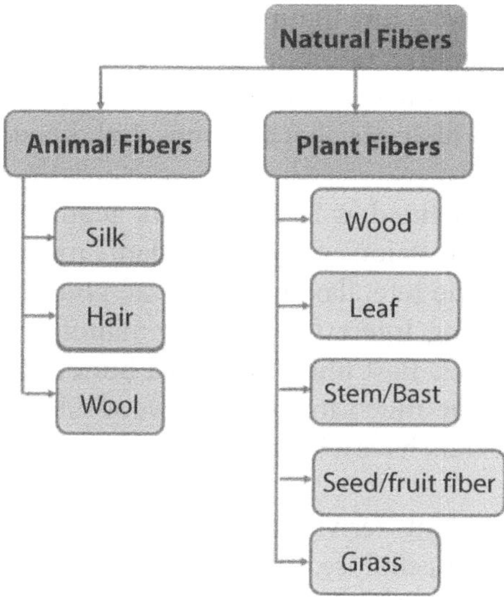

Figure 2.2 Classes of natural fibers.

2.2.1.1 Plant-Based Fibers

Fibers that are obtained from plants are known as plant fibers, e.g., cellulose, cotton, and jute. Lignocellulosic materials are recognized as plant-based fibers consisting of cellulosic materials, hemicellulose, and lignin. The available lignocellulosic materials such as wheat straw, rice straw, bamboo, fruit peels, cotton, jute, and others are considered a valuable source of cellulose. Properties of plant-based natural fibers include high strength, low density, low cost, low carbon footprint on the environment, and biodegradable in nature. Plant-based fibers utilize organic matter and renewable vegetal sources. Different types of plant-based fibers used for packaging are wood fibers, bast/stem fibers, seed fibers, stalk fibers, grass fibers, etc.

Wood Fiber: Wood is a naturally occurring renewable and biodegradable material. Among available wood fiber types, balsa wood has excellent stiffness-to-weight and strength-to-weight ratio [17]. The importance of wood as a raw material is continuously increasing as it is one of the most essential renewable resources used to meet the growing demand for bioenergy, construction material, paper, and composites.

Leaf Fiber: Leaf fiber is usually obtained by scraping away the nonfibrous materials from a leaf. The fiber which is produced can be coarser than other fibers. Leaf fiber can have cellulose contents as high as 70%.

As compared to wood, they have low lignin content, and the properties of the fibers may change depending upon the length. However, the leaves of pineapple can give fibers of much shorter length, which are also high in cellulose. Examples of leaf fiber are abaca, hemp, sisal, henequen, and cantala.

Bast Fiber: Bast fibers are defined as fibers that are acquired from the outer cell layers of the stems of various plants. The plants mostly used for the supply of bast fibers include jute, flax, hemp, kenaf, and ramie. These crops are gaining an increasing interest in various manufacturing processes of non-wood composites. Bast fibers contain a bundle of tube-like cell walls, and the fibers can be much longer than wood fibers. As compared to wood fibers, these fibers have a lower lignin content; however, the cellulose content is higher.

Cotton Fiber: Cotton is a plant-derived natural fiber. The cotton fiber contains approximately 90% of cellulose. The several noteworthy properties of cotton include nontoxic, biodegradable, absorbent, renewable, sustainable, and chemically modifiable. Cotton fiber can be defined as almost pure cellulose having the properties of soft and breathable nature, which have made it the world's most popular natural fiber. This fiber can absorb moisture readily and has high tensile strength. The development of blends of cotton fibers with other fibers such as wool, linen, nylon, and polyester can provide improved properties.

Coir: Coir is a form of natural fiber that can be obtained from the outer husk of coconut. The different types of coir fibers include curled fiber, bristol fiber, and mat fibers. Bristle coir is the longest coir fiber and contains high concentrations of lignin, which makes it stronger as compared to cotton fiber. As compared to fibers like abaca, the tensile strength of coir is lower. This fiber is resistant to microbial action and damage by salt water. The various uses of coir include packaging, agricultural uses, animal bedding, construction material, oil and fluid absorption, and doormats.

Seed Fibers: The fibers obtained from the seeds of plants are known as seed fibers such as cotton and coir.

2.2.1.2 Animal-Based Fibers

Animal fibers are considered the second most widely used fibers after vegetable or plant fibers. Animal fibers generally contain proteins, and thus, these fibers are potentially used for reinforcements in composites. There are different types of animal fibers such as wool and silk.

Wool: Wool fiber is considered a million-dollar textile fiber and is acquired from goats, sheep, rabbits, camels, alpacas, etc. The structure of wool is complex and consists of cortex, cuticle, and medulla tissues. Wool fibers have different properties based on their source, for instance, alpaca fiber is warmer compared to sheep fiber. Angora fiber, which is obtained from the Angora rabbit, is soft and thin, whereas cashmere fiber obtained from the Cashmere goat is a luxurious softwood fiber. Although different fibers possess different properties, sheep fiber is the most widely used fiber, as it is widely available and has a cheaper price.

Silk: Silk is an example of natural protein fiber, which is acquired from different varieties of sources. Most of the silk obtained from larvae of different species, such as mulberry silk is obtained from silkworms, whereas dragline silk is obtained from spiders. Silk fiber and its derivatives are extensively used in developing food packaging materials as nanocomposites.

2.2.2 Man-Made Fibers

Man-made fibers are fibers that are produced from various chemicals and can also be regenerated from plant fibers. Examples of man-made fibers include polyamide-nylon, polyester, acrylics, viscose, Kevlar, and Nomex. The structure, properties, and chemical compositions of these types are altered significantly during the manufacturing process. Man-made fibers can be classified into two types: synthetic fibers and regenerated fibers.

Synthetic fibers: These are the fibers that are synthesized from chemical processes. They were first developed in the 19th century to produce stronger fibers which can withstand high pressure. Nylon was the first man-made fiber and was discovered in the 1930s.

Polyester: Polyesters are the most widely used man-made fibers and exhibit properties similar to nylon. These fibers have low moisture absorption properties and are strong fibers. They can be blended with other fibers like cotton or wool.

Acrylic: These are synthetic fibers that are used as cheaper wool. They are manufactured as staple fibers and come in two forms: dry-spun and wet-spun. These fibers have advantages such as wrinkle resistance and sunlight fading resistance.

2.3 Plant Fiber–Based Composite for Food Packaging

The plant fibers include wood fibers (softwood and hardwood), stem fibers (jute, hemp, kenaf, ramie, and flax), leaf fibers (banana, sisal, abaca, pineapple, and palm), seed fibers (coconut and cotton), and grass (bamboo and rice) [18]. The lignocellulosic plant fibers are used as a reinforcement to develop biocomposite-based food packaging. However, the associated challenges with plant fiber–based composite include water sensitivity of fibers, poor adhesion to matrix materials, recycling of biocomposites, and dispersibility of fibers. Thus, the cellulose from plant fibers can be extracted to be used as potential fillers in biocomposite.

2.3.1 Wood-Based Composite for Food Packaging

Wood is widely used for collecting, storing, and transporting many food items such as fruits, vegetables, and grains. As an alternative to petrochemical plastics, wood-based composite plastic can be used as packaging materials [19]. The wood plastic composites are developed using hardwoods and softwoods as filler materials for versatile applications. Different matrix materials used for developing wood-reinforced plastic composite include thermoplastics such as polyvinyl chloride, polyethylene (PE), PLA, and poly(butylene adipate-co-terephthalate) (PBAT). The wood-based biocomposites with polymer matrices such as PLA and PBAT are used for manufacturing different prototypes for targeted food packaging including egg boxes, trays, and boxes [20]. Additionally, Bledzki *et al.* (2010) investigated the potential of bio-fillers (wheat husk and rye husk) in polypropylene base matrix with or without wood fibers, where the different properties such as bulk density, solubility, water absorption, and thermal degradation were reported [21]. It was observed that the wheat husk–based biocomposite has better impact strength (15%) and the rye husk–based biocomposite presented better elongation at break (110%) as compared to softwood biocomposites. However, the properties of wood-based composites can be improved using different types of additives such as colorants, UV stabilizers, foaming agents, and coupling agents.

2.3.2 Stem/Bast-Based Composite for Food Packaging

2.3.2.1 Kenaf Fiber

Kenaf fiber is used as a filler material for thermoplastics and other packaging materials, and to improve their properties, active compounds are added. In this context, antimicrobial films are being developed by introducing an active compound with antimicrobial activity in packaging material. The appropriate selection of an antimicrobial agent based on the type of food as well as preventing microbial growth to extend the shelf life of food products. However, it is required that the antimicrobial agent must be safe to utilize in food. Tawakkal *et al.* (2016) successfully fabricated the PLA/kenaf fiber–based biocomposite films containing thymol using melt blending followed by heat pressing for active food packaging applications [22].

2.3.2.2 Jute Fiber

Recently, a group in Bangladesh introduced Sonali Bag made from modified jute cellulose, which was biodegradable, and degraded in soil within 3 months [23]. Modified jute cellulose–based bags can be widely utilized in packaging industries where they can solve the problems associated with synthetic PE bags due to their low cost, biodegradability, and comparable or better physical and mechanical properties in comparison to PE bags. Additionally, jute fiber impregnated with red grape pomace extract (RGPE)–based sustainable food packaging material with antioxidant and antimicrobial properties has been developed, which was effective against various microorganisms such as *Escherichia coli*, *Pseudomonas aeruginosa*, and *Staphylococcus aureus* [24].

2.3.2.3 Ramie Fiber

Chitosan-based biocomposite film reinforced with lignin and ramie fiber has been fabricated using the casting method, and their mechanical, water absorption, and antioxidant properties were investigated [25]. The resultant food packaging films containing ramie fiber (20 wt%) showed improvement in tensile strength by 29.6%. Moreover, the incorporation of lignin (20 wt%) remarkably improved the antioxidant activity by up to 288% and decreased the water absorption by 41.2%. This film had a huge potential for the packaging of meat products. Thus, the ramie fiber can be a potential filler for chitosan-based biocomposites for food packaging applications.

2.3.2.4 Hemp Fiber

The lignocellulosic raw materials such as hemp fiber–based packaging can be used to replace single-use plastic packaging. It is biodegradable and compostable, which makes it a suitable choice for industrial-grade plastic and food contact materials. Additionally, hemp hurd powder, the byproduct of the hemp plant, can be processed to provide better antibacterial activity, and can be used as a filler in composite food packaging. The different types of hemp hurd powder include retted, semi-retted, and non-retted hemp hurd powder. Among the available, the heat-treated hemp hurd powder is efficient in antibacterial properties, where the 90% reduction in

CFU tested against *E. coli* can be obtained using heat-treated hemp hurd powder (160°C for 2 hours) [26].

2.3.3 Leaf-Based Composite for Food Packaging

2.3.3.1 Palm Fiber

The green composite based on PBAT as a matrix and peach palm as fiber has been developed using the melt mixing method for targeted food packaging applications [27]. The use of milled peach palm fibers with surface treatment using glycidoxypropyl-trimethoxy silane as a filler in PBAT-based biocomposite improved the mechanical properties. The development of cassava starch (CS)–based hybrid biocomposite with sugar palm fiber (SPF), cassava bagasse, and fructose as a plasticizer can be used as a food packaging material, where the incorporation of SPF remarkably improves the physical properties of the hybrid composite [28]. The addition of SPF led to significant changes in the resultant properties of the casava starch/cassava bagasse–based biocomposite films. Furthermore, the SPF incorporation reduces the density, water absorption, and water solubility of the hybrid composites, while the thickness and crystallinity are enhanced. The application of biocomposites of chitosan- and modified montmorillonite–based coating on the date palm fiber (DPF) trays can improve the mechanical and antibacterial properties against *E. coli*, *S. aureus*, and *P. aeruginosa* [29].

2.3.3.2 Sisal Fiber

Xie *et al.* (2018) developed a biodegradable starch/sisal fiber biocomposite with a nested structure to replace expandable polystyrene-based packaging [30]. Orue *et al.* (2018) studied the effect of linseed oil (epoxidized and non-epoxidized) and soyabean oil (epoxidized and non-epoxidized) as plasticizers in PLA/NaOH-treated sisal fiber–based biocomposite [31]. In this study, the mentioned biocomposite with epoxidized oil offers improved tensile modulus and elongation at break around 30 and 70%, respectively, as compared to pure PLA [31]. Additionally, it has been aimed to develop and characterize propionylated sisal fiber reinforced biocomposite using poly(3-hydroxybutyrate-co-valerate) (PHBV) matrix for household appliances and packaging [32]. The mechanical study has confirmed that the developed composite has better properties than that of the unmodified sisal/PHBV biocomposite, which indicated good interfacial adhesion between the PHBV matrix and sisal fiber. Furthermore, the bacterial growth test confirmed that the propionylated sisal fiber helped to reduce the colonization of *E. coli* and *S. aureus*.

Hybrid biocomposite was developed by melt mixing of PLA with granulated sisal and hemp fibers using extrusion followed by injection molding for different applications such as packaging, automotive, agriculture, and electronics [33]. The results showed significant improvement in the tensile and flexural properties of the hybrid biocomposite as compared to PLA. Additionally, the PLA-based hybrid biocomposite using flax and jute fiber can provide enhanced mechanical and biodegradable characteristics, which make them suitable candidates to be used in food packaging [34].

2.3.3.3 Pineapple Leaf Fiber

Aji *et al.* (2013) fabricated the high-density PE (HDPE)–based composite with pineapple leaf and kenaf fibers to study its mechanical properties [35]. The use of pineapple leaf fiber in HDPE-based composite has the potential to improve the flexural and tensile properties, whereas the kenaf fiber in HDPE composite reduced water uptake and impact strength. Also, PHB-coated pineapple leaf fiber–based paper has been utilized as a promising material for biodegradable paper packaging [36]. Additionally, PLA/pineapple leaf fiber–based paper packaging with an antimicrobial agent has been developed to promote food safety and improved shelf life [37]. To enhance the water resistance and mechanical properties, many biocoating solutions were introduced as coatings on pineapple leaf fiber–based paper. The application of biocoating of chitosan, beeswax, alginate/gellan gum, shellac, and beeswax/chitosan has been carried out on pineapple leaf pulp–based paper [38].

2.3.3.4 Banana Fiber

The effect of banana fiber on the mechanical properties of polyvinyl alcohol (PVA)–based composites was studied using the hand layup technique [39]. The tensile strength, percentage elongation, and flexural strength for PVA/banana fiber–based composite were 18.57 MPa, 9.1%, and 10.2 MPa, respectively, which can be utilized in food packaging applications due to its sufficient strength of handling and good biodegradability. According to Rodríguez *et al.* (2021), a hybrid matrix (mixture of HDPE and PLA) with banana fiber can be used to replace conventional bottle lids [40]. The surface modification of fibers using various methods such as chemical, physical, or biological methods is the possible solution to overcome most of the limitations associated with natural fiber, especially to promote the compatibility between polymer-based matrix and fiber. The incorporation of Cloisite 30B and chemically treated banana fiber into PLA-based bionanocomposites can be used to fabricate composite material for disposable applications [41]. Additionally, different types of coupling agents can be used to improve the interfacial compatibility between PLA matrix and filler and to improve packaging parameters such as crystallization temperature, crystallinity level, and flammability with better dispersion.

2.3.4 Seed/Fruit-Based Composite for Food Packaging

The various seed/fruit fibers are used as reinforcement materials in developing composites for targeted food packaging. The waste generated from different fruit products such as apples, bananas, olives, mangoes, cranberries, coconuts, blueberries, pineapples, and jackfruits is used for developing composite-based packaging. The various fruit parts are treated for surface modification to improve the compatibility and properties of formulated packaging materials. The seed/fruit fibers are used as filler materials in thermoplastic materials. In this context, Bajracharya *et al.* (2017) employed 10 to 30 wt% of cotton burrs (obtained from cotton gin waste) mixed with stem fibers to fabricate PLA-based hybrid biocomposite [42]. PLA, hybrid fiber (mixture of cotton burr and flax fiber) with maleic anhydride polypropylene (2 wt%) was dried and melted extruded to get composite granules, and the dried granules were compressed by hot press to produce biocomposite. Results indicated that the flexural modulus was enhanced by 42% due to the incorporation of 30 wt% cotton gin waste in comparison to pristine PLA. The modulus of elasticity of the biocomposites containing 30 wt% of cotton burr was comparable to a hybrid biocomposite loaded with various loadings of flax fibers.

2.3.5 Grass-Based Composite for Food Packaging

2.3.5.1 Bamboo Fiber

The bamboo fiber is lignocellulosic biomass, which is extensively utilized to extract cellulose for versatile applications. In Figure 2.3, the images of bamboo fiber, optical microscopy of bamboo fiber (untreated), and surface electron microscopy of bamboo fiber (untreated) have been represented. Bamboo waste is a type of agricultural waste that is processed using various methods such as chemical process and mechanical processes. The bamboo fiber is used for improving the texture of food products such as

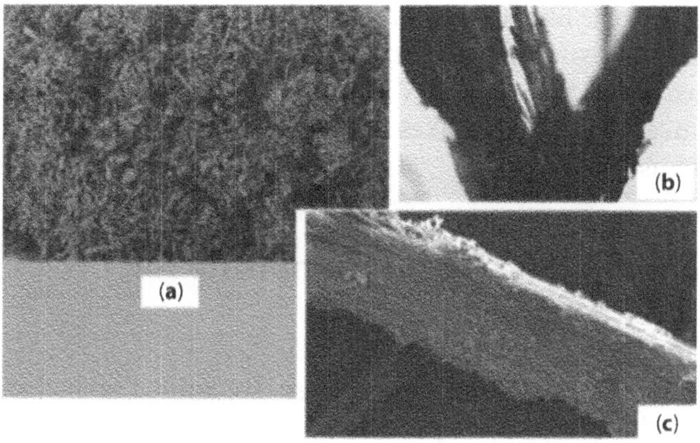

Figure 2.3 (a) Bamboo fiber, (b) optical microscopy of bamboo fiber (untreated), and (c) surface electron microscopy of bamboo fiber (untreated).

dairy products, bakery products, beverages, meat products, and fish products. However, in the current century, bamboo fiber–derived cellulose and nanocellulosic materials are used in developing sustainable packaging. The use of a mixture of bamboo fiber powder combined with silica aerogel, called TBFP, has been exploited as a reinforcing material in developing modified poly(lactic acid) (MPLA) biocomposite [43]. The resultant MPLA/TBFP composites possess high oxygen/water barrier proprieties, water resistance, and tensile strength as compared to neat PLA- and PLA/BFP-based biocomposites. The results also showed improvement in biodegradability of the MPLA/TBFP and PLA/BFP biocomposite with the increase in filler content. In addition, PLA/BFP and MPLA/TBFP showed good cytocompatibility. Some manufacturers such as BioHeroes already proposed biocomposite made of bamboo fiber and PLA to produce disposable food and beverage packaging [44]. The manufacturing of PLA/BF biocomposite starts with the stacking of BF and PLA layers alternatively and finally compressed to get the desired packaging materials. PLA/BF biocomposite can also be recycled using a melt extrusion process.

2.3.5.2 *Rice Husk and Wheat Straw*

Natural fibers such as rice husk, wheat straw, almond shell, and sea grass are used as fillers in developing composite packaging. The poly(hydroxybutyrate) (PHB) in the combination of three types of lignocellulosic waste such as rice husk, almond shells, and seagrass has been developed [45], which showed a reduction in thermal stability and water barrier properties.

Among these waste fibers, almond shell–based composite showed the most balance characteristics. Berthet *et al.* (2015) observed that the incorporation of wheat straw fiber to PHBV had a remarkable improvement in water vapor permeability, which makes it beneficial for the packaging of respiring food items [46].

2.4 Animal Fiber–Based Composite for Food Packaging

The second most widely used natural fibers are animal fibers [47]. Animal fibers are made up of certain proteins, the characteristics of animal fibers depend on the types, size, and structure of these proteins, providing a wide range of different characteristics, i.e., all the fibers do not have the same characteristics, and sometimes fibers are different even within a species [48]. Different types of animal fibers such as spider silk, mulberry silk (from *Bombyx mori*), tussah silk, wool fibers, cashmere wool, Yak fiber, chicken feathers, and hair have been reported for the use of human welfare and have shown the potential to outperform the most synthetic fibers [48]. The animal fibers are used as a filler material to develop composite food packaging. Animal fibers possess some unique properties and thus can be used as a potential reinforcement material in developing composite materials for food packaging. The most used animal fibers as filler in packaging are silk, wool, and feathers.

2.4.1 Silk-Based Composite for Food Packaging

Silk is generally made of highly structured proteins and shows a broad range of different characteristics such as extreme extensibility to high tensile strength and resistance to chemicals The toughness and compatibility of silk with biological systems can easily surpass most synthetic as well as natural fibers [48]. From the silk, silk fibroin (SF) is extracted following the degumming method, and based on the processing conditions, different morphological structures such as fibers, solutions, powders, films, and gels can be obtained [49]. Due to its nontoxicity, good biocompatibility, and degradability properties, SF is widely used in the food processing and pharmaceutical fields. The blending of SF, with some other suitable polymers, can enhance the desirable properties (mechanical and thermal) of the SF films, which is a remarkable research area to explore as innovative food packaging technology [50].

In the food packaging field, the application of silk and its derivatives with other polymer composites has been used for the design and development of various food packaging materials including active and intelligent packaging, packaging with good mechanical properties, and antimicrobial active packaging [49, 51, 52] that are summarized in Table 2.1. The combined use of SF materials with active components is beneficial for use as active food packaging for associated antimicrobial properties. The different combination of SF with active agents for packaging application includes SF with essential oils for longer shelf life of poultry meat [49], SF with eugenol and *Lycium barbarum* polysaccharide nanofibers [51], SF/PVA/silver nanoparticles and cellulose nanofibers of bacterial origin [52].

In natural form, SF is arranged in crystals of beta-sheet alternated by amorphous regions, which provides strength and resilience to the protein structure. Silk nanodisc–based edible chitosan nanocomposite [15], SF-PVA–based packaging [50], n-octadecane–incorporated silk hydrogel [53], SF-based edible coating [54], dialdehyde bacterial cellulose nanocrystal incorporated silk sericin films [55], SF/PVA/silver nanoparticles, and cellulose nanofiber–based composite have been developed that had improved mechanical, hydrophobic, and water barrier properties [52]. Thus, SF-based composite can be used for better shelf life of food products and are promising alternative material for the development of food packaging with a naturally derived material.

2.4.2 Wool-Based Composite for Food Packaging

Wool fibers are a potential source of inexpensive material and they are popular for their elasticity and abrasion resistance behavior and thus widely used in the production of various film/packaging materials. However, the poor mechanical properties of wool fibers lead to limited applications [56, 57]. The 95% weight of wool fibers is pure keratin, which is biocompatible, renewable, sustainable, and nontoxic in nature [58]. The use of keratin-enriched packaging is very less, as keratin has low mechanical properties [56]. The keratin-based fibers are used for developing value-added sustainable composite [57]. The application of wool and its extracts with other polymer composites has been used to design and develop various food packaging materials. Zhang *et al.* (2022) have developed moisture-permeable wool fabric with excellent antibacterial properties using phase-transited lysozyme as a secondary reaction platform [59]. Similarly, Ramirez *et al.* (2017) have developed wool keratin- and citric acid–based film for the shelf-life improvement of carrots, where the film

Table. 2.1 Animal fiber–based composite and their applications.

Composition of composite	Application	Reference
Silk fibroin nanoparticles, eugenol, and *Lycium barbarum* polysaccharide nanofibers	Successfully prepared nanofiber-based active food packaging, which showed as promising food packaging materials	[49]
Silkworm/silk sericin, chitosan, and carbon dots	Enhanced antimicrobial and antioxidant properties of film and keep the litchi fruit fresh for a longer time	[68]
Silk fibroin and polyvinyl alcohol	Significant enhancement of physical and thermal properties of the blended films with more than 100% elongation at break for a few blend films	[50]
Silk fibroin, polyvinyl alcohol, silver nanoparticles, and TEMPO-oxidized bacterial cellulose nanofibers	Developed active packaging has improved mechanical, hydrophobic, water barrier, UV blocking, and antimicrobial properties	[52]
Silkworm (*Bombyx mori*) and pomegranate peel powder	Pad with antimicrobial properties maintained microbial count	[69]
Silk nanodisc and edible chitosan	Edible packaging has improved thermal, optical, and mechanical properties and increased the shelf life of banana fruit	[15]
Silk sericin and dialdehyde bacterial cellulose nanocrystal	Enhanced mechanical, UV light-blocking, water-resistant, and antioxidant properties	[55]
Silk fibroin nanofibers and thyme essential oil	The developed membrane had a high antimicrobial nature and increased the shelf life of poultry meat	[51]

(Continued)

Table. 2.1 Animal fiber–based composite and their applications. (*Continued*)

Composition of composite	**Application**	**Reference**
n-Octadecane and silk hydrogel	Developed composite hydrogel film had high temperature sensitive	[53]
Silk fibroin	Developed thin membranes managed the freshness of food (strawberries and bananas)	[54]
Wool, phase-transited lysozyme, and silver nanoparticles	Composite had improved moisture permeability and excellent antibacterial properties	[59]
Wool keratin, carbon fibers, and graphene oxide	Developed functional composite had ultra-high sensitivity toward humidity sensing, and can be used for various food applications	[58]
Wool yarn wastes and shredded Tetra Pak wastes	Fabricated hybrid composites had better properties in terms of thickness swelling, and water absorption compared to commercial particleboards	[60]
Wool keratin and citric acid	Developed film had an excellent biocidal effect and high elongation value (600%), which improved the shelf life of the packed carrot	[56]
Chicken feather keratin and turmeric starch	Developed biocomposite showed improved tensile strength and surface smoothness	[64]
Hair fiber keratin	Ohmic heating-treated films showed distinct properties	[70]
Gelatin and chitosan	Developed films had excellent mechanical, surface hydrophobic, color, barrier, and thermal properties	[66]

(*Continued*)

Table. 2.1 Animal fiber–based composite and their applications. (*Continued*)

Composition of composite	Application	Reference
Chicken feathers keratin and microcrystalline cellulose	Keratin-based bioplastics had high structural strength and can be used for manufacturing food containers	[61]
Poultry feather keratin, edible dialdehyde carboxymethyl cellulose, and glycerol	Edible film had good UV-barrier properties and transmittance	[62]
Gelatin-chitosan blend and *Ferulago angulate* essential oil	Blended film had increased antimicrobial properties and enhanced the shelf life of turkey meat	[67]
Gelatin from bovine skin and chitosan	Composite film had enhanced mechanical strength up to five folds compared to neat gluten	[71]
Keratin	Keratin-based biofilms and biofibers can be used for food packaging	[72]
Chicken feather protein, gelatin, and clove oil	Composite film with 1.5% clove oil can be used as an active packaging material for smoked salmon	[63]
Feather keratin, carboxymethyl cellulose, and glycerol	The developed film showed a significant effect on the mechanical properties of blend films	[65]

has shown an excellent biocidal effect and high elongation value (600%) and enhanced the shelf life of carrots as compared to a commercial film [56]. Wool and its extracts are combined with other materials such as carbon fibers [58] and shredded Tetra Pak wastes [60], which can be used as a composite film for food packaging.

2.4.3 Other Animal Fiber-Based Composite

Several studies have been performed on the use of animal-based proteins including gelatin and keratin to develop food packaging [61, 62]. Apart from the silk and wool animal fibers, chicken feather protein is one of the potential sources to develop edible film due to its biodegradability, environmental-friendly nature, abundant availability, low toxicity, insolubility, hydrophobicity, and low density [63, 64]. Chicken feathers consist of 90% keratin proteins, which is a common inedible material produced in large quantities in the poultry industry [61]. Direct application of chicken feather protein as a food packaging material is less effective due to its brittleness in comparison to other proteins; therefore, feather keratin must be modified to meet the requirements for food packaging materials [62].

Chicken feather–based biocomposites are widely utilized in developing food packaging such as chicken feather keratin– and turmeric starch–based high mechanical strength film [64]; chicken feather keratin– and microcrystalline cellulose–based improved bioplastic for developing food containers [61]; poultry feather keratin–, edible dialdehyde carboxymethyl cellulose–, and glycerol-based edible food packaging [62]; and feather keratin–, carboxymethyl cellulose–, and glycerin-based higher strength packaging materials [65]. Additionally, materials such as carboxymethyl cellulose, glycerin, cellulose, and chitosan are used with animal fibers for enhanced mechanical and physical properties [65]; water vapor barrier properties [62]; surface hydrophobic property, color, and thermal properties [66]; and antimicrobial properties of composite films [67]. Composite films based on chicken feather protein, gelatin, and clove oil have been developed that showed antimicrobial properties and enhanced shelf life of smoked salmon [63].

Apart from chicken feathers, other materials such as gelatin [66, 67], hair keratin, and other poultry feather keratin [62] have been used for the development of food packaging materials. Naseri *et al.* (2020) developed gelatin, chitosan, and *Ferulago angulata* essential oil–based film with antimicrobial properties, which enhanced the shelf life of stored turkey meat [67]. Thus, over the years, animal fibers and vegetable fibers have been viewed as an attractive biomaterial for food packaging, but still, more research is needed for better insight [64, 73].

2.5 Nanomaterials from Natural Fiber

Natural fibers have several advantages over synthetic fiber; however, they also have several limitations, and thus, nanofiber from these natural fibers could be better alternatives for developing composite materials for food packaging applications. The techniques used for synthesis of nanofibers are physical methods such as ultrasonication, homogenization, and ball milling [76]; chemical methods such as sulfuric acid, hydrochloric acid hydrolysis, and alkaline treatment [75]; biological methods such as enzymes treatment using cellulase, endoglucanases, exoglucanases, and cellobiases [77]; and oxidation methods through oxidative radicals such as 2,2,6,6-tetramethylpiperidine-1-oxyl radical–mediated (TEMPO) oxidation [78]. In physical methods, the mechanical stress in terms of high RPM, high-intensity ultrasonic wave, high-pressure stress, and shear force is subjected to the larger fiber molecules/materials to obtain nanofibers of nanometer diameter, whereas in chemical methods, nanofibers are developed through alkali treatments and acid digestion, which resulted in a higher yield of nanocrystalline nanofibers. In biological methods, fibers are treated with different types of enzymes to cleave the fiber structures and make simpler ones, whereas in oxidation methods, nanofibers are obtained through the oxidation of fibers using different TEMPO-oxidized cellulose nanofibers and others [78, 79].

In recent years, several researchers have studied the production of nanofibers from natural fiber (Table 2.2) using wet-spinning [80, 81], dry-spinning [82, 83], electrospinning [49, 71, 84], high-pressure homogenization [85–87], ultrasonication [88–90], acid hydrolysis [15, 75, 91], enzymatic treatments [77, 78], etc. Also, synergetic methods by the combination of two or more techniques such as the combination of physical and chemical [92, 93], chemical-ultrasonic treatment [94, 95], ultrasound-electrospinning [96], and alkali-acid treatment [97] have been reported to synthesize nanofiber. Ample research studies are going on in the development of natural origin–derived nanofiber-based food packaging to overcome the problem associated with natural fibers. Nanofibers are also often combined with other biopolymers/materials to obtain desirable packaging materials. The technological innovations in biobased natural polymer technology combined with nanotechnology have become potential to develop bionanocomposites with superior properties [74, 98]. As compared to micro size, nanofibers confer some unique characteristics such as exceptional strength, lightweight, transparency, and better excellent biocompatibility, so it is considered a promising material for coatings, especially edible coating-based food packaging as well as paper processing [76, 99]. Ghosh *et al.* (2021) have developed silk nanostructured materials incorporated in edible food packages for the shelf-life enhancement of bananas [15]. The developed film showed better mechanical and physical properties. Apart from the improvement of mechanical and physical properties, nanofibers also have been used for the development of active packaging. Lin *et al.* (2019) developed cold plasma-treated SF nanofibers/thyme essential oils electrospinning materials that showed antibacterial properties against *S. typhimurium* on poultry meat and can be used for antimicrobial packaging [51].

Table 2.2 Nanofibers synthesized from natural fibers and their uses/advantages.

Sl. no	Materials	Technique	Advantages/outcomes	Reference
Chemical treatments				
1	Silk nanodisc	Acid hydrolysis	Edible film had superior thermal, hydrophobic, optical, mechanical, and physicochemical properties and enhanced the shelf life of food products	[15]
2	Cellulose nanofibers	Supercritical carbon dioxide–assisted mild acid hydrolysis	Used for developing cellulose nanopaper with tensile strength of 75.7 MPa	[91]
3	Cellulose nanomaterials	Sulfuric acid hydrolysis	Improved dewatering/drying process and improved the mechanical properties of tapes and yarns produce	[75]
4	Cellulose nanofibers from pomelo fruit	Alkali treatment and acid hydrolysis	Alkali treatment helped in eliminating the non-cellulosic constituents; however, acid hydrolysis was successfully used in individualizing cellulose nanofibers	[97]
Enzyme				
5	Cellulose nanofibrils	Mixed cellulase hydrolysis (exoglucanase and endoglucanase) and mechanical grinding	Controlled cellulose nanofiber length improved uniformity of cellulose nanofibers	[77]

(Continued)

Table 2.2 Nanofibers synthesized from natural fibers and their uses/advantages. (*Continued*)

Sl. no	Materials	Technique	Advantages/outcomes	Reference
Electrospinning				
6	Silk fibroin nanoparticles and eugenol	Electrospinning	Antibacterial activity and active food packaging	[49]
7	Nanofibers and gelatin from bovine skin	Nozzle-less electrospinning	Production of fine nanofibers on a large scale· and improved mechanical properties of edible films	[71]
8	Nanofibers	Electrospinning	Obtained homogeneous and ultrafine nanofibers	[84]
9	Silk fibroin nanofibers and thyme essential oil	Electrospinning	Modified properties such as stability, hydration, barrier, and mechanical properties	[51]
10	Cellulose nanofibers from sugar beet pulp	Steam explosion, high-speed blending, ultrasonic treatment, and bleaching	Environmentally friendly extraction process	[100]

(*Continued*)

Table 2.2 Nanofibers synthesized from natural fibers and their uses/advantages. (*Continued*)

Sl. no	Materials	Technique	Advantages/outcomes	Reference
Coprecipitation				
11	Magnetic cellulose nanofiber	Coprecipitation	Developed magnetic cellulose nanofiber (mgCNF) dispersed chitosan-based edible nanocoating with superior mechanical, thermal, optical, and texture properties and improved the texture of fresh-cut pineapple	[101]
High-pressure homogenization				
12	Cellulose nanofibers	Pretreatment and high-pressure homogenization	Pretreatment and high-pressure homogenization influence both fibrillation yield and morphology	[87]
13	Nanocellulosic fibers	Homogenizer	Good nanocellulose dye adhesion to the surface of the fabrics	[102]
14	Cellulose nanofibrils from microcrystalline cellulose	High-pressure homogenization	Cellulose nanofibrils reinforced hydroxypropyl cellulose films delivered high strength	[85]

(*Continued*)

Table 2.2 Nanofibers synthesized from natural fibers and their uses/advantages. (*Continued*)

Sl. no	Materials	Technique	Advantages/outcomes	Reference
Ultrasonication				
15	Cellulose nanofiber from rice straw	Ultrasonic-assisted method	Sonication time influences the disintegration of cellulose fibers	[103]
16	Chitin nanofibrils from crab shells	Ultrasonic-assisted method	The diameter and length of chitosan nanofibers are less than 100 nm and 3 μm, respectively	[90]
17	α-Chitin nanofibers	Ultrasonication	Highly transparent and flexible chitin films were prepared from chitin nanofiber hydrogels	[89]
Combination of various treatments				
18	Ginger cellulose nanofiber	Chemical-ultrasonication	Transparent film, high thermal stability, high antimicrobial activity	[95]
19	Nanofibers from bamboo fibers	Chemical-ultrasonication	Ultralong and highly uniform cellulose nanofibers (CNFs) with lengths >1 mm, diameters of 30 to 80 nm, and aspect ratios >10,000 were successfully prepared from bamboo fibers	[104]
20	Cellulose nanofibers	Chemical- ultrasonication process	Crystallinity of nanofibers increased using chemical treatments and can be used in green nanocomposites, filtration media, and optically transparent films	[105]

2.6 Natural Fiber–Based Composite for Circular Economy

Value-added biobased materials have gained significant importance in the recent decade due to the ever-expanding environmental issues and rising awareness of environmental safety. Specific interests have been laid on controlling the emissions of greenhouse gases and other undesirable waste materials, which may result in severe pollution problems. Thereby, the focus lies in the replacement of the existing fossil sources with biodegradable natural materials, thus paving the way for a circular economy with an infinite supply of resources. The usage of such bioresources should ensure that carbon footprint and water footprint are under perfect control [106, 107]. The materials that would be used for the preparation of such biofriendly products would play a vital role in contributing toward a circular economy. Based on this, various case studies for using natural fibers for versatile applications to move toward a circular economy have been discussed in this section.

(i) Keratin is a type of structural protein, which helps in forming the tissues of scales, nails, hair, feathers, etc. Keratin-based bioproducts can be obtained by the valorization of keratinous waste and are found to have a wide range of applications due to their natural properties, which enables a step toward a circular bioeconomy. Such keratin-based starting material is widely found in agricultural processing, wool processing, and grooming industry wherein the keratin needs to be extracted from the said materials in an eco-friendly manner and economically viable way to contribute toward the bioeconomy. Many extraction processes are available for the extraction of keratin which includes processes such as chemical hydrolysis, chemical-free hydrolysis, microbial hydrolysis, and enzymatic hydrolysis. The extracted keratin can be utilized in the preparation of keratin-based composites using methods such as lyophilization, casting, electrospinning, compression molding, fabrication of sponges, and 3D printing [108].

(ii) Sugar beet pulp residues are another important type of biomaterial that can be utilized for the preparation of biocomposites. Usually, the residue is produced during the preparation of sugar from beet which in turn is dumped for cattle feedstock. Although the use of beet for sugar production is still limited. The sugar beet pulp residue can be converted into useful products by using processes such as hydrolysis (enzyme/acid), pyrolysis, and nanofiber extraction. The modified waste residue could then be utilized in reinforcements for polymer composite, water treatment industry, bioenergy, etc. [109].

(iii) Cellulose sludge from wastewater plants has been utilized in the preparation of a cellulose plastic composite which falls in line with the requirements of circular bioeconomy wherein the aim is to replace wood with cellulose sludge. The improved property can be obtained by the incorporation of maleic anhydride and vinyltrimethoxysilane. The addition of maleic acid substantially improved the flexural and tensile strength of the composite along with better environmental and economic execution for the cellulose polymer composite [110].

(iv) Green PE composites using natural fibers (30% by weight) such as coir and agave are also widely used in the field of food packaging applications. Fibers are treated with maleated PE resulting in the improvement of several properties such as better wettability, tensile modulus as high as 345 MPa, tensile strength of 15.3 MPa, flexural strength of 100%, and flexural modulus of 34%. These composites can be used widely in automotive and packaging applications [111].

(v) Orchard wastes specifically from apple orchards have also gained interest among researchers to produce different bioproducts from the orchard waste using processes like anaerobic digestion and composting. The various products include biofuel, animal feed, biofertilizers, and biochemicals which are directed toward sustainable resource recovery and could progressively be linked with a circular bioeconomy [112].

(vi) Lignocellulose fibers are another set of natural materials that can be utilized in the building industry in the form of reinforcements known as natural fiber reinforcement polymers thereby decreasing the waste generation. The application can be in terms of structural abilities and architectural requirements like incorporation in furniture and acoustic panels [113]. Fiber-reinforced composites like the glass fiber–reinforced epoxy polymer can also be utilized for contributing toward circular bioeconomy. The aim is to extract the glass fibers from the epoxy matrix by using peroxide and tartaric acid as oxidizers to disintegrate the epoxy *via* the microwave-assisted chemical oxidation process. The process is considered to be a low-energy approach for recovering fibers from the end-of-life proceeding composites [114].

2.7 Conclusion and Future Perspective

Natural fibers having numerous benefits can be used for developing food packaging. Natural fibers are effective in dealing with existing environmental concerns. Utilization of natural materials to produce fibers that can be incorporated into the packaging material is considered one such approach that will gain more interest soon. Additionally, agricultural wastes have been identified as excellent candidates for the extraction of fiber to be used in packaging applications to replace fossil-based packaging. Apart from the packaging scenario, many countries are already focusing on the utilization of natural fibers to be incorporated in textile industries, automobile manufacturing, and dentistry. The life cycle of the products and the ratio of biobased material in the composite along with wastes generated and the end-of-life composites are important considerations for the product's full life cycle. At the same time, the introduction of the composite product into the market is beneficial in the replacement of already available products keeping in mind the aspects of economic viability and market expansion with biobased products.

Acknowledgment

This research was supported by the Seed Money (Research Grant) funded by Tezpur University, Tezpur, Assam, India (Memo No. TU/Fin/Conc./R/2021-22/410) and Start-up Research Grant (SRG) through Science and Engineering Research Board (SERB)-DST, New Delhi, India (SERB File no: SRG/2022/001406).

References

1. Katiyar, V. (Ed.), *Bio-Based Plastics for Food Packaging Applications*, Smithers Pira, United Kingdom, 2017.

2. Mondal, K., Ghosh, T., Bhagabati, P., Katiyar, V., Sustainable nanostructured materials in food packaging, in: *Dynamics of Advanced Sustainable Nanomaterials and Their Related Nanocomposites at the Bio-Nano Interface*, N. Karak (Ed.), pp. 171–213, Elsevier, St. Louis, Missouri, 2019.
3. Katiyar, V. and Ghosh, T., *Nanotechnology in Edible Food Packaging: Food Preservation Practices for A Sustainable Future*, Springer Nature, 2021.
4. Ghosh, T., Nakano, K., Katiyar, V., Curcumin doped functionalized cellulose nanofibers based edible chitosan coating on kiwifruits. *Int. J. Biol. Macromol.*, 184, 936, 2021.
5. Ghosh, T. and Katiyar, V., Nanochitosan functionalized hydrophobic starch/guar gum biocomposite for edible coating application with improved optical, thermal, mechanical, and surface property. *Int. J. Biol. Macromol.*, 211, 116, 2022.
6. Kim, Y.T., Min, B., Kim, K.W., General characteristics of packaging materials for food system, in: *Innovations in Food Packaging*, J.H. Han (Ed.), pp. 13–35, Academic Press, USA, 2014.
7. Ghosh, T., Borkotoky, S.S., Katiyar, V., Green composites based on aliphatic and aromatic polyester: Opportunities and application, in: *Advances in Sustainable Polymers*, V. Katiyar, R. Gupta, T. Ghosh (Eds.), pp. 249–275, Springer, Singapore, 2019.
8. Savadekar, N.R. and Mhaske, S.T., Synthesis of nano cellulose fibers and effect on thermoplastics starch based films. *Carbohydr. Polym.*, 89, 146–151, 2012.
9. Narmon, A.S., Dewaele, A., Bruyninckx, K., Sels, B.F., Van Puyvelde, P., Dusselier, M., Boosting PLA melt strength by controlling the chirality of co-monomer incorporation. *Chem. Sci.*, 12, 5672, 2021.
10. Ghosh, T., Dhar, P., Katiyar, V., 2. Nanocellulose: Extraction and fabrication methodologies, in: *Cellulose Nanocrystals: An Emerging Nanocellulose for Numerous Chemical Processes*, V. Katiyar and P. Dhar (Eds.), 23, De Gruyter, Berlin, Boston, 2020.
11. Obi, F.O., Ugwuishiwu, B.O., Nwakaire, J.N., Agricultural waste concept, generation, utilization and management. *Niger. J. Technol.*, 35, 957, 2016.
12. Verma, D. and Senal, I., Natural fiber-reinforced polymer composites: Feasibiliy study for sustainable automotive industries, in: *Biomass, Biopolymer-Based Materials, and Bioenergy*, D. Verma, E. Fortunati, S. Jain, X. Zhang (Eds.), pp. 103–122, Woodhead Publishing, 2019.
13. Ramesh, M., Palanikumar, K., Reddy, K.H., Plant fibre based bio-composites: Sustainable and renewable green materials. *Renew. Sustain. Energy Rev.*, 79, 558–584, 2017.
14. Ghosh, T., Dhar, P., Katiyar, V., Nanocellulose for food packaging applications, in: *Cellulose Nanocrystals*, V. Katiyar and P. Dhar (Eds.), p. 157, Walter de Gruyter, Berlin, Germany, 2020.
15. Ghosh, T., Mondal, K., Giri, B.S., Katiyar, V., Silk nanodisc based edible chitosan nanocomposite coating for fresh produces: A candidate with superior

thermal, hydrophobic, optical, mechanical and food properties. *Food Chem.*, 360, 130048, 2021.
16. Williams, G.I., and Wool, R.P., Composites from natural fibers and soy oil resins. *Appl. Compos. Mater.*, *7*, 421, 2000.
17. Da Silva, A. and Kyriakides, S., Compressive response and failure of balsa wood. *Int. J. Solids Struct.*, *44*, 8685, 2007.
18. Shetty, R., Pai, R., Barboza, A.B., Gandhi, V.P.K., Processing, mechanical charaterization and its tribological study of discontinously reinforced caryota urens fibre polyester composites. *ARPN J. Eng. Appl. Sci.*, 13, 3920, 2018.
19. Friedrich, D., Success factors of Wood-Plastic Composites (WPC) as sustainable packaging material: A cross-sector expert study. *Sustain. Prod. Consum.*, 30, 506, 2022.
20. Seggiani, M., Cinelli, P., Mihaela, G., Elen, P.M., Puccini, M., Lazzeri, A., Microbiological valorisation of bio-composites based on polylactic acid and wood fibres, *Chem. Eng. Trans.*, 49, 127–132, 2016.
21. Bledzki, A.K., Mamun, A.A., Volk, J., Physical, chemical and surface properties of wheat husk, rye husk and soft wood and their polypropylene composites. *Compos. Part A Appl. Sci. Manuf.*, *41*, 480, 2010.
22. Tawakkal, I.S., Cran, M.J., Bigger, S.W., Release of thymol from poly (lactic acid)-based antimicrobial films containing kenaf fibres as natural filler. *LWT-Food Sci. Technol.*, *66*, 629, 2016.
23. Shahinur, S., Sayeed, M.A., Hasan, M., Sayem, A.S.M., Haider, J., Ura, S., Current development and future perspective on natural jute fibers and their biocomposites. *Polymers*, 14, 1445, 2022.
24. Cejudo-Bastante, C., Arjona-Mudarra, P., Fernández-Ponce, M.T., Casas, L., Mantell, C., Martínez de la Ossa, E.J., Pereyra, C., Application of a natural antioxidant from grape pomace extract in the development of bioactive jute fibers for food packaging. *Antioxidants*, 10, 216, 2021.
25. Ji, M., Li, J., Li, F., Wang, X., Man, J., Li, J., Peng, S., A biodegradable chitosan-based composite film reinforced by ramie fibre and lignin for food packaging. *Carbohydr. Polym.*, 281, 119078, 2022.
26. Khan, B.A., Wang, J., Warner, P., Wang, H., Antibacterial properties of hemp hurd powder against E. coli. *J. Appl. Polym. Sci.*, 132, 10, 2015.
27. da Silva, J.S.P., da Silva, J.M.F., Soares, B.G., Livi, S., Fully biodegradable composites based on poly (butylene adipate-co-terephthalate)/peach palm trees fiber. *Compos. B. Eng.*, 129, 117, 2017.
28. Edhirej, A., Sapuan, S.M., Jawaid, M., Zahari, N.I., Cassava/sugar palm fiber reinforced cassava starch hybrid composites: Physical, thermal and structural properties. *Int. J. Biol. Macromol.*, 101, 75, 2017.
29. Hassani, F.Z.S.A., El Bourakadi, K., Merghoub, N., Bouhfid, R., Effect of chitosan/modified montmorillonite coating on the antibacterial and mechanical properties of date palm fiber trays. *Int. J. Biol. Macromol.*, 148, 316, 2020.

30. Xie, Q., Li, F., Li, J., Wang, L., Li, Y., Zhang, C., Chen, S., A new biodegradable sisal fiber–starch packing composite with nest structure. *Carbohydr. Polym.*, 189, 56, 2018.
31. Orue, A., Eceiza, A., Arbelaiz, A., Preparation and characterization of poly (lactic acid) plasticized with vegetable oils and reinforced with sisal fibers. *Ind. Crops Prod.*, 112, 170, 2018.
32. Kittikorn, T., Malakul, R., Stromberg, E., Ek, M., Karlsson, S., Enhancement of mechanical, thermal and antibacterial properties of sisal/polyhydroxybutyrate-co-valerate biodegradable composite. *J. Met. Mater. Miner.*, 28, 52, 2018.
33. Pappu, A., Pickering, K.L., Thakur, V.K., Manufacturing and characterization of sustainable hybrid composites using sisal and hemp fibres as reinforcement of poly (lactic acid) via injection moulding. *Ind. Crops Prod.*, 137, 260, 2019.
34. Sanivada, U.K., Mármol, G., Brito, F.P., Fangueiro, R., PLA composites reinforced with flax and jute fibers—A review of recent trends, processing parameters and mechanical properties. *Polymers*, 12, 2373, 2020.
35. Aji, I.S., Zainudin, E.S., Abdan, K., Sapuan, S.M., Khairul, M.D., Mechanical properties and water absorption behavior of hybridized kenaf/pineapple leaf fibre-reinforced high-density polyethylene composite. *J. Compos. Mater.*, 47, 979, 2013.
36. Vaithanomsat, P., Kongsin, K., Trakunjae, C., Boonyarit, J., Jarerat, A., Sudesh, K., Chollakup, R., Biosynthesized Poly (3-Hydroxybutyrate) on coated pineapple leaf fiber papers for biodegradable packaging application. *Polymers*, 13, 1733, 2021.
37. Todhanakasem, T., Panjapiyakul, S., Koombhongse, P., Novel pineapple leaf fibre composites coating with antimicrobial compound as a potential food packaging. *Packag. Technol. Sci.*, *35*, 97, 2022.
38. Iewkittayakorn, J., Khunthongkaew, P., Wongnoipla, Y., Kaewtatip, K., Suybangdum, P., Sopajarn, A., Biodegradable plates made of pineapple leaf pulp with biocoatings to improve water resistance. *J. Mater. Res. Technol.*, 9, 5056, 2020.
39. Rana, R.S., Rana, S., Nigrawal, A., Preparation and mechanical properties evaluation of polyvinyl alcohol and banana fibres composite. *Mater. Today: Proc.*, 26, 3145, 2020.
40. Rodríguez, L.J., Ospina, S., Ribeiro, I., Peças, P., Orrego, C.E., Banana fibre-biocomposite applied to bottle lid case-life-cycle engineering model for material selection. *Int. J. Sustain. Eng.*, 14, 1181, 2021.
41. Jandas, P.J., Mohanty, S., Nayak, S.K., Surface treated banana fiber reinforced poly (lactic acid) nanocomposites for disposable applications. *J. Cleaner Prod.*, *52*, 392, 2013.
42. Bajracharya, R.M., Bajwa, D.S., Bajwa, S.G., Mechanical properties of polylactic acid composites reinforced with cotton gin waste and flax fibers. *Proc. Eng.*, *200*, 370, 2017.

43. Wu, C.S., Wu, D.Y., Wang, S.S., Preparation and characterization of polylactic acid/bamboo fiber composites. *ACS Appl. Bio Mater.*, 5, 1038, 2022.
44. Darwin, S., Prajna, G., Tamba, T.A., Bamboo fiber-PLA composite materials for disposable food and beverages packaging tools: A brief review. *IOP Conf. Ser.: Earth Environ. Sci.*, 926, 1, 012087, IOP Publishing, 2021, November.
45. Sánchez-Safont, E.L., Aldureid, A., Lagarón, J.M., Gámez-Pérez, J., Cabedo, L., Biocomposites of different lignocellulosic wastes for sustainable food packaging applications. *Compos. B. Eng.*, 145, 215, 2018.
46. Berthet, M.A., Angellier-Coussy, H., Machado, D., Hilliou, L., Staebler, A., Vicente, A., Gontard, N., Exploring the potentialities of using lignocellulosic fibres derived from three food by-products as constituents of biocomposites for food packaging. *Ind. Crops Prod.*, *69*, 110, 2015.
47. Puttegowda, M., Rangappa, S.M., Jawaid, M., Shivanna, P., Basavegowda, Y., Saba, N., Potential of natural/synthetic hybrid composites for aerospace applications, in: *Sustainable Composites for Aerospace Applications*, pp. 315–351, Woodhead Publishing, 2018.
48. Stevens, C.V., *Industrial Applications of Natural Fibres: Structure, Properties And Technical Applications*, I.J. Mussig (Ed.), John Wiley & Sons, United Kingdom, 2010.
49. Lin, L., Luo, C., Li, C., Abdel-Samie, M.A., Cui, H., Eugenol/silk fibroin nanoparticles embedded Lycium barbarum polysaccharide nanofibers for active food packaging. *Food Packag. Shelf Life*, *32*, 100841, 2022.
50. Shetty, G.R. and Rao, B.L., Preparation and characterization of silk fibroin-poly-vinyl alcohol (PVA) blend films for food packaging materials. *Mater. Today: Proc.*, *55*, 194, 2022.
51. Lin, L., Liao, X., Cui, H., Cold plasma treated thyme essential oil/silk fibroin nanofibers against Salmonella Typhimurium in poultry meat. *Food Packag. Shelf Life*, *21*, 100337, 2019.
52. Wang, K., Hazra, R.S., Ma, Q., Jiang, L., Liu, Z., Zhang, Y., Han, G., Multifunctional silk fibroin/PVA bio-nanocomposite films containing TEMPO-oxidized bacterial cellulose nanofibers and silver nanoparticles. *Cellulose*, *29*, 1647, 2022.
53. Zhao, L., Li, Y., Wang, H., Luo, J., Song, G., Tang, G., Temperature buffering capacity of silk hydrogel: A useful packaging material. *Mater. Lett.*, 211, 110, 2018.
54. Marelli, B., Brenckle, M.A., Kaplan, D.L., Omenetto, F.G., Silk fibroin as edible coating for perishable food preservation. *Sci. Rep.*, *6*, 1, 2016.
55. Nam, J., Hyun, Y., Oh, S., Park, J., Jin, H.J., Kwak, H.W., Effect of cross-linkable bacterial cellulose nanocrystals on the physicochemical properties of silk sericin films. *Polym. Test.*, *97*, 107161, 2021.
56. Ramirez, D.O.S., Carletto, R.A., Tonetti, C., Giachet, F.T., Varesano, A., Vineis, C., Wool keratin film plasticized by citric acid for food packaging. *Food Packag. Shelf Life*, *12*, 100, 2017.

57. Ramesh, M., Rajeshkumar, L., Balaji, D., Bhuvaneswari, V., Keratin-based biofibers and their composites, in: *Advances in Bio-Based Fiber*, pp. 315–334, Woodhead Publishing, 2022.
58. Hammouche, H., Achour, H., Makhlouf, S., Chaouchi, A., Laghrouche, M., A comparative study of capacitive humidity sensor based on keratin film, keratin/graphene oxide, and keratin/carbon fibers. *Sens. Actuator A Phys.*, *329*, 112805, 2021.
59. Zhang, N., Deng, Z., Wang, Q., Zhou, M., Wang, P., Yu, Y., Phase-transited lysozyme with secondary reactivity for moisture-permeable antibacterial wool fabric. *Chem. Eng. J.*, *432*, 134198, 2022.
60. Hamouda, T., Hassanin, A.H., Saba, N., Demirelli, M., Kilic, A., Candan, Z., Jawaid, M., Evaluation of mechanical and physical properties of hybrid composites from food packaging and textiles wastes. *J. Polym. Environ.*, *27*, 489, 2019.
61. Alashwal, B.Y., Bala, M.S., Gupta, A., Sharma, S., Mishra, P., Improved properties of keratin-based bioplastic film blended with microcrystalline cellulose: A comparative analysis. *J. King Saud Univ. Sci.*, *32*, 853, 2020.
62. Dou, Y., Zhang, L., Zhang, B., He, M., Shi, W., Yang, S., Yin, G., Preparation and characterization of edible dialdehyde carboxymethyl cellulose crosslinked feather keratin films for food packaging. *Polymers*, 12, 158, 2020.
63. Song, N.B., Lee, J.H., Al Mijan, M., Song, K.B., Development of a chicken feather protein film containing clove oil and its application in smoked salmon packaging. *LWT-Food Sci. Technol.*, 57, 453, 2014.
64. Oluba, O.M., Osayame, E., Shoyombo, A.O., Production and characterization of keratin-starch bio-composite film from chicken feather waste and turmeric starch. *Biocatal. Agric. Biotechnol.*, 33, 101996, 2021.
65. Wang, H., Yin, G., Feng, G., Dou, Y., He, M., Deng, X., Preparation and properties of feather keratin and sodium carboxy methyl cellulose blend films. *Polym. Mater. Sci. Eng.*, 30, 139, 2014.
66. Wang, H., Ding, F., Ma, L., Zhang, Y., Edible films from chitosan-gelatin: Physical properties and food packaging application. *Food Biosci.*, 40, 100871, 2021.
67. Naseri, H.R., Beigmohammadi, F., Mohammadi, R., Sadeghi, E., Production and characterization of edible film based on gelatin–chitosan containing Ferulago angulate essential oil and its application in the prolongation of the shelf life of turkey meat. *J. Food Process. Preserv.*, 44, e14558, 2020.
68. Mei, S., Fu, B., Su, X., Chen, H., Lin, H., Zheng, Z., Yang, D.P., Developing silk sericin-based and carbon dots reinforced bio-nanocomposite films and potential application to litchi fruit. *LWT*, 164, 113630, 2022.
69. Giannelli, M., Lacivita, V., Posati, T., Aluigi, A., Conte, A., Zamboni, R., Del Nobile, M.A., Silk fibroin and pomegranate by-products to develop sustainable active pad for food packaging applications. *Foods*, 10, 2921, 2021.

70. Tinoco, A., Rodrigues, R.M., Machado, R., Pereira, R.N., Cavaco-Paulo, A., Ribeiro, A., Ohmic heating as an innovative approach for the production of keratin films. *Int. J. Biol. Macromol.*, 150, 671, 2020.
71. Ebrahimi, S., Fathi, M., Kadivar, M., Production and characterization of chitosan-gelatin nanofibers by nozzle-less electrospinning and their application to enhance edible film's properties. *Food Packag. Shelf Life*, 22, 100387, 2019.
72. McLellan, J., Thornhill, S.G., Shelton, S., Kumar, M., Keratin-based biofilms, hydrogels, and biofibers, in: *Keratin as a Protein Biopolymer*, pp. 187–200, Springer, Cham, 2019.
73. Nazri, W., Ezdiani, Z., Romainor, M., Erma, K.S., Jurina, J., Fadzlina, I.N., Effect of fibre loading on mechanical properties of durian skin fibre composite. *J. Trop. Agric. Food Sci.*, 42, 169, 2014.
74. Hosseini, S.B., A review: Nanomaterials as a filler in natural fiber reinforced composites. *J. Nat. Fibers*, 14, 311, 2017.
75. Ghasemi, S., Tajvidi, M., Bousfield, D.W., Gardner, D.J., Reinforcement of natural fiber yarns by cellulose nanomaterials: A multi-scale study. *Ind. Crops Prod.*, 111, 471, 2018.
76. Rana, S., Fangueiro, R., Thakur, V.K., Joshi, M., Thomas, S., Fiedler, B., Nanomaterials from natural products for industrial applications. *J. Nanomater.*, 2017.
77. Chen, Y., He, Y., Fan, D., Han, Y., Li, G., Wang, S., An efficient method for cellulose nanofibrils length shearing via environmentally friendly mixed cellulase pretreatment. *J. Nanomater.*, 2017.
78. Menon, M.P., Selvakumar, R., Ramakrishna, S., Extraction and modification of cellulose nanofibers derived from biomass for environmental application. *RSC Adv.*, 7, 42750, 2017.
79. Saito, I.M., Cabello, C., Fukushima, R.S., Análise da fibra residual do farelo de mandioca após tratamento hidrotérmico. *Rev. Raízes e Amidos Tropicais*, 2, 1, 2006.
80. Hakansson, K.M., Fall, A.B., Lundell, F., Yu, S., Krywka, C., Roth, S.V., Söderberg, L.D., Hydrodynamic alignment and assembly of nanofibrils resulting in strong cellulose filaments. *Nat. Commun.*, 5, 1, 2014.
81. Lundahl, M.J., Cunha, A.G., Rojo, E., Papageorgiou, A.C., Rautkari, L., Arboleda, J.C., Rojas, O.J., Strength and water interactions of cellulose I filaments wet-spun from cellulose nanofibril hydrogels. *Sci. Rep.*, 6, 1, 2016.
82. Hooshmand, S., Aitomäki, Y., Norberg, N., Mathew, A.P., Oksman, K., Dry-spun single-filament fibers comprising solely cellulose nanofibers from bioresidue. *ACS Appl. Mater. Interfaces*, 7, 13022, 2015.
83. Ghasemi, S., Tajvidi, M., Bousfield, D.W., Gardner, D.J., Gramlich, W.M., Dry-spun neat cellulose nanofibril filaments: Influence of drying temperature and nanofibril structure on filament properties. *Polymers*, 9, 392, 2017.
84. Fahmy, A., Zaid, H., Ibrahim, M., Optimizing the electrospun parameters which affect the preparation of nanofibers. *Biointerface Res. Appl. Chem.*, 9, 4463, 2019.

85. Lee, S.Y., Chun, S.J., Kang, I.A., Park, J.Y., Preparation of cellulose nanofibrils by high-pressure homogenizer and cellulose-based composite films. *J. Ind. Eng. Chem.*, 15, 50, 2009.
86. Davoudpour, Y., Hossain, S., Khalil, H.A., Haafiz, M.M., Ishak, Z.M., Hassan, A., Sarker, Z.I., Optimization of high pressure homogenization parameters for the isolation of cellulosic nanofibers using response surface methodology. *Ind. Crops Prod.*, 74, 381, 2015.
87. Balea, A., Tarrés, Q., Pèlach, M., Mutjé, P., Delgado-Aguilar, M., Blanco, A., Negro, C., Influence of pretreatment and mechanical nanofibrillation energy on properties of nanofibers from Aspen cellulose. *Cellulose*, 28, 9187, 2021.
88. Zhao, H.P., Feng, X.Q., Gao, H., Ultrasonic technique for extracting nanofibers from nature materials. *Appl. Phys. Lett.*, 90, 073112, 2007.
89. Lu, Y., Sun, Q., She, X., Xia, Y., Liu, Y., Li, J., Yang, D., Fabrication and characterisation of α-chitin nanofibers and highly transparent chitin films by pulsed ultrasonication. *Carbohydr. Polym.*, 98, 1497, 2013.
90. Wijesena, R.N., Tissera, N., Kannangara, Y.Y., Lin, Y., Amaratunga, G.A., de Silva, K.N., A method for top down preparation of chitosan nanoparticles and nanofibers. *Carbohydr. Polym.*, 117, 731, 2015.
91. Atiqah, M.N., Gopakumar, D.A., FAT, O., Pottathara, Y.B., Rizal, S., Aprilia, N.S., HPS, A.K., Extraction of cellulose nanofibers via eco-friendly supercritical carbon dioxide treatment followed by mild acid hydrolysis and the fabrication of cellulose nanopapers. *Polymers*, 11, 1813, 2019.
92. Fatah, I.Y.A., Abdul Khalil, H.P.S., Hossain, M.S., Aziz, A.A., Davoudpour, Y., Dungani, R., Bhat, A., Exploration of a chemo-mechanical technique for the isolation of nanofibrillated cellulosic fiber from oil palm empty fruit bunch as a reinforcing agent in composites materials. *Polymers*, 6, 2611, 2014.
93. Ren, S., Sun, X., Lei, T., Wu, Q., The effect of chemical and high-pressure homogenization treatment conditions on the morphology of cellulose nanoparticles. *J. Nanomater.*, 2014.
94. Xiao, S., Gao, R., Lu, Y., Li, J., Sun, Q., Fabrication and characterization of nanofibrillated cellulose and its aerogels from natural pine needles. *Carbohydr. Polym.*, 119, 202, 2015.
95. Abral, H., Ariksa, J., Mahardika, M., Handayani, D., Aminah, I., Sandrawati, N., Ilyas, R.A., Transparent and antimicrobial cellulose film from ginger nanofiber. *Food Hydrocoll.*, 98, 105266, 2020.
96. Nieminen, H.J., Laidmäe, I., Salmi, A., Rauhala, T., Paulin, T., Heinämäki, J., Hæggström, E., Ultrasound-enhanced electrospinning. *Sci. Rep.*, 8, 1, 2018.
97. Yongvanich, N., Isolation of nanocellulose from pomelo fruit fibers by chemical treatments. *J. Nat. Fibers*, 12, 323, 2015.
98. Jayakumar, A., Radoor, S., Kim, J.T., Rhim, J.W., Nandi, D., Parameswaranpillai, J., Siengchin, S., Recent innovations in bionanocomposites-based food packaging films–A comprehensive review. *Food Packag. Shelf Life*, 33, 100877, 2022.

99. Li, A., Xu, D., Luo, L., Zhou, Y., Yan, W., Leng, X., Fan, M., Overview of nanocellulose as additives in paper processing and paper products. *Nanotechnol. Rev.*, 10, 264, 2021.
100. Yang, W., Feng, Y., He, H., Yang, Z., Environmentally-friendly extraction of cellulose nanofibers from steam-explosion pretreated sugar beet pulp. *Materials*, 11, 1160, 2018.
101. Ghosh, T., Teramoto, Y., Katiyar, V., Influence of nontoxic magnetic cellulose nanofibers on chitosan based edible nanocoating: A candidate for improved mechanical, thermal, optical, and texture properties. *J. Agric. Food Chem.*, 67, 4289, 2019.
102. Minko, S., Sharma, S., Hardin, I., Luzinov, I., Wu Daubenmire, S., Zakharchenko, A., Kim, Y.S., Textile dyeing using nanocellulosic fibers, 2016.
103. Ratnakumar, A., Samarasekara, B.A., Amarasinghe, S.D., Karunanayake, L., Effect of sonication time on cellulose nanofiber disintegration from locally available rice straw. *Macromol. Symp.*, 402, 1, 2100425, 2022, April.
104. Chen, W., Yu, H., Liu, Y., Preparation of millimeter-long cellulose I nanofibers with diameters of 30–80 nm from bamboo fibers. *Carbohydr. Polym.*, 86, 453, 2011.
105. Chen, W., Yu, H., Liu, Y., Hai, Y., Zhang, M., Chen, P., Isolation and characterization of cellulose nanofibers from four plant cellulose fibers using a chemical-ultrasonic process. *Cellulose*, 18, 433, 2011.
106. Hassegawa, M., Karlberg, A., Hertzberg, M., Verkerk, P.J., Innovative forest products in the circular bioeconomy. *Open Res. Eur.*, 2, 19, 2022.
107. Karimah, A., Ridho, M.R., Munawar, S.S., Adi, D.S., Ismadi, Damayanti, R., Subiyanto, B., Fatriasari, W., Fudholi, A., A review on natural fibers for development of eco-friendly bio-composite: Characteristics, and utilizations. *J. Mater. Res. Technol.*, 13, 2442, 2021.
108. Ossai, I.C., Hamid, F.S., Hassan, A., Valorisation of keratinous wastes: A sustainable approach towards a circular economy. *Waste Manage.*, 151, 81, 2022.
109. Rana, S.K., Gupta, V.K., Newbold, J., Roberts, D., Rees, R.M., Krishnamurthy, S., Thakur, V.K., Sugar beet pulp: Resurgence and trailblazing journey towards a circular bioeconomy. *Fuel*, 312, 122953, 2022.
110. Zhou, Y., Stanchev, P., Katsou, P., Awad, S., Fan, M., A circular economy use of recovered sludge cellulose in wood plastic composite production: Recycling and eco-efficiency assessment. *Waste Manage.*, 42, 42–48, 2019.
111. Robledo-Ortíz, J.R., González-López, M.E., Rodrigue, D., Juan, F., Gutiérrez-Ruiz, Prezas-Lara, F., Pérez-Fonseca, A.A., Improving the compatibility and mechanical properties of natural fibers/green polyethylene biocomposites produced by rotational molding. *J. Polym. Environ.*, 28, 1040, 2020.
112. Qin, S., Giri, B.S., Patel, A.K., Sar, T., Liu, H., Chen, H., Juneja, A., Kumar, D., Zhang, Z., Awasthi, M.K., Taherzadeh., M.J., Resource recovery and biorefinery potential of apple orchard waste in the circular bioeconomy. *Bioresour. Technol.*, 321, 124496, 2021.

113. Dahy, H., Natural fibre-reinforced polymer composites (NFRP) fabricated from lignocellulosic fibres for future sustainable architectural applications, Case Studies: Segmented-Shell construction, acoustic panels, and furniture. *Sensors*, 19, 738, 2019.
114. Zabihi, O., Ahmadi, M., Liu, C., Mahmoodi, R., Li, Q., Ghandehari Ferdowsi, M.R., Naebe, M., A sustainable approach to the low-cost recycling of waste glass fibres composites towards circular economy. *Sustainability*, *12*, 641, 2020.

3

Corncob Waste for Food Packaging

Antony Catherine Flora Louis[1] **and Sivakumar Venkatachalam**[2]*

[1]*St. Joseph's College of Engineering, OMR, Chennai, Tamil Nadu, India*
[2]*Food Process Engineering Laboratory, Department of Chemical Engineering, A.C. Tech, Anna University, Chennai, Tamil Nadu, India*

Abstract
Plastic packaging materials synthesized from petrochemicals are inexpensive and profusely used causing serious environmental concerns owing to their nondegradable and nonrenewable nature. These synthetic plastics are majorly utilized by the food packaging industry for packing a variety of food products. Recently, a number of research studies have focussed on developing a biodegradable substitute for synthetic polymers from low-cost materials. Biodegradable polymers are the best solution for protecting the environment from the increasing threat of petroleum-based synthetic plastics. These biodegradable polymers are present in large amounts in agro-waste generated enormously from agricultural activities and food processing. These lignocellulose-rich agro-wastes are disposed of indiscriminately or are burned off, creating an environmental threat. Among agro-waste, waste generated in maize cultivation in the form of corncobs occupies a large proportion, as 200 million tonnes of corncobs are produced annually due to extensive cultivation of maize worldwide. These corncob wastes can be effectively valorized through different routes and applied in various applications. In recent decades, corncob waste has been valorized for use in biodegradable food packaging materials. This chapter presents recent activities on the valorization of corncobs to create sustainable, biodegradable food packaging materials.

Keywords: Corncob, cellulose, hemicellulose, polyhydroxy butyrate, nanocomposite

**Corresponding author*: drvsivakumar@gmail.com

Santosh Kumar, Avik Mukherjee and Vimal Katiyar (eds.) Agro-Waste Derived Biopolymers and Biocomposites: Innovations and Sustainability in Food Packaging, (69–90) © 2024 Scrivener Publishing LLC

3.1 Introduction

The production of agricultural commodities has significantly increased, reaching approximately 23.7 million tonnes of food per day to serve the growing population [1], which is projected to exceed 9.1 billion by 2050 [2]. An increase in agricultural activity results in an increased accumulation of agricultural waste, which is estimated to be 140 billion metric tonnes. Many developing countries dispose of the generated waste by burning it in open places, which eminently increases soil and air pollution, resulting in global warming. Therefore, to avoid this environmental nuisance, an innovative approach must be adopted to valorize these agro-waste residues in an eco-friendly manner. In order to generate value-added products from agro-waste, residual agro-waste can be used to produce high-performance materials that have commercial applications. This century has witnessed remarkable achievements in developing high-performance materials like biocomposites from natural resources. Among the agricultural products, corn (*Zea mays*) generates enormous amounts of waste. The corncob residue generated globally has reached 1114.63 million metric tonnes [3, 4].

Corn is a cereal crop grown globally with an annual global production of approximately 520 billion kilograms, which is an important resource for the agro-food industry and contributes to 39% of global grain production. It was first domesticated from the Mexican grass *Teosinte* in Central America approximately 9000 years ago and is now grown in the tropical and temperate regions of the world [5]. The annual production of corn in different countries is shown in Table 3.1. The harvested corn is processed through dry milling or wet milling to separate the maize kernels and further processed into frozen corn, canned corn, or directly consumed, resulting in the production of enormous waste residues in the form of stems, leaves, fastballs, silk, husks, and cob. Corncob is the main waste constituting 15 to 30% of the corn harvest, and for every 100 kg of corn kernels produced, around 18 kg of corncob waste is generated. According to a survey conducted by FAO (2022), Canada and the United States produced 63,787,820 metric tonnes of corncobs, and 40 to 50 million tonnes of corncobs is produced annually in China. Corncobs are the fibrous main axes of the female inflorescence of corn [6]. Although it is a potential lignocellulosic biomass, the majority of corncobs produced are used for animal feed and fertilizer. They are also used in the production of furfural, used as abrasives and polishers in cleaning products [7], and the remaining unused cobs are burned in the open air, creating severe risks to human health and

Table 3.1 Annual production of corn in the year 2022 [10].

Country	Production of corn (metric tonnes)
USA	383,943,000
China	272,552,000
Brazil	115,000,000
European Union	69,960,000
Argentina	54,000,000
Ukraine	42,000,000
India	30,000,000
Mexico	27,600,000
South Africa	17,000,000
Russia	15,000,000

the environment, or disposed of in landfills. Disposal in landfills leads to the growth of pathogens, contamination of underground water through seepage, and release of foul odors, whereas burning of corncobs produces large quantities of greenhouse gasses, air pollutants, and volatile particulate matter [8]. Recently, owing to the depletion of fossil resources, several studies have focused on utilizing lignocellulosic biomass rich in agro-waste residues and have gained economic significance for industrial applications. Lignocellulosic resources are abundantly available and are the most abundant material on Earth, largely present in the cell walls of woody plants, agricultural and forest residues. The composition of lignocellulosic species can vary from one source to another depending on the type of plant species, plant parts, and growth conditions. Studies have focused on utilizing corncobs to produce bioethanol, biochar, xylo-oligosaccharides, and adsorbents to produce fine chemicals. In some cases, corncob residues are used in the production of sugar/ethanol by enzymatic hydrolysis; however, large-scale utilization is avoided because of the high production cost. Corncob is rich in cellulose (40 to 44%) and has a hemicellulose content of 31 to 33% and 16 to 18% lignin, which is responsible for its sponge structures that maintain a low density and good bearing capacity [9]. Therefore, corncob can be effectively valorized to produce cellulose and nanocellulose through chemical and mechanical processes.

3.2 Isolation of Cellulose from Corncob

Cellulose plays a vital role in human life and is a landmark in human evolution. It has great economic importance, as it can be used for manufacturing of broad spectrum of products and materials from packaging to the biomedical fields. Cellulose is the most abundant polymer with a wide range of applications in several fields, owing to its high surface area, high strength, modulus, good optical properties, biodegradability, and renewability. It comprises a large portion of lignocellulosic biomass (30 to 35%) and is found exclusively in the cell walls of plants. In certain cases, they are also produced by a few bacterial species. Owing to its outstanding properties, it has been identified as a suitable alternative to nonrenewable and nonbiodegradable packaging materials that pose a threat to the environment. Cellulose is an insoluble, high-molecular-weight polysaccharide with a polymerization degree of 10,000. It is made up of repeating units of anhydro-D-glucopyranose units, and each repeating unit is connected by a β-1,4 glycosidic linkage [11], as shown in Figure 3.1, with a chemical formula ($C_6H_{10}O_5$). Cellulose units are interlinked by a network of inter- and intramolecular hydrogen bonds and Van der Waals forces to form flat sheets. Interactions of these bonds among the cellulose fibrils create crystalline and noncrystalline cellulose structures. Although glucose is soluble in water, cellulose is insoluble owing to its high molecular weight and low chain flexibility [12].

The main role of cellulose in corn is to provide rigidity and structural integrity to plant material. The properties of isolated cellulose largely depend on the source of extraction, processing factors, solvents used, and the technique for extraction [13]. Nanocellulose is a natural fiber that is extracted/synthesized from cellulose and is generally less than 100 nm in diameter and several micrometers in length. They have several reactive surface hydroxyl groups, which can be modified to impart various surface properties. Owing to its enhanced properties and biodegradability, nanocellulose can be used in many fields such as nanocomposite materials and surface-modified materials. Nanocellulose being nontoxic and

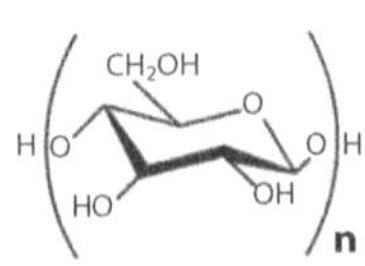

Figure 3.1 Structure of cellulose.

biocompatible can be potentially used as reinforcement material as they improve the mechanical property. The high surface area and crystallinity of the nanoparticles aid in forming strong hydrogen bonds creating a strong barrier property against water vapor and oxygen molecules.

Several methods have been used for the extraction of cellulose fibers from lignocellulosic agro-waste residues, such as grinding, cryo-crushing, steam explosion, alkali extraction, ultrasonication, high-speed blending, and ionic liquid–based extraction. Cellulose obtained by various methods can self-assemble into defined structures of varying sizes from macro, micro, and nano sizes. Isolated cellulose and nanocellulose can be potentially used to replace a variety of nonrenewable or synthetic materials to produce food additives, nanocomposites, biomedical scaffolds, sensors, and food packaging applications.

3.2.1 Pretreatment of Corncob Waste Residues

Different types of pretreatment strategies are carried out on lignocellulosic biomass to open the fibrous structure of corncob to facilitate greater release of hemicellulose and lignin and subsequently make cellulose more accessible for extraction. Isolation/synthesis of cellulose nanoparticles involves two main steps: The first step is the fragmentation of the biomass to recover cellulose from lignocellulosic resources and the second step involves the synthesis of cellulose nanoparticles to achieve nanoscale dimensions. Mechanical pretreatment, such as grinding and chipping, is also carried out to increase the surface area for further chemical pretreatment. These chemical treatments are acid hydrolysis, alkaline hydrolysis, ionic liquid treatment, enzymatic treatment, and organo-solvent treatment. Alkali extraction of cellulose is the most used low-cost method for isolating cellulose from agro-waste residues.

3.2.1.1 Acidic Hydrolysis

One of the commonly and commercially used methods for producing corncob pulp is acidic hydrolysis, which is generally carried out with dilute acids, such as hydrochloric acid or sulfuric acid at a particular temperature and pressure. The acidic hydrolysis steps loosened the lignocellulosic matrix and aided in increasing the accessibility of lignin and hemicellulose and their removal. Lignin and hemicellulose are hydrolyzed into their monomeric or oligomeric forms and are easily dissolved in acid liquor and washed away during the washing step. The concentration of acid, reaction time, and temperature are important parameters that play a vital role in the isolation of cellulose. A higher concentration of acid is also used for producing cellulose nanocrystals (CNCs) from cellulose fiber. During the acid hydrolysis, the acid diffuses into the noncrystalline region of the fiber and breaks the glycosidic bonds. The commonly used acid for nanocellulose production is sulfuric acid. Other acids like hydrochloric acid, hydrobromic acid, and phosphoric acid can also be used, but comparatively, they form lesser stable colloidal suspension than sulfuric acid. Hydrolysis using sulfuric acid was optimized by several researchers and the commonly optimized method was using 64 wt % sulfuric acid at 45°C for 45 to 60 min which is followed by quenching with 10-fold cold water. The major setback here is the generation of acid water during the washing process to neutralize the nanocellulose suspension. Maiti *et al.* (2013) isolated nanocellulose using 47% sulfuric acid, after the hydrolysis the acid was washed, centrifuged, and neutralized using 0.5 N NaOH [14].

3.2.1.2 Alkaline Treatment

The main aim of the pulping process is to separate the interlinked cellulose fibers into individual fibers, wherein the lignin that holds the fibers together is solubilized along with hemicellulose, liberating the cellulose fibers. Soda pulping is the preferred pulping process that utilizes sodium hydroxide as the active chemical under high pressure and temperature between 150°C and 170°C. As the alkali penetrates the raw material, the fibers swell, breaking the β-O-4 linkages within the lignin resulting in the delignification of lignin. Although various pretreatments are available, alkaline pretreatment for the extraction of cellulose is preferred because it is operated at a relatively lower temperature and pressure than other processes. It causes less sugar degradation when compared to acidic hydrolysis, and most of the caustic salts used in the process can be recovered [15].

Jia *et al.* (2016) extracted cellulose from corncobs using 20 to 70% NaOH at 170°C for 90 min at a pressure of 0.7 MPa [16]. The basic chemistry involved in this process is solvation and saponification, which result in the depolymerization and cleavage of ester bonds that link hemicellulose and lignin. The alkali treatment also results in the cleavage of α- and β-ether bonds in the phenolic structure of lignin and the β-ether bonds in the non-phenolic units, which leads to the delignification process. Simultaneously, the glycosidic group present in hemicellulose is also hydrolyzed by alkali treatment to release sugar monomers and soluble oligomers from the fiber into the hydrolysate. In addition, the hydroxyl groups present in the material are ionized, and the OH bonding is disrupted owing to the alkali reaction. This combination leads to the separation of interfibrillar regions from the cellulose fibers [15].

3.2.2 Bleaching Process

The alkaline pulping process is followed by bleaching using sodium chlorite and hydrogen peroxide to remove undesirable residues such as lignin, wax, and lipids. Several bleaching agents have been used, such as chlorine gas, ozone, chlorine dioxide, oxygen, hydrogen peroxide, and hypochlorite. Some of them react with aromatic lignin units, whereas others react with the free phenolic hydroxyl groups in the lignin structures. This method, also termed the delignification process, is widely used in the pulp industry. Ibrahim *et al.* (2010) isolated cellulose from corncobs using steam explosion technology followed by alkaline peroxide bleaching [17]. Steam explosion technology involves the use of high-pressure steam for a short duration, which is subsequently followed by a sudden explosion that results in the defiberization or mulching of biomass. The vital parameters that must be monitored during the steam explosion process are the reaction temperature and retention time, and in this study, the steam explosion was performed at 220°C for 1 to 4 min. The fibers were further water extracted, followed by alkali extraction using 20 wt % NaOH at 80°C for 1 hour and then peroxide bleaching. This process resulted in the fractionation of corncob biomass into polymeric fractions with reasonable purity. The steam explosion process degraded a major portion of hemicellulose, whereas the alkali process removed all the remaining lignin. The cellulose obtained exhibited a higher degree of crystallinity.

Singh *et al.* (2019) isolated microcrystalline cellulose from corncobs using alkali treatment with varying concentrations of NaOH, followed by bleaching using sodium chlorite [11]. The bleached corncob biomass was further acid-hydrolyzed to obtain microcrystalline cellulose. Under the optimized conditions of 2% w/v NaOH and 2% w/v $NaClO_2$, 29.60% cellulose was obtained from corncobs. After the acid hydrolysis treatment using 10% v/v H_2SO_4, cellulose of sizes 1.3 to 2.5 μm was obtained. Andrade *et al.* (2019) isolated cellulose from corncobs and fabricated a chitosan-based film incorporating the isolated nanocellulose [18]. Cellulose was isolated using an alkali treatment (2% NaOH), followed by bleaching with 1.7% sodium hypochlorite. Nanocellulose was obtained using 45% H_2SO_4, and the obtained nanocellulose had a length/diameter ratio of 32.19. Chitosan films were fabricated using 5% nanocellulose by solvent-casting method and the addition of which showed the highest tensile strength of 11.43 MPa and lower moisture content of 18.32%.

3.2.3 Extraction of Nanocellulose

Acid hydrolysis is the primary process used to extract nanocellulose from cellulose. During hydrolysis, the disordered regions of the cellulosic materials are hydrolyzed by acid, leaving only the ordered parts. Acid hydrolysis is conducted using strong acids like sulfuric acid, hydrochloric acid, phosphoric acid, maleic acid, nitric acid, formic acid, and hydrobromic acid. The hydrolysis is conducted at controlled temperature, time, and cellulose/acid ratio conditions. Sulfuric acid is a commonly used acid that forms a stable colloid system owing to the esterification of hydroxyl groups by sulfate ions. Sulfuric acid is also the most used mineral acid for removing hemicellulose, lignin, and other extracts present in biomass. The use of sulfuric acid also generates negative charges on the surface of the cellulose, which produces an enhanced colloidal suspension. The CNCs formed in the acid are diluted with water to neutralize the acid and washed. Mechanical treatments such as high-pressure homogenization, ultrasonication, and ball milling can also be employed to produce cellulose nanofibrils that produce high shear forces to cleave the cellulose fibers along the longitudinal axis [19]. Among the mechanical methods for size reduction, planetary ball milling is one of the simplest, safest, and most cost-effective methods for producing nano-sized particles. CNCs isolated by chemical method produced a nanoscale range of 0.3 to 7.0 nm, and nanofiber isolated by ball milling produced particles in the range of 45 to 108 nm. Nanocellulose is potentially used as a filler in several applications and has been found to improve the mechanical strength of materials and their barrier properties when combined with polymers such as carboxymethyl cellulose, chitosan, alginate, and other biodegradable polymers. It has also been used in enzyme immobilization, sensor devices, and electronic appliances [20]. Ali *et al.* (2020) isolated CNCs and nanofibers from corncobs using chemical and mechanical methods [21].

Kumar *et al.* (2010) isolated cellulose nanoparticles from corncob by alkali treatment in a rotary digester. The corncob was delignified using varying concentrations of NaOH (0 to 24%) at 165°C [22]. Furthermore, it was treated with sodium chlorite and was disintegrated to obtain homogeneous nanoparticles with an average particle size of 24 to 200 μm with a crystallinity of 48%. Silverio *et al.* (2013) extracted cellulose from corncob using 2% NaOH for 4 hours at 100°C under mechanical stirring followed by bleaching using 1.7% sodium chlorite at 80°C for 6 hours. The obtained cellulose particles were further hydrolyzed using concentrated sulfuric acid (9.7 M) at a temperature of 45°C for 60 min with an acid/cellulose ratio of 15:1 v/w. CNCs with an average length of 210 nm and diameter of 4 nm with a crystallinity of 83.7% were obtained [7]. Advanced methods like treatment with ionic liquids, enzymatic hydrolysis, and TEMPO-mediated oxidation can also be used for CNC preparation but these methods are usually applied along with other chemical or mechanical methods.

3.3 Isolation of Hemicellulose from Corncob

Hemicelluloses are heteropolysaccharides with different types of sugar units, such as D-xylose, D-mannose, D-glucose, L-arabinose, and varying amounts of uronic acids, and are the second most abundant polymer in lignocellulosic biomass. They are branched polymers with shorter chains and are amorphous in nature, which binds cellulose fibrils non-covalently. Different types of hemicellulose are available in corncob residues, such as xylans, which are mainly made up of D-xylose units linked by β-1,4 linkages, mannans made up of D-mannose units, galactans comprising D-galactose units, and arabinans made up of L-arabinose units. Xylan is usually made up of three components 4-O-methyl-D-glucuronic acid, L-arabinose, and D-xylose in the ratio of 2:7:19. The structure of xylan is shown in Figure 3.2. They comprise approximately 25 wt % lignocellulosic biomasses. Hemicellulose has a molecular weight lower than that of cellulose, with a low degree of polymerization of approximately 150 to 200. Owing to their noncrystalline nature, they are more susceptible to depolymerization. They are also water-soluble at higher temperatures. Several studies have been conducted to effectively isolate hemicellulose from corncob residues. The major challenge in isolating the hemicellulose is the strong linkage of hemicellulose with lignin. The close association of hemicellulose with lignin must be ruptured to facilitate the separation of cellulose and hemicellulose.

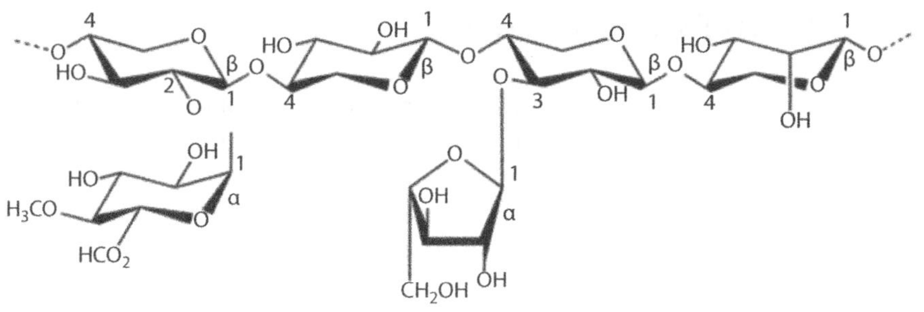

Figure 3.2 Structure of typical hemicellulose.

In order to completely valorize the agricultural waste, hemicellulose can be extracted from biomass for film production as this hemicellulose are renewable, low-cost material with good gas barrier properties due to its dense macromolecular structure [23]. The characteristics of the developed packaging material depend on the chemical structure of hemicellulose, its molecular weight, crystallinity, polymerization degree, and the condition in which it is processed. The extracted hemicellulose is also chemically modified by esterification, carboxy-methylation, hydroxypropylsulfonation, and acetylation to improve the film-forming properties. Other than packaging materials, they are also used in bioethanol production, foams, surfactants, as additives in paper and flocculation aids, and as coating color components. The increasing demand for xylans has led to the identification of various methods to extract hemicellulose from agro-waste [24].

Samanta *et al.* (2012) extracted xylan from corncob using two different alkalis NaOH and KOH [25]. Extraction using alkali allows for hydrolysis of ester linkage in the biomass followed by extraction in an aqueous medium. The concentration of alkali was varied and the extraction was carried out by incubating overnight or by application of steam at 121°C, 15 lb for 45 min. Results showed that NaOH was better than KOH in recovering xylan and steam application was better than overnight incubation for maximizing the yield of xylan. NaOH (12%) under steam application resulted in a maximum recovery of 32.38% yield of xylan. Louis and Venkatachalam (2020) isolated cellulose and hemicellulose from corncob using a sequential process of microwave irradiation and ultrasound-assisted alkali extraction, resulting in the extraction of 0.445 g/g of cellulose and 0.327 g/g of hemicellulose [26]. The extracted particles were converted into nanoparticles of size 131.4 nm for nanocellulose and 130.4 nm for nanohemicellulose.

3.4 Microbial Biosynthesis of Polyhydroxy Butyrate (PHB) from Corncobs

In recent decades, the depletion of petroleum deposits has initiated an upsurge in the search for alternative resources, and thus, more interest has been focused on the production of bioplastics using microorganisms. One such type of bioplastic is polyhydroxyalkanoate (PHA). When excess carbon is present, more than 250 species of bacteria under stress growth conditions can produce PHA as inclusion bodies accumulated as a reserve material. They are produced by microorganisms by assimilating carbon sources and are used as stored energy when other energy sources are unavailable. They are stored in the bacterial cytoplasm as granules and can be observed by transmission electron microscopy (TEM). Polyhydroxy butyrate (PHB) is a common polymer in PHA, which degrades faster in natural environmental conditions, making it a suitable alternative to synthetic plastics. They also possess properties like those of synthetic plastics such as polypropylene, and thus, PHB has food packaging applications by replacing synthetic plastics with the future of commercial mass production.

PHB is a linear polyester of the homopolymer 3-hydroxybutyric acid of high molecular weight that was discovered in 1926 by Maurice Lemoigne and is the best-characterized member of the PHA family. Currently, PHB is industrially produced by several companies including Zeneca Bio Products and Fine Chemicals Company. Although they have a wide range of applications, commercialization of PHB is a major issue owing to its high production cost, which is almost 10 times higher than the production cost of synthetic polymers. The major contributor to the increase in production cost was found to be the carbon substrate cost, comprising 50% of the total production cost. Strategies such as developing efficient bacterial strains and optimizing the fermentation and recovery process were found to decrease the cost of production.

The efficient utilization of agro-waste, especially lignocellulosic biomass, has recently gained global attention. Value-added products such as bioethanol, biofuels, succinic acid, enzymes, and PHAs can be produced using these lignocellulosic resources, which could benefit both rural and agricultural economies. Agricultural waste materials are economically feasible carbon substrates for microbial growth and PHB production. Lignocellulosic waste that is abundantly available in agricultural waste is a potential source of PHB production. Corncob agro-waste is a vital resource that can potentially be used as a carbon substrate. Synthesis of PHB by microorganisms uses a biosynthetic pathway involving three genes: phaA, phaB, and phaC. The enzyme β-ketothiolase, encoded by phaA, is responsible for the condensation of two molecules of acetyl-CoA to produce acetoacetyl-CoA. Acetoacetyl-CoA is converted to (R)-3-hydroxybutyrate–CoA by the enzyme acetoacetyl-CoA reductase, encoded by phaB. Polymerization of (R)-3-hydroxybutyrate–CoA is further performed by PHA synthase, which is encoded by the phaC gene. The three enzymes involved in the biosynthetic pathway for the synthesis of PHB are present in the cytosol of bacterial cells, where PHB accumulation occurs.

The carbon substrate in corncobs can be hydrolyzed using enzymatic or chemical methods. Acid hydrolysis is the preferred method because of its high efficiency, even if microbial-inhibiting byproducts are produced.

Dilute acid hydrolysis rapidly depolymerizes the hemicellulose present in corncob to xylose, glucose, arabinose, mannose, and galactose. Stoica *et al.* (2018) developed a fermentation strategy to produce PHB using bacteria from corncob, as a single carbon source [27]. Corncobs were hydrolyzed using dilute acid and used as a carbon source in fermentation to cultivate spore-forming bacteria to achieve biosynthesis of PHB. Dilute acid hydrolysis resulted in a total sugar content of 35.84 g/L, which was found to be the optimum fermentative condition for the growth of *Bacillus megatherium* BM 37 for PHB bioproduction. A yield of 36.16% w/w dry biomass of PHB was obtained by submerged aerobic fermentation of corncob hydrolysate by selected bacteria. Getachew and Woldesenbet (2016) isolated bacterial strains from different localities that were identified as *Bacillus* spp., and they were found to be the best for producing PHB at an optimum pH of 7 and 37°C for 48 hours at 150 rpm [28]. The strain produced 51.6% PHB on pretreated corncobs as a carbon source. Corncob was hydrolyzed using the zinc chloride method, and the reducing sugar content was 4465 μg/ml.

3.5 Biopolymers-Based Food Packaging Reinforced with Corncob Fibers

Environmentalists have begun to look for a potential substitute for synthetic plastics to resolve the ongoing long-term environmental crisis due to the use of plastics. The term biodegradation was defined by Albertsson *et al.* (1994) as a biological process that occurs due to the chemical decomposition of the polymer by the action of enzymes or living organisms (bacteria, fungi, etc.), and the decomposable polymers are referred to as biodegradable polymers [29]. Biopolymers are derived from bioresources such as plants and animals (cellulose, starch, chitin, casein, and gluten), as well as a mixed source of biomass and petroleum. Sometimes, biopolymers such as PHA and PHB are also synthesized from small molecules by microorganisms. The utilization of such biodegradable polymers promises to reduce the volume of garbage/waste generated, improve compostability in the natural cycle, and to a certain extent, preserve fossil-based raw materials. Biodegradable polymers have major drawbacks such as poor mechanical properties, low barrier properties, and low heat distortion temperatures that limit their use in a wide range of applications [30]. Biopolymers, such as starch, are low-cost materials that are surplus in nature, recyclable, renewable, nontoxic, and biodegradable. But as such in their native form, they exhibit poor barrier and mechanical properties.

Many strategies are being researched to enhance the properties of biodegradable packaging materials like starch, such as the usage of plasticizers like glycerol, formamide, and sorbitol that increases the flexibility of the materials by forming hydrogen bonds with the polymeric chain and blending or addition of other polymers to produce material with intermediate properties. However, these strategies failed to meet the properties of synthetic plastics cost-effectively. Recent developments in the field of nanotechnology opened the window of opportunity to create biodegradable materials reinforced with nanofillers. Many researchers have utilized lignocellulosic fibers as reinforcements in biopolymer matrices to enhance their mechanical and barrier properties. Several studies have been conducted on agro-waste to extract fibers to prepare biocomposite materials because of their abundant availability, biodegradability, nontoxicity, etc. Nanofiber-reinforced biocomposite consists of two phases, a continuous phase of biopolymer matrix and a discontinuous phase of nanofiller. The addition of fibers in the form of nanofillers increases the surface area of the filler between the biopolymer and the filler. The commonly investigated properties of fiber-reinforced composites are the tensile, flexural, impact properties, and barrier properties. Other than these, the long-term performance of the composite, dynamic mechanical behavior, and compressive properties are also analyzed. In certain cases, there is poor compatibility between the fiber and thermoplastic matrix, which raises problems that can be rectified using various physical and chemical methods. Similarly, several methods have been proposed to improve or modify the fiber surface properties to enhance their adhesion to the polymer matrix. Chemical methods such as alkali treatment, the addition of compatibilizers, the addition of cross-linking agents, and physical methods such as thermal treatment and stretching are carried out to improve the surface properties of the fibers. The surface-treated fibers are reinforced in biopolymer by using various methodologies like mixing, compounding, injection molding, compression molding and extrusion [31].

Corncobs are a rich source of fiber that can be used as fillers in biopolymer matrices to improve their physicochemical, mechanical, and barrier properties. In addition, these fibers are nontoxic, inexpensive, widely available, have high strength and stiffness, and have low weight. The use of corncobs as a reinforcement adds value to crop residue, providing macroeconomic benefits to rural communities. The influence of fiber reinforcement in different biopolymer matrix and their effect on mechanical properties is tabulated in Table 3.2. The effect of fiber addition from different natural resources can be observed from the given data. Corncob filler reinforced with chitosan improved the mechanical properties, but the

Table 3.2 Properties of fiber reinforced composite.

Biopolymer matrix	Fillers/fibers	Improvement in properties of composite		Reference
		Tensile strength (MPa)	Elongation at break (%)	
Kappa carrageenan	-	25	13	[34]
	Nanocellulose from cotton	37	10	
Corn starch	-	-	31	[35]
	Bleached *E. urograndis* pulp	-	11	
Corn starch	-	33	20	[36]
	Chitosan	44	30	
Rice starch	-	0.3	18	[37]
	Cotton fiber	0.8	23	
Corn starch	-	6	52	[38]
	Birch cellulose	14	3	
Chayote tubers	Cellulose from cotton	0.85	27	[39]
	Nanocellulose from cotton	1.05	45	
Potato starch	Cellulose from cotton	0.8	27	
	Nanocellulose from cotton	1.1	30	
Cassava starch	-	2	81	[40]
	Cellulose fibers from softwood	23	15	
Chitosan	-	54	14	[41]
	Corncob	33	11	
	Corncob treated with acrylic acid	37	13	

(Continued)

Table 3.2 Properties of fiber reinforced composite. (*Continued*)

Biopolymer matrix	Fillers/fibers	Improvement in properties of composite		Reference
		Tensile strength (MPa)	Elongation at break (%)	
Corn starch	-	2.9	37	[26]
	Nanocellulose from corncob	8.2	21	
	Nanohemicellulose from corncob	5.45	38	
	Nanocellulose from rice straw	5.2	23	[42]
	Nanohemicellulose from rice straw	6.2	25	

excess addition of fillers resulted in a decrease in the tensile strength and elongation at break [32]. The addition of corncob filler directly into chitosan leads to a decrease in mechanical properties as filler matrix adhesion was affected due to the roughness of the filler surface, hydrophilicity, and surface wettability. Therefore, surface treatment of the biomass was carried out using acrylic acid and the property improvement is tabulated below (Table 3.2). Similarly, the addition of corncob filler to polypropylene composites also showed a decrease in mechanical properties as the filler loading increased from 10, 20, 30, and 40 wt % [33]. The decrease in mechanical properties is due to the irregular shape of the corncob particles, resulting in a poor interlocking matrix. The decreasing trends in mechanical properties can be improved by surface modification, adding compatibilizers or silane coupling agents, grafting polymers with hydrophilic functional groups, or plasma treatment on the filler surface. Such methods promise to increase the interfacial adhesion between the filler and the polymer matrix.

3.6 Hybrid Nanocomposite of Corncob for Food Packaging

Bionanocomposites are composed of two or more constituents, divided into a continuous phase or matrix and a discontinuous nanofiller layer.

The development of bionanocomposites plays a significant role in improving the physicochemical, mechanical, barrier, and functional properties of food packaging materials. The addition of different types of nanofillers adds to functions, such as antimicrobial agents, biosensors, and oxygen scavengers, and various agricultural residues like rice straw, wheat straw, and rice husk are been used as fillers in biocomposites. Recently corncob has gained attention due to its renewability and biodegradability and has shown potential application as reinforcements in bionanocomposite. Corncob facilitates efficient processing as it is low weight with high strength and stiffness and is nontoxic in nature. Arabinoxylan was extracted from corncobs and used for film preparation by the solvent-casting method described by Egues *et al.* (2014) [43]. The films were fabricated using corncob arabinoxylan, washed arabinoxylan, and bleached corncob arabinoxylan. Purification of arabinoxylan increased tensile strength from 9 to 53 MPa and elastic modulus increased from 293 to 1600 MPa, while the elongation at break remained constant at 8% for the untreated and purified corncob-derived arabinoxylan films. Film fabrication using bleached arabinoxylan showed further improvements in tensile strength (67 MPa), elongation at break (13.4%), and elastic modulus (2241 MPa).

Corncob powder has also been used as a filler in the fabrication of PHB/polylactic acid (PLA) biopolymer matrices. Films developed using pure PHB are soft, have low rigidity, and lack the qualities required for use in bending applications. These properties can be improved by using biodegradable materials as fillers in the PHB matrix, which not only lowers its cost but also expands its industrial usage. The properties of the PHB/PLA composite with corncob fillers were assessed by varying the filler loading from 1, 2, 4, 6, to 8 wt % [44]. The findings showed improved tensile, flexural moduli, and the glass transition temperature as the filler loading increased. Louis *et al.* (2022) developed starch-based food packaging materials reinforced with nanocellulose and nanohemicellulose isolated from corncobs, and the result showed that the addition of nanofillers significantly enhanced tensile strength, Young's modulus, and barrier properties [42]. The developed films were found to be thermally stable, with onset degradation at 260°C and 280°C, and were successfully used to pack button mushrooms.

Gairola *et al.* (2022) developed polypropylene-based composites reinforced with corncob fiber [4]. Fillers were added at different concentrations and the property of the developed composite is reported in Table 3.3. The composite prepared with a higher filler size showed higher modulus

Table 3.3 Mechanical properties of composites.

Composites	Tensile strength (MPa)	Tensile modulus (MPa)	Elongation at break (%)	Flexural strength (MPa)	Flexural modulus (MPa)	Reference
Polypropylene	35.77	826.33	--	65.23	2075.08	[4]
Polypropylene + 10% corncob filler	29.26	813.59	10.19	62.47	2257.91	
Polypropylene + 20% corncob filler	27.42	813.32	08.02	60.89	2397.64	
PVA	21	-	-	-	-	[7]
PVA + 3% CNC from corncob	32	-	-	-	-	
PVA + 6% CNC from corncob	41	-	-	-	-	
PVA + 9% CNC from corncob	50	-	-	-	-	

with a decrease in tensile and flexural strength. The developed composite also showed better water resistance with a water absorption of only 1.1%. Silverio *et al.* (2013) extracted CNCs from corncob and the reinforcing capability of the CNC was analyzed using polyvinyl alcohol (PVA) [7]. The tensile strength of the PVA composite increased with an increase in the concentration of CNC in the matrix. The CNC improved the tensile strength by 140.2% when added at 9%. This confirms a close relationship between filler and matrix due to hydrogen bonds between PVA and CNC.

3.7 Conclusion and Future Perspectives

Petroleum-based synthetic plastics are a major concern worldwide, as they create enormous environmental problems, such as climate change, global warming, and resource depletion. To compete with these synthetic plastics, biodegradable materials must attain the required mechanical and barrier properties. The use of bio-renewable fillers in the preparation of bionanocomposites is of great interest, as it improves the properties of biopolymer-based packaging materials that are promising alternatives to synthetic plastic packaging. Corncob is rich in carbon, a low-cost, and abundantly available source of raw material for producing cellulose, hemicellulose, and fermentative PHB. The use of agro-waste such as corncob creates a path toward sustainable development and a circular economy. Thus, corncob is a potential source for the production of cellulose, hemicellulose, and their nano-derivatives and their subsequent use in developing food packaging materials. However, there is a need to formulate a biotechnological route to valorize corncobs at an industrial scale. Valorization of corncob biomass offers numerous opportunities for biorefineries to develop sustainable bioproducts, thus improving the biotechnological management of corncobs.

References

1. Duque-Acevedo, M., Belmonte-Urena, L.J., Cortes-Garcia, F.J., Camacho-Ferre, F., Agricultural waste: Review of the evolution, approaches and perspectives on alternative uses. *Glob. Ecol. Conserv.*, 22, e00902, 2020.
2. Leisner, C.P., Climate change impacts on food security focus on perennial cropping systems and nutritional value. *Plant Sci.*, 293, 110412, 2020.
3. Gandam, P.K., Chinta, M.L., Prashanth, N.P., Veliandi, A., Sharma, M., Kuha, R.C., Corncob based biorefinery: A comprehensive review of pretreatment methodologies, and biorefinery platforms. *J. Energy Inst.*, 101, 290–08, 2022.

4. Gairola, S., Naik, T.P., Sinha, S., Singh, I., Corncob waste as a potential filler in biocomposite: A decision towards sustainability. *Compos. Part C*, 9, 100317, 2022.
5. Stitzer, M.C. and Ross-Ibarra, J., Maize domestication and gene interaction. *New Phytol.*, 220, 395–408, 2018.
6. Vasudeva Nayaka, K. and Rangaswamy, B., Pulverized maize cobs as a recycled agro waste – a feasibility study. *Int. J. Waste Resour.*, 10, 376, 2020.
7. Silverio, H.A., Neto, W.P.F., Dantas, N.O., Pasquini, D., Extraction and characterization of cellulose nanocrystal from corncob for application as reinforcing agent in nanocomposite. *Ind. Crops Prod.*, 44, 427–436, 2013.
8. Elegbede, J.A., Ajayi, V.A., Lateef, A., Microbial valorization of corncob:Novel route for biotechnological products for sustainable bioeconomy. *Environ. Technol. Innov.*, 24, 102073, 2021.
9. Sun, Y. and Cheng, J., Hydrolysis of lignocellulosic materials for ethanol production: A review. *Bioresour. Technol.*, 83, 1–11, 2002.
10. Statista, Global corn production by countries, 2022, htpps://www.statista.com/statistics/254292/global-corn-production-by-country/.
11. Singh, H.K., Patil, T., Vineeth, S.K., Das, S., Pramanil, A., Mhaske, S.T., Isolation of microcrystalline cellulose from corn stover with emphasis on its constituents: Corn cover and corn cob. *Mater. Today: Proc.*, 27, 589–594, 2019.
12. Brandt, A., Grasvik, J., Hallet, J.P., Welton, T., Deconstruction of lignocellulosic biomass with ionic liquids. *Green Chem.*, 15, 3, 550–583, 2013.
13. Khalil, H.A., Bhat, A., Yusra, A.I., Green composites from sustainable cellulose nanofibrils: A review. *Carbohydr. Polym.*, 87, 963–79, 2014.
14. Maiti, S., Jayaramuu, J., Das, K., Rey, S.M., Saiku, R., Ray, S.S., Liu, D., Preparation and characterization of nano-cellulose with new shape from different precursor. *Carbohydr. Polym.*, 98, 562–567, 2013.
15. Pinto, E., Aggrey, W.N., Boakye, P., Amenuvor, G., Sokama-Neuyam, Y.A., Fokuo, M.K., Karimaie, H., Sarkodie, K., Adenutsi, C.D., Erzuah, S., Rockson, M.A.D., Cellulose processing from biomass and its derivitization into carboxymethylcellulose: A review. *Sci. Afr.*, 15, e01978, 2022.
16. Jia, F., Liu, H., Zhang, G., Preparation of carboxymethyl cellulose from corn cob. *Proc. Environ. Sci.*, 31, 98–102, 2016.
17. Ibrahim, M.M., Agblevor, F.A., El-Zawawy, W.K., Isolation and characterization of cellulose and lignin from steam exploded lignocellulosic biomass. *Bioresources*, 5, 397–418, 2010.
18. Andrade, M.R., Nery, T.B.R., Santana, T.I.S., Leal, I.L., Rodriques, L.A.P., Reis, J.H.O., Druzian, J.I., Machado, B.A.S., Effect of cellulose nanocrystals from different lignocellulosic residues to chitosan/glycerol films. *Polymers*, 11, 658, 2019.
19. Phanthong, P., Reubroycharoen, P., Hao, X., Xu, G., Abudula, A., Guan, G., Nanocellulose: Extraction and application. *Carbon Resour. Convers.*, 1, 32–43, 2018.

20. Gupta, G.K. and Shukla, P., Lignocellulosic biomass for the synthesis of nano-cellulose and its eco-friendly advanced applications. *Front. Chem.*, 8, 601256, 2020.
21. Ali, J.B., Danladi, A., Bukhari, M.M., Nyakuma, B.B., Mamza, P., Mohamad, Z.B., Musa, A.B., Inuwa, I.M., Extraction and characterization of cellulose nanofibers and cellulose nanocrystals from Sammaz-14 Maize Cobs. *J. Nat. Fibers*, 19, 7, 2020.
22. Kumar, S., Negi, Y.S., Upadhyaya, J.S., Studies on characterization of corn cob based nanoparticles. *Adv. Mater. Lett.*, 1, 3, 246–253, 2010.
23. Egues, I., Eceiza, A., Labidi, J., Effect of different hemicelluloses characteristics on film forming properties. *Ind. Crops Prod.*, 47, 331–338, 2013.
24. Sousa, S., Ramos, A., Evtuguin, D.V., Gamelas, J.A.F., Xylan and xylan derivatives-their performance in bio base films and effect of glycerol addition. *Ind. Crops Prod.*, 94, 682–689, 2016.
25. Samanta, A.K., Senani, S., Kolte, A.P., Sridhar, M., Sampath, K.T., Jayapal, N., Devi, A., Production and *in vitro* evaluation of xylooligosaccharides generated from corn cobs. *Food Bioprod. Process.*, 90, 466–474, 2012.
26. Louis, A.C.F. and Venkatachalam, S., Energy efficient process for valorization of corn cob as a source for nanocrystalline cellulose and hemicellulose production. *Int. J. Biol. Macromol.*, 163, 260–69, 2020.
27. Stoica, I., Petrovici, A.R., Silion, M., Varganici, C., Dinica, R., Bahrim, G., Corn cob hydrolyzates used for microbial biosynthesis of polyhydroxybutyrate. *Cellul. Chem. Technol.*, 52, 65–74, 2018.
28. Getachew, A. and Woldesenbet, F., Production of biodegradable plastic by poly-hydroxybutyrate (PHB) accumulating bacteria using low cost agricultural waste material. *BMC Res. Notes*, 9, 509, 2016.
29. Albertsson, A.C., Barenstedt, C., Karlsson, S., Abiotic degradation products from enhanced environmentally degradable polyethylene. *Acta Polym.*, 45, 97–103, 1994.
30. Sinha, R.S. and Bousmina, M., Biodegradable polymers and their layered silicate nanocomposites: In greening the 21st century materials world. *Prog. Mater. Sci.*, 50, 962–1079, 2005.
31. Faruk, O., Blezki, A.K., Fink, H., Sain, M., Biocomposites reinforced with natural fibers: 2000-2010. *Prog. Polym. Sci.*, 37, 1552–1596, 2012.
32. Yeng, C.M., Husseinsyah, S., Ting, S.S., Modified corn cob filled chitosan biocomposite films. *Polym. Plast. Technol. Eng.*, 52, 1496–1502, 2013.
33. Fuqua, M.A., Chevali, V.S., Ulven, C.A., Lignocellulosic byproducts as filler in polypropylene comprehensive study on the effects of compatibilization and loading. *J. Appl. Polym. Sci.*, 127, 862–868, 2013.
34. Savadekar, N.R., Karande, V.S., Vigneshwaran, N., Bharimalla, A.K., Mhaske, S.T., Preparation of nanocellulose fibers an it application in kappa-carrageenan base film. *Int. J. Biol. Macromol.*, 51, 1008–1013, 2012.

35. Curvelo, A.A., de Carvalho, A.J.F., Agnelli, J.A.M., Thermoplastic starch-cellulosic fibers composites: Preliminary results. *Carbohydr. Polym.*, 45, 183–188, 2001.
36. Zhai, M., Zhao, L., Yoshii, F., Kume, T., Study on antibacterial starch/chitosan blend film formed under the action of irradiation. *Carbohydr. Polym.*, 57, 83–88, 2004.
37. Prachayawarakorn, J., Sangnitiej, P., Boonpasith, P., Properties of thermoplastic rice starch compoites reinforce by cotton fiber or low density polyethylene. *Carbohydr. Polym.*, 81, 425–433, 2010.
38. Teaca, C., Bodirlau, R., Spiridon, I., Effect of cellulose reinforcement on the properties of organic acid modified tarch micorparticles/plasticized starch biocomposite films. *Carbohydr. Polym.*, 93, 307–315, 2013.
39. Aila-Suarez, S., Palma-Rodriguez, H.M., Rodriguez-Hernanez, A.I., Hernanez-Uribe, J.P., Bello-Perez, L.A., Vargas-Torres, A., Characterization of films made with chayote tuber and potato starches blending with cellulose nanoparticles. *Carbohydr. Polym.*, 98, 102–107, 2013.
40. Muller, C.M.O., Laurindo, J.B., Yamashita, F., Effect of celluloe fibers addition on the mechanical properties and water vapour barrier of starch-based films. *Food Hydrocoll.*, 23, 1328–1333, 2009.
41. Yeng, C.M., Husseinsyah, S., Ting, S.S., Modified corn cob fille chitosan biocomposite films. *Polym.-Plast. Technol. Eng.*, 52, 1496–1502, 2013.
42. Louis, A.C.F., Venkatachalam, S., Gupta, S., Innovative strategy for rice straw valorization into nanocellulose and nanohemicellulose and its application. *Ind. Crops Prod.*, 179, 114695, 2022.
43. Egues, I., Stepan, A.M., Eceiza, A., Toriz, G., Gateholm, P., Labid, J., Corncob arabinoxylan for new materials. *Carbohydr. Polym.*, 102, 12–20, 2014.
44. Ohaeri, O. and Cree, D., Development and characterization of PHB-PLA/corn cob composite for fused filament fabrication. *J. Compos. Sci.*, 6, 249, 2022.

4

Coir Fibers for Sustainable Food Packaging

Sweety Kalita[1], Manashi Das Purkayastha[2], Avik Mukherjee[1*] and Santosh Kumar[1†]

[1]Department of Food Engineering and Technology, Central Institute of Technology Kokrajhar, Kokrajhar, Assam, India
[2]Food Science and Technology Programme Department of Horticulture, Assam Agricultural University, Jorhat, Assam, India

Abstract

Coconuts, scientifically known as *Cocos nucifera*, thrive in tropical regions. The rising need for environmentally friendly, biodegradable, renewable, and sustainable materials has driven the increasing utilization of biocomposites reinforced with coir fibers sourced from coconut husks. These husks and shells, once considered waste, are now highly regarded resources for environmentally friendly biocomposite production. Coir fibers, well known for their strength and stiffness, are combined with various thermoplastic and thermosetting materials to produce biocomposites with superior mechanical, thermal, and physical properties compared to synthetic alternatives. The mechanical performance can be further improved through surface pretreatment. Coir fibers can be employed as reinforcement alongside biodegradable polymers or plastics to create biocomposites suitable for robust applications such as packaging. This chapter provides an extensive exploration of coir fiber and its composites, addressing surface treatments and properties, and as a result, this advancement paves the way for the mass production of biocomposites reinforced with coir fibers, making them suitable for various uses, particularly within the realm of food packaging.

***Keywords*:** Coir fiber, biocomposites, food packaging, sustainability, sustainable development goal

**Corresponding author*: ak.mukherjee@cit.ac.in
†*Corresponding author*: s.kumar@cit.ac.in

Santosh Kumar, Avik Mukherjee and Vimal Katiyar (eds.) Agro-Waste Derived Biopolymers and Biocomposites: Innovations and Sustainability in Food Packaging, (91–114) © 2024 Scrivener Publishing LLC

4.1 Introduction

Natural fiber–reinforced composite materials are emerging as low-cost, renewable nature, biodegradability, and environment-friendly materials in the packaging sector of food industries. The frequently employed natural biodegradable fillers encompass lignocellulosic fibers like coir, hemp, ramie, rice husk, jute, and kenaf [1]. When compared to synthetic fibers, natural fibers have abundant availability, affordability, and overall machinability that have made them appealing for various applications. Natural fiber–reinforced composite materials are used in a wide range of sectors, such as automobiles, construction and building, consumer goods, biomedicine, and food packaging. Nonetheless, owing to the absence of appropriate technology, investigation, and scientific progress for the industrialization of natural fibers, synthetic fiber–enhanced goods are still being utilized in the production of composite materials [2–4]. The types of natural fibers include animal, and plant fibers such as seed, stalk, grass/reed, wood, and leaf fibers. Coconut coir is a dense, stiff, and biodegradable lignocellulosic fiber made from the fibrous mesocarp of coconuts [5, 6]. Longevity represents a noteworthy benefit in composites reinforced with natural fibers. Nevertheless, coir fiber (CF) stands out as exceptionally durable among natural fiber options due to its higher lignin content. CF-reinforced composites also exhibit elasticity up to their elastic threshold without breaking, thanks to increased elongation at the point of rupture. Consequently, researchers and manufacturers are currently focusing their efforts on developing more convenient approaches to harness the potential of coconut husk–derived fibers [6, 7].

CFs, typically extracted from coconut husks, are naturally hydrophilic due to the presence of polarized hydroxyl groups in cellulose and hemicellulose. This hydrophilicity limits their use in industrial applications, as it can lead to moisture absorption and poor bonding with hydrophobic polymer matrices, affecting the mechanical properties of biocomposites. Various treatments, including physical methods like plasma and corona discharge, chemical methods like acid or alkali treatment, enzymatic treatment, and steam explosion treatment have been employed to enhance the adhesion between fibers and polymers in biocomposites [6, 7]. Various studies have reported promising findings in producing CF-reinforced biocomposites having superior thermal, mechanical, and morphological characteristics. Rejeesh *et al.* (2017) proposed that CF could serve as a flame-retardant material to conventional plywood [8]. Furthermore, in the case of food biocomposite packaging, the inclusion of fibers is associated with benefits such as improved gas barrier qualities, increased heat stability, and, in some circumstances, improved mechanical capabilities. Fibers, due to their hydrophilic qualities, also contribute to increased swelling thickness and water absorption, which is especially beneficial in the case of high-respiring food packages [9]. The CF has low density, high thermal insulating qualities, mechanical properties, low pricing, wide availability, nontoxicity, problem-free disposal, and superior compatibility with diverse polymeric materials. Thus, the purpose of this chapter is to provide an overview of the chemical, morphological, and thermomechanical properties of CF-reinforced biocomposites, which have been developed in collaboration with thermoplastics, thermosetting polymers, and biopolymers, as well as their potential applications and economic benefits.

4.2 Coir Fibers as Reinforcement Material for Synthetic Polymers

4.2.1 Polyethylene-Based Composites Reinforced with Coir Fibers

CFs possess significant promise as strengthening elements in thermoplastic matrices such as polyethylene (PE), where the fatty layer of CFs establishes robust linkages with thermoplastic polymers, thereby enhancing the robustness and thermomechanical attributes of the composite [3]. The interaction between phases is a critical factor in composite performance, involving methods such as combining with an external enhancer or chemical bonding, followed by purification and copolymerization. Babu and Narayanankutty, in their 2021 study, introduced an improved method for creating chemically functionalized and self-compatible composites (HDPE-TCF5%-MA2%). This was achieved through the *in situ* grafting of maleic anhydride (MA) in the polymer melt [high-density PE (HDPE)], followed by the reactive compatibilization of treated short CF (TCF) [10]. Examination of the composites' microstructure revealed enhanced bonding between fibers and matrix, leading to upgraded mechanical, viscoelastic, and thermal characteristics, as validated by FTIR investigations. HDPE-TCF5%-MA2% demonstrated roughly 74 and 25% enhancements in tensile strength compared to unprocessed and processed fiber composites without MA, respectively. Dynamic mechanical analysis of the composites demonstrated a corresponding pattern, with the highest storage modulus observed in HDPE-TCF5%-MA2% [10].

With the goal of improving the compatibility between CFs and the HDPE matrix, three chemical processes, namely, silane, sodium hydroxide (NaOH), and dodecane bromide (C12) treatments were carried out. The results indicated that composites using treated fibers displayed superior mechanical properties compared to those using untreated fibers. The Young's modulus computed in the composites was greater for the fibers treated with alkali compared to the untreated ones, and the C12 composite also displayed a significant tensile modulus compared to the polymer reinforced with untreated fibers. Furthermore, the torsional test revealed that the NaOH treatment increased the flexibility of the fibers, while the silane treatment made them stiffer [11]. Rather than the typical method of altering CFs, this research explored the potential of plasma-treated PE (PPE) as a matrix for CF composites *via* rotational molding. Through assessments of water absorption, scanning electron microscopy (SEM), and mechanical properties, the authors established four distinct interfaces (PE/CF, PE/bleached CF, PPE/CF, and PPE/bleached CF) and assessed their interfacial connections. The PPE/bleached CF composite demonstrated reduced water absorption owing to enhanced fiber/matrix bonding [12]. This study also utilized the rotational molding method to determine the optimum level of reinforcement of coir with linear low-density PE (LLDPE) to provide its required process capability. In this experiment, powdered coir was mixed with varying concentrations of LLDPE, ranging from 3 to 20%, and the FTIR spectra showed that 3 to 12% by weight ratios were acceptable for flow abilities; however, 10 wt % coir was suitable in terms of material processability, shear, and heat parameters [13]. A similar kind of work has successfully incorporated NaOH-treated long fibers up to 15 wt % of both CF and Tampico as the reinforcement factors into linear low-density PE (LLDPE) composites using a biaxial rotomolding machine [14]. The fractured surface microstructure of the molded specimens revealed a uniform blend of CF and LDPE matrix. Nevertheless, there was evidence of limited adhesion at the interface between the CF and LDPE matrix. Examination of water absorption demonstrated that the resulting composite materials exhibited minimal water uptake at lower fiber concentrations, but at higher fiber concentrations, water absorption markedly escalated [15].

Awodi *et al.* (2021) conducted another study involving CF-reinforced recycled HDPE composites, focusing on eight properties, including impact resistance, denseness, tensile strength, tensile modulus, elongation percentage, flexural modulus, flexural strength, and water absorption proportions. The outcomes demonstrated the feasibility of producing CF-reinforced recycled HDPE composites with relatively modest fiber content, achieving an optimal gray relational grade of 0.8286. The most influential factor impacting the various performance attributes was the fiber loading, with fiber length also playing a significant role. This suggests that optimizing the performance of CF-reinforced recycled HDPE composites with lower fiber content can be attained by regulating both fiber loading and length [16]. Kakou *et al.* (2015) fabricated hybrid composites by combining palm fibers, CFs, and HDPE for a fixed fiber content of 40% by weight using a coupling agent (maleated PE) *via* twin-screw extrusion and compression molding. The findings indicated that the treated composites exhibited enhanced fiber dispersion and distribution, leading to a more uniform composite structure due to the chemical cross-linkage, resulting in favorable interfacial attachment [17].

4.2.2 Polypropylene-Based Composites Reinforced with Coir Fibers

Among thermoplastics, polypropylene consists of a linear hydrocarbon chain containing carbon atoms, and due to its low water vapor and gas permeability, high mechanical strength, excellent resistance to abrasion and chemical, and durability at high temperatures, it has been profusely for packaging applications [18]. Currently, there is a lot of research and development being focused on environmentally benign, sustainable materials, for example, the reinforcement of natural fibers in synthetic plastics like polypropylene to improve composite properties [19]. To enhance the interaction between polypropylene and fibers, an investigation of chemically and mechanically processed coconut fiber (CF) composites was conducted by Gelfuso *et al.* (2011). Injection molding was used to fabricate composite with CF contents of 5, 10, 15, and 20%, and the result showed that mechanically treated 5 vol % CF had improved tensile strength compared to untreated fiber [20]. Surface treatment of CF with various solvents and their efficiency was evaluated and the outcomes proclaimed that inclusion of fibers in polypropylene caused a decrease in thermal stability and hydrophilicity [21]. Another study employed basic chromium sulfate and sodium bicarbonate salt to treat raw CFs before blending them with polypropylene, which resulted in improved interfacial adhesion between the coir and polypropylene matrix and improved mechanical qualities of the developed composite [22].

For a better understanding of the chemical composition and natural filler surface to develop natural filler–reinforced composites, Essabir *et al.* (2016) used coir shell and CFs (20 wt %) as bio-fillers in polypropylene-based matrix by combining extrusion and injection molding methods using a coupler, i.e., SEBS-g-MA (8%). The results demonstrated that the used coupler accelerated interfacial bonding between filler and matrix and thus mechanical properties (improvement between 35 and 50%) [23]. Morandim-Giannetti *et al.* (2012) exhibited how the inclusion of lignin as a coupling agent impacts the mechanical, thermal, and morphological attributes of polypropylene/coir composites. They conducted this study both with and without polypropylene grafted with MA as a compatibilizer and the findings indicated that the introduction of lignin did not impact tensile strength when modified polypropylene was not present, but tensile strength decreased in its presence [24]. A hybrid composite derived from polypropylene-treated CF and yam peel particulate (YPP) was blended and processed through compression molding. CFs were analyzed with 1.5 M sodium hydroxide, and yam peel was altered using 1 M solution of sodium hydroxide and hydrochloric acid in 30 and 70%, respectively. Composites were created utilizing processed and unprocessed reinforcements with a consistent CF ratio (15%) and different quantities of yam peel particles (2, 4, 6, and 8 wt %). The outcomes indicated that composites reinforced with processed yam peel and CF exhibited favorable mechanical and thermal characteristics [25].

4.2.3 Polyester-Based Composites Reinforced with Coir Fibers

The utilization of thermosetting polymers, like polyester, offers promising prospects for CF-reinforced biocomposites. The natural waxy surface layer of CF enhances interfacial bonding with the polyester matrix, leading to increased hydrophobic behavior, tensile strength, and resistance to fungal attack in coir-polyester composites [26]. Polyesters are robust polymers derived from diols and dicarboxylic acids, with applications determined by their synthesis and polymer chain arrangement [27]. Polyester-based composites typically use polymer resins as the matrix and incorporate fillers to meet specific objectives [28]. CFs with their curved structure, on the other side, are challenging to orient as desired in composite development. Consequently, CFs are commonly used in applications requiring randomly oriented fiber-reinforced composites and are considered one of the most suitable biofibers for reinforcing polyester matrices [29]. Several CF-polyester–based composites have been developed for various applications. Prasad *et al.* (2017) studied the energy absorption capability of CF-polyester composites using untreated and treated CFs where they discovered increased impact strength with an increase in treated fiber thickness [29]. The treated CFs have the potential to enhance the mechanical characteristics of the CF-reinforced thermosetting polymeric matrix by promoting stronger bonding between the fiber and polymeric matrix

(Figure 4.1) [1], for instance, CFs mixed with polyester matrix after being treated with NaOH resulted in increased impact strength of the overall composite [7, 29]. Another research even used a higher concentration of NaOH to treat CF with respect to different periods of soaking time that later concluded that the soaking time of the coir in NaOH solution decreased the amount of hemicellulose and cellulose, making the surface of the fiber coarser and thus maximized the tensile strength and interfacial shear strength of the developed composite [30].

Moreover, polyester and resins were enhanced with CFs subsequent to their surface modification using sodium carbonate. The density of the fibers increased after alkali treatment, but their thermal stability was reduced. Despite the fact that the properties of the polyester-coir composites were inferior to those of the neat matrix (aside from impact strength), the proposal was to employ sodium carbonate treatment to safeguard against corrosion and potential hazards linked to conventional chemicals frequently used for surface fiber modification [31]. Aiming to give information on the tensile properties of coconut coir single fiber treated with alkali and to evaluate the effects of reinforcement on unsaturated polyester polymer, this particular study took a data-driven approach. By treating coconut coir with NaOH alkali, the final fiber had an average tensile strength of 130.9 MPa, a failure strain of 22.4%, and an elastic modulus of 681.4 MPa [32]. A research project involving a drilling test of woven coir-polyester composites treated with NaOH led to absolute percentage errors of 5.4, 6.2, and 6.5% for thrust force, torque, and tool wear, respectively [33]. Coconut flower cover fiber was further utilized to reinforce polyester composites through hot compression molding, yielding a composite with high specific

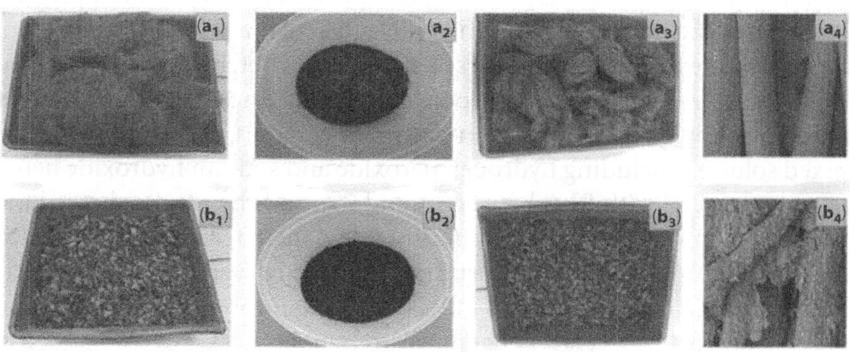

Figure 4.1 (a_1 and b_1) Coir fibers before pretreatment; (a_2 and b_2) treatment of coir fibers in alkaline solutions; (a_3 and b_3) coir fibers after treatment; and (a_4 and b_4) SEM images of coir fibers showing morphological view [1].

strength and excellent impact objection. This composite is also considered cost-effective for potential use in the automotive sector [34]. Furthermore, to obtain high flexibility in composites with exclusive properties that are difficult to manage by using individual fiber-reinforced composites, Balaji *et al.* (2020) hybridized Zea and CFs with polyester resin through compression sheet molding. The resulting hybrid composite demonstrated the highest tensile strength, flexural strength, and impact strength along with good thermal stability [35]. CF-reinforced polyester composites were further analyzed and found to have significant improvement by hybridizing them with Napier grass fibers, as demonstrated in this study. As the loading of Napier grass fibers increased, the flexural strength and impact strength values raised as well. A SEM examination revealed that the fiber and matrix had the strongest adhesion [36].

4.3 Coir Fibers as Reinforcement Material in Biopolymers

4.3.1 Composites of Coir and Polylactic Acid (PLA)

Despite being classified as biodegradable, PLA's degradation rate is the lowest among similar polymers, and therefore, it has been reinforced with other biopolymers for food packaging applications [37, 38]. CFs with a density of 1.15 g/cm^3 were amalgamated with polylactic acid (PLA), which has a density of 1.25 g/cm^3, to form biocomposites. The fibers were segmented into lengths ranging from 6 to 8 mm and underwent treatment with a 2% sodium hydroxide solution at 70°C for 1 hour, followed by heating for 15 min at 180°C before being compressed under 1.5-MPa pressure for 15 min. The fiber composition and NaOH treatment had an impact on the composites' tensile properties, particularly stretchability, and 10% of the CFs treated with alkali incorporation in PLA increased the biodegradation level by 35% [39]. In a similar manner, short CFs were treated by mixed solution including hydrogen peroxide and sodium hydroxide before being reinforced with PLA between 1 and 7%, and the results showed that the composites' tensile strength declined but their tensile modulus escalated as the amount of fiber material increased; however, the addition of the fiber made the composites stiff [40]. Using alkali (6%), slaked lime (6%), and silicon hydride (2%) as chemical treatments of CF were studied to evaluate the effects of these treatments on the tensile, morphological, thermal, and structural properties of CFs and pineapple leaf fibers (PALF) in order to enhance their interfacial bonding with PLA. The alkali treatment enhances the interfacial shear strength between the fibers and PLA matrix, suggesting the successful fabrication of CF/PALF-reinforced PLA hybrid biocomposites [41]. The surface of CFs was further modified by incorporating polymethyl vinyl ether altered with MA that further grafted with a PLA matrix, leading to improved tensile modulus and flexural modulus in addition to a notable reduction in water absorption in the biocomposites when compared to untreated fibers [42]. Moreover, the enhancement of PLA with unidirectionally aligned CFs resulted in improved tensile properties, with single-layer coir/PLA composites exhibiting an 11% acceleration in tensile strength and elongation percentage when differentiated with double-layer coir/PLA composites. The application of chemical treatments such as NaOH and a silane coupling agent can further enhance mechanical properties by strengthening the bond between the fiber and matrix [43]. Coskun *et al.* (2019) used enzymatic treatments of CF, particularly lactase, lipase, pectinase, and cellulase, and reinforced in PLA, and the findings suggested that employing enzyme-treated CFs led to changes in the mechanical properties, with the exception of flexural characteristics. Tensile strength increased with lipase and lactase treatments but had no effect with pectinase and cellulase treatments [44]. Even amylase treatment on a PLA composite with CF significantly increased the composite's strength and stiffness, with tensile strength, bending strength, and impact strength increasing by 5-fold, 3.6-fold, and 2-fold, respectively [45]. The enzymatic treatments enhanced the interfacial adhesion between CF and PLA by removing waxes and fatty acids and increasing surface roughness [44, 45]. The hybridization of PALF and CFs in equal proportions (1:1), along with their integration into a matrix of PLA to create complex materials, yielded a favorable combination of mechanical characteristics and enhanced thermal stability. This renders the material well-suited for various applications, including environmentally friendly food packaging and structural components, aimed at reducing environmental impact [46]. Nonetheless, when CFs were combined with thermoplastic starch (TPS) and PLA, the resulting biocomposites displayed decreased extensibility, tensile strength, and impact strength in contrast to the pristine TPS/PLA blend. This reduction was further counterbalanced by enhanced stiffness and hardness achieved through reinforced hydrogen bonding interactions between starch, PLA, and CF, along with increased PLA crystallinity [47].

4.3.2 Composites of Coir and Protein

Proteins are natural condensation polymers involved in biological processes, offering versatility for creating various biodegradable materials.

Plant proteins, including soy proteins, wheat protein, wheat gluten, and corn zein, are explored as alternatives to synthetic polymers for biocomposites and films. These plant proteins offer advantages such as oxygen barrier properties, making them preferable for applications like food packaging. However, protein-based polymers often have low elongation and can be brittle, leading to the use of plasticizers like glycerol to enhance film flexibility. However, biocomposites from protein-based polymers, like wheat gluten, are often brittle and susceptible to water absorption post-processing. Incorporating natural reinforcements like CFs have been effective in enhancing the mechanical properties [48–50]. In a study, for instance, CF-reinforced wheat gluten biocomposites treated with saline exhibited significant enhancements in composite strength, with a remarkable 80% increase, along with improved fiber-matrix interfacial adhesion, as observed through SEM [48]. Varying lengths of CFs (2.5, 5, 10, and 20 mm) were reinforced in cement-albumen composite at a ratio of 65:35 (w/w) increased flexural strength to 6.719 MPa by using 5-mm coir could be due to efficient bridging of cracks by longer fibers and proper withstanding of higher load than the shorter fibers [51]. Pereira *et al.* (2021) incorporated coconut kernel fiber in myofibrillar protein (MP), and increasing the concentration of CF from 0 to 1.5% improved the water-holding capacity and hardness of the protein gel ($p < 0.05$). This was evidenced by the gel micrographs, showing a more homogeneous and stable structure in the presence of CF, which acted as filler material in the protein matrix [52]. Khanam *et al.* (2009) prepared a hybrid composite of mercerized coir and silk fibers (1:1) by incorporating them into polyester, in which both the type of fiber (coir or silk) of varying length, i.e., 1, 2, and 3 cm, were combined with an unsaturated polyester resin that resulted in significant improvement in tensile, flexural, and compressive strength [53]. Another similar study utilizing coconut coir (C), regenerated silk microparticles (S), and construction-demolition waste wood (W), to form W-C, S-C, and W-S biocomposites, in which a binary mixture of the above components was blended by hot-pressing (170°C to 180°C) at ratios of 80:20, 60:40, 50:50, 40:60, and 20:80 (w/w). Among the acquired biocomposites, the addition of protein-based S enhanced the mechanical characteristics and water repellency in W-S. In the case of W-C and S-C, there was a modest improvement in these properties [54]. Gluten is basically composed of two individual proteins, namely, glutenin and gliadin, and the mixing of these two protein components with water forms a stretchy mass called gluten. Muensri *et al.* (2011) fabricated wheat gluten–based amalgamate in the presence of treated CFs improved the mechanical properties of the composite [55]. In another study, a coir-reinforced gluten matrix was evaluated for interfacial adhesion, and alkali-silane–treated fiber-reinforced gluten biocomposite exhibited reduced pullout length [48]. Since this gluten biocomposite showed matrix cracking beyond a certain stress level, the same group of authors further experimented with toughening the wheat gluten by molecular cross-links *via* thiolated polyvinyl alcohol (PVA) resulting in minimal cracking, increased stiffness, and strength [49]. Gypsum-based tiles are ubiquitously used for various applications; the interaction of biodegradable fibers like sheep wool and coir (25 to 75% each) with gypsum, at a ratio of 30% fiber and 70% gypsum, gave increased strength (up to 90% rise) and modulus (7% enhancement) to the fiber-reinforced tiles [56].

4.3.3 Composites of Coir with Starch

Blended coir with TPS has also been studied by many scientists for food packaging applications [57]. Corradini *et al.* (2006) added 10% mercerized (alkali-treated) coir to TPS, resulting in a 52.2 and 84.3% rise in elastic modulus and tensile strength of the composite, respectively [58]. Meanwhile, the addition of 10% untreated natural coir in TPS increased the elastic modulus and tensile strength by 100 and 76%, respectively, against a pristine matrix containing no coir. Likewise, a composite made from mercerized coir and starch/ethylene vinyl alcohol (EVOH) blend through melt extrusion presented an enhancement in tensile strength (53%) and tensile modulus (17%), when differentiated from the same formulation made with unprocessed coir [59]. Moura *et al.* (2021) mercerized and bleached babassu CFs amalgamated with TPS and glycerol (40 wt %), and a composite made with fibers having no chemical pretreatment had 15-MPa tensile strength, 1.8% elongation, and 1600-MPa elastic modulus, whereas those with treated fibers, the respective values were 18 MPa, 2.2%, and 1500 MPa [60]. In addition to the abovementioned chemical interactions, the treatment of coirs with plasma (using oxygen or air) is successful in eliminating the amorphous layer rich in lignin and amplifying the adhesion of treated coirs with the biopolymer. This phenomenon occurs by etching and ablating the fiber surface with plasma. Plasma-treated CFs of size 125 to 250 μm were melt compounded with starch-glycerol (7:3) mixture, which was then compressed at 60°C into discs, and the composites made from such plasma-treated coir showed high strength (approximately 300% rise) and elasticity (approximately 20 times higher) [61]. Corn starch–based biocomposite by amalgamating carnauba wax and CFs has also been reported to result in higher tensile strength and modulus of elasticity with increasing amounts of wax and fibers into starch [62].

Chitosan-starch blended film having jute and CFs was designed by incorporating 1, 2, 3, and 4% of the two kinds of fibers. Unlike the thermoplastic condition, the authors cast the blend in film after cross-linking with $CaCl_2$. Prior to incorporation, both the fibers were treated with NaOH solution for alkalization and then ground to pass through a 60 BSS sieve. The results showed that the blended films had higher antibacterial properties against *Escherichia coli* and found no bacterial growth on the film surface [63]. Gupta *et al.* (2022) prepared carboxymethyl cellulose (CMC) from coconut coir and groundnut shell by mercerization followed by an etherification process for making biocomposite with either starch or commercial CMC. The composite produced with CMC displayed higher tensile strength and elongation at break than its counterpart made from groundnut shell [64]. Iovino *et al.* (2008) fabricated a composite using a blend of PLA and TPS (75:25 w/w), in the presence or absence of MA (coupler) and short CFs, and showed that the incorporation of CFs and MA seemed to increase the rate of biodegradation of PLA [65]. Starch composite with coir was evaluated by Rabe *et al.* (2019), who applied feather keratin protein fibers (from tannery waste) and CFs as fillers in a TPS-polyester blend that resulted in increased Young's modulus; however, tensile strength, strain at break, tenacity, and impact resistance were reduced [66].

4.3.4 Hybrid Composites of Coir

Hybrid composites consist of a minimum of two fillers within a unified matrix. The attractiveness of this type of composite arises from the potential to achieve a multifaceted material with a combination of performance attributes while also reducing costs and mitigating the disadvantages linked to using a single filler. The volume percentage of the related fibers in hybrid composites has a substantial influence on the mechanical properties of the blend as well as stress migration between the reinforcements that are polymers and fibers in the matrix structure. Natural fibers like CFs have tremendous promise for replacing synthetic fibers in the development of hybrid composites with improved mechanical and functional qualities while reducing material and production costs [7, 9, 67]. As per the study conducted by Tran *et al.* in 2019, enhancing bamboo with CF can have a beneficial impact on strain failure. Consequently, the incorporation of bamboo fiber materials can enhance the rigidity of polymeric composites reinforced with CFs [68]. The superficial stiffness of the CFs was further enhanced by their surface modification with chemicals like sodium carbonate that led to the effective removal of hemicellulosic compounds which promoted swelling and the heightened surface roughness

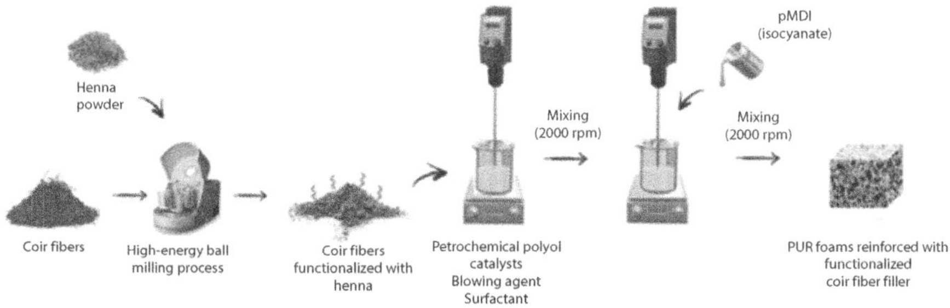

Figure 4.2 Schematic procedure for coir fiber- and henna powder-based PUR composite preparation [69].

of the fibers led to enhancements in the mechanical characteristics of the synthetic composites reinforced with coir [31]. Successful modification of CFs with henna (*Lawsonia inermis*) resulted in enhanced physicomechanical properties, particularly with a 1 wt % coir-fiber-henna filler, as shown in Figure 4.2 [69]. In another research, coir was utilized as fibrous reinforcement and rice husk was employed as particulate filler to formulate vinyl ester–based composites. The addition of rice husk particles enhanced adhesion and consequently imparted better characteristics. When 30-mm fibers and 15% particles were utilized, a high impact strength of 43 kJ/m^2 was produced [70]. An even more innovative composite was created by combining HDPE with oil palm, coir, and a coupling agent (maleated PE), through twin-screw extrusion and compression molding to create hybrid composites with CF (40%). Composite materials were subjected to assessments regarding their morphological, structural, and thermal attributes. Findings indicated that the composites with improved compatibility exhibited enhanced fiber arrangement and scattering (uniformity) owing to chemical binding, resulting in improved interfacial attachment [17]. Table 4.1 depicts the mechanical characteristics of the hybrid composite, achieved by fortifying CFs with various natural fibers or synthetic polymers.

4.4 Biodegradable Package/Container from Coconut Coir

The issue of ecological contamination resulting from non-decomposable artificial plastics and the disposal and incineration of agricultural residues are of primary concern globally. In the meantime, biodegradable

Table 4.1 Mechanical properties of coir fiber blended composites of natural and synthetic polymers.

Hybrid composites	Composite formulation	Tensile strength (MPa)	Tensile modulus (GPa)	Flexural strength (MPa)	Flexural modulus (GPa)	Impact strength (kJ m^{-2})	Elongation at break (%)	Reference
Pineapple leaf fiber (P)/ coir fiber (C)/epoxy resin (E)	P20:C20:E60	71.39	61.29	75.4	58.87	86.63	-	[71]
Coir (C)/sugarcane leaf sheath (S)/polyester resin (P)	C15:S15:P70	16.88	1.20	28.41	3.16	0.95	-	[72]
Banana stem fiber (B)/ coir fiber (C)/maleic anhydride grafted polypropylene	B10:C10	36.2	1.09	32.6	-	9.3	-	[73]
Coir fiber/glass/polyester	-	29.3948	-	73.1751	-	40.97	8.85	[74]
Coir (C)/carbon fiber/epoxy	C30	285.74		215.79	-	-	-	[75]
Date palm fiber (D)/coir fiber (C)/epoxy	D50:C50	76.51	7.54	-	-	832.2	2.77	[76]
Coir fiber (C)/jute fiber (J)/polypropylene	C25:J75	21.63	1.25	31.97	1.44	360	-	[77]
Coir fiber (C)/Caeser weed fiber (CW)/polyester (P)	C-CW30:P70	55.53	-	90.8	-	6.8	-	[78]

(Continued)

Table 4.1 Mechanical properties of coir fiber blended composites of natural and synthetic polymers. (*Continued*)

Hybrid composites	Composite formulation	Tensile strength (MPa)	Tensile modulus (GPa)	Flexural strength (MPa)	Flexural modulus (GPa)	Impact strength (kJ m^{-2})	Elongation at break (%)	Reference
Starch (S)/polylactic acid (P)/coir fiber (C)	S60:P40:C10.9	26.6 - 30.7	-	-	-	-	2.1 -2.4	[47]
Thermoplastic starch (TPS)/coconut fibers (CF)/ammonium polyphosphate (APP)	CF20:APP10	14	-	387	-	95	19	[66]
Coir fiber (CF)/banana fiber (CF)	E70:SF10: BF10:CF10	0.92	2.23	30.0	-	-	12.8	[79]
Sisal fiber (SF)/coir fiber (CF)/polylactic acid (PLA)	SF7:CF3	50	-	95	-	2.1	-	[80]
Banana fiber (BF)/coir fiber (F)	BF10:CF10	20	-	48	-	-	-	[81]
Coir fiber (CF)/chitosan (CH)	CF90:CH10	1.8	-	-	-	-	-	[82]
Coir fiber (CF)/yam peel particulate (YPP)/ polypropylene (PP)	CF15:YPP4	22	290	18	190	33	-	[25]

polymer–based packaging can overcome these problems due to its biodegradability and environment-friendly nature. However, biodegradable polymer–based packaging films have inferior mechanical properties compared with synthetic nonbiodegradable polymers. Organic fillers made from natural resources are the most beneficial because they are renewable, nontoxic, biodegradable, low-density, and mechanically durable [83]. In this context, researchers are focusing on utilizing agro-wastes such as coconut waste to produce useful byproducts like CF as raw filler. In a research endeavor, CF served as the primary filler for producing cellulose nanofibers (CNF) derived from coconut shells. This was accomplished using a combination of methods such as ball milling, acid hydrolysis, and ultrasonication before integrating them into a PVA matrix in conjunction with lemon oil and linseed oil. The resulting bionanocomposite films demonstrated superior mechanical and melting characteristics when compared to a plain PVA film. Furthermore, these films displayed remarkable biodegradability, with an 87.34 ± 0.91% degradation rate over a 45-day period. By adding essential oil and coconut shell nanofibers, the hydrophobic properties of the PVA-based film were enhanced as well as its antimicrobial status [84]. To explore the biodegradability, color, and textural features in comparison to plain PLA, this particular study conducted accelerated weathering and soil burial tests on biocomposites of varied ratios of coir CF/PALF with PLA. Consequently, exposure to weather conditions induced alterations in the physical and morphological characteristics of the biocomposites following 250 hours, resulting in a reduced degradation rate of PLA compared to the developed biocomposites. Post-weathering, it was reported that biocomposites degraded through photoradiation, thermal breakdown, oxidation, and hydrolysis, with water expediting the process by inducing fiber swelling, leading to increased light scattering. The CF/PALF/PLA biocomposites appeared to have good biodegradability according to soil burial studies [85].

CF is abundant in nature and rarely used in value-added products, which makes it an economical and environmentally friendly alternative. Afzal and team (2021) investigated the reinforcement of CF into a biodegradable composite film based on corn starch and PVA. Prior to fabrication, CF was modified with chemical agents (H_2O_2, NaOCl, and $KMnO_4$) at concentrations of 0.5 and 1.5% (v/v). Among the treatments used, it was discovered that NaOCl treatment made the fiber surface clean, rough, and active for better bonding with matrix molecules, and composite film developed with NaOCl-treated fiber particles exhibited high tensile strength, moderate elongation, high water absorbency, and good degradation rate. The hydrophilicity of fiber particles improves as the surface is modified, which causes degradation and loss of tensile qualities when it comes into contact with water and soil, and thus, the produced films might be employed in food packaging requiring a short shelf life [86]. Corncob powder and CFs were incorporated into biocomposite films based on maize starch, and the results indicated that the produced biocomposite films were entirely environmentally degradable, as the degradation rate increased over time, reaching 87.37% in only 10 days [87].

Polyhydroxyalkanoate (PHA) was impregnated with 10% CFs treated with 3% oregano essential oil (OEO) to improve the antibacterial activity of the composite. Using OEO-treated CFs, the addition of OEO resulted in a bacteriostatic effect that could be the ability of CF to entrap active compounds such as essential oil, as well as their ability to withstand typical processing conditions of thermoplastic materials used in food packaging [88]. Furthermore, antimicrobial agents added to packaging had a vital role in regulating the absorption and release of chemicals in packaged food [89]. Biodegradable composite foam made from cassava starch mixed with CF loaded with boric acid (cross-linking agent) using a conventional compression molding technique. The morphology of biocomposite foam showed a high interaction between the starch phase and the coconut phase; moreover, boric acid as a cross-linking agent improved the tensile strength, hardness, and water absorption properties, while the density of composite foam decreased [90]. The authors extended their prior research, with a focus on how coconut residual fibers (CRFs) and their modified counterparts affect the properties of eco-friendly composite foam made from cassava starch. CMC was used to modify the structure of CRF, and biodegradable composite foam was loaded with CRF and modified CRF at various doses of 0, 2, 4, 6, and 8 wt % and 0, 2, 4, and 6 wt %, respectively. As a result, the density of the composite was reduced with increased CRF (modified) concentrations from 2 wt %, and mechanical qualities such as hardness and tensile strength were further enhanced, and modified CRF (2 wt %) mixed with CRF (6 wt %) created best-performing biodegradable composite foam, which had good qualities for foam packaging applications [91].

4.5 Conclusion and Future Perspective

This chapter offers an overview of CF as a viable sustainable filler material for biopolymer-based composites and their applications in food packaging. Additionally, it discusses the outstanding mechanical and thermal stability of the CF-reinforced composites. The interfacial adhesion of the coir and biopolymer matrix in composite materials was shown to be significantly improved by modifying the coir surface. The pretreatment of CFs lessens the number of voids in the composites, which is a major issue in the production of composite materials. CFs combined with other natural or synthetic fibers in a hybrid composite can also be used to customize the thermomechanical properties. The expansion of scientific advancements and technology has also led to more applications of CF-reinforced composites, such as biodegradable food packaging materials, making them a more sustainable choice than conventional materials.

References

1. Hasan, K.M.F. *et al.*, Thermo-mechanical properties of pretreated coir fiber and fibrous chips reinforced multilayered composites. *Sci. Rep.*, **11**, 1, 3618, 2021.
2. Keya, K.N. *et al.*, Natural fiber reinforced polymer composites: History, types, advantages, and applications. *Mater. Eng. Res.*, **1**, 2, 69–87, 2019.
3. S, J. *et al.*, Recent development and future trends in coir fiber-reinforced green polymer composites: Review and evaluation. *Polym. Compos.*, **37**, 11, 3296–3309, 2015.
4. Mahmud, S. *et al.*, Comprehensive review on plant fiber-reinforced polymeric biocomposites. *J. Mater. Sci.*, **56**, 12, 7231–7264, 2021.
5. Verma, D. and Gope, P.C., The use of coir/coconut fibers as reinforcements in composites, in: *Biofiber Reinforcements in Composite Materials*, pp. 285–319, Elsevier Ltd, Cambridge, UK, 2015.
6. Adeniyi, A.G. *et al.*, A review of coir fiber reinforced polymer composites. *Compos. Part B: Eng.*, **176**, 107305, 2019.
7. Hasan, K.M.F. *et al.*, A state-of-the-art review on coir fiber-reinforced bio-composites. *RSC Adv.*, **11**, 18, 10548–10571, 2021.
8. Rejeesh, C.R. and Saju, K.K., Methods and materials for reducing flammability behaviour of coir fibre based composite boards: A review. *Mater. Today: Proc.*, **4**, 9, 9399–9407, 2017.
9. Sydow, Z. and Bieńczak, K., The overview on the use of natural fibers reinforced composites for food packaging. *J. Nat. Fibers*, **16**, 8, 1189–1200, 2018.
10. Babu, S.A. and Narayanankutty, S.K., Preparation of chemically functionalized and self compatibilized short coir fiber - high density polyethylene composites. *J. Appl. Polym. Sci.*, **138**, 42, 10, e51254, 2021.
11. Arrakhiz, F.Z. *et al.*, Mechanical properties of high density polyethylene reinforced with chemically modified coir fibers: Impact of chemical treatments. *Mater. Des.*, **37**, 379–383, 2012.

12. Sari, P.S. *et al.*, Effect of plasma modification of polyethylene on natural fibre composites prepared via rotational moulding. *Compos. Part B: Eng.*, **177**, 107344, 2019.
13. Gupta, N. and Ramkumar, P.L., Experimental investigation of linear low density polyethylene composites based on coir for rotational molding process. *Polym. Polym. Compos.*, **29**, 8, 1114–1125, 2020.
14. Abhilash, S.S. and Lenin Singaravelu, D., A comparative study of mechanical, dynamic mechanical and morphological characterization of tampico and coir fibre-reinforced LLDPE processed by rotational moulding. *J. Ind. Text.*, **51**, 285S–310S, 2020.
15. Mohammed Bukar, A., Mohammed El-Jummah, A., Hammajam, A.A., Development and evaluation of the mechanical properties of coconut fibre reinforced low density polyethylene composite. *Open J. Compos. Mater.*, **12**, 03, 83–97, 2022.
16. Awodi, E. *et al.*, Experimentally predicted optimum processing parameters assisted by numerical analysis on the multi-physicomechanical characteristics of coir fiber reinforced recycled high density polyethylene composites. *Res. Sq.*, 1–21, 2021. https://doi.org/10.21203/rs.3.rs-591200/v1.
17. Kakou, C.A. *et al.*, Hybrid composites based on polyethylene and coir/oil palm fibers. *J. Reinf. Plast. Compos.*, **34**, 20, 1684–1697, 2015.
18. Arman Alim, A.A. *et al.*, A review of nonbiodegradable and biodegradable composites for food packaging application. *J. Chem.*, **2022**, 1–27, 2022.
19. Kenechi, N.-O., Linus, C., Kayode, A., Utilization of rice husk as reinforcement in plastic composites fabrication- a review. *Am. J. Mater. Synth. Process. (AJMSP)*, **1**, 3, 32–36, 2016.
20. Gelfuso, M.V., Silva, P.V.G.d., Thomazini, D., Polypropylene matrix composites reinforced with coconut fibers. *Mater. Res.*, **14**, 3, 360–365, 2011.
21. Leão, R.M. *et al.*, Surface treatment of coconut fiber and its application in composite materials for reinforcement of polypropylene. *J. Nat. Fibers*, **12**, 6, 574–586, 2015.
22. Mir, S.S. *et al.*, Improvement of physico-mechanical properties of coir-polypropylene biocomposites by fiber chemical treatment. *Mater. Des.*, **52**, 251–257, 2013.
23. Essabir, H. *et al.*, Structural, mechanical and thermal properties of bio-based hybrid composites from waste coir residues: Fibers and shell particles. *Mech. Mater.*, **93**, 134–144, 2016.
24. Morandim-Giannetti, A.A. *et al.*, Lignin as additive in polypropylene/coir composites: Thermal, mechanical and morphological properties. *Carbohydr. Polym.*, **87**, 4, 2563–2568, 2012.
25. Adediran, A.A. *et al.*, Effect of surface modification on the properties of polypropylene matrix reinforced with coir fibre and yam peel particulate. *Sci. World J.*, **2021**, 8891563, 2021.
26. Monteiro, S.N., Terrones, L.A.H., D'Almeida, J.R.M., Mechanical performance of coir fiber/polyester composites. *Polym. Test.*, **27**, 5, 591–595, 2008.

27. Kobayashi, S. and Uyama, H., Synthesis of polyesters I: Hydrolase as catalyst for polycondensation (condensation polymerization), in: *Enzymatic Polymerization towards Green Polymer Chemistry*, pp. 105–163, Springer Nature Singapore Pte Ltd., Singapore, 2019.
28. Bagherpour, S., Fibre reinforced polyester composites, in: *Polyester*, InTech, London, UK, 2012.
29. Prasad, G.L.E., Gowda, B.S.K., Velmurugan, R., A study on impact strength characteristics of coir polyester composites. *Proc. Eng.*, **173**, 771–777, 2017.
30. Arsyad, M., Soenoko, R., Arman, A., Influence of the soaking time on the mechanical properties of coir as a natural composite reinforcement. *Fibres Text. East. Eur.*, **28**, 6, 98–103, 2020.
31. dos Santos, J.C. *et al.*, Effects of sodium carbonate on the performance of epoxy and polyester coir-reinforced composites. *Polym. Test.*, **67**, 533–544, 2018.
32. Widnyana, A. *et al.*, Tensile properties of coconut coir single fiber with alkali treatment and reinforcement effect on unsaturated polyester polymer. *Mater. Today: Proc.*, **22**, 300–305, 2020.
33. Jayabal, S. *et al.*, Mechanical and machinability behaviors of woven coir fiber-reinforced polyester composite. *Fibers Polym.*, **14**, 9, 1505–1514, 2013.
34. Palani Kumar, K. *et al.*, Evaluation of mechanical properties of coconut flower cover fibre-reinforced polymer composites for industrial applications. *Prog. Rubber Plast. Recycl. Technol.*, **37**, 1, 3–18, 2020.
35. Balaji, N.S. *et al.*, Study of mechanical and thermal behaviours of zea-coir hybrid polyester composites. *Mater. Today: Proc.*, **27**, 2048–2051, 2020.
36. Karthi, S. *et al.*, Effect of relative content on mechanical properties of coir and Napier grass fibers reinforced hybrid polyester composites. *Mater. Today: Proc.*, **33**, 2929–2933, 2020.
37. Castro-Aguirre, E. *et al.*, Poly(lactic acid)—Mass production, processing, industrial applications, and end of life. *Adv. Drug Deliv. Rev.*, **107**, 333–366, 2016.
38. Azlin, M.N.M. *et al.*, Natural polylactic acid-based fiber composites: A review, in: *Advanced Processing, Properties, and Applications of Starch and Other Bio-Based Polymers*, pp. 21–34, Elsevier Inc., Amsterdam, Netherlands, 2020.
39. Dong, Y. *et al.*, Polylactic acid (PLA) biocomposites reinforced with coir fibres: Evaluation of mechanical performance and multifunctional properties. *Compos. Part A: Appl. Sci. Manuf.*, **63**, 76–84, 2014.
40. Sun, Z. *et al.*, Mechanical and thermal properties of PLA biocomposites reinforced by coir fibers. *Int. J. Polym. Sci.*, **2017**, 1–8, 2017.
41. Siakeng, R. *et al.*, Effects of surface treatments on tensile, thermal and fibre-matrix bond strength of coir and pineapple leaf fibres with poly lactic acid. *J. Bionic Eng.*, **15**, 6, 1035–1046, 2018.
42. González-López, M.E. *et al.*, Effect of surface treatment on the physical and mechanical properties of injection molded poly(lactic acid)-coir fiber bio-composites. *Polym. Compos.*, **40**, 6, 2132–2141, 2018.

43. Madueke, C.I. *et al.*, Investigations into the tensile properties and microstructural features of Coconut fibre (Coir) reinforced Polylactic acid (PLA) biodegradable composites. *UNIZIK J. Eng. Appl. Sci. (UJEAS)*, **2**, 2, 301–310, 2023.
44. Coskun, K. *et al.*, Effect of various enzymatic treatments on the mechanical properties of coir fiber/poly(lactic acid) biocomposites. *J. Thermoplast. Compos. Mater.*, **34**, 8, 1066–1079, 2019.
45. Sun, Z. *et al.*, Properties of coir fibre/polylactic acid composites based on amylase processing. *J. Thermoplast. Compos. Mater.*, 37, 426–439, 2023.
46. Siakeng, R. *et al.*, Mechanical, dynamic, and thermomechanical properties of coir/pineapple leaf fiber reinforced polylactic acid hybrid biocomposites. *Polym. Compos.*, **40**, 5, 2000–2011, 2018.
47. Chotiprayon, P., Chaisawad, B., Yoksan, R., Thermoplastic cassava starch/poly(lactic acid) blend reinforced with coir fibres. *Int. J. Biol. Macromol.*, **156**, 960–968, 2020.
48. Hemsri, S. *et al.*, Wheat gluten composites reinforced with coconut fiber. *Compos. Part A: Appl. Sci. Manuf.*, **43**, 7, 1160–1168, 2012.
49. Diao, C. *et al.*, Toughened wheat gluten and treated coconut fiber composite. *Compos. Part A: Appl. Sci. Manuf.*, **58**, 90–97, 2014.
50. Jagadeesh, D., Kanny, K., Prashantha, K., A review on research and development of green composites from plant protein-based polymers. *Polym. Compos.*, **38**, 1504–1518, 2015.
51. Zuraida, A. *et al.*, Effect of fiber length variations on mechanical and physical properties of Coir Fiber Reinforced Cement-albumen Composite (CFRCC). *IIUM Eng. J.*, **12**, 1, 65–77, 2011.
52. Pereira, J. *et al.*, Insight into the effects of coconut kernel fiber on the functional and microstructural properties of myofibrillar protein gel system. *LWT*, **138**, 110745, 2021.
53. Noorunnisa Khanam, P. *et al.*, Tensile, flexural, and compressive properties of coir/silk fiber-reinforced hybrid composites. *J. Reinf. Plast. Compos.*, **29**, 14, 2124–2127, 2009.
54. Alharbi, M.A.H. *et al.*, Effect of coconut coir and regenerated silk microparticles as blends and natural binders for construction and demolition waste (CDW) wood on the mechanical, thermal, and structural properties of bio-microcomposites prepared by hot-pressing. *Biomass Convers. Biorefin.*, **13**, 4, 3209–3226, 2021.
55. Muensri, P. *et al.*, Effect of lignin removal on the properties of coconut coir fiber/wheat gluten biocomposite. *Compos. Part A: Appl. Sci. Manuf.*, **42**, 2, 173–179, 2011.
56. Guna, V. *et al.*, Wool and coir fiber reinforced gypsum ceiling tiles with enhanced stability and acoustic and thermal resistance. *J. Build. Eng.*, **41**, 102433, 2021.

57. Lomelí Ramírez, M.G. *et al.*, Study of the properties of biocomposites. Part I. Cassava starch-green coir fibers from Brazil. *Carbohydr. Polym.*, **86**, 4, 1712–1722, 2011.
58. Corradini, E. *et al.*, A preliminary study for the use of natural fibers as reinforcement in starch-gluten-glycerol matrix. *Macromol. Symp.*, **245-246**, 1, 558–564, 2007.
59. Rosa, M.F. *et al.*, Effect of fiber treatments on tensile and thermal properties of starch/ethylene vinyl alcohol copolymers/coir biocomposites. *Bioresour. Technol.*, **100**, 21, 5196–202, 2009.
60. Moura, C.V.R.d. *et al.*, New biodegradable composites from starch and fibers of the babassu coconut. *Polímeros*, **31**, 1, e2021007, 2021.
61. de Farias, J.G. *et al.*, Surface lignin removal on coir fibers by plasma treatment for improved adhesion in thermoplastic starch composites. *Carbohydr. Polym.*, **165**, 429–436, 2017.
62. Gomesa, Á.V.R. *et al.*, Effect of carnauba wax and coconut fiber contents on tensile properties of corn starch-based biocomposites. *Mater. Res.*, **22**, e20190053, 2019.
63. Singh, N. and Maitra, J., Exploring the antibacterial activity of jute and coir fiber added chitosan-starch blended films. *Der Pharma Chem.*, **14**, 7, 21–28, 2022.
64. Gupta, H. *et al.*, Preparation and characterization of bio-composite films obtained from coconut coir and groundnut shell for food packaging. *J. Mater. Cycles Waste Manage.*, **24**, 569–581, 2022.
65. Iovino, R. *et al.*, Biodegradation of poly(lactic acid)/starch/coir biocomposites under controlled composting conditions. *Polym. Degrad. Stab.*, **93**, 147–157, 2008.
66. Rabe, S. *et al.*, Natural keratin and coconut fibres from industrial wastes in flame retarded thermoplastic starch biocomposites. *Materials*, **12**, 3, 344, 2019.
67. Reddy, N., Composites from coir fibers, in: *Sustainable Applications of Coir and Other Coconut By-Products*, pp. 141–185, 2019.
68. Tran, L.Q.N. *et al.*, Tensile behavior of unidirectional bamboo/coir fiber hybrid composites. *Fibers*, **7**, 7, 62, 2019.
69. Czlonka, S., Strakowska, A., Kairyte, A., Coir fibers treated with henna as a potential reinforcing filler in the synthesis of polyurethane composites. *Materials*, **14**, 5, 1128, 2021.
70. Ramprasath, R. *et al.*, Investigation on impact behavior of rice husk impregnated coir-vinyl ester composites. *Macromol. Symp.*, **361**, 1, 123–128, 2016.
71. Mittal, M. and Chaudhary, R., Experimental investigation on the mechanical properties and water absorption behavior of randomly oriented short pineapple/coir fiberreinforced hybrid epoxy composites. *Mater. Res. Express*, 6, 1–26, 2018.

72. Arul, M. *et al.*, Mechanical and fracture study of hybrid natural fiber reinforced composite – Coir and sugarcane leaf sheath. *Mater. Today: Proc.*, **33**, 2795–2797, 2020.
73. Arifuzzaman Khan, G.M. *et al.*, Influence of chemical treatment on the properties of banana stem fiber and banana stem fiber/coir hybrid fiber reinforced maleic anhydride grafted polypropylene/low-density polyethylene composites. *J. Appl. Polym. Sci.*, **128**, 2, 1020–1029, 2012.
74. Jayabal, S., Natarajan, U., Sekar, U., Regression modeling and optimization of machinability behavior of glass-coir-polyester hybrid composite using factorial design methodology. *Int. J. Adv. Manuf. Technol.*, **55**, 263–273, 2011.
75. Singh, Y. *et al.*, Fabrication and characterization of coir/carbonfiber reinforced epoxy based hybrid composite for helmet shells and sports-good applications: Influence of fiber surface modifications on the mechanical, thermal and morphological properties. *J. Mater. Res. Technol.*, **9**, 6, 15593–15603, 2020.
76. Dixit, S. *et al.*, Novel hybrid structural biocomposites from alkali treated-date palm and coir fibers: Morphology, thermal and mechanical properties. *J. Polym. Environ.*, **28**, 9, 2386–2392, 2020.
77. Siddika, S. *et al.*, Effect of reinforcement and chemical treatment of fiber on the properties of jute-coir fiber reinforced hybrid polypropylene composites. *Fibers Polym.*, **15**, 5, 1023–1028, 2014.
78. Dagwa, I., Mudiare, E., Nelson, E., Effect of fiber length on the mechanical properties of hybrid ceasar weed/coir reinforced polyester composite. *Usak Univ. J. Mater.*, **4**, 2, 29–29, 2015.
79. Wang, J. and Hu, Y., Novel particleboard composites made from coir fiber and waste banana stem fiber. *Waste Biomass Valorization*, **7**, 6, 1447–1458, 2016.
80. Duan, J. *et al.*, Mechanical properties of hybrid sisal/coir fibers reinforced polylactide biocomposites. *Polym. Compos.*, **39**, 2017. https://doi.org/10.1002/pc.24489
81. Okafor, C.E. *et al.*, Optimal route to robust hybridization of banana-coir fibre particulate in polymer matrix for automotive applications. *Materialia*, **16**, 101098, 2021.
82. V, B. *et al.*, Novel coir/chitosan composites for packaging applications. *J. Mines Met. Fuels*, **70**, 222–227, 2022.
83. Sarwar, M.S. *et al.*, Preparation and characterization of PVA/nanocellulose/Ag nanocomposite films for antimicrobial food packaging. *Carbohydr. Polym.*, **184**, 453–464, 2018.
84. Arun, R. *et al.*, Biodegradable nano composite reinforced with cellulose nano fiber from coconut industry waste for replacing synthetic plastic food packaging. *Chemosphere*, **291**, 132786, 2022.
85. Siakeng, R. *et al.*, Accelerated weathering and soil burial effect on biodegradability, colour and textureof coir/pineapple leaf fibres/PLA biocomposites. *Polymers*, **12**, 2, 458, 2020.

86. Afzal, A. *et al.*, Development and characterization of biodegradable composite film. *Environ. Technol. Innovation*, **23**, 101664, 2021.
87. Singh, N. and Nanda, A., Preparation and antimicrobial activity of corn cob and coir reinforced biodegradable starch biocomposite films for food packaging application. *Asian Food Sci. J.*, **21**, 7, 53–66, 2022.
88. Torres-Giner, S. *et al.*, Melt processability, characterization, and antibacterial activity of compression-molded green composite sheets made of poly(3-hydroxybutyrate-co-3-hydroxyvalerate) reinforced with coconut fibers impregnated with oregano essential oil. *Food Packag. Shelf Life*, **17**, 39–49, 2018.
89. Wangprasertkul, J., Siriwattanapong, R., Harnkarnsujarit, N., Antifungal packaging of sorbate and benzoate incorporated biodegradable films for fresh noodles. *Food Control*, **123**, 107763, 2021.
90. Nansu, W. *et al.*, Effect of crosslinking agent on the physical and mechanical properties of a composite foam based on cassava starch and coconut residue fiber. *Mater. Today: Proc.*, **17**, 2010–2019, 2019.
91. Nansu, W. *et al.*, Coconut residue fiber and modified coconut residue fiber on biodegradable composite foam properties. *Mater. Today: Proc.*, **47**, 3594–3599, 2021.

5

Sugarcane Bagasse for Sustainable Food Packaging

Sweety Kalita, Avik Mukherjee and Santosh Kumar*

Department of Food Engineering and Technology, Central Institute of Technology Kokrajhar, Kokrajhar, Assam, India

Abstract

Waste management in agriculture is a growing concern due to increased food processing and production resulting in the generation of huge amounts of solid waste. Agricultural byproducts and residues, known as agro-wastes, are valuable resources often unutilized. Sugarcane bagasse, one of the agro-wastes because of its chemical composition and biodegradability, is gaining attention for various applications. Sugarcane bagasse contains a range of components that have demonstrated the potential to serve as raw materials for manufacturing composite materials with diverse properties and performance characteristics. Research has explored the impacts, performance, and applications of sugarcane bagasse in its natural state, as well as in transformed forms, after treatment with various techniques and chemicals, and by blending with other materials having different characteristics. The current chapter aims to deliver a contemporary and comprehensive review of the current literature, summarizing the extensive studies conducted to explore the possible applications and capabilities of sugarcane bagasse as a composite material in the food packaging sector.

Keywords: Sugarcane bagasse, pretreatment, sustainable packaging, food container, circular economy

5.1 Introduction

Sugarcane (*Saccharum officinarum*) is a significant agricultural crop primarily grown in Brazil, India, and other tropical and subtropical regions

**Corresponding author*: s.kumar@cit.ac.in

Santosh Kumar, Avik Mukherjee and Vimal Katiyar (eds.) Agro-Waste Derived Biopolymers and Biocomposites: Innovations and Sustainability in Food Packaging, (115–152) © 2024 Scrivener Publishing LLC

of the world. Brazil is the largest producer of sugarcane globally, whereas India is the second-largest producer. In India, the estimated sugarcane production stands at approximately 306 Mt, whereas Brazil produces 758 Mt. Sugarcane plays a pivotal role in fulfilling nearly 80% demand for sugar in the world. The process of harvesting sugarcane, its transportation to mills, and subsequent processing including juice extraction are important steps in sugar production [1–3]. This industrial process yields several valuable byproducts, including sugarcane bagasse (SB), mud cake, sugarcane trash, and fly ash. Among these byproducts, SB is particularly noteworthy, as it is typically obtained after the cleaning and juice extraction from sugarcane [4]. SB is a substantially dry, pulpy, and fibrous material, which is composed of approximately 32 to 45% cellulose, 20 to 32% hemicellulose, 17 to 32% lignin, 1.0 to 9.0% ash, and several other minor constituents [5]. In 2020, the global production of sugarcane amounted to approximately 1869.71 Mt, and approximately 280 kg of bagasse was generated as a byproduct for every ton of sugarcane processed [6]. The excessive production of bagasse has led to its disposal through burning, resulting in environmental pollution. Consequently, waste recycling has emerged as a pivotal focus in scientific research, driven by environmental concerns [7]. Utilizing this byproduct of the sugar industry holds significant importance not only from an environmental perspective but also as a valuable substrate for the large-scale production of innovative and valuable products, as it contains structural sugars in the form of cellulose and hemicellulose [8]. However, lignin presence in bagasse poses a challenge to the saccharification process, as lignin can nonproductively bind to enzymes, hindering their effectiveness in breaking down cellulose and hemicellulose into fermentable sugars [9, 10]. To overcome the challenges posed by lignin and to improve the saccharification process, an effective pretreatment strategy is required. This strategy usually aims to solubilize hemicellulose, expose cellulose to cellulolytic enzymes, and reduce the size of particles, crystallinity, and the polymerization of cellulose. These pretreatment steps lead to increased surface area that eventually enhances the enzymatic digestibility of biomass [11]. SB, with its high cellulose content, presents an attractive opportunity for manufacturing composite materials due to its abundant availability and cost-effective processing. SB composites offer several advantages, including flexibility, high strength, tensile strength (TS), ease of shaping, customization, and biodegradability, and have applications in diverse areas like food packaging, automotive, construction, furniture, and healthcare [12]. This chapter offers insights into the structural properties of SB, discusses important pretreatment methods, and explores diverse applications of SB, with a specific focus on its utilization in the food packaging sector.

5.2 Chemical Composition and Characteristics of Sugarcane Bagasse (SB)

SB represents a residue of lignocellulose fibers derived from the sugarcane culm, following the milling and juice extraction process. The bagasse comprises two distinct regions, i.e., the outer rind and the inner pith. The pith is primarily made up of parenchyma cells and contains short fibers, whereas the rind is constituted of sclerenchyma cells, with a significant presence of longer and finer bundles. These inherent structures within the culm/bagasse, each possessing unique characteristics and containing vascular bundles, which contribute to its heterogeneous nature [5]. The chemical composition of bagasse is contingent upon factors such as the age of the fibers, varieties of sugarcane, and extraction conditions [13]. The usual composition of sugarcane includes water (65 to 75%), sugars (11 to 18%), fibers (8 to 14%), and soluble solids (12 to 23%); however, the bagasse consists of 32 to 45%, 20 to 32%, 17 to 32%, and 1.0 to 9.0% of cellulose, hemicellulose, lignin, and ash, respectively [14–20]. The relatively low ash content in bagasse (1.0 to 9.0%) has several benefits in microbial bioconversion in comparison with the other biomass such as rice and wheat straw, which have ash contents of 17.5 and 11.0%, respectively [21]. Bagasse ash is characterized by carbonized black-colored particles, primarily containing cliftonite (C), moissanite (SiC), quartz (SiO_2), and titanium oxide (Ti_6O), which are generated by pyrolysis of bagasse powder [22, 23]. In bagasse, a significant portion of the cellulose is present in a crystalline structure (47%), whereas hemicellulose is an amorphous structure that includes constituents like xylose and glucose. Such cellulose polymers are arranged within fibrils, which are encased by lignin and hemicellulose, and the lignin acts as a binding agent among hemicellulose and cellulose, contributing to the rigidity and strength of the material [5]. This structural arrangement in bagasse with the intertwined roles of cellulose, hemicellulose, and lignin results in a complex yet effective combination that imparts strength and structural integrity to the material.

In general, bagasse possesses a diameter of 0.3 to 0.4 mm, a density ranging from 344 to 360 kg/m^3, a Young's modulus of 4500 MPa, a TS varying between 70 and 107 MPa, and an elongation at break (EB) falling within the range of 3 to 7% [24]. The incorporation of bagasse into composites has garnered significant attention in the mechanical properties as one of the influencing factors. Key properties of SB encompass TS, impact strength, flexural strength, weathering, durability, and flame resistance, making it a significant contender to reinforce in the composite.

As per a study, higher mechanical values have been reported, such as a high TS of 222 MPa, Young's modulus of 27.1 GPa, and an elongation at a break of 1.1% [25]. Some reports indicate that the TS of composites enhances with increasing bagasse% [26]. The flexural strength of bagasse is dependent on several factors including fiber length, presence of the rind, powder size, and content. Initially, as the bagasse content increases, the flexural strength tends to rise; however, beyond a certain point, further addition of bagasse content can lead to a decline in flexural strength [27]. Notably, there is a study that contradicts this pattern, presenting a situation where added bagasse results in stable flexural strength [28]. When it comes to the impact strength of bagasse composites, an increase in bagasse content generally leads to a reduction in impact strength [29].

5.3 Cellulosic and Hemicellulosic Fractions of Sugarcane Bagasse

Cellulose represents a linear natural polymer made up of anhydro-glucose units linked together through β-glycosidic bonds at the C-1 and C-4 atoms, which leads to the development of a crystalline pattern, due to the existence of three hydroxyl groups, i.e., a primary OH at C-6 and two secondary OH at C-2 and C-3. Consequently, this arrangement fosters the creation of robust intra- and intermolecular H-bonds, playing a pivotal character in facilitating fibril formation [30, 31]. The analysis displayed that untreated bagasse cellulose has an average crystallinity index (CrI) of 56.7% [30, 32, 33]. The customary molecular weight of cellulose sourced from SB lies between 157,800 and 168,400 g/mol, and the dimensions of cellulose fibers vary from 1.0 to 1.5 mm. On a more detailed scale, fundamental cellulose units termed elementary fibrils undergo a self-assembly process, culminating in the emergence of microfibrils. These microfibrils can then establish connections through hemicellulose matrices, ultimately giving rise to macrofibrils. This hierarchical arrangement significantly bolsters cellulose's resilience against both chemical and enzymatic degradation procedures. Cellulose I, the primary polymorph of cellulose, can be detected through x-ray diffraction (XRD). The degree of polymerization (DPw) refers to the number of glucose units within a polymer molecule and can impact the rate of enzymatic hydrolysis. The average DPw typically falls within the range of 974 to 1039.50 in the case of cellulose of SB, and the variation depends on the method of cellulose isolation and is closely linked to the intrinsic viscosity (η) [32]. For instance, when SB undergoes potassium hydroxide isolation and delignification, the DPw increases to 1406.5, whereas the combination of both the acetic and nitric acid reduced the DPw to 822.5, demonstrating that this mixture degrades cellulose macromolecules more than potassium hydroxide [30]. Cellulose usually contains crystalline, para crystalline (disordered), and amorphous structures, with the amorphous portion being more reactive than the crystalline one, and such complication influences cellulose thermal decomposition [34–36]. To explore this further, SB cellulose was subjected to extremely low acid hydrolysis using 0.07% sulfuric acid at 190°C, 210°C, and 225°C for varying durations, and thermal analysis, SEM, and XRD results indicated that cellulose residues (microcrystalline celluloses) got at these conditions have alike behavior of thermal decomposition, crystallite size, CrI, and particle size as Avicel PH-101 [37]. Nanocellulose refers to cellulose that exists in fiber or crystal forms typically ranging from a few micrometers and diameters falling within the <100-nm range. These nanocellulose materials are obtained from natural cellulose fibers and are of three primary types, i.e., nanofibrillated cellulose, bacterial nanocellulose, and cellulose nanocrystals (CNCs). Recent research efforts have been focused on extracting nanocellulose or nanocrystals from SB. Various chemical treatments have been explored by different researchers, including the use of different chemicals, ratios, and conditions [38]. For instance, Sofla *et al.* (2016) employed a ball milling technique in their research [39], whereas Feng *et al.* (2018) utilized ultrasonic treatment with acid hydrolyzation [40]. The results demonstrated that the ball milling technique did not effectively remove the amorphous portions of cellulose, whereas acid hydrolysis was found to remove these amorphous parts and enhance the crystalline content of CNCs. In addition, CNCs were obtained through sulfuric acid extraction from cellulose that had undergone purification via steam explosion and enzyme treatment. Findings from FTIR, XRD, thermal stability, and cytotoxicity assessments indicated the absence of noncellulosic components, enhanced crystallinity in the CNCs, reduced thermal stability of the CNCs, and no signs of cytotoxicity [41]. Also, chemical composition analysis revealed a whiteness index of 92.59 ± 0.12 and an α-cellulose content of 87%.

Hemicellulose, a heteropolysaccharide with a relatively low molecular weight of 7380 g mol^{-1}, is composed of a β-(1,4)-xylopyranose backbone, which contains approximately 200 β-xylopyranose residues linked with 1,4-glycosidic bonds. It encompasses various constituents such as glucomannans, xyloglucans, galactomannans, β-glucans, and small quantities of uronic acids [31, 42, 43]. In SB, the sugar composition of hemicellulose mainly includes 11.5 to 39.9% galactose, 15.5 to 28.9% xylose, and 17.5 to 50.5% glucose, along with smaller proportions of 5.35 to 14.31% arabinose, 0.0 to 14.0% mannose, 2.5 to 10.6% rhamnose, and 1.0 to 2.3% uronic acids [44]. The kind and quantity of hemicelluloses can vary depending on the plant species and the specific tissue being considered. In bagasse, different sections of the bagasse contain varying amounts of hemicelluloses, i.e., 16, 27, and 29% of bagasse mass are located on the external fraction, node, and internode, respectively [45]. The degree of branching in hemicellulose is indicated by the arabinose/xylose ratio, and a ratio of around 0.2 indicates that the presence of arabinose is relatively limited in comparison to xylose [43]. Hemicellulose forms covalent bonds with lignin and H-bonds with cellulose, leading to integration between these components that contribute to stability and flexibility [46].

There are several methods employed for extracting hemicellulose, and the method depends on reasons like the DPw, branching or substitution, impurities like residual lignin content, reactivity, and solubility. Depending on whether the goal is to recover hemicellulose in its polysaccharide or monosaccharide form, different approaches involving either acidic or alkaline conditions can be effectively employed. When hemicellulose extraction is carried out in an acidic medium and the process is tailored accordingly, it results in a final product primarily composed of monosaccharides and oligosaccharides. In contrast, the extraction of hemicelluloses in an alkaline medium, along with a process designed for this purpose, yields a final product primarily consisting of polysaccharides having high DPw [45, 47]. Generally, the process of isolating hemicellulose involves multiple alkaline extraction steps that break ester linkages as well as extract hemicellulose from the lignocellulosic matrix. Notably, KOH and NaOH with H_2O_2 pretreatments for isolation of hemicellulose have demonstrated high yields of xylose, ranging from 83.1 to 85.02% [48, 49]. In contrast, procedures involving dewaxed and distilled H_2O, as well as H_2O with solid alkali precooking resulted in significantly lower yields of xylose, ranging from 55.20 to 57.43% [42]. FTIR analysis showed the complete breakdown of ester bonds during alkali pretreatment, specifically with 10% KOH at temperatures of 20°C, 35°C, and 50°C [49]. Furthermore, NMR was also able to identify strong signals confirming the existence of 1,4-linked β-D-xylopyranose groups and the 4-O-methoxyl units of glucuronic acid residues in xylan, and thus, it was deduced that β-D-xylopyranose represents the primary constituent in bagasse hemicellulose [42]. Another investigation presents an efficient alkaline procedure for extracting high-quality hemicellulose using a combination of NaOH and NH_4OH from SB, specifically to produce xylooligosaccharides (XOs). The comparative analysis revealed that the use of mixed alkalis led to a notably elevated recovery of hemicellulose (68% by weight of xylose), which was relatively 1.3 times higher compared to individual alkali protocols. Moreover, the hemicellulose extracted through this method was primarily composed of xylose (with a xylose/arabinose ratio of 3.94) in the form of xylan [50]. Synergistic interaction between different alkalis (involving weak and strong bases) played a significant role in selectively breaking the linkages between lignin and carbohydrates, which facilitated a greater release of hemicellulose under the relatively mild reaction conditions (with a 10% alkali concentration and a temperature of 120°C [50]). A recent study explored the use of ultrasound-assisted alkaline extraction to obtain hemicellulose from SB pith, in which the optimal conditions were 28-min ultrasonic treatment, a KOH mass concentration of 3.7%, and an extraction temperature of 53°C that led to a maximum total hemicellulose yield of 23.05%. The researchers characterized the monosaccharide composition, and findings revealed that the predominant constituent in both water-soluble hemicellulose (69.05%) and alkali-soluble hemicellulose (85.83%) is xylose [51].

5.4 Pretreatment Approaches for SB

The initial and crucial steps in maximizing the utilization of complex lignocellulosic biomass render it amenable to subsequent transformations. The effectiveness of biomass conversion hinges on factors including its surface, structure, particle size, and composition of cellulose, hemicellulose, and lignin. In the case of SB, its structural composition presents challenges in terms of recalcitrance primarily due to the protective lignin-hemicellulose structure that shields the valuable cellulose, cellulose crystallinity, as well as the interconnections between hemicellulose and lignin, which hinder its full utilization potential. Pretreatment methods are imperative to enhance the bioavailability of SB, thereby increasing its potential value. Pretreatment techniques must not only enhance sugar yields or make the biomass more accessible for biological processes but also be economically viable and scalable. Given that this is the initial step, it significantly impacts the cost of all downstream biological processing, and consequently, the pretreatment process should be advanced and integrated into the overall process to fully exploit the potential of lignocellulosic biomass [4, 31, 33, 52]. Various pretreatment strategies have been documented for SB (Figure 5.1); however, a single pretreatment approach cannot be universally applicable to all types of lignocellulosic biomass under all conditions [53–55]. Therefore, when selecting a pretreatment method, the key criteria associated with the highest recovery of crucial compounds and polymers should be considered [56]. In conclusion,

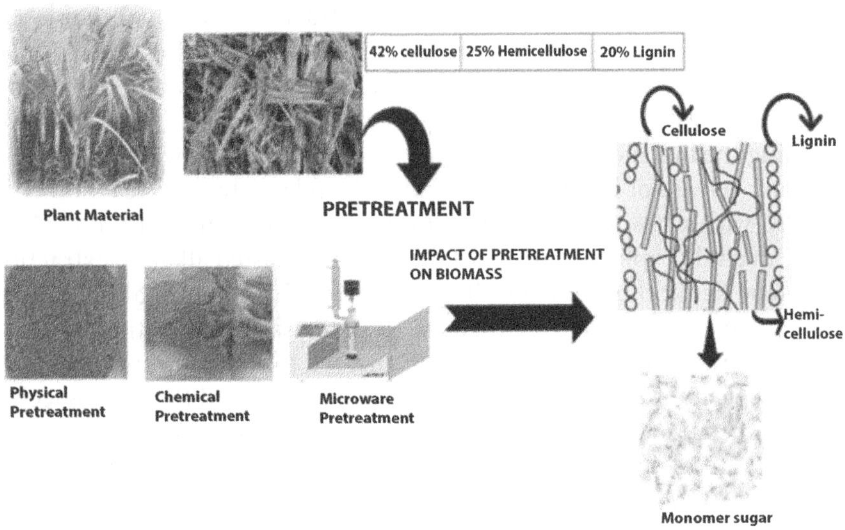

Figure 5.1 Some pretreatment techniques for sugarcane bagasse adopted with permission from [57].

before choosing an appropriate pretreatment method for lignocellulosic biomass, a comprehensive consideration of all these factors is essential. Pretreatment methods can be either physical, chemical, biological, or a combination of two or more methods.

5.4.1 Physical Pretreatments

5.4.1.1 Mechanical Pretreatment

Comminution stands as the prevailing physical pretreatment method, involving the milling of SB to achieve the desired particle size range. This milling process serves a dual purpose, i.e., reduces particle size, augments specific surface area, and diminishes crystallinity of cellulose, which subsequently contributes to an improved enzymatic hydrolysis process, thereby enhancing both the rate and yield of saccharification. Consequently, this enhancement fosters stronger binding of enzymes to the cellulose fractions, further accelerating the saccharification rate and yield. This increased surface area facilitates more efficient interactions between enzymes and substrates, thereby promoting effective bioconversion processes [58–60]. The effectiveness of milling and the amount of energy needed for milling are influenced by various factors including the moisture content of SB, the desired reduction in particle size (expressed as the comminution ratio), and the type of mill being used. Higher moisture content leads to an increased energy requirement for milling and can also result in clogging or blockages in the milling process. Similarly, when aiming for a greater reduction in particle size (higher comminution ratio), the specific milling energy also increases. For example, in a study conducted by Miao *et al.* (2011) using a bench scale knife mill, approximately 720 kWh/tonne of dry weight was needed to grind bagasse with around 40% moisture content. In this study, the initial particle size ranged from 20 to 25.4 mm, and the final particle size targeted was 2 mm [58, 59, 61]. When aiming for a larger final particle size of 8 mm, the energy needed for milling decreased to around 390 kWh/tonne of dry weight. However, an alternative approach involving wet milling has also been explored in research for pretreating SB. For instance, da Silva *et al.* (2010) utilized wet disk milling as a pretreatment of SB, which was found to enhance the yield of saccharification during subsequent processes [58]. Mechanical activation serves to alter both the structure and physicochemical characteristics of bagasse through the disruption of chemical bonds, which subsequently leads to a reduction in particle size. Consequently, this pretreatment method has demonstrated its significance as a pivotal step in converting the feedstock into diverse biomaterials, useful in different applications [62]. Despite the benefits of mechanical activation, the intricate composition of bagasse necessitates the support of supplementary pretreatment methods, which could be chemical, biological, or physiochemical in nature [63]. Consequently, this approach is recognized as an effective and environmentally friendly physical pretreatment strategy, and its success lies in reducing the recalcitrance of bagasse through modifications in the structure of lignin, addressing the complexity of the bagasse matrix [64, 65].

5.4.1.2 Microwave Pretreatment (MWP)

MWP is a convenient and energy-efficient process, which breaks down the framework of cellulosic bundles in SB with minimal formation of inhibitors [66]. This technique utilizes electromagnetic radiation as a nonconventional method for breaking down the multifaceted construction of SB. A significant advantage of microwave pretreatment is its ability to uniformly distribute heat within biomass, whereas conventional heating methods often create temperature gradients. The dielectric polarization during pretreatment using a microwave creates molecular collisions, effectively disassembling the multifaceted construction of SB [67, 68]. Typically, MWP is used in conjunction with other pretreatment methods to achieve improved results, for example, de Souza Moretti *et al.* (2014)

conducted MWP of SB (particle size: 1 to 3 mm) by submerging it in distilled H_2O and subjecting the slurry to microwave radiation for 5 min at 2450 MHz [69]. Similarly, Portero-Barahona *et al.* (2019) investigated microwave-assisted sulfolane pretreatment of SB (at 10% concentration) at various temperatures and reaction times [68]. Furthermore, continuous-flow microwave irradiators and large-scale batch types, with capacities up to 50 liters, have been developed for biomass pretreatments and have been successfully employed in processes aimed at producing bioethanol from lignocellulosic materials, marking significant progress in the field [70, 71].

5.4.1.3 Ultrasound Pretreatment (USP)

Ultrasound frequencies emit radiation that initiates the phenomenon of acoustic cavitation that leads to the creation of numerous microbubbles, each with a diameter of around 100 μm, which undergo rapid implosion in less than a microsecond, triggering compression of the enclosed gasses and vapors. The intense compression generates substantial heat, swiftly elevating the temperature of the liquid in the immediate vicinity of the cavity. Similarly, in the case of biomass, ultrasound pretreatment involves the disruption of the cell wall structure, which in turn increases the available surface area that further facilitates the formation of a more reactive cellulosic fraction, enhancing the production of fermentable sugars by effectively dissociating the lignin and xylan linkages [72–74]. USP shows both chemical and mechanical effects on bagasse, particularly at lower frequencies, and this pretreatment approach lies in its ability to extract substantial amounts of cellulose, lignin, and hemicellulose, ultimately enhancing enzymatic hydrolysis efficiency [75]. In a study by Velmurugan and Muthukumar *et al.* (2012), the researchers conducted ultrasound-supported alkali pretreatment of SB at 24 kHz, 400 W, and temperatures ranging from 30°C to 50°C for 5 to 50 min, achieving an impressive 99.6% delignification of the biomass [76]. In a comparative study that involves alkali-steam and only steam explosion pretreatments, sono-assisted alkaline pretreatment demonstrated superior results with higher glucose recovery and fermentation efficiencies. Specifically, it achieved a remarkable 95.8% recovery of total glucose from the original bagasse, based on the glucan content, and a fermentation efficiency of 91.4%. Sono-assisted alkaline pretreatment of SB emerged as an efficient, straightforward, and sustainable process for the production of bioethanol, especially when considering the principal costs associated with the steam explosion [77].

5.4.2 Chemical Pretreatments

5.4.2.1 Acidic Pretreatment (AP)

AP stands out as an efficient method that proficiently breaks down hemicellulose, facilitating the release of sugars from bagasse, which has two variations i.e., concentrated acid and diluted acid pretreatment [78]. In concentrated acid hydrolysis, the process can occur at relatively low temperatures (30°C to 60°C) utilizing acid concentrations ranging from 40 to 80%. Although this technique yields high sugar quantities, it demands substantial amounts of acid, which can be both toxic and corrosive [79, 80]. To address this concern, it is important to explore alternative reagents such as organic acids (e.g., oxalic acid and acetic acid) that can yield XOs during acidic pretreatment [81]. Moreover, the acid must be recovered after hydrolysis to ensure that the process remains economically viable, and efficient acid recovery technologies have revitalized interest in this method [82]. Conversely, dilute acid hydrolysis represents one of the most widely adopted approaches, which involves the utilization of diluted acids within the range of 0.5 to 6%, at elevated temperatures ranging from 120°C to 170°C with varying periods. Dilute acid pretreatment serves the critical purpose of disrupting the glycosidic bonds within hemicellulose, as well as the bonds between lignin and hemicellulose, and lignin itself. Because of these alterations, sugar dissolution within hemicellulose is achieved, and the cell wall porosity is enhanced to promote effective enzyme digestibility [83, 84]. However, note that acid pretreatments can yield unintended effects, including the development of furan compounds and derivatives of short-chain aliphatic acids that are recognized as potent obstructers in microbial fermentation processes [85, 86]. A variety of acids have been employed for pretreating SB, including dilute H_2SO_4 [87–89], dilute HCl [90], and dilute H_3PO_4 [91]. Among these options, dilute sulfuric acid has demonstrated particularly high hydrolysis yields, and thus, it is the most utilized acid in the pretreatment of bagasse, which outperforms hydrochloric, phosphoric, and nitric acids [79, 80, 92].

5.4.2.2 Alkaline Treatment

In addition to acid pretreatment, alkaline pretreatment has emerged as a prominent chemical technology garnering significant attention for pretreating SB, primarily involving the process of delignification, which is typically conducted at lower temperatures and pressures. This method disrupts the structural integrity of the cell wall of bagasse through several mechanisms, i.e., hydrolyzing uronic and acetic esters, dissolving lignin, hemicelluloses, and silica, as well as swelling cellulose under mild conditions. This series of processes leads to the division of the material into two distinct fractions; a liquid fraction containing hemicellulose oligomers and lignin and a solid fraction consisting of cellulose [20, 93–95]. Alkaline pretreatments stand out from acid pretreatments due to their higher effectiveness in removing lignin, and even when only a portion of the lignin is removed, alkaline pretreatments substantially enhance the digestibility of cellulose. The hydrolysis of ester linkages between lignin and hemicellulose residues during this process contributes to an increase in biomass porosity, and thus, both the hemicellulose and cellulose become more susceptible to enzyme activity [31, 96]. This pretreatment approach retains a significant portion of both cellulose and hemicellulose, thereby offering the potential for hydrolyzing more amount of the pretreated biomass leading to the release of glucose, and additional pentose sugars from cellulose and hemicellulose, respectively. Notably, this hydrolysis occurs in an environment devoid of strong acids and fermentation inhibitors, resulting in minimal sugar degradation [97]. This process employs a range of bases, including NaOH [98–103], $Ca(OH)_2$ [104], KOH [105], aqueous ammonia [106], and NaOH in combination with calcium hydroxide [107]. Among various alkalis, NaOH is notably effective in alkaline pretreatment, showcasing the highest degree of lignin degradation.

5.4.2.3 *Ionic Liquids Pretreatment*

Ionic liquids are unique solvents made of poorly organized cations and anions of organic compounds that give them a boiling point below 100°C. The ionic liquids are characterized by their high thermal stability and lack of volatility, making them exceptional catalysts for the delignification of bagasse [108, 109]. They are effective in reducing the crystallinity of cellulose and can lead to the removal of up to 60% of lignin [110]. For instance, 1-ethyl-3-methylimidazolium acetate has been recognized as a potent ionic solvent capable of dissolving cellulose from bagasse [32]. Ionic liquids, which consist of cations and anions of various organic compounds, also serve as catalysts for lignin removal from bagasse [63], for instance, Bahrani *et al.* (2015) employed the ionic liquid (1,3-dimethylimidazolium dimethyl phosphate) for pretreating SB at a ratio of 1:20 and temperatures ranging from 100°C to 140°C for 120 min for achieving effective pretreatment [111]. Other solvents utilized for bagasse pretreatment include 1-butyl-3-methylimidazolium chloride,

1-allyl-3-methylimidazolium chloride, 1-butyl-3-methylimidazolium acetate, and 1,3-dimethylimidazolium dimethyl phosphate [111, 112].

5.4.3 Physiochemical Pretreatment

5.4.3.1 Organosolv Pretreatments

Organic solvent pretreatments, often referred to as organosolv pretreatments, constitute a delignification process that is capable of fractionating SB into its components (cellulose, lignin, and hemicellulose) in comparatively pure forms and allow efficient recovery of the solvent [113]. This pretreatment involves the use of organic solvent, sometimes in combination with alkali or acid catalysts, for extracting lignin from the lignocellulosic biomass, such as SB. Organic solvents enable the recovery of most of the cellulose in solid form, whereas a significant portion of lignin and hemicellulose liquify in these solvents. This selective dissolution of hemicellulose and lignin in organic solvents contributes to reducing the recalcitrance of the biomass, making cellulose more accessible for enzymatic hydrolysis and subsequent bioethanol production [114, 115]. Organic solvents used in organosolv pretreatment could be high boiling point liquids such as glycerol or glycol, and low boiling point alcohols such as methanol or ethanol. Low boiling point alcohols are often preferred due to their cost-effectiveness and easier recovery during the process; however, they may require special equipment to operate at high pressure, which can increase the overall process cost [116]. In this approach, a mixture of ethanol and water is used as the organic solvent that is effective in breaking down the α-aryl ether linkages in lignin, causing their hydrolysis and solubilization. The process is particularly suitable for smaller mills due to its straightforward recovery method and the absence of sulfur in the discharge [55, 117]. For example, Zhang and Wu (2014) conducted pretreatment of bagasse using a mixture of ethanol and water (60:40 ratio) along with acetic acid (5%) for 45 min at 190°C [118]. Another study by Zhang *et al.* (2018) involved pretreating SB with a mixture of FeCl3 (0.025%) and ethanol (60%) for 72 hours at 160°C [119]. Moreover, aqueous ethanol and organic acid catalysts were used to pretreat lignocellulosic biomass, investigating various factors such as type of catalyst (including acetic, oxalic, and citric acids), temperature (from 150°C to 190°C), time (up to 60 min), amount of CO_2 (ranging from 25 to 50 g), ethanol titer in water (ranging from 0 to 80 vol %), and catalyst concentration (between 0.5 and 1.5%, w/v), and the most effective results were attained for 15 min at 170°C using a mixture of ethanol in water (60:40 ratio), 25 g of CO_2, and 1 wt % oxalic acid as catalyst. Under these conditions, delignification reached 86%, whereas glucan retention was 89% by weight [120].

On the other hand, high boiling point alcohols, while operating at lower temperatures and pressures, may consume more energy during solvent recovery [113]. Glycerol is also a promising solvent for organosolv pretreatment for sugar bagasse [121]. The glycerol and acid pretreatment of SB was systematically studied using a factorial design that considered various factors including temperature (ranging from 80°C to 120°C), solid loading (2.5 to 7.5%), glycerol concentration (85 to 95%), acid concentration (0.5 to 3%), and reaction time (10 to 60 min) to optimize cellulose digestibility. The optimal conditions were 130°C with an acid/SB mass ratio of 15% and a reaction time of 57.5 min, at which an impressive 68.5% digestibility of cellulose was achieved [122]. Various organic solvents, including chloroethanol, acetone, alcohols, phenol, formaldehyde, amines, and propionic acid, have been explored for the pretreatment of diverse lignocellulosic biomass sources. For instance, Li *et al.* (2022) conducted a mild organosolv pretreatment of SB using a solution of acetone, phenoxyethanol, and water, aiming to enhance subsequent sugar manufacturing in enzymatic hydrolysis. Through optimization, they identified the best conditions for the pretreatment as 125°C and 120 min, using 0.17 M H_2SO_4 and a liquid-solid ratio of 15. Under these conditions, an impressive 98.1% of lignin was removed from the biomass, and the cellulose digestibility was notably increased to 74.5%, in contrast to the low digestibility of 9.3% observed with raw SB [123].

5.4.3.2 Steam Explosion Pretreatment

Steam explosion is widely regarded as one of the most efficient, cost-effective, and environment-friendly pretreatment methods, particularly in the sugarcane industry [117]. It is a combined physical and chemical process that involves subjecting the lignocellulosic materials within SB to high-temperature steam and pressure, followed by rapid decompression, which leads to several beneficial effects, such as rearrangement of lignin due to partial hemicellulose hydrolysis and the solubilization of lignin at high temperatures [124]. Additionally, steam explosion induces changes in cellulose crystallinity and DPw, which in turn facilitate enzymatic hydrolysis [125]. Several key factors influence the effectiveness of steam explosion pretreatment, including the source of biomass, particle size, temperature, residence time, and moisture content [126], for instance, Silveira *et al.* (2018) conducted pilot-scale steam explosion pretreatment of SB to check the effect of different combined severity factors in steam explosion (autocatalyzed and sulfuric acid–catalyzed) on xylose and oligosaccharides recovery, using a 65-L steam gun reactor with 50% moisture content, enhancing saccharification [127]. The study by Pitarelo *et al.* (2016) involved steam explosion pretreatment of SB, both in the presence (autohydrolysis) and absence of H_3PO_4. The objective was to assess the impacts of various parameters including acid concentration (0 and 19 mg g^{-1}, dry basis), temperature (180°C and 210°C), and pretreatment time (5 and 10 min) on reactivity and structure of SB. The results indicated that autohydrolysis for 10 min at 210°C showed better pretreatment performance; however, the addition of phosphoric acid as a catalyst (at a concentration of 19 mg g^{-1}) resulted in even better yields, and this was achieved under considerably milder conditions, specifically at 180°C for 5 min [128].

5.4.3.3 *Hot Water Pretreatments*

Hot water pretreatment is an appealing technique that involves using a substantial volume of water at high temperatures, typically within the range of 160°C to 240°C, for durations spanning 1 to 80 min to maintain water in its liquid state and achieves theoretical enzymatic digestibility of cellulose, while concurrently recovering 55 to 84% of hemicellulose [55, 129–131]. During this water pretreatment at high temperatures, water molecules permeate from the cell wall of biomass and hydrate cellulose, leading to the fractional removal of hemicellulose and a slight reduction in lignin content [96]. The near-neutral pH prevents corrosion, minimizes excessive chemical use, and reduces the creation of extra furans during the degradation of sugar [132]. However, compared to diluted acid pretreatments, the sugar release yields from liquid hot water pretreated biomass tend to be lower, often requiring higher pretreatment temperatures and longer residence times to achieve comparable performance [133, 134]. Researchers have examined liquid hot water pretreatment of bagasse at different solid-to-liquid ratios and temperatures and demonstrated effective hemicellulose removal and improved saccharification at a solid-to-liquid ratio of 1:20, 180°C for 30 min [135].

5.4.4 Biological Treatment

Biological pretreatment of lignocellulosic materials is indeed an environmentally friendly and sustainable approach to enhance the enzymatic digestion of polysaccharides. This method involves microorganisms, such as fungi or bacteria, to break down complex lignocellulosic structures, primarily lignin and hemicellulose, without the need for excessive energy or harsh chemicals [136–138]. Indeed, biological pretreatment offers several advantages over physical and chemical methods, including lower energy requirements, simple operating conditions, and reduced inhibitor formation. However, the success of biological pretreatment largely depends on the choice of fungal or microbial strains and their ability to enhance polysaccharide digestibility while minimizing lignin degradation. Numerous fungal species can enhance the breakdown of polysaccharides, but only a limited selection of species possess the capability to break down substantial quantities of lignin, thus safeguarding cellulose, and a portion of hemicellulose [139, 140]. The selection of a specific microorganism for biological pretreatment depends on the type of biomass being treated, when it comes to SB, white-rot fungi are often preferred for their ability to effectively delignify the biomass. White-rot fungi are equipped with lignolytic enzymes such as laccases and peroxidases, which play a key role in breaking down lignin and hemicellulose in SB. Lignolytic enzymes, through effective delignification, serve a dual role in the breakdown of lignocellulosic materials: first, enhance cellulose accessibility, making it easier to convert cellulose into sugars and, second, break down lignin [136, 141, 142]. In a study conducted by Machado and Ferraz (2017), the researchers investigated the pretreatment of SB using wood-decay fungi resulting in diverse patterns of biodegradation of the biomass and revealed that a specific type of white-rot fungus was essential to achieve optimal glucose yields. The chosen white-rot fungus in this study was *Ceriporiopsis subvermispora*, which exhibited selectivity for lignin degradation within the lignocellulosic material and remarkably preserved a significant portion of the glucan fraction while enhancing the digestibility of cellulose [139]. In a study, white-rot fungus, Ceriporiopsis subvermispora, was employed as an effective biological pretreatment method for SB at 27°C for 60 days, with the aim of achieving a high recovery of sugars [139]. Another researcher utilized laccase obtained from Pycnoporus cinnabarinus as part of an effective delignification process for SB, in which the enzyme played a crucial role in breaking down lignin, thereby facilitating the subsequent utilization of the bagasse for various applications [143].

5.5 Sugarcane Bagasse in Biopolymer Matrix as Reinforcement Filler

SB, historically used primarily as a fuel source in sugar factories and, to a lesser extent, in the production of paper and board, has found new and promising applications in recent times. One of the most effective and attractive uses for SB is as a reinforcement filler in polymeric composites [25, 144, 145]. SB consisting of cellulose, hemicellulose, and lignin renders it an exceptional candidate for use as a reinforcing fiber in composite materials, which improves the physical, chemical, barrier, and mechanical properties of the composite [146, 147]. In terms of cost-effectiveness, SB fibers (SBFs) hold a significant commercial advantage compared to other fibrous materials due to their low production and pretreatment costs [148]. Starch is often used in conjunction with SB to create composite materials, for example, biocomposite films of SB cellulose nanofibers (SBCN) and tapioca starch (TS) were developed using a casting method, exhibited promising properties, i.e., the highest TS (20.84 MPa) [149]. The interaction between CNCs and starch exhibited reduced solubility, altered contact angles, and water sorption isotherms compared to starch-only films [150]. In another study, researchers assessed the thermal and mechanical performance of SBF-reinforced thermoplastic cassava starch (TPCS), which had been previously modified with beeswax (BW). The addition of SBF into the TPCS/BW led to substantial improvements in flexural strength, TS, and impact strength, and, notably, the sample with a 20 wt % SBF loading exhibited the highest TS (12.2 MPa) [151]. Researchers have also developed biodegradable films using a combination of polyvinyl alcohol (PVOH), starch, and nanocellulose derived from SB [152].

Apart from starch, SBF has also been incorporated with other biopolymers, for instance, Singha *et al.* (2023) developed a biodegradable film using sweet lime peel powder (SLPP) and PVOH as the base materials, and the results showed that incorporating 25% SBF into the film enhanced TS, EB, and water vapor permeability compared to the film without SBF. One hundred percent SBF blended film had the least solubility due to the higher fiber content; however, films with 50, 75, and 100% SBF exhibited lower TS and EB, possibly due to phase separation between the fiber, SLPP, and PVOH matrix [6]. Another research modified polylactic acid (PLA) with SBFs, resulting in a significant increase in stiffness, nearly constant TS, and a slight improvement in impact resistance, which led to the development of a material with a balanced set of properties suitable for structural applications [153]. Gan and Chow (2018) employed a similar approach, utilizing acid hydrolysis for nanocellulose preparation and the casting technique to create a nanocellulose-reinforced PLA composite incorporating SB, and observed that the presence of hydrogen bonding between cellulose and PLA in the composite contributed to its improved thermal stability compared to pure PLA [154]. Table 5.1 summarizes the study on sugarcane as reinforcement material in biopolymer composites for food packaging applications.

Table 5.1 Sugarcane as reinforcement material in biopolymer-based composites.

Biopolymer blend	Composition of composite	Major findings	Reference
Sugarcane bagasse fiber (SBF)/polylactic acid (PLA)	SBF- 0, 5, 10, 15, and 20 wt %	Tensile testing showed that pure PLA film had a TS of 25.63 MPa, while the inclusion of 7 wt % fiber increased it to 45.27 MPa.	[155]
Cocoa pod husk cellulose (CPHC)/sugarcane bagasse fiber (SBF)	CPHC/SBF- 100/0, 75/25, 50/50, 25/75, and 0/100	The most appropriate bioplastic film for food packaging appears to be the blend consisting of 25% fiber bioplastic and 75% cellulose, as it exhibited the lowest levels of water vapor permeability and water absorption.	[156]
Polyhydroxybutyrate (PHB)/sugarcane bagasse (SB)	PHB/SB- 100/0, 80/20, and 70/30 w/w %	The composite showed considerable improvement compared to the pure polymer, with the tensile modulus increasing upon the addition of the fiber. Moreover, the developed material exhibited noteworthy antimicrobial properties against *E. coli*, *S. typhi*, *S. aureus*, and *S. pyogenes*.	[157]
Whey protein isolates (WPI)/sugarcane cellulose nanocrystals (SCNCs)	SCNSs- 2, 5, and 8% WPI- 5%	A film containing 8% cellulose nanocrystals demonstrated suitable mechanical and barrier properties, making it a promising candidate as an alternative material for packaging films.	[158]
Starch (S)/maize hull (MH)/sugarcane bagasse (SB)	S- 50 to 65 wt % MH- 23 wt % SB- 23 wt %	The mechanical properties of the thermoplastic starch composite improved when both maize hull and SB were added, but only the latter achieved results comparable to the reference HDPE.	[159]

(*Continued*)

Table 5.1 Sugarcane as reinforcement material in biopolymer-based composites. (*Continued*)

Biopolymer blend	Composition of composite	Major findings	Reference
Sugarcane bagasse fiber (SBF)/starch	SBF- 20, 40, 60, and 80% Starch- 30%	With increasing fiber content from 0 to 60%, both the flexural strengths and tensile increased, water uptake and thermal degradation were maintained	[160]
Tapioca starch (TS) and SBCN	TS- 2 wt % SBCN- 0 to 8 wt %	The combination of SBCN and tapioca starch led to strong adhesion and bonding, resulting in the highest tensile strength observed in the TS matrix with 4% SBCN content.	[149]
Sugarcane nanocrystalline cellulose (SNCC)/polylactic acid (PLA)	SNCC- 0, 0.2, 0.3, 0.4, and 0.5%	The rise in tensile strength of the films affirmed the heightened and effective positive interaction occurring at the interfaces between the matrix and the filler.	[161]
Polyurethane (PU)/cellulose and lignin from sugarcane bagasse	Cellulose- 10, 20, 30, and 40 % w/w Lignin- 10, 20, 30, 40, 50, and 60 % w/w	The incorporation of cellulose fibers and lignin in the polyurethane resulted in decreased thermal stability and increased mechanical properties in comparison with the pure matrix.	[162]
Tamarind shell powder (TSP)/sugarcane fiber (SF)/epoxy resin (ER)	TSP- 5 to 50 wt % SF- 25 to 50 wt % ER- 50 wt %	Epoxy composites containing 35 wt % SF and 15 wt % TSP filler demonstrated improved mechanical and barrier properties.	[163]
Palm sheath fiber (PSF)/bagasse fiber (BF)/epoxy resin (ER)	Ratio of matrix and fiber- 70/30, 65/35, and 60/40 Ratio of PSF and BF- 40/60	The most favorable combination was found to be a 60/40 ratio of matrix to fiber treatment, which resulted in superior mechanical and barrier properties for the developed composite.	[164]

(*Continued*)

Table 5.1 Sugarcane as reinforcement material in biopolymer-based composites. (*Continued*)

Biopolymer blend	Composition of composite	Major findings	Reference
Tapioca starch (TS)/SBF	SBF- 1 wt % TS- 10 wt %	The sample subjected to 15 min of ultrasonication at 2.5 MPa exhibited the highest tensile strength. Furthermore, an increase in ultrasonication time correlated with a reduction in the moisture absorption rate.	[165]
Rice husk (RH)/sugarcane bagasse (SB)	RH (starch/CMC)- 50/50 wt % SB (starch/CMC)- 50/50 wt %	With the increase in the degree of substitution of carboxymethyl cellulose (CMC), the moisture content, tensile strength, solubility, and opacity of the film have also increased. Notably, CMC derived from SB outperformed CMC obtained from rice husks in these aspects.	[166]
Sugarcane cellulose nanocrystals (SCNCs)/k-carrageenan	SCNCs- 1, 3, 5, and 8 wt %	A nanocomposite film based on k-carrageenan with an 8 wt % content of SCNCs exhibited strong mechanical properties and flexibility and has potential for use in food packaging applications.	[167]
Polyvinyl alcohol (PVA)/ carboxymethyl cellulose (CMC)/ sugarcane cellulose nanocrystals (SCNCs)	PVA/CMC- 50/50 w/w SCNCs- 0.5, 2.5, 5, and 10 wt %	Incorporating 5 wt % SCNCs into a PVA/CMC blend results in the enhancement of tensile strength while simultaneously reducing water vapor permeability.	[168]
Sugarcane hemicellulose (SHC)/chitosan (CH)	SHC- 1.6, 2.4, 3.2, 4.0, and 4.8 wt % CH- 20, 25, 30, 35, and 40%	As the hemicellulose concentration increased from 1.6 to 4.0%, there was a noticeable improvement in the film's elongation, tensile strength, and water vapor barrier properties.	[169]

(*Continued*)

Table 5.1 Sugarcane as reinforcement material in biopolymer-based composites. (*Continued*)

Biopolymer blend	Composition of composite	Major findings	Reference
All-sugarcane cellulose nanocomposite (ASCNC)	ASCNC- 0.5 wt %	ASCNC was deemed advantageous because of its promising attributes, which include toughness, biobased origin, biodegradability, and acceptable levels of water vapor permeability.	[170]
Sugarcane nanocellulose (SNC)/polyvinyl alcohol (PVA)/	2.5, 5, 7.5, and 10 wt % PVA- 5 wt %	Nanocomposite containing 5 and 7.5% nanocellulose had the highest tensile strength, elongation at break, and toughness.	[171]
polyvinyl alcohol (PVA)/ starch/sugarcane nano cellulose (SNC)	PVA- 3, 5.5, and 8% Starch- 1, 2.5, and 4% SNC- 1 to 9%	A combination of 3% polyvinyl alcohol (PVA), 1% starch, and 1% nanocellulose (SNC) demonstrates robust adhesion and bonding among the polar molecules within the PVA-starch blend, leading to minimal water absorption.	[146]

5.6 Food Containers and Trays Made From SB

In recent years, there has been a growing interest in the development of bioplastics as a sustainable alternative to petrochemical-based synthetic plastics, and it has been a widely accepted strategy aimed at creating a more sustainable and environmentally friendly society [172–176]. Recent advancements in SB processing hold great promise for the sugarcane industry, which led to the creation of numerous value-added bioproducts and technologies that utilize SB, particularly in the production of food packaging materials [177]. Numerous works have been carried out on the use of SB as reinforcement in biopolymer-based composite for food packaging applications. In a recent study, Farhadi and Javanmard (2023) developed a novel cellulose fibrous casing using SB as a base material loaded with extracts from rosemary and thyme, known for their antioxidant properties, with the aim of creating active food packaging that could prolong the shelf life of sausages. Different percentages of thyme and rosemary extracts (0, 2, 4, and 6%) were applied to the SB-based fibrous casings, and the results showed a significant improvement in the oxidative stability of sausages ($p < 0.05$) with the fibrous casing loaded with 6% thyme [178]. Nida *et al.* (2022) developed customized 3D food packaging materials from mixtures of banana peel (BP) and SB in different ratios (BP:SB) with the addition of guar gum (GG) to assess their printability. Successful extrusion of BP:SB (1:1) and BP:SB (9:1) with GG occurred at printing speeds of 400 and 600 mm/min, along with nozzle heights of 1.34 and 0.54 mm, respectively. Ultimately, this research showcases the conversion of nonprintable agro-wastes into printable materials that may serve as an eco-friendly substitute for traditional synthetic plastic–based packaging materials [179].

Researchers have successfully developed decomposable trays from sugarcane fibers, cassava starch, and Na-montmorillonite (Na-MMT) through the baking process, and the result showed that the incorporation of Na-MMT and sugarcane fibers led to less rigid and dense trays. Notably, the equilibrium moisture content increased, when the samples were stored at relative humidity (RH) levels above 75% and thus offer a sustainable alternative for packaging foods with low water contents [180]. By incorporating sugarcane fibrous wastes into biopolymer-based packaging composites, along with sweet potato starch and asparagus peel, researchers have developed foam trays suitable for packaging dry foods. The addition of fibrous waste resulted in yellowish foam trays characterized by reduced luminosity, increased porosity, enhanced mechanical resistance, greater deformability, and improved water absorption capability, when compared to foam trays made solely from sweet potato starch, without the inclusion of fibers [181]. Liu *et al.* (2020) successfully developed environmentally friendly tableware using SBF and bamboo fiber through a scalable pulp-molding method, which exhibited hygienic properties, remained stable when exposed to water and oil, demonstrated mechanical durability, and was cost-effective to produce [182]. Azmin *et al.* (2020) identified the optimal bioplastic film for food packaging by blending 75% cellulose and 25% fiber derived from cocoa pod husk and SB, which exhibited the lowest levels of water absorption and water vapor permeability, making it highly suitable for food packaging applications [156]. Different forms of food packaging containers and trays made from SB are summarized in Table 5.2.

Table 5.2 Sugarcane bagasse–based containers and trays for food packaging applications.

Biopolymer blend	Content of biopolymers in packaging	Packaging form	Major findings	Reference
Sweet potato starch (SPS)/ sugarcane bagasse (SB)/ asparagus peel (AP)	SPS- 60 to 100% SB- 0 to 40% AP- 0 to 40%	Baked foam trays	The trays exhibited reduced porosity, lower water absorption capacity, and higher tensile strength when asparagus peel fibers were not added.	[181]
Sugarcane fibers (SF)/ cassava starch (CS)/ polyvinyl Alcohol (PVA)	SB- 0 to 40% CS- 0 to 100% PVA- 0 to 40%	Baked foam trays	The sample composed of 66.6% starch, 6.7% fibers, and 26.7% PVA demonstrated the most favorable balance of mechanical and expansion properties, making it suitable for use as loose-fill packaging material.	[183]
Oca starch (OS)/SB asparagus peel (AP)	OS/SB and OS/AP- 100/0, 95/5, 90/10, 85/15, 80/20, 70/30, and 60/40	Foam trays	The trays showed superior physical properties, including greater density and thickness, when compared to the control polystyrene trays.	[184]

(Continued)

Table 5.2 Sugarcane bagasse–based containers and trays for food packaging applications. (*Continued*)

Biopolymer blend	Content of biopolymers in packaging	Packaging form	Major findings	Reference
Cassava starch (CS)/ sugarcane fibers (SF)/ Na-montmorillonite (Na-MMT)	CS- 75 to 100% SF- 10 to 20% Na-MMT- 0.5%	Baked foam trays	The obtained sample having SF as 20% showed the maximum yield production with less rigid and dense trays in comparison with the control samples (only starch).	[180]
Sweet potato starch (SPS)/ sugarcane bagasse fiber (SBF)	SPS/SBF- 90 to 100% Glycerol- 7.5% Magnesium stearate- 5%	Foam trays	Using fibers with a size range of 75 to 45 μm and a 2.5% ratio led to the production of trays with reduced thicknesses and densities.	[152]
Sugarcane cellulose nanofibrils (CNFs)/nisin	CNFs- 0.8% (w/w) Nisin- 640, 1280, 1920, 2560, or 3200 mg/L	Packaging for ready-to-eat ham	Hybrid films exhibited higher tensile strength, light transmission, and oxygen barrier properties and also inhibited bacterial growth during 7 days of storage at 4°C.	[125, 185]

5.7 Conclusion and Future Perspective

SB is produced in sugar industries during sugarcane processing as a byproduct, and a large portion of which is burnt causing environmental pollution. SB consists of all the necessary properties that enable it as a sustainable source to produce various value-added products, and bioplastics that can be used as an alternative to synthetic plastic–based packaging materials in food industries. SB, the primary solid waste produced during the sugar manufacturing process from sugarcane, mainly consists of cellulose and hemicellulose, which are closely bound to lignin. Due to this intricate structure, significant pretreatment is required for efficient utilization. The pretreatment of SB can be broadly classified into four groups: physical, physicochemical, chemical, and biological. This pretreatment is essential to overcome the recalcitrance of the biomass, allowing for increased productivity and reduced costs associated with microorganisms and enzymes during processes like enzymatic hydrolysis. SB has found applications in various biopolymer-based composites, including films, foams, and trays, expanding its use in diverse fields, particularly in food packaging. However, the water sensitivity of these materials limits their suitability for food packaging applications, particularly for products having high moisture content. Therefore, there is a need for the development of new materials and innovative technologies that can broaden the application of environment-friendly biodegradable composites in the packaging sector. In conclusion, utilizing SB as a renewable and sustainable material in food packaging can support various engineering endeavors aimed at transforming sugarcane waste into environmentally friendly decomposable materials.

References

1. Meghana, M. and Shastri, Y., Sustainable valorization of sugar industry waste: Status, opportunities, and challenges. *Bioresour. Technol.*, **303**, 122929, 2020.
2. Silalertruksa, T. and Gheewala, S.H., Competitive use of sugarcane for food, fuel, and biochemical through the environmental and economic factors. *Int. J. Life Cycle Assess.*, **25**, 7, 1343–1355, 2019.
3. Dotaniya, M.L. *et al.*, Use of sugarcane industrial by-products for improving sugarcane productivity and soil health. *Int. J. Recycl. Org. Waste Agric.*, **5**, 3, 185–194, 2016.
4. Konde, K.S. *et al.*, Sugarcane bagasse based biorefineries in India: Potential and challenges. *Sustain. Energy Fuels*, **5**, 1, 52–78, 2021.

5. Mahmud, M.A. and Anannya, F.R., Sugarcane bagasse - A source of cellulosic fiber for diverse applications. *Heliyon*, **7**, 8, e07771, 2021.
6. Singha, P., Rani, R., Badwaik, L.S., Influence of sugarcane bagasse fibre on the properties of sweet lime peel- and polyvinyl alcohol-based biodegradable films. *Sustainable Food Technol.*, **1**, 4, 610–620, 2023.
7. Mokhena, T.C. *et al.*, Sugarcane bagasse and cellulose polymer composites, in: *Sugarcane - Technology and Research*, 2018.
8. Rabelo, S.C., da Costa, A.C., Vaz Rossel, C.E., Industrial waste recovery, in: *Sugarcane*, pp. 365–381, 2015.
9. Azar, R. *et al.*, Impact of protein blocking on enzymatic saccharification of bagasse from sugarcane clones. *Biotechnol. Bioeng.*, **116**, 7, 1584–1593, 2019.
10. Florencio, C., Badino, A.C., Farinas, C.S., Addition of soybean protein improves saccharification and ethanol production from hydrothermally pretreated sugarcane bagasse. *Bioenergy Res.*, **12**, 1, 81–93, 2019.
11. Kim, D., Physico-chemical conversion of lignocellulose: Inhibitor effects and detoxification strategies: A mini review. *Molecules*, **23**, 2, 309, 2018.
12. Kusuma, H.S. *et al.*, Sugarcane bagasse as an environmentally friendly composite material to face the sustainable development era. *Biomass Convers. Biorefin.*, 2190–6823, 2023. https://doi.org/10.1007/s13399-023-03764-2
13. de Sousa, M.V., Monteiro, S.N., d'Almeida, J.R.M., Evaluation of pretreatment, size and molding pressure on flexural mechanical behavior of chopped bagasse–polyester composites. *Polym. Test.*, **23**, 3, 253–258, 2004.
14. Rocha, G.J.d.M. *et al.*, Influence of mixed sugarcane bagasse samples evaluated by elemental and physical–chemical composition. *Ind. Crops Prod.*, **64**, 52–58, 2015.
15. Yao, S. *et al.*, Efficient extraction of bagasse hemicelluloses and characterization of solid remainder. *Bioresour. Technol.*, **185**, 21–7, 2015.
16. Arni, S.A., Extraction and isolation methods for lignin separation from sugarcane bagasse: A review. *Ind. Crops Prod.*, **115**, 330–339, 2018.
17. Haghdan, S., Renneckar, S., Smith, G.D., Sources of lignin, in: *Lignin in Polymer Composites*, pp. 1–11, 2016.
18. Ibrahim, M.I.J. *et al.*, Preparation and characterization of cornhusk/sugar palm fiber reinforced Cornstarch-based hybrid composites. *J. Mater. Res. Technol.*, **9**, 1, 200–211, 2020.
19. Kordkheili, H.Y., Hiziroglu, S., Farsi, M., Some of the physical and mechanical properties of cement composites manufactured from carbon nanotubes and bagasse fiber. *Mater. Des.*, **33**, 395–398, 2012.
20. Sabiha-Hanim, S. and Asyikin Abd Halim, N., Sugarcane bagasse pretreatment methods for ethanol production, in: *Fuel Ethanol Production from Sugarcane*, 2019.
21. Pandey, A. *et al.*, Biotechnological potential of agro-industrial residues. I: Sugarcane bagasse. *Bioresour. Technol.*, **74**, 69–80, 2000.

22. Aigbodion, V.S., Hassan, S.B., Agunsoye, J.O., Effect of bagasse ash reinforcement on dry sliding wear behaviour of polymer matrix composites. *Mater. Des.*, **33**, 322–327, 2012.
23. Aigbodion, V.S. and Hassan, S.B., The study of the precipitation process in Al–Cu–Mg/bagasse ash particulate composites. *J. Alloys Compd.*, **501**, 1, 104–109, 2010.
24. Frollini, E. *et al.*, Poly(butylene succinate) reinforced with different lignocellulosic fibers. *Ind. Crops Prod.*, **45**, 160–169, 2013.
25. Hajiha, H. and Sain, M., The use of sugarcane bagasse fibres as reinforcements in composites, in: *Biofiber Reinforcements in Composite Materials*, pp. 525–549, 2015.
26. Chaharmahali, M. *et al.*, Effects of nano-graphene on the physico-mechanical properties of bagasse/polypropylene composites. *Polym. Bull.*, **71**, 2, 337–349, 2013.
27. Hemmasi, A.H. *et al.*, Studying the effect of size of bagasse and nanoclay particles on mechanical properties and morphology of bagasse flour/recycled polyethylene composites. *Bioresources*, **8**, 3, 3791–3801, 2013.
28. Cao, Y., Goda, K., Shibata, S., Development and mechanical properties of bagasse fiber reinforced composites. *Adv. Compos. Mater.*, **16**, 4, 283–298, 2007.
29. Saini, G. *et al.*, Effect of particle size and alkali treatment of sugarcane bagasse on thermal, mechanical, and morphological properties of PVC-bagasse composites. *J. Reinf. Plast. Compos.*, **29**, 5, 731–740, 2009.
30. Sun, J.X. *et al.*, Isolation and characterization of cellulose from sugarcane bagasse. *Polym. Degrad. Stab.*, **84**, 2, 331–339, 2004.
31. Khattab, S.M.R. and Watanabe, T., Bioethanol from sugarcane bagasse: Status and perspectives, in: *Bioethanol Production from Food Crops*, pp. 187–212, 2019.
32. Bian, J. *et al.*, Effect of [Emim]Ac pretreatment on the structure and enzymatic hydrolysis of sugarcane bagasse cellulose. *Carbohydr. Polym.*, **100**, 211–217, 2014.
33. Bezerra, T.L. and Ragauskas, A.J., A review of sugarcane bagasse for second-generation bioethanol and biopower production. *Biofuels Bioprod. Biorefin.*, **10**, 5, 634–647, 2016.
34. Laluce, C. *et al.*, Effects of pretreatment applied to sugarcane bagasse on composition and morphology of cellulosic fractions. *Biomass Bioenergy*, **126**, 231–238, 2019.
35. Shen, D.K. and Gu, S., The mechanism for thermal decomposition of cellulose and its main products. *Bioresour. Technol.*, **100**, 24, 6496–504, 2009.
36. Jain, A., Wei, Y., Tietje, A., Biochemical conversion of sugarcane bagasse into bioproducts. *Biomass Bioenergy*, **93**, 227–242, 2016.
37. Gurgel, L.V.A. *et al.*, Characterization of depolymerized residues from extremely low acid hydrolysis (ELA) of sugarcane bagasse cellulose: Effects

of degree of polymerization, crystallinity and crystallite size on thermal decomposition. *Ind. Crops Prod.*, **36**, 1, 560–571, 2012.
38. Sharma, A. *et al.*, Commercial application of cellulose nano-composites - A review. *Biotechnol. Rep.*, 21, e00316, 2019.
39. Sofla, M.R.K. *et al.*, A comparison of cellulose nanocrystals and cellulose nanofibres extracted from bagasse using acid and ball milling methods. *Adv. Nat. Sci.: Nanosci. Nanotechnol.*, **7**, 3, 035004, 2016.
40. Feng, Y.-H. *et al.*, Characteristics and environmentally friendly extraction of cellulose nanofibrils from sugarcane bagasse. *Ind. Crops Prod.*, **111**, 285–291, 2018.
41. Lam, N.T. *et al.*, Characterization of cellulose nanocrystals extracted from sugarcane bagasse for potential biomedical materials. *Sugar Tech.*, **19**, 5, 539–552, 2017.
42. Shi, J.B. *et al.*, The structural changes of the bagasse hemicelluloses during the cooking process involving active oxygen and solid alkali. *Carbohydr. Res.*, **359**, 65–69, 2012.
43. Sukhbaatar, B. *et al.*, Optimization of hot-compressed water pretreatment of bagasse and characterization of extracted hemicelluloses. *Carbohydr. Polym.*, **101**, 196–202, 2014.
44. Peng, F. *et al.*, Comparative studies on the physico-chemical properties of hemicelluloses obtained by DEAE-cellulose-52 chromatography from sugarcane bagasse. *Food Res. Int.*, **43**, 3, 683–693, 2010.
45. Brienzo, M. *et al.*, Sugarcane bagasse hemicellulose properties, extraction technologies and xylooligosaccharides production, in: *Food Waste*, G.L. Riley (Ed.), Nova Science Publishers, NY, USA, 2016.
46. Hamzeh, Y. *et al.*, Pre-extraction of hemicelluloses from bagasse fibers: Effects of dry-strength additives on paper properties. *Ind. Crops Prod.*, **43**, 365–371, 2013.
47. Carvalho, A.F.A. *et al.*, Xylo-oligosaccharides from lignocellulosic materials:Chemical structure, health benefits and production by chemical and enzymatic hydrolysis. *Food Res. Int.*, **51**, 1, 75–85, 2013.
48. Sun, J.X. *et al.*, Fractional extraction and structural characterization of sugarcane bagasse hemicelluloses. *Carbohydr. Polym.*, **56**, 2, 195–204, 2004.
49. Bian, J. *et al.*, Isolation of hemicelluloses from sugarcane bagasse at different temperatures: Structure and properties. *Carbohydr. Polym.*, **88**, 2, 638–645, 2012.
50. Kundu, P., Kansal, S.K., Elumalai, S., Synergistic action of alkalis improve the quality hemicellulose extraction from sugarcane bagasse for the production of xylooligosaccharides. *Waste Biomass Valorization*, **12**, 6, 3147–3159, 2020.
51. Xie, Y. *et al.*, Efficient extraction and structural characterization of hemicellulose from sugarcane bagasse pith. *Polymers*, **12**, 3, 608, 2020.
52. Alokika, *et al.*, Cellulosic and hemicellulosic fractions of sugarcane bagasse: Potential, challenges and future perspective. *Int. J. Biol. Macromol.*, **169**, 564–582, 2021.

53. Wang, B. *et al.*, Enhancement of sugar release from sugarcane bagasse through NaOH-catalyzed ethylene glycol pretreatment and water-soluble sulfonated lignin. *Int. J. Biol. Macromol.*, **221**, 38–47, 2022.
54. Dionísio, S.R. *et al.*, Second-generation ethanol process for integral use of hemicellulosic and cellulosic hydrolysates from diluted sulfuric acid pretreatment of sugarcane bagasse. *Fuel*, **304**, 121290, 2021.
55. Vallejos, M.E. *et al.*, Low liquid-solid ratio fractionation of sugarcane bagasse by hot water autohydrolysis and organosolv delignification. *Ind. Crops Prod.*, **65**, 349–353, 2015.
56. Jönsson, L.J., Alriksson, B., Nilvebrant, N.O., Bioconversion of lignocellulose: Inhibitors and detoxification. *Biotechnol. Biofuels*, **6**, 16, 2015.
57. Shangdiar, S. *et al.*, Pretreatment of lignocellulosic biomass from sugar bagasse under microwave assisted dilute acid hydrolysis for biobutanol production. *Bioresour. Technol.*, **361**, 127724, 2022.
58. da Silva, A.S. *et al.*, Milling pretreatment of sugarcane bagasse and straw for enzymatic hydrolysis and ethanol fermentation. *Bioresour. Technol.*, **101**, 19, 7402–7409, 2010.
59. Park, S. *et al.*, Cellulose crystallinity index: Measurement techniques and their impact on interpreting cellulase performance. *Biotechnol. Biofuels*, **3**, 10, 2010.
60. Barakat, A. *et al.*, Mechanical pretreatments of lignocellulosic biomass: Towards facile and environmentally sound technologies for biofuels production. *RSC Adv.*, **4**, 89, 48109–48127, 2014.
61. Miao, Z. *et al.*, Energy requirement for comminution of biomass in relation to particle physical properties. *Ind. Crops Prod.*, **33**, 2, 504–513, 2011.
62. Barakat, A., de Vries, H., Rouau, X., Dry fractionation process as an important step in current and future lignocellulose biorefineries: A review. *Bioresour. Technol.*, **134**, 362–73, 2013.
63. Silveira, M.H. *et al.*, Current pretreatment technologies for the development of cellulosic ethanol and biorefineries. *ChemSusChem*, **8**, 20, 3366–90, 2015.
64. Gan, T. *et al.*, Esterification of bagasse cellulose with metal salts as efficient catalyst in mechanical activation-assisted solid phase reaction system. *Cellulose*, **24**, 12, 5371–5387, 2017.
65. Zhao, X. *et al.*, Effect of mechanical activation on structure changes and reactivity in further chemical modification of lignin. *Int. J. Biol. Macromol.*, **91**, 1081–1089, 2016.
66. Kumar, A.K. and Sharma, S., Recent updates on different methods of pretreatment of lignocellulosic feedstocks: A review. *Bioresour. Bioprocess.*, **4**, 7, 2017.
67. Sarker, T.C., Azam, S.M.G.G., Bonanomi, G., Recent advances in sugarcane industry solid by-products valorization. *Waste Biomass Valorization*, **8**, 2, 241–266, 2016.
68. Portero-Barahona, P. *et al.*, Sugarcane bagasse hydrolysis enhancement by microwave-assisted sulfolane pretreatment. *Energies*, **12**, 9, 2019.

69. de Souza Moretti, M.M. *et al.*, Pretreatment of sugarcane bagasse with microwaves irradiation and its effects on the structure and on enzymatic hydrolysis. *Appl. Energy*, **122**, 189–195, 2014.
70. Watanabe, T. and Mitani, T., Microwave technology for lignocellulosic biorefinery, in: *The Role of Green Chemistry in Biomass Processing and Conversion*, H. Xie and N. Gathergood (Eds.), John Wiley & Sons, New Jersey, USA, 2013.
71. Hasegawa, N. *et al.*, Pilot-plant scale 12 kW microwave irradiation reactor for woody biomass pretreatment. *IEICE Trans. Electron.*, **E97.C**, 10, 986–993, 2014.
72. Zheng, Y. *et al.*, Pretreatment of lignocellulosic biomass for enhanced biogas production. *Prog. Energy Combust. Sci.*, **42**, 35–53, 2014.
73. Ramadoss, G. and Muthukumar, K., Mechanistic study on ultrasound assisted pretreatment of sugarcane bagasse using metal salt with hydrogen peroxide for bioethanol production. *Ultrason. Sonochem.*, **28**, 207–217, 2016.
74. Yaldagard, M., Mortazavi, S.A., Tabatabaie, F., The effect of ultrasound in combination with thermal treatment on the germinated barley's alpha-amylase activity. *Korean J. Chem. Eng.*, **25**, 3, 517–523, 2008.
75. Esfahani, M.R. and Azin, M., Pretreatment of sugarcane bagasse by ultrasound energy and dilute acid. *Asia-Pac. J. Chem. Eng.*, **7**, 2, 274–278, 2012.
76. Velmurugan, R. and Muthukumar, K., Ultrasound-assisted alkaline pretreatment of sugarcane bagasse for fermentable sugar production: Optimization through response surface methodology. *Bioresour. Technol.*, **112**, 293–299, 2012.
77. Torres da Silva, G. *et al.*, Sono-assisted alkaline pretreatment of sugarcane bagasse for cellulosic ethanol production. *Catal. Today*, **269**, 21–28, 2016.
78. Kumar, S. *et al.*, Kinetic studies of two-stage sulphuric acid hydrolysis of sugarcane bagasse. *Renewable Energy*, **83**, 850–858, 2015.
79. Canilha, L. *et al.*, A study on the pretreatment of a sugarcane bagasse sample with dilute sulfuric acid. *J. Ind. Microbiol. Biotechnol.*, **38**, 9, 1467–75, 2011.
80. Zhao, X., Zhou, Y., Liu, D., Kinetic model for glycan hydrolysis and formation of monosaccharides during dilute acid hydrolysis of sugarcane bagasse. *Bioresour. Technol.*, **105**, 160–168, 2012.
81. Qin, L. *et al.*, Mass balance and transformation of corn stover by pretreatment with different dilute organic acids. *Bioresour. Technol.*, **112**, 319–326, 2012.
82. Moe, S.T. *et al.*, Saccharification of lignocellulosic biomass for biofuel and biorefinery applications – A renaissance for the concentrated acid hydrolysis? *Energy Proc.*, **20**, 50–58, 2012.
83. Zhang, H., Xu, Y., Yu, S., Co-production of functional xylooligosaccharides and fermentable sugars from corncob with effective acetic acid prehydrolysis. *Bioresour. Technol.*, **234**, 343–349, 2017.

84. Sritrakul, N., Nitisinprasert, S., Keawsompong, S., Evaluation of dilute acid pretreatment for bioethanol fermentation from sugarcane bagasse pith. *Agric. Nat. Resour.*, **51**, 6, 512–519, 2017.
85. Hendriks, A.T. and Zeeman, G., Pretreatments to enhance the digestibility of lignocellulosic biomass. *Bioresour. Technol.*, **100**, 1, 10–18, 2009.
86. Kumar, R. *et al.*, Physical and chemical characterizations of corn stover and poplar solids resulting from leading pretreatment technologies. *Bioresour. Technol.*, **100**, 17, 3948–3962, 2009.
87. Soares, I.B. *et al.*, Evaluation of the effects of operational parameters in the pretreatment of sugarcane bagasse with diluted sulfuric acid using analysis of variance. *Chem. Eng. Commun.*, **204**, 12, 1369–1390, 2017.
88. Timung, R. *et al.*, Effect of subsequent dilute acid and enzymatic hydrolysis on reducing sugar production from sugarcane bagasse and spent citronella biomass. *J. Energy*, **2016**, 1–12, 2016.
89. Dussán, K.J. *et al.*, Dilute-acid hydrolysis of cellulose to glucose from sugarcane bagasse. *Chem. Eng. Trans.*, **38**, 433–438, 2014.
90. Hernandez-Salas, J.M. *et al.*, Comparative hydrolysis and fermentation of sugarcane and agave bagasse. *Bioresour. Technol.*, **100**, 3, 1238–1245, 2009.
91. Rueda, S.M.G. *et al.*, Pretreatment of sugarcane bagasse with phosphoric and sulphuric diluted acid for fermentable sugars production by enzymatic hydrolysis. *Chem. Eng. Trans.*, **20**, 321–326, 2010.
92. Alvira, P. *et al.*, Pretreatment technologies for an efficient bioethanol production process based on enzymatic hydrolysis: A review. *Bioresour. Technol.*, **101**, 13, 4851–4861, 2010.
93. Wu, L. *et al.*, Efficient conversion of sugarcane stalks into ethanol employing low temperature alkali pretreatment method. *Bioresour. Technol.*, **102**, 24, 11183–11188, 2011.
94. S, S.-H. *et al.*, Reducing sugar production from oil palm fronds and rice straw by acid hydrolysis. *IEEE Colloq. Humanit. Sci. Eng. Res.*, 2013.
95. Xu, Z. and Huang, F., Pretreatment methods for bioethanol production. *Appl. Biochem. Biotechnol.*, **174**, 1, 43–62, 2014.
96. Jonsson, L.J. and Martin, C., Pretreatment of lignocellulose: Formation of inhibitory by-products and strategies for minimizing their effects. *Bioresour. Technol.*, **199**, 103–112, 2016.
97. Visser, E.M. *et al.*, Production and application of an enzyme blend from Chrysoporthe cubensis and Penicillium pinophilum with potential for hydrolysis of sugarcane bagasse. *Bioresour. Technol.*, **144**, 587–594, 2013.
98. Yu, N. *et al.*, Bioethanol from sugarcane bagasse: Focused on optimum of lignin content and reduction of enzyme addition. *Waste Manage.*, **76**, 404–413, 2018.
99. Ye, G. *et al.*, Ethanol production from mixtures of sugarcane bagasse and Dioscorea composita extracted residue with high solid loading. *Bioresour. Technol.*, **257**, 23–29, 2018.

100. Carvalho, D.M.d., Queiroz, J.H.d., Colodette, J.L., Assessment of alkaline pretreatment for the production of bioethanol from eucalyptus, sugarcane bagasse and sugarcane straw. *Ind. Crops Prod.*, **94**, 932–941, 2016.
101. Guilherme, A.A. *et al.*, Evaluation of composition, characterization and enzymatic hydrolysis of pretreated sugar cane bagasse. *Braz. J. Chem. Eng.*, **32**, 1, 23–33, 2015.
102. Nadeem, M. *et al.*, Production of ethanol from alkali-pretreated sugarcane bagasse under the influence of different process parameters. *Front. Life Sci.*, **8**, 4, 358–362, 2015.
103. Wunna, K. *et al.*, Effect of alkali pretreatment on removal of lignin from sugarcane bagasse. *Chem. Eng. Trans.*, **56**, 1831–1836, 2017.
104. Grimaldi, M.P. *et al.*, Evaluation of lime and hydrothermal pretreatments for efficient enzymatic hydrolysis of raw sugarcane bagasse. *Biotechnol. Biofuels*, **8**, 205, 2015.
105. Paixão, S.M. *et al.*, Sugarcane bagasse delignification with potassium hydroxide for enhanced enzymatic hydrolysis. *RSC Adv.*, **6**, 2, 1042–1052, 2016.
106. Hedayatkhah, A. *et al.*, Improvement of hydrolysis and fermentation of sugarcane bagasse by soaking in aqueous ammonia and methanolic ammonia. *Biosci. Biotechnol. Biochem.*, **77**, 7, 1379–1383, 2013.
107. Ahmadi, F. *et al.*, Pre-treatment of sugarcane bagasse with a combination of sodium hydroxide and lime for improving the ruminal degradability: Optimization of process parameters using response surface methodology. *J. Appl. Anim. Res.*, **44**, 1, 287–296, 2015.
108. Silveira, M.H. *et al.*, Supercritical carbon dioxide combined with 1-butyl-3-methylimidazolium acetate and ethanol for the pretreatment and enzymatic hydrolysis of sugarcane bagasse. *Bioresour. Technol.*, **192**, 389–396, 2015.
109. Long, J. *et al.*, Catalytic delignification of sugarcane bagasse in the presence of acidic ionic liquids. *Catal. Today*, **200**, 99–105, 2013.
110. Siqueira, G. *et al.*, Enhancement of cellulose hydrolysis in sugarcane bagasse by the selective removal of lignin with sodium chlorite. *Appl. Energy*, **102**, 399–402, 2013.
111. Bahrani, S., Raeissi, S., Sarshar, M., Experimental investigation of ionic liquid pretreatment of sugarcane bagasse with 1,3-dimethylimadazolium dimethyl phosphate. *Bioresour. Technol.*, **185**, 411–415, 2015.
112. Moghaddam, L. *et al.*, Characterisation of lignins isolated from sugarcane bagasse pretreated with acidified ethylene glycol and ionic liquids. *Biomass Bioenergy*, **70**, 498–512, 2014.
113. Zhang, K., Pei, Z., Wang, D., Organic solvent pretreatment of lignocellulosic biomass for biofuels and biochemicals: A review. *Bioresour. Technol.*, **199**, 21–33, 2016.
114. Koo, B.-W. *et al.*, Organosolv pretreatment of Liriodendron tulipifera and simultaneous saccharification and fermentation for bioethanol production. *Biomass Bioenergy*, **35**, 5, 1833–1840, 2011.

115. Zhao, X., Cheng, K., Liu, D., Organosolv pretreatment of lignocellulosic biomass for enzymatic hydrolysis. *Appl. Microbiol. Biotechnol.*, **82**, 5, 815–27, 2009.
116. Borand, M.N. and Karaosmanoğlu, F., Effects of organosolv pretreatment conditions for lignocellulosic biomass in biorefinery applications: A review. *J. Renew. Sustain. Energy*, **10**, 033104, 2018.
117. Silva Ortiz, P. and de Oliveira, S., Exergy analysis of pretreatment processes of bioethanol production based on sugarcane bagasse. *Energy*, **76**, 130–138, 2014.
118. Zhang, H. and Wu, S., Efficient sugar release by acetic acid ethanol-based organosolv pretreatment and enzymatic saccharification. *J. Agric. Food Chem.*, **62**, 48, 11681–7, 2014.
119. Zhang, H. *et al.*, FeCl(3)-catalyzed ethanol pretreatment of sugarcane bagasse boosts sugar yields with low enzyme loadings and short hydrolysis time. *Bioresour. Technol.*, **249**, 395–401, 2018.
120. Hermsdorff, G.B. *et al.*, Ethanol organosolv pretreatment of sugarcane bagasse assisted by organic acids and supercritical carbon dioxide. *Carbohydr. Polym.*, **300**, 120263, 2023.
121. Sun, F.F. *et al.*, Industrially relevant hydrolyzability and fermentability of sugarcane bagasse improved effectively by glycerol organosolv pretreatment. *Biotechnol. Biofuels*, **9**, 59, 2016.
122. Terán Hilares, R. *et al.*, Organosolv pretreatment of sugar cane bagasse for bioethanol production. *Ind. Eng. Chem. Res.*, **56**, 14, 3833–3838, 2017.
123. Li, W. *et al.*, Mild organosolv pretreatment of sugarcane bagasse with acetone/phenoxyethanol/water for enhanced sugar production. *Green Chem.*, **25**, 3, 1169–1178, 2023.
124. Wanderley, M.C. *et al.*, Increase in ethanol production from sugarcane bagasse based on combined pretreatments and fed-batch enzymatic hydrolysis. *Bioresour. Technol.*, **128**, 448–453, 2013.
125. Aguiar, R.S. *et al.*, Kinetics of enzyme-catalyzed hydrolysis of steam-exploded sugarcane bagasse. *Bioresour. Technol.*, **147**, 416–423, 2013.
126. Rabemanolontsoa, H. and Saka, S., Various pretreatments of lignocellulosics. *Bioresour. Technol.*, **199**, 83–91, 2016.
127. Silveira, M.H.L. *et al.*, Production of hemicellulosic sugars from sugarcane bagasse via steam explosion employing industrially feasible conditions: Pilot scale study. *Bioresour. Technol. Rep.*, **3**, 138–146, 2018.
128. Pitarelo, A.P. *et al.*, Ethanol production from sugarcane bagasse using phosphoric acid-catalyzed steam explosion. *J. Braz. Chem. Soc.*, **27**, 10, 1889–1898, 2016.
129. Gao, Y. *et al.*, Effects of different pretreatment methods on chemical composition of sugarcane bagasse and enzymatic hydrolysis. *Bioresour. Technol.*, **144**, 396–400, 2013.

130. da Cruz, S.H. *et al.*, Hydrothermal pretreatment of sugarcane bagasse using response surface methodology improves digestibility and ethanol production by SSF. *J. Ind. Microbiol. Biotechnol.*, **39**, 3, 439–47, 2012.
131. Zeng, M. *et al.*, Tissue-specific biomass recalcitrance in corn stover pretreated with liquid hot-water: Enzymatic hydrolysis (part 1). *Biotechnol. Bioeng.*, **109**, 2, 390–7, 2012.
132. Geddes, C.C. *et al.*, Optimizing the saccharification of sugar cane bagasse using dilute phosphoric acid followed by fungal cellulases. *Bioresour. Technol.*, **101**, 6, 1851–7, 2010.
133. Yang, B., Tao, L., Wyman, C.E., Strengths, challenges, and opportunities for hydrothermal pretreatment in lignocellulosic biorefineries. *Biofuels Bioprod. Biorefin.*, **12**, 1, 125–138, 2017.
134. Ladeira-Azar, R.I.S. *et al.*, Inhibitors compounds on sugarcane bagasse saccharification: Effects of pretreatment methods and alternatives to decrease inhibition. *Appl. Biochem. Biotechnol.*, **188**, 1, 29–42, 2019.
135. Hongdan, Z., Shaohua, X., Shubin, W., Enhancement of enzymatic saccharification of sugarcane bagasse by liquid hot water pretreatment. *Bioresour. Technol.*, **143**, 391–6, 2013.
136. Liguori, R. and Faraco, V., Biological processes for advancing lignocellulosic waste biorefinery by advocating circular economy. *Bioresour. Technol.*, **215**, 13–20, 2016.
137. Sindhu, R., Binod, P., Pandey, A., Biological pretreatment of lignocellulosic biomass–An overview. *Bioresour. Technol.*, **199**, 76–82, 2016.
138. Vasco-Correa, J., Ge, X., Li, Y., Fungal pretreatment of non-sterile miscanthus for enhanced enzymatic hydrolysis. *Bioresour. Technol.*, **203**, 118–123, 2016.
139. Machado, A.D.S. and Ferraz, A., Biological pretreatment of sugarcane bagasse with basidiomycetes producing varied patterns of biodegradation. *Bioresour. Technol.*, **225**, 17–22, 2017.
140. Salvachua, D. *et al.*, Fungal pretreatment: An alternative in second-generation ethanol from wheat straw. *Bioresour. Technol.*, **102**, 16, 7500–7506, 2011.
141. Masran, R. *et al.*, Harnessing the potential of ligninolytic enzymes for lignocellulosic biomass pretreatment. *Appl. Microbiol. Biotechnol.*, **100**, 12, 5231–5246, 2016.
142. Chen, S. *et al.*, Biological pretreatment of lignocellulosics: Potential, progress and challenges. *Biofuels*, **1**, 1, 177–199, 2010.
143. Rencoret, J. *et al.*, Delignification and saccharification enhancement of sugarcane byproducts by a laccase-based pretreatment. *ACS Sustain. Chem. Eng.*, **5**, 8, 7145–7154, 2017.
144. Loh, Y.R. *et al.*, Sugarcane bagasse—The future composite material: A literature review. *Resour. Conserv. Recycl.*, **75**, 14–22, 2013.
145. Devadiga, D.G. *et al.*, Sugarcane bagasse fiber reinforced composites: Recent advances and applications. *Cogent Eng.*, **7**, 1823159, 2020.
146. Asem, M., Nawawi, W.M.F.W., Jimat, D.N., Evaluation of water absorption of polyvinyl alcohol-starch biocomposite reinforced with sugarcane bagasse

nanofibre: Optimization using two-level factorial design. *IOP Conf. Ser.: Mater. Sci. Eng.*, **368**, 012005, 2018.
147. Zafeer, M.K. *et al.*, Mechanical characteristics of sugarcane bagasse fibre reinforced polymer composites: A review. *Cogent Eng.*, **10**, 2200903, 2023.
148. Gilfillan, W.N., Moghaddam, L., Doherty, W.O.S., Preparation and characterization of composites from starch with sugarcane bagasse nanofibres. *Cellulose*, **21**, 4, 2695–2712, 2014.
149. Asrofi, M. *et al.*, Improvement of biocomposite properties based tapioca starch and sugarcane bagasse cellulose nanofibers. *Key Eng. Mater.*, **849**, 96–101, 2020.
150. Slavutsky, A.M. and Bertuzzi, M.A., Water barrier properties of starch films reinforced with cellulose nanocrystals obtained from sugarcane bagasse. *Carbohydr. Polym.*, **110**, 53–61, 2014.
151. Jumaidin, R. *et al.*, Effect of sugarcane bagasse on thermal and mechanical properties of thermoplastic cassava starch/beeswax composites. *Phys. Sci. Rev.*, 9, 1–15, 2022.
152. Cruz-Tirado, J.P., Tapia-Blácido, D.R., Siche, R., Influence of proportion and size of sugarcane bagasse fiber on the properties of sweet potato starch foams. *IOP Conf. Ser.: Mater. Sci. Eng.*, **225**, 012180, 2017.
153. Bartos, A. *et al.*, Biobased PLA/sugarcane bagasse fiber composites: Effect of fiber characteristics and interfacial adhesion on properties. *Compos. Part A: Appl. Sci. Manuf.*, **143**, 106273, 2021.
154. Gan, I. and Chow, W.S., Synthesis of phosphoric acid-treated sugarcane bagasse cellulose nanocrystal and its thermal properties enhancement for poly(lactic acid) nanocomposites. *J. Thermoplast. Compos. Mater.*, **32**, 5, 619–634, 2018.
155. Nampitch, T. *et al.*, Mechanical and morphological properties of poly(lactic acid)/bagasse fiber composite foams. *Appl. Mech. Mater.*, **851**, 31–36, 2016.
156. Azmin, S.N.H.M., Hayat, N.A.b.M., Nor, M.S.M., Development and characterization of food packaging bioplastic film from cocoa pod husk cellulose incorporated with sugarcane bagasse fibre. *J. Bioresour. Bioprod.*, **5**, 4, 248–255, 2020.
157. Sabapathy, P.C. *et al.*, Active polyhydroxybutyrate (PHB)/sugarcane bagasse fiber-based anti-microbial green composite: Material characterization and degradation studies. *Appl. Nanosci.*, **13**, 2, 1–13, 2021.
158. Sukyai, P. *et al.*, Effect of cellulose nanocrystals from sugarcane bagasse on whey protein isolate-based films. *Food Res. Int.*, **107**, 528–535, 2018.
159. Dogossy, G. and Czigany, T., Thermoplastic starch composites reinforced by agricultural by-products: Properties, biodegradability, and application. *J. Reinf. Plast. Compos.*, **30**, 21, 1819–1825, 2011.
160. Mehanny, S. *et al.*, Fabrication and characterization of starch based bagasse fibre composite, in: *International Mechanical Engineering Congress & Exposition 2012*, Houston, Texas, USA.

161. Panicker, A.M., Rajesh, K.A., Varghese, T.O., Mixed morphology nano-crystalline cellulose from sugarcane bagasse fibers/poly(lactic acid) nano-composite films: Synthesis, fabrication and characterization. *Iran. Polym. J.*, **26**, 2, 125–136, 2017.
162. Miléo, P.C. *et al.*, Thermal and chemical characterization of sugarcane bagasse cellulose/lignin-reinforced composites. *Polym. Bull.*, **73**, 11, 3163–3174, 2016.
163. Jayaraman, R., Viknesh, M., Girimurugan, R., Experimental investigations on mechanical and water absorption properties of epoxy resin matrix treated sugarcane and tamarind shell powder reinforced bio-composites. *Mater. Today: Proc.*, **74**, 636–641, 2023.
164. Marichelvam, M.K. *et al.*, A novel palm sheath and sugarcane bagasse fiber based hybrid composites for automotive applications: An experimental approach. *Polym. Compos.*, **42**, 1, 512–521, 2020.
165. Asrofi, M. *et al.*, Characteristic of composite bioplastics from tapioca starch and sugarcane bagasse fiber: Effect of time duration of ultrasonication (Bath-Type). *Mater. Today: Proc.*, **46**, 1626–1630, 2021.
166. Gupta, H. *et al.*, Synthesis of biodegradable films obtained from rice husk and sugarcane bagasse to be used as food packaging material. *Environ. Eng. Res.*, **25**, 4, 506–514, 2019.
167. Kassab, Z. *et al.*, Improved mechanical properties of k-carrageenan-based nanocomposite films reinforced with cellulose nanocrystals. *Int. J. Biol. Macromol.*, **123**, 1248–1256, 2019.
168. El Achaby, M. *et al.*, Processing and properties of eco-friendly bio-nanocomposite films filled with cellulose nanocrystals from sugarcane bagasse. *Int. J. Biol. Macromol.*, **96**, 340–352, 2017.
169. Liu, Y. *et al.*, Mechanical and water vapor barrier properties of bagasse hemicellulose-based films. *BioResources*, **11**, 4226–4236, 2016.
170. Ghaderi, M. *et al.*, All-cellulose nanocomposite film made from bagasse cellulose nanofibers for food packaging application. *Carbohydr. Polym.*, **104**, 59–65, 2014.
171. Mandal, A. and Chakrabarty, D., Studies on the mechanical, thermal, morphological and barrier properties of nanocomposites based on poly(vinyl alcohol) and nanocellulose from sugarcane bagasse. *J. Ind. Eng. Chem.*, **20**, 2, 462–473, 2014.
172. Majumder, R. *et al.*, Biopolymers, in: *Biopolymer-Based Food Packaging*, pp. 29–65, 2022.
173. Kumar, S. *et al.*, Current progress in valorization of food processing waste and by-products for pectin extraction. *Int. J. Biol. Macromol.*, **239**, 124332, 2023.
174. Kumar, S. *et al.*, Recent progress in pectin extraction and their applications in developing films and coatings for sustainable food packaging: A review. *Int. J. Biol. Macromol.*, **239**, 124281, 2023.

175. Basumatary, I.B. *et al.*, Edible films and coatings, in: *Biopolymer-Based Food Packaging*, pp. 445–475, 2022.
176. Kumar, S. *et al.*, An overview of natural biopolymers in food packaging, in: *Biopolymer-Based Food Packaging*, pp. 1–28, 2022.
177. Singh, R.V., Sharma, P., Sambyal, K., Application of sugarcane bagasse in chemicals and food packaging industry: Potential and challenges. *Circ. Econ. Sust.*, **2**, 1479–1500, 2022.
178. Farhadi, S. and Javanmard, M., Development of fibrous casings based on sugarcane bagasse with natural antioxidant using rosemary and thyme extract in dried sausages. *J. Food Meas. Charact.*, **17**, 2475–2487, 2023.
179. Nida, S., Moses, J.A., Anandharamakrishnan, C., 3D extrusion printability of sugarcane bagasse blended with banana peel for prospective food packaging applications. *Sugar Tech.*, **24**, 764–778, 2022.
180. Vercelheze, A.E.S. *et al.*, Properties of baked foams based on cassava starch, sugarcane bagasse fibers and montmorillonite. *Carbohydr. Polym.*, **87**, 2, 1302–1310, 2012.
181. Cruz-Tirado, J.P. *et al.*, The addition of sugarcane bagasse and asparagus peel enhances the properties of sweet potato starch foams. *Packag. Technol. Sci.*, **32**, 5, 227–237, 2019.
182. Liu, C. *et al.*, Biodegradable, hygienic, and compostable tableware from hybrid sugarcane and bamboo fibers as plastic alternative. *Matter*, **3**, 6, 2066–2079, 2020.
183. Mali, S. *et al.*, Starch, sugarcane bagasse fibre, and polyvinyl alcohol effects on extruded foam properties: A mixture design approach. *Ind. Crops Prod.*, 32, 3, 353–359, 2010.
184. Cruz-Tiradoa, J.P. *et al.*, Properties of baked foams from oca (Oxalis tuberosa) starch reinforced with sugarcane bagasse and asparagus peel fiber, in: *3rd International Conference on Natural Fibers: Advanced Materials for a Greener World*, Procedia Engineering, Braga, Portugal, 2017.
185. Yang, Y. *et al.*, Bio-based antimicrobial packaging from sugarcane bagasse nanocellulose/nisin hybrid films. *Int. J. Biol. Macromol.*, **161**, 627–635, 2020.

6

Husk and Straw of Cereals Grains for Sustainable Food Packaging

Pooja Baiju, Maria Jose Mukkadan and Preetha Radhakrishnan*

Department of Food Process Engineering, School of Bioengineering, Faculty of Engineering and Technology, SRM IST, Kattankulathur, Tamil Nadu, India

Abstract
Cellulose is one of the major abundantly available polymers on earth. Agro-residues such as husk and straw of cereals are rich in cellulose content that can be extracted using various physical and chemical methods. The extracted cellulose can be converted into its derivatives with different processes such as esterification, oxidation, and etherification and are explained elaborately. Nanocrystalline cellulose confers additional stabilization and reinforcement properties to the cellulose, making them a noteworthy component in packaging applications such as compostable tableware and containers. There are various companies around the globe manufacturing packaging components from cereal grains byproducts, especially husk and straw for the development of packaging materials, which are discussed in detail in this chapter. The packaging materials thus produced were reported to have excellent tensile and compressive strength. This book chapter discusses the applicability of cellulose obtained from husk and straw of cereal grains for sustainable food packaging, which helps to replace synthetic plastics and thus reduce environmental pollution and waste disposal problems.

***Keywords*:** Cellulose, cellulose nanocrystals, biodegradable, packaging, tableware, containers

**Corresponding author*: preethar@srmist.edu.in

Santosh Kumar, Avik Mukherjee and Vimal Katiyar (eds.) Agro-Waste Derived Biopolymers and Biocomposites: Innovations and Sustainability in Food Packaging, (153–186) © 2024 Scrivener Publishing LLC

Abbreviations

AC_2O	Acetic Anhydride
BPA	Bisphenol A
CA	Cellulose Acetate
CH_3COOH	Acetic Acid
ClO_2	Chlorine Dioxide
CMC	Carboxymethyl Cellulose
CN	Cellulose Nitrate
CNC	Cellulose Nano Crystals
CNF	Cellulose Nano Fibrils
CS	Cellulose Sulfate
EC	Ethyl Cellulose
H_2O_2	Hydrogen Peroxide
H_2SO_4	Sulfuric Acid
HCl	Hydrochloric Acid
HEC	Hydroxy Ethyl Cellulose
HPC	Hydroxy Propyl Cellulose
HPH	High-Pressure Homogenization
HPMC	Hydroxypropyl Methyl Cellulose
LB	Lignocellulosic Biomass
MC	Methyl Cellulose
Na_2CO_3	Sodium Bicarbonate
NaCl	Sodium Chloride
NaClO	Sodium Hypochlorite
$NaClO_2$	Sodium Chlorite
NaOH	Sodium Hydroxide
NCO	Isethionate
OH	Hydroxyl
TEMPO	2,2,6,6-Tetramethylpiperidine-1-oxyl

6.1 Introduction

The most commonly cultivated cereals across the world are rice, wheat, millet, barley, rye, sorghum, and maize [1]. Agricultural wastes produced from these cereals can be categorized as *in situ* residues, which are the wastes produced on the field itself (rice straw), and *ex situ* residues, which are produced at the processing units, for example, sugarcane bagasse obtained after the juice extraction. These residues are high in cellulose, carbohydrates, proteins, lipids, etc., that can be used for a wide range of applications [2]. Cellulose is an abundantly available biopolymer in the world, which consists of β-1,4-linked D-glucose residues. It has good film-forming performance and is chemically stable. It is an easily available material having suitable mechanical properties and great demand in the food industry. Cellulose derivatives such as ethers and esters are used for the preparation of edible food packaging [3]. Cellulose and its derivatives are biodegradable, edible, and can be used to supplement the nutritional values of foods. Cellulose and its derivatives can be extracted from different agricultural residues. Renewable cellulose–based biomaterials are used for the development of biodegradable packaging materials nowadays, thereby contributing to environmental sustainability by reducing nondegradable waste disposal problems [4]. Recently, synthetic plastics are being replaced by cellulose nanomaterials reinforced biodegradable polymers in food packaging industries. Nanocellulose is of three types, i.e., cellulose nanocrystals (CNCs), bacterial cellulose (BC), and nanofibrillated cellulose (NFC). CNC has a diameter of 10 to 20 nm and a length ranging from 100 to 600 nm, whereas NFC consists of a stretched chain of cellulose molecules having 1- to 100-nm size. These nanocellulose have been used as reinforcing elements in polymer composites [5]. Biodegradable polymers generally have low water barrier properties, and so the incorporation of cellulosic nanomaterials helps to preserve the freshness of food because of its excellent water barrier properties [6].

The development of transparent packaging films from cellulose and its derivatives is gaining much importance among researchers [7]. Cereal manufacturing involves different types of processes, which result in the production of different byproducts such as husk, bran, and straw. The byproducts of paddy processing include husk, bran, and germ representing 20, 10, and 2%, respectively [8]. The application of these products as a source for the development of bioplastics helps in resolving the problems of disposal and environmental pollution [9]. The low-cost easily available raw materials for the development of sustainable packaging can be classified into different categories: cereal crop residues (straw, leaves, and stalks), cereal processing byproducts (bran, husk, and hulls), sugarcane processing waste (sugarcane bagasse), and dairy byproducts (whey) [10].

Agro-wastes are pretreated or modified by physical, chemical, or biological methods before using them for the development of packaging materials. Agro-waste and wood have many similar properties, which enable the use of agricultural wastes in the production of paper. They can be pulped and used for the development of paper and packaging materials such as tableware, dishes, cartons, and containers [11]. Biodegradable packaging products made from agro-residues degrade faster than synthetic plastics [12]. Molded pulp packaging is gaining importance because of its environmental sustainability [13]. Cellulose nanofibers synthesized from oat hulls and corn husks were used for the production of nanocomposite packaging materials [14]. CNCs obtained from rice husk were used as reinforcement fillers in biopolymers such as polylactic acid and polyvinyl alcohol, which resulted in bionanocomposites with good thermal strength, tensile strength, and morphological properties [15]. The biodegradable packaging films reinforced with nanofibers synthesized from rice husk and bleached rice husk showed good barrier and mechanical properties [16].

6.2 Extraction and Purification of Cellulose from Husk and Straw

Cellulose comprises around 50% of carbon in the biosphere and is found plenteous in lignocellulosic biomass (LB). Cellulose consists of repeated units of cellobiose disaccharide with extensive intermolecular and intramolecular hydrogen bonding [17–19]. LB usually consists of 35 to 50% cellulose, and this content varies based on the raw material. The extraction, modification pretreatment, and purification of cellulose consist of various steps, which involve an amalgam of different methods with the aim of getting the polymer in its purest form α-cellulose. Conventional pretreatment steps include washing, mechanical refining along with soaking in dilute acid solutions, and steam explosion for specific LB, while the purification step involves the elimination of undesirable components such as hemicellulose, lignin, waxes, and other extractives through the processes of alkali treatment and bleaching [20]. Extraction of cellulose could be as nanocrystals, nanofibers, antimicrobial microcrystalline cellulose (MCC), and cellulose nanoyarn through the process of controlled hydrolysis of amorphous regions in the cellulose preferably by an acid (Figure 6.1);

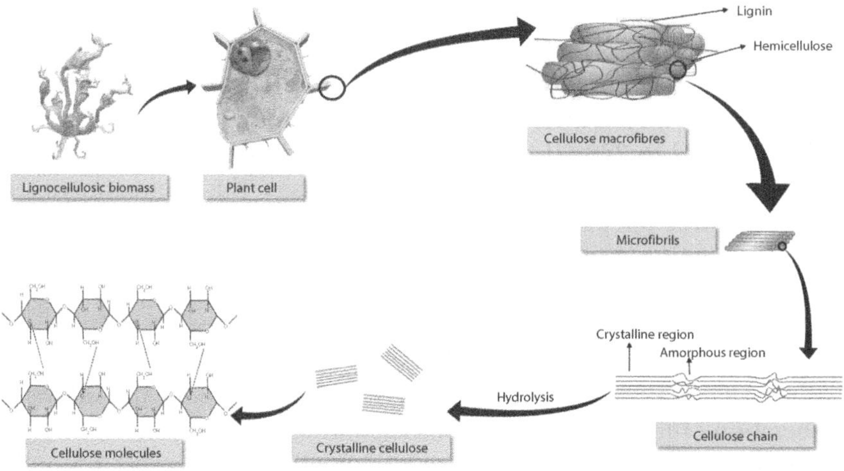

Figure 6.1 Schematic diagram showing extraction of crystalline cellulose from lignocellulosic biomass.

at the same time, a few studies reported other extraction methods which involve ultrasonication and enzyme technology [21–23]. Cellulose fibers can be extracted from agricultural residues especially husk and straw in either micro or nano forms. The extraction can be done by means of chemical or mechanical techniques. Cellulose fiber extraction mainly involves three important steps such as prehydrolysis (alkali or acid treatment), pulping (cooking the fiber with alkali), and bleaching (extraction of the pure bleached form of cellulose with hydrogen peroxide, ozone, etc.).

6.2.1 Pretreatment Methods

These methods are performed to reduce the complexity of LB to facilitate the facile extraction of cellulose as well as to confer better yield. An effective pretreatment method should conserve the cellulose fraction (hexose) of the biomass while adversely hydrolyzing hemicellulose (pentose) and depolymerizing the lignin fraction. The initial step is the appropriate cleaning to remove dust, debris, and other undesirable impurities followed by grinding or different physical treatments. High-pressure homogenization (HPH) of 50 cycles as a pretreatment of sugarcane bagasse has been found to reduce the size of CNCs to a greater extent with an improvement in physical properties [24]. Pretreatment of barley straws involved the size reduction by chopping followed by dewaxing, by boiling the biomass in a mixture of toluene and ethanol (2:1 v/v, 6 hours) [25]. Alternatively, the

Soxhlet extraction method can also be adopted for the dewaxing process [26]. Steam explosion is another popular pretreatment method, which results in breakage in lignin and hemicellulose bonds causing its depolymerization. This method utilizes high pressure/high temperature (1 to 3.5 MPa/180°C to 240°C), followed by rapid decompression causing the mechanical breakage and defiberation of biomass. This pretreatment simplifies the extraction process by increasing the vulnerability of the substrate toward upcoming steps. Nonetheless, cellulose fibers get swollen during the process, weakening the bonds, whereas hemicellulose gets converted into uronic and acetic acid [27]. Stream explosion is a useful technique for the preparation of strawboards [28].

The enzymatic pretreatment method is another technique that allows the defiberation of lignin and the deconstruction of cellulose fibers into fibrils (cellulose nanofibrillation). Cellulase treatments have often been found to remove the amorphous parts of the cellulose, reducing the size of the fibers, and therefore making the substrate receptive to the following alkali treatment. This also leads to the improved thermal stability of the CNF obtained when compared to groups without enzymatic pretreatment. To date, cellulase is the most common enzyme used for pretreatment. The enzymatic pretreatment of barley husk, straw, and cocoa pod husk was reported in the literature [25, 29, 30]. Other enzymes that are being used include canonical endoglucanase (EG), xylanase, and other auxiliary enzymes such as lytic polysaccharide monooxygenases (LPMO), laccases, polygalacturonase, alpha-amylase, and hemicellulase [31–33]. Other pretreatments include thermal processes, ammonia/urea treatment, organo-solvent, kraft method, sulfite treatment, soaking in ethanol and lime, ensilage, microbial degradation, and so forth [34, 35]. Thermal processes such as conventional heating, microwave treatment, and hyperthermophilic have high energy consumption and are less desirable because of reduced crystallinity and reduction in the polymerization of the cellulose [36]. The usage of microbial degradation is noteworthy due to the high degree of cellulose polymerization as well as the reduced usage of chemicals. However, from an industrial point of view, this technique is not economical and the hydrolysis rate happened to be much lower compared to other treatment methods. In all the processes, the treatment severity and parameters such as temperature, pH, and treatment time will notably affect the yield of the final product formed.

6.2.2 Purification Methods

Purification of cellulose includes chemical treatment of LB to remove predominantly a larger portion of lignin and hemicellulose. The former method includes alkali treatment and bleaching. Alkali treatment improves the moisture resistance and surface topography of the LB with minimum thermal degradation, whereas the latter method is considered the crucial step of the purification process.

6.2.2.1 *Alkali Treatment*

Literature has shown that alkalization is an effective and inexpensive process for the maximum removal of undesirable components, modification of cellulose, and detachment of cellulose fibers from the bundle (defibrillation). Alkaline treatment of rice straws combined with high temperature (80°C) has been found to reduce the lignin content present and improve moisture resistance [37, 38]. The varying alkali concentration has different effects on the substrate, as a low concentration effectively removes hemicellulose and retains the surface properties of cellulose, whereas high concentrations weaken the fibers and increase the surface roughness. The practical treatments utilize concentration ranges from 2 to 10% to efficaciously remove the disagreeable contents. At times, the cellulose shows a tendency to pack or cluster together due to hydrogen bonding particularly when the concentration of NaOH is 5% [39]. Rice husk treated with different concentrations of NaOH was reported better interfacial adhesion and enhanced fiber resin bonding in packaging composites but only up to a concentration of 8% [37, 40]. The alkalization process always brings a favorable effect on the chemical constituents of cellulose in terms of its functional and tribological properties, contributing toward its efficiency as a reinforcement in the polymer matrices or composites [41].

6.2.2.2 *Bleaching*

The bleaching process is mainly carried out for the delignification of the substrate, thereby increasing the whiteness index. Lignin has a brownish color due to its phenolic components, and this process aids in the removal of the chromatophore group from the substrate, proffering a whitish color [42]. Different works of literature illustrate various techniques for bleaching, but the basis is the oxidation of water-insoluble lignin to water-soluble aromatic aldehydes and carboxylic acids [43, 44]. Various bleaching agents include individual or a combination of H_2O_2, NaCl, Na_2CO_3, NaClO, $NaClO_2$, and acidified chlorination treatment using glacial acetic acid [45]. Hydrogen peroxide (H_2O_2) and sodium chlorite ($NaClO_2$) are the most widely used bleaching agents for corncob, husk, and stalk; rice husk, and wheat straw [46]. H_2O_2 is known as a green bleaching agent as peroxide-based methods are eco-friendly [47]. H_2O_2 produces hydroperoxyl radicals (HO_2-) in an alkaline environment (pH 10-11) which oxidates the quinone structure of lignin to aliphatic compounds. Sorghum straws treated with H_2O_2 gave good delignification in a short period of time compared to other bleaching agents [48]. Sodium chlorite, in the presence of an acid buffer, liberates chlorine dioxide (ClO_2) which reacts with lignin and precipitates lignin chloride. In $NaClO_2$-treated rice straw, it has been observed that this treatment has an insignificant effect on the amount of silica ash present [49]. The usage of organo-solvents and ozone has also been discussed in some literature reports. Organo-solvents include the usage of organic solvents in the presence of catalysts to hydrolyze the lignin-carbohydrate bonds of the LB, wherein in ozonation, the hydroxyl radicles formed attack lignin bonds. Peroxyformic acid (a mixture of formic acid and hydrogen peroxide) also induces the oxidative depolymerization of lignin.

6.2.3 Extraction Methods

Once the cellulose is purified, the succeeding step is the extraction of cellulose in a relevant form such as nanofiber, nanocrystal, or antimicrobial MCC, corresponding to the application. There are several extraction methods available, but acid hydrolysis is the most routinely used technique. Generally, the extraction methods are categorized into biological, chemical, and mechanical processes [50]. Biological treatment comprises enzyme hydrolysis, which was earlier discussed as one of the pretreatment methods. Application of enzymes onto purified samples could deliver a high yield of pure cellulose in pertinent form. Nanocellulose is often extensively produced by chemical methods such as acid hydrolysis, TEMPO oxidation, ionic liquid (IL) treatment, and eutectic solvent treatment. Mechanical methods include HPH, ultrasonication, ball milling, cryo-crushing, and extrusion [51].

6.3 Cellulose Nanocrystals

CNCs, alternatively known as nanocrystalline cellulose (NCC) and cellulose nanowhiskers (CNWs), are a product of the revalorization of LB, especially cellulose. These nanomaterials have excellent physical and chemical properties contributing to their various biobased applications. The usage of CNC-based materials is multifarious according to the modification and functionalization, as well as depending upon the extraction method. The well-known extraction method of CNC is through acid hydrolysis of the

LB, which was discussed in the previous section. They are also used as organic nanofillers in different composite packaging materials. CNC provides enhanced flexural and tensile stress to the composite. Besides the aspect ratio, crystallinity, large surface area, elastic modulus, and tensile strength delivered by CNC enhance their mechanical stability to establish stable structures with excellent molecular assembly [52]. CNC is a rigid, crystal-like rod of varying lengths with a diameter of 10 to 30 nm [53]. The dimensions rely on the parameters maintained during the extraction process. CNC is hydrophilic, but various surface modifications and functionalization can be executed on them for an extended application as packaging material [54].

6.3.1 Modifications and Functionalization of CNC

Modifications and functionalization of CNC are possible to a larger extent by virtue of free hydroxyl moieties on the surface of CNC. Generally, the modifications are performed by the disintegration of the supramolecular structure of cellulose followed by the substitution of the functional group in place of the hydroxyl group of nanocellulose, mostly at the C6 position as they act as the primary alcohol among the three hydroxyl groups. Generally, the modifications are performed to increase hydrophobicity and dimensional stability, reduce aggregation of particles, improve miscibility, and enhance interfacial compatibility of the CNC [55]. Due to the formation of hydrogen bonds between and within the molecules, nanocellulose is well known for its ability to self-assemble. This characteristic is not always desirable as this aggregation can weaken the interfacial adhesion between CNC and the matrix. Furthermore, this property also interferes with the transfer of mechanical stress from the extrinsic environment to the intrinsic reinforcement [56–58]. Therefore, surface modifications ameliorate CNC as an excellent reinforcement filler with the hosting matrices. Modifications are categorized into two types: functionalization and polymer grafting (Figure 6.2) [59].

Functionalization is often defined in the context of the degree of substitution of a particular functional group on the CNC. This is carried out by processes such as amidation, carbamation, etherification, esterification, TEMPO-mediated oxidation, and silylation. Esterification is a modification method in which an organic functional group is substituted at the hydroxyl moieties on the CNC surface. Acetylation is the most regular esterification reaction performed in nanocellulose. This process is simple and convenient to increase the hydrophobic nature of CNC by lessening the hydroxyl density on the surface by partially replacing it with an

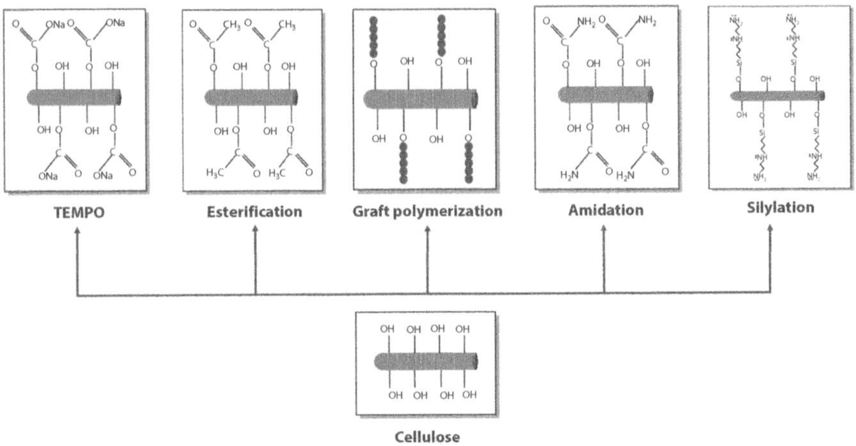

Figure 6.2 Major modifications of CNC.

acetyl group [60]. Acetylation is performed by the addition of acetic acid (CH_3COOH) and acetic anhydride (AC_2O) catalyzed by perchloric or sulfuric acid. AC_2O acts as an acyl donor, and the degree of substitution is tailored by the reaction time and the ratio between CH_3COOH and AC_2O. As reaction time proceeds, more hydroxyl groups are substituted by acetyl moieties, and the degree of substitution is expressed in terms of DS, which points out how many hydroxyl groups are substituted per anhydro-glucose unit. Carboxylic acids are also used in other esterification reactions with catalysts like acid anhydrides or acid chlorides [61, 62].

Etherification reactions involve the introduction of quaternary ammonium salts in the CNC chain at places of hydroxyl groups [63]. Literature has shown several etherification reactions involving the usage of azetidinium *viz* N,N-hexyl 3-methoxyazetidinium, N,N-diallyl-3- methoxyazetidinium, and N-morpholino-3-methoxyazetidinium [64]; (2,3-epoxypropyl) trimethylammonium chloride (EPTMAC), 3-isocyanatopropyltriethoxysilane (ICPTS), glycidyl trimethylammonium chloride, and tetra-n-butylammonium bromide (TBAB) for the cationic surface functionalization [65–67]. This leads to a nucleophilic addition reaction under basic conditions, causing the hydroxyl moieties replaced by a highly unstable functional group. Several substituent groups along with numerous substituent patterns are achievable by etherification reaction. Carbamation reaction, also known as urethanization, inserts covalently bonded urethane linkage on the surface of CNC due to the reaction between hydroxyl moieties with the isethionate (-NCO) group. This reaction is highly specific and unique owing to the high reaction rate, exceptional stability of the functional group, and the absence of other undesirable products [68]. Aliphatic and aromatic NCO groups can be utilized for the functionalization, whereas the mono-NCO group acts as a surface modifier, while di-NCO moieties act as chemical linkers and surface modifiers [69, 70].

Silanization, also known as silylation, involves the insertion of silane (amine or amido) groups onto the CNC using coupling agents to upgrade the hydrophobicity and enhance dispersibility in the solvent or polymer matrices [71]. This process happens more swiftly than other functionalization methods, with increased values of DS and high degrees of silylation. However, this could lead to solubilization and structural degradation of CNC in the medium due to the blizzard association between coupling agents and hydroxyl moieties. Typically, a DS value below 1 is preferred for the preservation of CNC structure and fine dispersibility in the medium [72]. Other functionalization methods include amidation, nucleophilic substitution reactions, and ion exchange methods. Graft polymerization is a modification technique in which organic molecules or "grafts" are introduced onto the CNC surface mainly *via* three approaches: graft onto, graft through, and graft from [52, 67]. Polymer grafting helps to increase the mechanical strength, stability, tensile strength, and thermal properties due to the inclusion of the polymer with a reactive end chain. This tailors the characteristics of the functional group to the cellulose chain, thereby tuning the properties and interaction of CNC with solvent and polymer matrices [73]. Literature has reported the modification of cellulose packaging materials for durable applications [74]. The graft-onto approach entails the covalent bonding of a linker on the surface of CNC, followed by the attachment of the preformed polymer, whereas the graft-from approach includes the growth of the polymer right from the CNC surface where the initiators were anchored earlier. In the graft-through approach, the functionalization followed by the *in situ* polymerization occurs, given that the initiator and the monomers are provided externally [75–78].

6.3.2 Applications of CNC in Packaging Films

The main application of CNC in packaging films is as a reinforcement filler for food preservation. It is also used in edible films and coatings for shelf-life extension of the food. Besides providing excellent mechanical properties, certain functionalizations of cellulose can endow antimicrobial properties to the matrices. Certain studies have showcased the antimicrobial property of CNC etherified by dodecyl-trimethyl-ammonium bromide (DDTMABr) in which even the lowest functionalization ratio of 1:0.2012 provided an inhibition rate of 100% against *Candida albican, Escherichia coli*, and *Staphylococcus aureus* [79]. This is possible due to the interaction between cationic quaternary ammonium salts and the anionic cell wall of the microorganism followed by the penetration of the former into the latter [80]. This disrupts the cells and denatures the proteins, causing destabilization. The nanocellulose cationized by epoxypropyl trimethylammonium chloride [EPTMAC] has also shown similar results by inhibiting Gram-negative bacteria, whereas the polymer grafted NC by poly(2-methacryloyloxyethyl trimethyl ammonium iodide) (PDMQ) exhibited inhibition against both Gram-positive and Gram-negative bacteria [81]. This might be due to the penetration of the grafts through the cell membrane. Similarly, the incorporation of other biocidal agents such as metal nanoparticles and ligands also produces antimicrobial NC. Successful attempts to dope NC with copper, iron, and zinc terpyridine metal complexes to impart antimicrobial properties against some microorganisms have been reported whereupon copper exhibited maximum antimicrobial attribute [82]. New investigations are being conducted for the usage of CNC as a pH sensor in packaging, where the color changes with the quality parameters of the food product [83]. However, these works have been performed in bacterial nanocellulose hitherto.

6.4 Use of Cellulose and Its Derivatives in Food Packaging

Cellulose and its derivatives are being broadly used in the food processing industry as an ingredient in formulations and a reinforcement agent in sustainable packaging (Table 6.1). The range of substituent groups/derivatives upgrades the applicability and versatility of cellulose in packaging development. Cellulose and its derivatives can be used alone or in combination with other polymers in the fabrication of packaging material. This depends upon the polymer's molecular geometry and physical structure, interconnectivity, and interaction of functionalized cellulose with the polymer. The major chemical derivatizations of cellulose are as follows.

6.4.1 Cellulose Ethers

6.4.1.1 Ethyl Cellulose (EC) and Methyl Cellulose (MC)

EC and MC are nontoxic and biodegradable polymers obtained by comparable functionalization of cellulose. EC is a linear polysaccharide obtained by the controlled substitution of hydroxyl groups by ethoxy ($-OC_2H_5$)

Table 6.1 The derivatizations of cellulose extracted from husk and straw of cereal grains for food packaging applications.

Type of derivatization	Raw material	Copolymer/other ingredients used	Packaging material	Procedure	Properties	References
Methyl cellulose	Rice straw	-	Paper sheet	Handmade into paper sheets	Increase in water absorption rate by 230%	[119]
Carboxymethyl cellulose	Rice straw	-	Paper sheet	SCA standard method	Increase in water absorption rate by 290%	[120]
	Rice husk	Starch and citric acid	Film	Solvent casting	Increased tensile strength, Maximum elongation	[121]
Hydroxypropyl cellulose	Rice straw	OXCNC	Film	Solvent casting	Increased tensile strength, WVP	[122]
Cellulose acetate	Corncob	-	Film	Solvent casting	Noteworthy mechanical properties, with DS 2.68 and 60% yield.	[123]

moieties in the reaction of alkalized cellulose with ethyl chloride [84, 85]. EC is mostly semicrystalline or amorphous due to the inter- and intra-hydrogen bonding between free hydroxyl groups. It is inert, colorless, odorless, and tasteless, with excellent film-forming capacity and thermoplasticity. It can also be used as a rheology modifier in films at room temperature; however, the strength of EC is inversely proportional to temperature [86]. Its solubility in different solvents is influenced by the degree of substitution of ethyl groups in the repetitive anhydrous glucose units, with water solubility and organic solvent solubility achieved at the DS range of 1.0 to 1.5 and 2.4 to 2.5, respectively [87]. MC is produced by the substitution of hydroxy groups by methoxy (-OCH_3) moieties when alkaline cellulose is reacted with methyl chloride [88].

In a recent study, coating EC on paper surfaces with a thickness of around 10 μm has been reported to make the coated paper water-resistant and biodegradable [89]. Active bi-films prepared from EC by electrospinning showed high hydrophobicity, which is important for packaging applications [90]. Works on the incorporation of active compounds such as tannins and cinnamaldehyde into these films render them with antioxidant and antimicrobial properties [90, 91]. A composite film prepared by ethyl(hydroxyethyl) cellulose and methyl nanocellulose has significant oxygen barrier properties due to the interaction between the components [92]. Similarly, hydroxypropyl derivatization of MC along with a lipid component has enhanced water vapor permeability. MC in its derivatizations has more packaging applications than the former. MC combined with components with good cross-linking ability such as chitosan showed better results compared to pristine MC [93].

6.4.1.2 *Carboxymethyl Cellulose (CMC)*

CMC is a widely used anionic polymer produced by the esterification reaction of cellulose in alkaline conditions provided by NaOH, followed by the treatment using monochloroacetic acid or sodium monochloroacetate in an alcohol-containing medium [94]. CMC consists of both carboxymethyl and hydroxyl groups in its structure. It is nontoxic and can be consumed safely. The highest purity grade of CMC is called cellulose gum, used as a food additive. It is also used in bakery products to preserve the structural constancy of finished products. Due to its viscous properties, gelation ability, pH-sensitive swelling, and thixotropic rheology, this linear long polymer is frequently utilized as a matrix material [95, 96]. CMC has been widely used in food packaging applications as hydrogels, composites with active compounds, nanoparticles, polyvinyl chloride (PVC), and much more.

The important criteria here are to tackle the hygroscopic nature of CMC for better packaging applications. Biopolymer film prepared from starch and CMC of sugarcane bagasse showed high mechanical properties compared to rice husk CMC [96]. A composite of BC, CMC, and yeast was successful in preserving the shelf life of oranges and tomatoes [97]. Also, another work conducted has prepared a breathable hydrogel out of CMC and PVC to prevent the dehydration of grapes [98].

6.4.1.3 Hydroxyethyl Cellulose (HEC)

Hydroxyethyl cellulose (HEC) is a nonionic cellulose ester produced by the reaction between cellulose and ethylene oxide in the presence of NaOH as a base catalyst. The production of HEC involves opening the ethylene oxide ring nucleophilically and replacing hydroxyl anions on the cellulose anhydroglucose ring [99]. It is viscous, nontoxic, soluble in water and organic solvents, and can be used as a thickening agent in the food industry. This non-immunogenic polymer is also widely used for the fabrication of packaging material, paper making, cosmeceutical, and pharmaceutical industries [100, 101]. However, it has poor water barrier features by virtue of the hydroxyl groups but can be easily subjected to modifications due to the same reason [102]. Antimicrobial films prepared from HEC have widely been used in food packaging. HEC-polyaniline electroactive film was developed to detect food deterioration by sensing pH changes [103]. These pH-sensitive films are suitable as packaging material for fish and other muscle foods. Similarly, another pH-sensing HEC-guar gum anthocyanin–based film was developed to detect the spoilage of pork and soy milk [104]. Oxygen scavenging films are also fabricated from HEC-PVA, in which the respiratory activity of the entrapped microorganisms will scavenge the oxygen from the headspace of the package [105]. Oxygen indicator films have also been prepared using HEC-polyviologen which could detect oxygen concentration above 0.5% [106].

6.4.1.4 Hydroxypropyl Cellulose (HPC)

Hydroxypropyl cellulose is the linear chained cellulose esterified by hydroxypropyl groups *via* partial or complete substitution. Due to a profuse number of hydroxyl groups in the backbone cellulose chain, it can also be modified, just as HEC [95]. HPC is mesophasic and exhibits chiral behavior as a result of the helicoidal molecular configuration [96]. This edible derivatization of cellulose has wide food and pharmaceutical industry applications. However, the low tolerance to other electrolytes and the high viscosity due to inadequate cross-linking hinder its applicability in many fields [97, 98]. HPC-based packaging films loaded with nanofillers, chitosan, lipids, antimicrobial compounds, etc., have been addressed in packaging applications. Paperboards with HPC coating have improved barrier properties and oil resistance when incorporated with gelatin, latex, and talc [99]. Guavas coated with HPC starch maintained the physiochemical parameters of the fruit for up to 13 days [100]. An intelligent package based on hydroxypropyl methyl cellulose (HPMC)-MCC-anthocyanin was a successful freshness indicator for fish [101]. Similarly, HPMC-beewax-nanoclay composites were able to minimize the oxygen and water barrier properties of the film [102]. Another work has incorporated nisin in HPMC-based films to increase the antimicrobial activity of the film, thereby targeting postprocess contamination [103].

6.4.2 Cellulose Esters

6.4.2.1 Cellulose Acetate (CA)

CA is the end product of the cellulose acetylation process. The final product of acetylation is cellulose triacetate, which is subjected to other treatment methods to lower the substitution pattern to yield cellulose diacetate, and then cellulose acetate. CA is being used widely in active packaging, as they are appropriate for the controlled delivery of actives and antimicrobial agents into food [104, 124]. Biobased packaging of CA is widely used in the food industry. Active compounds and nanoparticles are often incorporated in CA-based films to improve their functional properties. CA films prepared by dry phase inversion technique and added with lysozymes provided enhanced antimicrobial activity and controlled release of the components [105]. Organoclays associated with silver nanoparticles when added to CA-based films along with thymol and gelatin have provided the film with a wide range of functionalities owing to each component of the film [106]. Colorimetric sensor based on polydiacetylene (PDA) was successful in its color transitioning based on pH changes [107]. Another such film incorporating pyranoflavylium dye as a bio-sensitive pH indicator was also efficient in monitoring food deterioration [108].

6.4.2.2 Cellulose Nitrate (CN)

CN is also known as nitrocellulose or gun cotton, and it is made by replacing hydroxyl groups with nitrate groups in the presence of strong nitrating agents during the esterification of cellulose. Due to increased reaction speed, nitric acid (HNO_3) is employed for its production. Like other derivatives, the application of CN depends upon the degree of substitution, which relies on parameters such as reaction conditions, preliminary treatment applied, and properties of the substrate utilized [109]. A final nitrogen level of 13.5% or more is considered to be the threshold for complete nitration of cellulose [110]. Certain works suggest that a nitrogen content of 12.6% is optimum for the fine biocompatibility of this poly-nitric ester [122]. Gupta *et al.* (2022) also observed that the nitration yield increased when the substrate was crystalline [125]. CN as a packaging material is not widely accepted due to some safety concerns [111]. Antioxidant release from CN-based films containing antioxidant extracts was studied by Licciardello *et al.* (2015) who concluded that CN demonstrated a quicker release of antioxidant substances [112].

6.4.2.3 *Cellulose Sulfate (CS)*

CS is produced by the esterification reaction of hydroxyl groups by sulfate groups ($-SO_3^-$) with sulfating agents under homogeneous, heterogeneous, or quasi-homogeneous conditions. Homogeneous reactions provide a high degree of substitution of sulfate groups, along with the formation of an intermediate product cellulose nitrite [109]. Chlorosulfonic acid, sulfur trioxide, and sulfuric acid are commonly used sulfating agents. CS is a water-soluble, biodegradable polymer having broad-spectrum antimicrobial properties [113]. The microbicidal effect is conferred by the sulfate groups, and this property is comparable to that of sulfated polysaccharides found in nature. In addition, the anticoagulation and biocompatible characteristics make them an ideal candidate for fabrication of film. The CS film displays transparency and high dissolubility in aqueous solutions when the sulfonic acid groups are raised to 528 to 689 mol/g [114]. Numerous packaging materials based on CS are being used for food applications. Sodium CS film added with chitosan and curcumin was productive in surveilling pork deterioration, where curcumin acted as a pH sensor [115]. Edible film from CS was prepared and was successful in extending the shelf life of bananas. This film is resistant to fats and oils and therefore can be used for such packaging [116]. Another work conducted on CS edible film loaded with β-cyclodextrin and mustard essential oil showed enhanced antimicrobial properties contributed by each component in the film. The organoleptic properties of the film after the addition of actives remained unchanged and the taste was unaffected [117]. Another study conducted has fabricated a film with good transparency from BC sulfate. This film has good potential in the food industry, and can even be made into an edible film with optimum modifications [118].

6.5 Paper-Based Package from Straw and Husk

Currently, a huge emphasis has been given to innovations and research related to converting waste natural fibers into useful products. The properties and chemical composition of rice straw fiber are similar to the properties exhibited by wood-based sources like eucalyptus. The chemical composition of some of the byproducts of cereal grains is given in Table 6.2. The use of these waste natural fibers helps in reducing greenhouse gas emissions. Rice straw contains a lesser amount of lignin when compared to conventional wood; therefore, it requires only milder chemical pretreatments. Rice straw pulp can be used for producing paperboard and paper, which involves the following steps: stock preparation, papermaking, finishing, and packaging. The usage of rice straws to produce paper has many advantages and disadvantages. Rice straw is biodegradable, recyclable, reduces carbon emissions, and is a suitable replacer for polystyrene containers. It also possesses some demerits such as the product price being higher than the conventional synthetic plastic–based products and the material availability being seasonal [124].

Wheat is another widespread crop produced in the world, and its main byproducts are bran and straw. On average, 1.3 kg of wheat straw will be generated as a byproduct for every 1 kg of wheat grain. These byproducts are commonly used in paper production or to feed animals [126]. Research on improving the properties of paperboard packaging is gaining importance because of its recyclability and biodegradability. Bideau *et al.* (2018) developed a paperboard with (2,2,6,6-tetramethylpiperidin-1-yl)oxyl-oxidized cellulose nanofibers and polypyrrole coating. The developed paperboard had reduced gas permeability and improved mechanical properties because of the dense network formed by the TEMPO-oxidized cellulose nanofibers and polypyrrole coating. The results proved that polypyrrole coating can be used for the production of multilayer paperboard packaging [127]. Paper and boards are of more than 3000 types and consist of different categories. They are used for packing different products like soft drinks and milk. Straw

Table 6.2 Cereal grains byproducts and composition [125].

Cereal grains byproducts	Cellulose (%)	Hemicellulose (%)	Lignin (%)
Wheat straw	33.4–41.93	25.2–30.2	14.55–18.0
Rice husk	30.0–51.0	18.5–25.71	20.0–25.42
Cornstalk	41.7	47	7.4

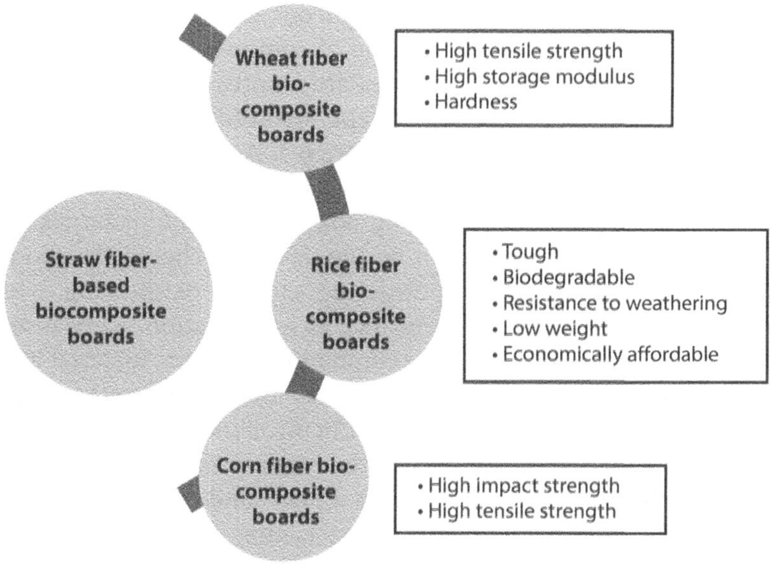

Figure 6.3 Properties of straw fiber–based biocomposite boards [131].

fiber–based biocomposite boards can be produced from the straw of cereals such as wheat, rice, and corn, which has excellent properties (Figure 6.3). A variety of papers such as vegetable parchment and greaseproof are used for direct food contact. Packaging paper can be shaped and molded based on the shape and size of the foods or can be modified depending on the type of food whether it is solid or liquid. Uncoated paper is not used for liquid foods because of the low water vapor and oxygen barrier properties and is commonly used for dry foods [128]. Food packaging paper is coated with waxes, ethylene vinyl alcohol, and polyolefins to improve grease, oil, and water resistance [129]. Cellulosic fibers can also be used as a coating material in addition to their use as a material for paper and board production. HPMC is one of the cellulose derivatives that is commonly used as a coating material in papers to improve the barrier properties [130].

6.6 Tableware and Food Containers from Straw and Husk

6.6.1 Compostable and Biodegradable Tableware and Containers

Compostable tableware is produced from cereal byproducts with the help of compressive processing. High pressure and temperature are applied to the material as a result of which the ligneous cell structure of the plant breaks. The feasibility of the production of compostable tableware will be good only if its mechanical properties are comparable with those made from plastic [132]. The processing steps involved in the production of compostable tableware are given in Figure 6.4. Agricultural residues such as rice straw, wheat straw, sugarcane bagasse, and coconut shells can be used for the production of packaging boxes, where the materials can be used in their natural form or can be physically or chemically modified. Physical modifications include powdering, cutting, or shearing, whereas chemical modifications can be performed by esterification, acetylation, etc. Packaging material is produced by combining a fibrous material as reinforcement and a binder as a matrix. Protein-based resins, soy resin, gluten resin, and carbohydrates like cellulose or hemicellulose are some examples of binders. Rice husk was combined with protein powder and the obtained slurry was kept overnight, poured onto a plate, and dried, and the developed packaging material was used for packaging cookies [133]. Thermal insulating plates using agricultural wastes such as rice straw and flax boon fibers composites were developed by Bakatovich *et al.* (2018) who reported lower moisture absorption [134].

A company named Biotrem started manufacturing wheat bran–based tableware in the 1990s. They claim that disposable tableware from wheat bran is a suitable alternative to tableware from paper and plastics. Their products are fully biodegradable by composting within 1 month. They produce about 10,000 tons of bowls or plates from one ton of edible wheat bran. It can be used to serve both hot and cold foods [136]. The list of some manufacturers producing cellulose-based packaging materials is enlisted in Table 6.3.

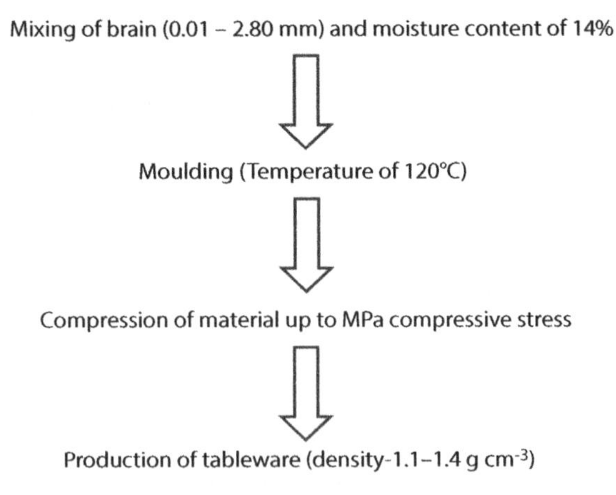

Figure 6.4 Steps involved in the production of compostable tableware using wheat bran [135].

Table 6.3 List of a few manufacturers producing cellulose-based packaging materials [137, 138].

Manufacturer	Packaging material	Properties	Applications in food
Innovia Films	NatureFlex™ (cellulose film)	• Long shelf life • Good gas barrier properties	• Dried foods • Confectionary • Dairy • Meat • Tea packaging
Bilerud Korsnas	FibreForm®	• High elasticity • High strength	Food and beverage packaging
Elopak	Pure-Pak® cartons	Paperboard coated with a polyethylene (PE) layer acts as a liquid barrier	• Milk • Juices • Other liquid foods
Twinnings	Films	• Biodegradable • Approved for direct food contact	Tea
Genpak	Films	• Barrier for oxygen and water • Coatable with aluminum	Crisps
Amcor	Film laminate	• Approved for direct food contact • Compostable • Biodegradable	Dry foods
Bio4pack	Top covering film	• Transparent • Sealable • Barrier to oxygen and water	Meat and fish

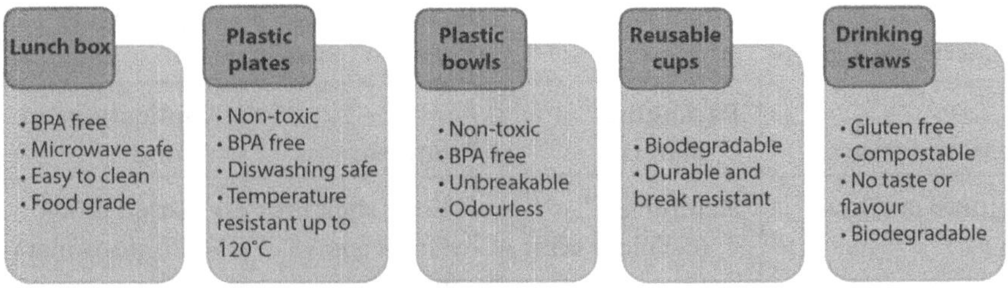

Figure 6.5 Properties of packaging materials produced from wheat straw plastic [139].

6.6.2 Wheat Straw Plastic

Wheat straw plastic is a food-grade material produced from wheat straw, which is bisphenol A (BPA)–free, and accepted by the U.S. Food and Drug Administration (FDA). Wheat straw is mainly used for manufacturing food containers, plastic plates, reusable coffee cups, drinking straws, etc., having different properties (Figure 6.5). Lignin present in wheat straw is broken down by the bacteria, *Rhodococcus jostii*, by acid production which is then properly mixed with sugar and results in the formation of a substance similar to plastic. The resultant product can be then molded into plates, cups, and containers. Wheat straw plastic possesses many benefits such as complete biodegradability (3 to 6 months to compost), easy to clean, nonallergenic, and BPA-free, which is the commonly used chemical in the production of plastics and microwave-safe food-grade material.

6.7 Conclusion and Future Perspective

The development of cellulose-based packaging materials from agricultural residues such as straw and husk from cereals contributes to environmental sustainability. Agro-residue–based packaging materials are low cost, eco-friendly, nontoxic, biodegradable, and abundantly available, for example, husk and straw of various cereals can be converted to sustainable packaging materials. Agro-waste needs to be pretreated before using it for the development of any packaging material. Pretreatments can be done by physical, chemical, or mechanical methods. Physical methods include milling, grinding, and defibrillation, whereas chemical methods involve acid or alkali treatment. Packaging films from husk and straw of cereal grains are developed by different methods like solution casting, suspension casting, extrusion, wet mixing, and casting. Cellulose extracted from agro-waste is used to produce biodegradable packaging materials and films with improved mechanical strength and barrier properties. Straws and husks can also be used for producing biodegradable tableware and containers which serve as an alternative to synthetic plastic containers. In short, these findings will eliminate synthetic plastic–based packaging, tableware, and containers, thereby preventing environmental hazards from plastic waste.

References

1. Leff, B., Ramankutty, N., Foley, J.A., Geographic distribution of major crops across the world. *Global Biogeochem. Cycles*, 18, 1, 607–620, 2004.
2. Candido, R.G. and Gonçalves, A.R., Synthesis of cellulose acetate and carboxymethylcellulose from sugarcane straw. *Carbohydr. Polym.*, *152*, 679–686, 2016.
3. Ferreira, A.R., Alves, V.D., Coelhoso, I.M., Polysaccharide-based membranes in food packaging applications. *Membranes*, 6, 2, 22, 2016.
4. Arun, R., Shruthy, R., Preetha, R., Sreejit, V., Biodegradable nano composite reinforced with cellulose nano fiber from coconut industry waste for replacing synthetic plastic food packaging. *Chemosphere*, 291, 132786, 2022.
5. Hazwan Hussin, M., Trache, D., Chuin, C.T.H., Nurul Fazita, M.R., Mohamad Haafiz, M.K., Hossain, M., Extraction of cellulose nanofibers and their eco-friendly polymer composites, in: *Sustainable Polymer Composites and Nanocomposites*, pp. 653–691, Springer, Cham, 2019.
6. Shruthy, R. and Preetha, R., Cellulose nanoparticles from agro-industrial waste for the development of active packaging. *Appl. Surf. Sci.*, *484*, 1274–1281, 2019.
7. Ramesh, S. and Radhakrishnan, P., Areca nut fiber nano crystals, clay nano particles and PVA blended bionanocomposite material for active packaging of food. *Appl. Nanosci.*, **12**, 295–307, 2022.
8. Galanakis, C.M., Sustainable applications for the valorization of cereal processing by-products. *Foods*, *11*, 2, 241, 2022.
9. Wojnowska-Baryła, I., Kulikowska, D., Bernat, K., Effect of bio-based products on waste management. *Sustainability*, *12*, 5, 2088, 2020.
10. Deshwal, G.K., Alam, T., Panjagari, N.R., Bhardwaj, A., Utilization of cereal crop residues, cereal milling, sugarcane and dairy processing by-products for sustainable packaging solutions. *J. Polym. Environ.*, *29*, 7, 2046–2061, 2021.
11. Vigneswaran, C., Pavithra, V., Gayathri, V., Mythili, K., Banana fiber: Scope and value added product development. *J. Text. Appar. Technol. Manage.*, *9*, 1–7, 2015.
12. Shruthy, R., Jancy, S., Preetha, R., Cellulose nanoparticles synthesized from potato peel for the development of active packaging film for enhancement

of shelf life of raw prawns (Penaeus monodon) during frozen storage. *Int. J. Food Sci. Technol.*, *56*, 8, 3991–3999, 2021.
13. Pulp2Pack, Eco-friendly packaging, paper pulp molding, 2021, https://www.pulp-2pack.com/product.php.
14. Valdebenito, F., Pereira, M., Ciudad, G., Azocar, L., Briones, R., Chinga-Carrasco, G., On the nanofibrillation of corn husks and oat hulls fibres. *Ind. Crops Prod.*, *95*, 528–534, 2017.
15. Bhardwaj, A., Alam, T., Sharma, V., Alam, M.S., Hamid, H., Deshwal, G.K., Lignocellulosic agricultural biomass as a biodegradable and eco-friendly alternative for polymer-based food packaging. *J. Packag. Technol. Res.*, *4*, 2, 205–216, 2020.
16. Nascimento, P., Marim, R., Carvalho, G., Mali, S., Nanocellulose produced from rice hulls and its effect on the properties of biodegradable starch films. *Mater. Res.*, *19*, 167–174, 2016.
17. Rana, A.K., Frollini, E., Thakur, V.K., Cellulose nanocrystals: Pretreatments, preparation strategies, and surface functionalization. *Int. J. Biol. Macromol.*, *182*, 1554–1581, 2021.
18. Isikgor, F.H. and Becer, C.R., Lignocellulosic biomass: A sustainable platform for the production of bio-based chemicals and polymers. *Polym. Chem.*, *6*, 25, 4497–4559, 2015.
19. Nagarajan, K.J., Balaji, A.N., Basha, K.S., Ramanujam, N.R., Kumar, R.A., Effect of agro waste α-cellulosic micro filler on mechanical and thermal behavior of epoxy composites. *Int. J. Biol. Macromol.*, *152*, 327–339, 2020.
20. Melikoğlu, A.Y., Bilek, S.E., Cesur, S., Optimum alkaline treatment parameters for the extraction of cellulose and production of cellulose nano-crystals from apple pomace. *Carbohydr. Polym.*, *215*, 330–337, 2019.
21. Prado, K.S., Jacinto, A.A., Spinacé, M.A., Cellulose nanostructures extracted from pineapple fibres, in: *Pineapple Leaf Fibers*, pp. 185–234, Springer, Singapore, 2020.
22. Shokrkar, H., Ebrahimi, S., Zamani, M., A review of bioreactor technology used for enzymatic hydrolysis of cellulosic materials. *Cellulose*, *25*, 11, 6279–6304, 2018.
23. Radotić, K. and Mićić, M., Methods for extraction and purification of lignin and cellulose from plant tissues, in: *Sample Preparation Techniques for Soil, Plant, and Animal Samples*, pp. 365–376, Humana Press, New York, NY, 2016.
24. Rana, A.K., Frollini, E., Thakur, V.K., Cellulose nanocrystals: Pretreatments, preparation strategies, and surface functionalization. *Int. J. Biol. Macromol.*, *182*, 1554–1581, 2021.
25. Fortunati, E., Benincasa, P., Balestra, G.M., Luzi, F., Mazzaglia, A., Del Buono, D., Torre, L., Revalorization of barley straw and husk as precursors for cellulose nanocrystals extraction and their effect on PVA_CH nanocomposites. *Ind. Crops Prod.*, *92*, 201–217, 2016.

26. Nanayakkara, M.P.A., Pabasara, W.G.A., Samarasekara, A.B., Amarasinghe, D.A.S., Karunanayake, L., Synthesis and characterization of cellulose from locally available rice straw. *Moratuwa Engineering Research Conference (MERCon)*, IEEE, pp. 176–181, 2017.
27. Chadni, M., Grimi, N., Bals, O., Ziegler-Devin, I., Brosse, N., Steam explosion process for the selective extraction of hemicelluloses polymers from spruce sawdust. *Ind. Crops Prod.*, 141, 111757, 2019.
28. Chen, H., Yang, Y., Zhang, J., Biotechnological potential of cereal (wheat and rice) straw and bran residues, in: *Biotechnology for Agro-Industrial Residues Utilisation: Utilisation of Agro-Residues*, pp. 327–340, 2009.
29. Tao, P., Zhang, Y., Wu, Z., Liao, X., Nie, S., Enzymatic pretreatment for cellulose nanofibrils isolation from bagasse pulp: Transition of cellulose crystal structure. *Carbohydr. Polym.*, *214*, 1–7, 2019.
30. Chen, Y.W., Lee, H.V., Juan, J.C., Phang, S.M., Production of new cellulose nanomaterial from red algae marine biomass Gelidium elegans. *Carbohydr. Polym.*, 151, 1210–1219, 2016.
31. Nie, S., Zhang, K., Lin, X., Zhang, C., Yan, D., Liang, H., Wang, S., Enzymatic pretreatment for the improvement of dispersion and film properties of cellulose nanofibrils. *Carbohydr. Polym.*, 181, 1136–1142, 2018.
32. Bian, H., Dong, M., Chen, L., Zhou, X., Ni, S., Fang, G., Dai, H., Comparison of mixed enzymatic pretreatment and post-treatment for enhancing the cellulose nanofibrillation efficiency. *Bioresour. Technol.*, 293, 122171, 2019.
33. Perzon, A., Jørgensen, B., Ulvskov, P., Sustainable production of cellulose nanofiber gels and paper from sugar beet waste using enzymatic pretreatment. *Carbohydr. Polym.*, 230, 115581, 2020.
34. Long, L., Tian, D., Hu, J., Wang, F., Saddler, J., A xylanase-aided enzymatic pretreatment facilitates cellulose nanofibrillation. *Bioresour. Technol.*, 243, 898–904, 2017.
35. Singh, S.S., Lim, L.T., Manickavasagan, A., Ultrasound-assisted alkaliurea pre-treatment of Miscanthus× giganteus for enhanced extraction of cellulose fiber. *Carbohydr. Polym.*, 247, 116758, 2020.
36. Koupaie, E.H., Dahadha, S., Lakeh, A.B., Azizi, A., Elbeshbishy, E., Enzymatic pretreatment of lignocellulosic biomass for enhanced biomethane production-A review. *J. Environ. Manage.*, 233, 774–784, 2019.
37. Kathirselvam, M., Kumaravel, A., Arthanarieswaran, V.P., Saravanakumar, S.S., Characterization of cellulose fibers in Thespesia populnea barks: Influence of alkali treatment. *Carbohydr. Polym.*, 217, 178–189, 2019.
38. Shanmugasundaram, N., Rajendran, I., Ramkumar, T., Characterization of untreated and alkali treated new cellulosic fiber from an Areca palm leaf stalk as potential reinforcement in polymer composites. *Carbohydr. Polym.*, 195, 566–575, 2018.
39. Loganathan, T.M., Sultan, M.T.H., Ahsan, Q., Jawaid, M., Naveen, J., Shah, A.U.M., Hua, L.S., Characterization of alkali treated new cellulosic fibre from Cyrtostachys renda. *J. Mater. Res. Technol.*, 9, 3, 3537–3546, 2020.

40. Senthamaraikannan, P. and Kathiresan, M., Characterization of raw and alkali treated new natural cellulosic fiber from Coccinia grandis. L. *Carbohydr. Polym.*, 186, 332–343, 2018.
41. Ganapathy, T., Sathiskumar, R., Senthamaraikannan, P., Saravanakumar, S.S., Khan, A., Characterization of raw and alkali treated new natural cellulosic fibres extracted from the aerial roots of banyan tree. *Int. J. Biol. Macromol.*, 138, 573–581, 2019.
42. de Souza Fonseca, A., Panthapulakkal, S., Konar, S.K., Sain, M., Bufalinof, L., Raabe, J., Tonoli, G.H.D., Improving cellulose nanofibrillation of non-wood fiber using alkaline and bleaching pre-treatments. *Ind. Crops Prod.*, 131, 203–212, 2019.
43. Hafemann, E., Battisti, R., Marangoni, C., Machado, R.A., Valorization of royal palm tree agroindustrial waste by isolating cellulose nanocrystals. *Carbohydr. Polym.*, 218, 188–198, 2019.
44. Wang, T. and Zhao, Y., Optimization of bleaching process for cellulose extraction from apple and kale pomace and evaluation of their potentials as film forming materials. *Carbohydr. Polym.*, 253, 117225, 2021.
45. Dai, J., Patti, A.F., Saito, K., Recent developments in chemical degradation of lignin: Catalytic oxidation and ionic liquids. *Tetrahedron Lett.*, 57, 45, 4945–4951, 2016.
46. Li, K., Tang, B., Zhang, W., Shi, Z., Tu, X., Li, K., Zhang, H., Formation mechanism of bleaching damage for a biopolymer: Differences between sodium hypochlorite and hydrogen peroxide bleaching methods for shellac. *ACS Omega*, 5, 35, 22551–22559, 2020.
47. Shi, S.C. and Liu, G.T., Cellulose nanocrystal extraction from rice straw using a chlorine-free bleaching process. *Cellulose*, 28, 10, 6147–6158, 2021.
48. Alves, J.A.A., Dos Santos, M.D.L., Morais, C.C., Ascheri, J.L.R., Signini, R., Dos Santos, D.M., Ascheri, D.P.R., Sorghum straw: Pulping and bleaching process optimization and synthesis of cellulose acetate. *Int. J. Biol. Macromol.*, *135*, 877–886, 2019.
49. Gu, J. and Hsieh, Y.L., Alkaline cellulose nanofibrils from streamlined alkali treated rice straw. *ACS Sustain. Chem. Eng.*, *5*, 2, 1730–1737, 2017.
50. Salimi, S., Sotudeh-Gharebagh, R., Zarghami, R., Chan, S.Y., Yuen, K.H., Production of nanocellulose and its applications in drug delivery: A critical review. *ACS Sustain. Chem. Eng*, 7, 19, 15800–15827, 2019.
51. Zhang, Y., Li, T., Shen, Y., Wang, L., Zhang, H., Qian, H., Qi, X., Extrusion followed by ultrasound as a chemical-free pretreatment method to enhance enzymatic hydrolysis of rice hull for fermentable sugars production. *Ind. Crops Prod.*, 149, 112356, 2020.
52. Du, H., Liu, W., Zhang, M., Si, C., Zhang, X., Li, B., Cellulose nanocrystals and cellulose nanofibrils based hydrogels for biomedical applications. *Carbohydr. Polym.*, 209, 130–144, 2019.
53. Du, H., Liu, C., Mu, X., Gong, W., Lv, D., Hong, Y., Li, B., Preparation and characterization of thermally stable cellulose nanocrystals via a sustainable

approach of FeCl3-catalyzed formic acid hydrolysis. *Cellulose*, 23, 4, 2389–2407, 2016.
54. George, J. and Sabapathi, S.N., Cellulose nanocrystals: Synthesis, functional properties, and applications. *Nanotechnol. Sci. Appl.*, 8, 45, 2015.
55. Garg, M., Linares, M., Zozoulenko, I., Theoretical rationalization of self-assembly of cellulose nanocrystals: Effect of surface modifications and counterions. *Biomacromolecules*, 21, 8, 3069–3080, 2020.
56. Nascimento, D.M., Nunes, Y.L., Figueirêdo, M.C., de Azeredo, H.M., Aouada, F.A., Feitosa, J.P., Dufresne, A., Nanocellulose nanocomposite hydrogels: Technological and environmental issues. *Green Chem.*, 20, 11, 2428–2448, 2018.
57. Dhuiège, B., Pecastaings, G., Sèbe, G., Sustainable approach for the direct functionalization of cellulose nanocrystals dispersed in water by transesterification of vinyl acetate. *ACS Sustain. Chem. Eng.*, 7, 1, 187–196, 2018.
58. Trinh, B.M. and Mekonnen, T., Hydrophobic esterification of cellulose nano-crystals for epoxy reinforcement. *Polymer*, 155, 64–74, 2018.
59. Li, B., Xu, C., Liu, L., Yu, J., Fan, Y., Facile and sustainable etherification of ethyl cellulose towards excellent UV blocking and fluorescence properties. *Green Chem.*, 23, 1, 479–489, 2021.
60. Zhou, X., Lin, X., White, K.L., Lin, S., Wu, H., Cao, S., Chen, L., Effect of the degree of substitution on the hydrophobicity of acetylated cellulose for production of liquid marbles. *Cellulose*, 23, 1, 811–821, 2016.
61. Barbosa, R.F.S., Souza, A.G., Ferreira, F.F., Rosa, D.S., Isolation and acetylation of cellulose nanostructures with a homogeneous system. *Carbohydr. Polym.*, 218, 208–217, 2019.
62. Chin, K.M., Sung Ting, S., Ong, H.L., Omar, M., Surface functionalized nanocellulose as a veritable inclusionary material in contemporary bioinspired applications: A review. *J. Appl. Polym. Sci.*, 135, 13, 46065, 2018.
63. Sahlin, K., Forsgren, L., Moberg, T., Bernin, D., Rigdahl, M., Westman, G., Surface treatment of cellulose nanocrystals [CNC]: Effects on dispersion rheology. *Cellulose*, 25, 1, 331–345, 2018.
64. Bonardd, S., Robles, E., Barandiaran, I., Saldías, C., Leiva, Á., Kortaberria, G., Biocomposites with increased dielectric constant based on chitosan and nitrile-modified cellulose nanocrystals. *Carbohydr. Polym.*, 199, 20–30, 2018.
65. Anžlovar, A., Krajnc, A., Žagar, E., Silane modified cellulose nanocrystals and nanocomposites with LLDPE prepared by melt processing. *Cellulose*, 27, 10, 5785–5800, 2020.
66. Ghasemlou, M., Daver, F., Ivanova, E.P., Habibi, Y., Adhikari, B., Surface modifications of nanocellulose: From synthesis to high-performance nanocomposites. *Prog. Polym. Sci.*, 119, 101418, 2021.
67. Zhang, X., Wang, L., Dong, S., Zhang, X., Wu, Q., Zhao, L., Shi, Y., Nanocellulose 3, 5-dimethylphenylcarbamate derivative coated chiral stationary phase: Preparation and enantioseparation performance. *Chirality*, 28, 5, 376–381, 2016.

68. Guo, J., Du, W., Gao, Y., Cao, Y., Yin, Y., Cellulose nanocrystals as water-in-oil Pickering emulsifiers via intercalative modification. *Colloids Surf. A: Physicochem. Eng. Asp.*, 529, 634–642, 2017.
69. Abushammala, H. and Mao, J., A review of the surface modification of cellulose and nanocellulose using aliphatic and aromatic mono-and di-isocyanates. *Molecules*, 24, 15, 2782, 2019.
70. Zhu, W., Ji, M., Zhang, Y., Wang, Z., Chen, W., Xue, Y., Synthesis and characterization of aminosilane grafted cellulose nanocrystal modified formaldehyde-free decorative paper and its CO2 adsorption capacity. *Polymers*, 11, 12, 2021, 2019.
71. Wu, Y., Zhang, Y., Chen, N., Dai, S., Jiang, H., Wang, S., Effects of amine loading on the properties of cellulose nanofibrils aerogel and its CO2 capturing performance. *Carbohydr. Polym.*, 194, 252–259, 2018.
72. Wohlhauser, S., Delepierre, G., Labet, M., Morandi, G., Thielemans, W., Weder, C., Zoppe, J.O., Grafting polymers from cellulose nanocrystals: Synthesis, properties, and applications. *Macromolecules*, 51, 16, 6157–6189, 2018.
73. Kang, H., Liu, R., Huang, Y., Graft modification of cellulose: Methods, properties and applications. *Polymer*, 70, A1–A16, 2015.
74. Zhu, G. and Lin, N., Surface chemistry of nanocellulose, in: *Nanocellulose: From Fundamentals to Advanced Materials*, pp. 115–153, 2019.
75. Kedzior, S.A., Zoppe, J.O., Berry, R.M., Cranston, E.D., Recent advances and an industrial perspective of cellulose nanocrystal functionalization through polymer grafting. *Curr. Opin. Solid State Mater. Sci.*, 23, 2, 74–91, 2019.
76. Macchione, M.A., Biglione, C., Strumia, M., Design, synthesis and architectures of hybrid nanomaterials for therapy and diagnosis applications. *Polymers*, 10, 5, 527, 2018.
77. Rol, F., Belgacem, M.N., Gandini, A., Bras, J., Recent advances in surface-modified cellulose nanofibrils. *Prog. Polym. Sci.*, 88, 241–264, 2019.
78. Peng, N., Wang, Y., Ye, Q., Liang, L., An, Y., Li, Q., Chang, C., Biocompatible cellulose-based superabsorbent hydrogels with antimicrobial activity. *Carbohydr. Polym.*, 137, 59–64, 2016.
79. Tripathi, A. and Melo, J.S. (Eds.), *Advances in Biomaterials for Biomedical Applications*, vol. 66, Springer, Singapore, 017.
80. Littunen, K., Snoei de Castro, J., Samoylenko, A., Xu, Q., Quaggin, S., Vainio, S., Seppälä, J., Synthesis of cationized nanofibrillated cellulose and its anti-microbial properties. *Eur. Polym. J.*, 75, 116–124, 2016.
81. Hassan, E.A. and Hassan, M.L., Rice straw nanofibrillated cellulose films with antimicrobial properties via supramolecular route. *Ind. Crops Prod.*, 93, 142–151, 2016.
82. Taherkhani, E., Moradi, M., Tajik, H., Molaei, R., Ezati, P., Preparation of on-package halochromic freshness/spoilage nanocellulose label for the visual shelf life estimation of meat. *Int. J. Biol. Macromol.*, 164, 2632–2640, 2020.

83. Nguyen, L.H., Naficy, S., McConchie, R., Dehghani, F., Chandrawati, R., Polydiacetylene-based sensors to detect food spoilage at low temperatures. *J. Mater. Chem.*, 7, 7, 1919–1926, 2019.
84. Altunina, L., Tikhonova, L.D., Yarmukhametova, E.G., Method for deriving carboxymethyl cellulose. *Eurasian Chem.-Technol. J.*, 3, 49, 2016.
85. Dai, L., Cheng, T., Duan, C., Zhao, W., Zhang, W., Zou, X., Ni, Y., 3D printing using plant-derived cellulose and its derivatives: A review. *Carbohydr. Polym.*, 203, 71–86, 2019.
86. Hossieni-Aghdam, S.J., Foroughi-Nia, B., Zare-Akbari, Z., Mojarad-Jabali, S., Motasadizadeh, H., Farhadnejad, H., Facile fabrication and characterization of a novel oral pH-sensitive drug delivery system based on CMC hydrogel and HNT-AT nanohybrid. *Int. J. Biol. Macromol.*, 107, 2436–2449, 2018.
87. Sarkar, C., Chowdhuri, A.R., Kumar, A., Laha, D., Garai, S., Chakraborty, J., Sahu, S.K., One pot synthesis of carbon dots decorated carboxymethyl cellulose-hydroxyapatite nanocomposite for drug delivery, tissue engineering and Fe3+ ion sensing. *Carbohydr. Polym.*, 181, 710–718, 2018.
88. Alatawi, F.S., Monier, M., Elsayed, N.H., Amino functionalization of carboxymethyl cellulose for efficient immobilization of urease. *Int. J. Biol. Macromol.*, 114, 1018–1025, 2018.
89. Hosen, M.D., Hossain, M.S., Islam, M.A., Haque, A.N.M.A., Naebe, M., Utilisation of natural wastes: Water-resistant semi-transparent paper for food packaging. *J. Clean. Prod.*, *364*, 132665, 2022.
90. Martins, V.D., Cerqueira, M.A., Fuciños, P., Garrido-Maestu, A., Curto, J.M., Pastrana, L.M., Active bi-layer cellulose-based films: Development and characterization. *Cellulose*, *25*, 6361–6375, 2018.
91. Olejar, K.J., Ray, S., Ricci, A., Kilmartin, P.A., Superior antioxidant polymer films created through the incorporation of grape tannins in ethyl cellulose. *Cellulose*, *21*, 4545–4556, 2014.
92. Lyytikäinen, J., Morits, M., Österberg, M., Heiskanen, I., Backfolk, K., Skin and bubble formation in films made of methyl nanocellulose, hydrophobically modified ethyl (hydroxyethyl) cellulose and microfibrillated cellulose. *Cellulose*, *28*, 787–797, 2021.
93. Kamari, A. and Phillip, E., Chitosan, gelatin and methylcellulose films incorporated with tannic acid for food packaging. *Int. J. Biol. Macromol.*, *120*, 1119–1126, 2018.
94. Khairuddin, N., Muhamad, I.I., Abd Rahman, W.A.W., Siddique, B.M., Physicochemical and thermal characterization of hydroxyethyl cellulose—Heat starch-based films incorporated thymol intended for active packaging. *Sains Malays.*, 49, 2, 323–333, 2020.
95. Zare-Akbari, Z., Farhadnejad, H., Furughi-Nia, B., Abedin, S., Yadollahi, M., Khorsand-Ghayeni, M., PH-sensitive bionanocomposite hydrogel beads based on carboxymethyl cellulose/ZnO nanoparticle as drug carrier. *Int. J. Biol. Macromol.*, *93*, 1317–1327, 2016.

96. Gupta, H., Kumar, H., Kumar, M., Gehlaut, A.K., Gaur, A., Sachan, S., Park, J.W., Synthesis of biodegradable films obtained from rice husk and sugarcane bagasse to be used as food packaging material. *Environ. Eng. Res.*, *25*, 4, 506–514, 2020.
97. Atta, O.M., Manan, S., Ahmed, A.A.Q., Awad, M.F., Ul-Islam, M., Subhan, F., Yang, G., Development and characterization of yeast-incorporated antimicrobial cellulose biofilms for edible food packaging application. *Polymers*, *13*, 14, 2310, 2020.
98. Saha, N., Benlikaya, R., Slobodian, P., & Saha, P., Breathable and polyol based hydrogel food packaging. *J. Biobased Mater. Bio.*, *9*, 2, 136–144, 2015.
99. Leminen, V., Ovaska, S.S.C., Tanninen, P., Varis, J., Convertability and oil resistance of paperboard with hydroxypropyl-cellulose-based dispersion barrier coatings. *J. Appl. Packag. Res.*, *7*, 3, 6, 2015.
100. Francisco, C.B., Pellá, M.G., Silva, O.A., Raimundo, K.F., Caetano, J., Linde, G.A., Dragunski, D.C., Shelf-life of guavas coated with biodegradable starch and cellulose-based films. *Int. J. Biol. Macromol.*, *152*, 272–279, 2020.
101. Boonsiriwit, A., Lee, M., Kim, M., Inthamat, P., Siripatrawan, U., Lee, Y.S., Hydroxypropyl methylcellulose/microcrystalline cellulose biocomposite film incorporated with butterfly pea anthocyanin as a sustainable pH-responsive indicator for intelligent food-packaging applications. *Food Biosci.*, *44*, 101392, 2021.
102. Ortega-Toro, R., Jiménez, A., Talens, P., Chiralt, A., Properties of starch-hydroxypropyl methylcellulose-based films obtained by compression molding. *Carbohydr. Polym.*, *109*, 155–165, 2014.
103. Imran, M., El-Fahmy, S., Revol-Junelles, A.M., Desobry, S., Cellulose derivative based active coatings: Effects of nisin and plasticizer on physico-chemical and antimicrobial properties of hydroxypropyl methylcellulose films. *Carbohydr. Polym.*, *81*, 2, 219–225, 2010.
104. Gouvêa, D.M., Mendonça, R.C.S., Soto, M.L., Cruz, R.S., Acetate cellulose film with bacteriophages for potential antimicrobial use in food packaging. *LWT - Food Sci. Technol.*, 63, 1, 85–91, 2015.
105. Gemili, S., Yemenicioğlu, A., Altınkaya, S.A., Development of cellulose acetate based antimicrobial food packaging materials for controlled release of lysozyme. *J. Food Eng.*, *90*, 4, 453–462, 2009.
106. Dairi, N., Ferfera-Harrar, H., Ramos, M., Garrigós, M.C., Cellulose acetate/AgNPs-organoclay and/or thymol nano-biocomposite films with combined antimicrobial/antioxidant properties for active food packaging use. *Int. J. Biol. Macromol.*, *121*, 508–523, 2019.
107. Ardila-Diaz, L.D., Oliveira, T.V.D., Soares, N.D.F., Development and evaluation of the chromatic behavior of an intelligent packaging material based on cellulose acetate incorporated with polydiacetylene for an efficient packaging. *Biosensors*, *10*, 6, 59, 2020.
108. Gomes, V., Pires, A.S., Mateus, N., de Freitas, V., Cruz, L., Pyranoflavylium-cellulose acetate films and the glycerol effect towards the development of

pH-freshness smart label for food packaging. *Food Hydrocoll.*, *127*, 107501, 2022.
109. Oprea, M. and Voicu, S.I., Recent advances in composites based on cellulose derivatives for biomedical applications. *Carbohydr. Polym.*, *247*, 116683, 2020.
110. Wsoo, M.A., Shahir, S., Bohari, S.P.M., Nayan, N.H.M., Abd Razak, S.I., A review on the properties of electrospun cellulose acetate and its application in drug delivery systems: A new perspective. *Carbohydr. Res.*, *491*, 107978, 2020.
111. Tajeddin, B., Cellulose-based polymers for packaging applications, in: *Lignocellulosic Polymer Composites: Processing, Characterization, and Properties*, pp. 477–498, 2014.
112. Licciardello, F., Wittenauer, J., Saengerlaub, S., Reinelt, M., Stramm, C., Rapid assessment of the effectiveness of antioxidant active packaging—Study with grape pomace and olive leaf extracts. *Food Packag. Shelf Life*, *6*, 1–6, 2015.
113. Espinha, A., Dore, C., Matricardi, C., Alonso, M.I., Goñi, A.R., Mihi, A., Hydroxypropyl cellulose photonic architectures by soft nanoimprinting lithography. *Nat. Photonics*, 12, 6, 343–348, 2018.
114. Ovaska, S.S., Backfolk, K., Optimizing grease resistance of dispersion barrier coatings for heterogeneous grease mixtures. 26th PTS Coating Symposium. Munich, Germany. PTS - ST 1301, 411–423, 2013.
115. Tang, C., Zhao, Z., Yang, M., Lu, X., Fu, L., Jiang, G., Preparation and characterization of sodium cellulose sulfate/chitosan composite films loaded with curcumin for monitoring pork freshness. *Curr. Res. Food Sci.*, *5*, 1475–1483, 2022.
116. Chen, G., Zhang, B., Zhao, J., Chen, H., Development and characterization of food packaging film from cellulose sulfate. *Food Hydrocoll.*, *35*, 476–483, 2014.
117. Chen, G. and Liu, B., Cellulose sulfate based film with slow-release antimicrobial properties prepared by incorporation of mustard essential oil and β-cyclodextrin. *Food Hydrocoll.*, *55*, 100–107, 2016.
118. Palaninathan, V., Chauhan, N., Poulose, A.C., Raveendran, S., Mizuki, T., Hasumura, T., Kumar, D.S., Acetosulfation of bacterial cellulose: An unexplored promising incipient candidate for highly transparent thin film. *Mater. Express*, *4*, 5, 415–421, 2014.
119. Maha, M. and TA M, A.R., Preparation of environment friendly cellulose derivatives and paper from rice straw. *J. Environ. Sci.*, 40, 1, 51–74, 2017.
120. Gupta, H., Kumar, H., Kumar, M., Gehlaut, A.K., Gaur, A., Sachan, S., Park, J.W., Synthesis of biodegradable films obtained from rice husk and sugarcane bagasse to be used as food packaging material. *Environ. Eng. Res.*, *25*, 4, 506–514, 2020.
121. El-Wakil, N.A., Kassem, N.F., Hassan, M.L., Hydroxypropyl cellulose/rice straw oxidized cellulose nanocrystals nanocomposites and their use in paper coating. *Ind. Crops Prod.*, 93, 186–192, 2016.

122. Araujo, D., Castro, M.C.R., Figueiredo, A., Vilarinho, M., Machado, A., Green synthesis of cellulose acetate from corncob: Physicochemical properties and assessment of environmental impacts. *J. Clean. Prod.*, 260, 120865, 2020.
123. Jia, F., Liu, H.J., Zhang, G.G., Preparation of carboxymethyl cellulose from corncob. *Proc. Environ. Sci.*, 31, 98–102, 2016.
124. Ibrahim, R., Sapuan, S.M., Ilyas, R.A., Atikah, M.S.N., Utilization of rice straw as a raw material for food packaging, in: *Bio-based Packaging: Material, Environmental and Economic Aspects*, pp. 205–224, 2021.
125. Gupta, N., Mahur, B.K., Izrayeel, A.M.D., Ahuja, A., Rastogi, V.K., Biomass conversion of agricultural waste residues for different applications: A comprehensive review. *Environ. Sci. Pollut. Res.*, 29, 73622–73647, 2022.
126. Gan, I. and Chow, W.S., Antimicrobial poly (lactic acid)/cellulose bionanocomposite for food packaging application: A review. *Food Packag. Shelf Life*, *17*, 150–161, 2018.
127. Bideau, B., Loranger, E., Daneault, C., Nanocellulose-polypyrrole-coated paperboard for food packaging application. *Prog. Org. Coat.*, 123, 128–133, 2018.
128. Markevičiūtė, Z. and Varžinskas, V., Plant-origin feedstock applications in fully green food packaging: The potential for tree-free paper and plant-origin bio-plastics in the Baltic Sea region. *Sustainability*, *14*, 12, 7393, 2022.
129. Sönmez, S. and Özden, Ö., Barrier properties of paper and cardboard, in: *Academic Researches in Architecture, Engineering Planning and Design*, pp. 171–183, 2018.
130. Rastogi, V.K. and Samyn, P., Bio-based coatings for paper applications. *Coatings*, *5*, 4, 887–930, 2015.
131. Aladejana, J.T., Wu, Z., Fan, M., Xie, Y., Key advances in development of straw fibre bio-composite boards: An overview. *Mater. Res. Express*, *7*, 1, 012005, 2020.
132. Olt, J., Maksarov, V.V., Soots, K., Leemet, T., Technology for the production of environment friendly tableware. *Environ. Clim. Technol.*, *24*, 2, 57–66, 2020.
133. N. Reddy, V. Krishna, B. Santhosh, Kishore, Natural fiber or fibrous material based completely biodegradable food containers. International Patent PCT/IB2016/000218, 2017.
134. Bakatovich, A., Davydenko, N., Gaspar, F., Thermal insulating plates produced on the basis of vegetable agricultural waste. *Energy Build.*, *180*, 72–82, 2018.
135. W. Jerzy, Method for manufacturing biodegradable moldings in particular tableware and packages. US Patent 9517578-B2, 2016.
136. Biotrem, Innovative products from wheat bran http://biotrem.pl/en/, 2016.
137. Sid, S., Mor, R.S., Kishore, A., Sharanagat, V.S., Bio-sourced polymers as alternatives to conventional food packaging materials: A review. *Trends Food Sci. Technol.*, *115*, 87–104, 2021.

138. Khalil, H.A., Davoudpour, Y., Saurabh, C.K., Hossain, M.S., Adnan, A.S., Dungani, R., Haafiz, M.K.M., A review on nanocellulosic fibres as new material for sustainable packaging: Process and applications. *Renew. Sustain. Energy Rev.*, *64*, 823–836, 2016.
139. Eco World, *Wheat straw plastic*, 2020, https://ecoworldonline.com.

7

Sericulture Waste for Edible Films and Coating of Fruits and Vegetables

Sweety Kalita, Avik Mukherjee and Santosh Kumar*

Department of Food Engineering and Technology, Central Institute of Technology, Kokrajhar, Kokrajhar, Assam, India

Abstract

Sericulture represents an agricultural sector primarily focused on cultivating silkworms, with the primary goal of producing silk. The most commonly utilized form of silk originates from the larvae of *Bombyx mori* moths, which construct cocoons using a single thread of raw silk, finding widespread application in the textile industry. In addition to silk production, a variety of secondary products, including proteins known as fibroin and sericin, have emerged, boasting significant economic value. These by-products have found diverse applications in various sectors, including tissue engineering and drug delivery systems. Notably, the food industry has recently integrated this biomaterial into its packaging systems. In a recent time, various types of silk-based active food packaging have been proposed, demonstrating the manifold potential applications of silk in the food industry. This chapter delves into the potential use of sericulture waste as a coating for food and in the production of film-based packaging to extend the freshness of perishable food items.

***Keywords*:** Sericulture waste, fibroin, sericin, edible films and coatings, sustainable food packaging

7.1 Introduction

Sericulture is an agricultural sector that primarily focuses on growing silkworms (*Bombyx mori*) mainly for the purpose of production of silk. Silk cocoon is recognized as the primary product of sericulture, and silk industries use them as raw materials. Besides cocoons, the byproducts of

**Corresponding author*: s.kumar@cit.ac.in

Santosh Kumar, Avik Mukherjee and Vimal Katiyar (eds.) Agro-Waste Derived Biopolymers and Biocomposites: Innovations and Sustainability in Food Packaging, (187–204) © 2024 Scrivener Publishing LLC

silk production are moths and silkworm fibrous waste including proteins that can be used for various purposes. Generally, fibrous silk waste is a heterogeneous waste consisting of rejected cocoons, non-reeling parts of the cocoon, and other byproducts that are not used by textile industries. Silkworms also serve as a biofactory to produce valuable proteins such as fibroin and sericin utilizing the silk gland, and these proteins may be obtained as high-value products from low-value byproducts or wastes of sericulture and silk industries [1, 2]. A total of 609,332 tonnes of silk is produced annually worldwide, among which China, India, Korea, Uzbekistan, Thailand, and Brazil are the leading producers, and sericulture has been recognized as a significant cottage industry in these countries. China produces approximately 403,021 tonnes of silk annually contributing more than 60% of the global production, followed by India, the second-largest silk producer with an annual production of around 161,127 tonnes [3].

About one-third of foods grown are squandered in the distribution chain from the farm to the table. The wastage is more prominent in fruits and vegetables, which are highly perishable and may suffer rapid physical, chemical, and microbiological deterioration resulting in limited postharvest shelf life [4]. Reduction in the rate of moisture loss, microbial growth, and physiological deteriorations to extend the shelf life of fresh fruits and vegetables can be achieved by using sustainable technologies such as edible films and coatings. Besides being effective protection of perishable food items, such biodegradable films and coatings have shown minimal alterations in nutritional and sensory qualities for a longer duration without using any synthetic chemicals [5, 6]. Biopolymers such as polysaccharides, proteins, lipids, etc. have been extensively used as the primary ingredients for the fabrication of edible films and coatings [7–11]. Silk fibroin (SF), among various proteins, has garnered increasing attention for its potential applications in addressing critical challenges within agriculture, food safety, and security. Notably, it is considered a Generally Recognized as Safe (GRAS) substance by the U.S. Food and Drug Administration. One of the key advantages of SF is its capacity to effortlessly form thin, conformal films which are characterized by their transparency, devoid of taste or odor, flexibility, biodegradability, and effective barrier properties against oxygen and water vapors. SF boasts remarkable toughness (80 to 78 MJ m^{-3}), extraordinary tensile strength (300 to 740 MPa), elongation at break ranging from 4 to 26%, and Young's modulus between 10 and 17 GPa. The toughness and mechanical strength of SF are heavily influenced by the presence of β sheet crystals, and these structures play a crucial role in determining the polymer's overall stability [12]. Moreover, the primary structure of SF has numerous hydrophobic groups that prevent water molecules from permeating through the structure while simultaneously facilitating the formation of a densely packed material [13, 14].

The primary structure of fibroin mainly consists of three essential amino acids: glycine, alanine, and serine. SF exhibits a structural composition that includes both crystalline and amorphous regions. Additionally, SF can exist in three distinct polymorphic forms, labeled as I, II, and III. Silk I represents the fibroin structure before the spinning process and belongs to the orthorhombic crystal system, whereas silk II is the solid state of the protein obtained after the spinning process and is classified within the monoclinic crystal system. Notably, treating silk I with potassium phosphate can induce its conversion into the second polymorph, silk II [15–17]. SF has a characteristic known as polymorphism that allows it to be processed using water to produce either water-soluble or water-insoluble end products, depending on the molecular makeup of the protein (i.e., random coil or β sheet, respectively). The polymorphism offers the chance to adjust the material's properties such as surface energy, molecular structure, water solubility, adhesion, thermal reflow, and self-assembly while developing new material fabrication techniques, leading to the development of proof-of-concept designs that can be used alone or with other structural biopolymers [14]. This remark makes an attempt to shed light on the use of SF as a structural biopolymer having potential applications as food packaging or coating materials that can improve food security and safety. The chapter also shows how SF can support new biopolymer-based technologies as an advanced material having beneficial impacts on numerous fields, including food preservation and packaging, and includes particular interest in the toxicological studies of SF upon oral consumption. With this context, detailed documentation of sericulture waste, SF extraction, and SF-based edible films and coating are provided in this chapter.

7.2 Sericulture Wastes

India holds the rare distinction of producing four different types of silk, namely, mulberry, eri, tasar, and muga silk. Among these four, the mulberry silkworm produces roughly 75% of the total silk production in India, and the remaining 25% is shared by the wild silks, i.e., eri, tasar, and muga silk. The term "silk" is typically described as "mulberry silk," whereas the other three are referred to as "non-mulberry silk" or "vanya silk". In sericulture, the development of host plants and the development of cocoons are interlinked, and plant growth/cultivation affects cocoons' development. The production of cocoons in all forms of silk generates a large amount of waste and secondary byproducts, major portions of which come from mulberry farms. Sericulture has pre-cocoon and post-cocoon activities, in which host plant cultivation and silkworm rearing lead to the production of waste materials including silkworm litter, leaves, dead larvae, moths, and other insects are the pre-cocoon activities. On the other hand, post-cocoon activities include raw silk extraction and its conversion into valued items. Pulling or unwinding silk fibers from the cocoon is a process known as "reeling", and to obtain a single thread of silk, reeling a number of cocoons is needed. Eri cocoons are the only type of silkworm cocoon that is open-ended and cannot be reeled due to a lack of continuous filament, and thus, it is utilized to make spun yarns (machine or handspun) that use 100% of the fiber material without any waste generation. Mulberry, tasar, and muga cocoons, on the other hand, are made of continuous silk threads and must be reeled to extract continuous filament. "Pupal waste" and "fibrous silk waste" are the principal, unavoidable byproducts of the reeling process [18]. Mala and Chandrashekhar (2020) reported that in the rearing process, 1000 kg of mulberry leaves is needed to raise 100 disease-free silkworms, and this yielded about 300 kg of litter and 500 kg of mulberry trash, which includes dried leaves, veins from leaves, and leaf stalks [19]. Silkworm only digests about 55% of the ingested leaf and excretes the remainder as litter. During the post-cocoon period, 1 kg of raw silk yields 8 kg of wet and 2 kg of dry pupa that are rich in proteins, fats, carbohydrates, and vitamins, but neither of which are useful in the sericulture sector. Thus, a proper strategy for the utilization of these byproducts to obtain valuable products is important. Among these, fibrous silk waste can be effectively exploited to extract silk protein, i.e., fibroin for its use in foods, biomedical, cosmetic, and pharmaceutical applications [20, 21].

7.3 Extraction and Purification of Silk Protein/Fibroin

Silk is composed of two proteins organized in a core-shell configuration, in which two fibers of fibroin protein are adhered by an outer layer of a globular protein known as sericin. Depending on the type of silk, sericin accounts for 25 to 30% of the cocoon's weight, and the rest is made up of fibroin. Studies showed that sericin combined with fibroin can lead to significant allergic reactions, and thus, sericin must therefore be separated if silk is to be used as a material for biological applications. The steps involved in the extraction of SF from silk cocoons are summarized in Figure 7.1 [22]. Degumming is a process by which these two proteins can be separated. Degumming is carried out by several methods, among which the

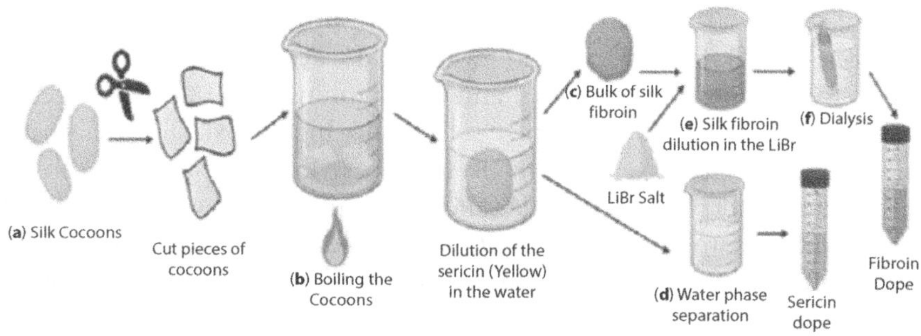

Figure 7.1 Procedure of silk fibroin extraction from silk cocoons [22].

alkali degumming process using sodium carbonate (Na_2CO_3) is the most used and studied method [23]. Calcium hydroxide ($Ca(OH)_2$), another alkali, was also investigated as another degumming agent. Degumming of Chinese *B. mori* male and female cocoons was done by boiling them in 0.01 to 0.1% $Ca(OH)_2$ solution at a ratio of 1:60 for up to 60 min. The fibers produced by using degumming calcium hydroxide have stronger mechanical strength and elongation compared to those produced by degumming using sodium carbonate. Additionally, molecular weight analysis showed that degumming using calcium hydroxide instead of sodium carbonate significantly reduced the damage of protein [24]. The degumming effluents produced by this traditional approach contain not only a significant amount of sericin but also alkali, soap, and other chemical compounds. Thus, the environmental impacts of such effluents and the sustainability of the degumming process are major concerns that have generated interest in exploring greener, cleaner, and more facile degumming methods to produce uniform, undamaged, improved silk fibers. Various environment-friendly techniques such as degumming via steam [25], ionic liquids [26, 27], enzymes such as cocoonase [28], proteases [29], trypsin [30], and microwave-assisted degumming [31] have been explored by researchers. Degummed SF is often dissolved in inorganic salt systems to produce aqueous solutions, and the most common solvents used are concentrated aqueous lithium bromide (LiBr) and ternary solvent, i.e., $CaCl_2$/CH_3CH_2OH/H_2; however, additional inorganic salt systems, CH_3OH, lithium thiocyanate (LiSCN), sodium thiocyanate (NaSCN), and $ZnCl_2$ can also be used. In inorganic salt systems, disruption of hydrogen bonds in the crystalline SF occurs leading to dissolution of the biopolymer. SF dissolved in a salt solution requires dialysis and concentration against aqueous polyethylene glycol (20 wt %), and thus, the dissolving, dialysis, and concentration processes for SF extraction take a longer time and are quite unstable.

Nonaqueous solvents such as hexafluoro-2-propanol (HFIP), hexafluoroacetone (HFA), formic acid, and formic acid/$CaCl_2$ are typically employed for electrospinning, wet spinning, and dry spinning of SF to yield regenerated fibers and films. However, these solvents are highly toxic, flammable, and corrosive. Attempts have been made to use environment-friendly solvents like N-methylmorpholine N-oxide (NMMO) and ionic liquids like 1-allyl-3-methylimidazolium chloride (AMIMCl), 1-butyl-3 methylimidazolium acetate (BMIMAc), and 1-ethyl-3-methylimidazolium chloride (EMIMCl) to dissolve SF [32], which require high temperatures resulting in breakdown of silk protein. Recently, Wang *et al.* (2020) proposed a green and environment-friendly approach to prepare aqueous SF solution using LiBr solution for dialysis over acetone, and the results showed that in the obtained SF, there was no alteration in stability, peptide chains, and secondary structure. Also, acetone and salt from the resultant effluents could be easily separated and recovered [33].

7.4 Silk Protein–Based Active Food Packaging

7.4.1 Silk Protein/Fibroin–Based Active Films

SF-based materials have been used for various applications in agriculture, food security, and food safety. A significant part of perishable foods particularly fruits and vegetables are wasted due to premature deterioration before consumption. Because SF has been classified as a GRAS material, it can be used as a promising biopolymer to fabricate edible and active films for food packaging applications due to its flavorless, odorless, mechanically strong, flexible, transparent, biodegradable, and hydrophobic nature [14]. SF nanofibers and thyme oil–based composite films treated by cold plasma were applied as antibacterial packaging for poultry meat, and the results showed that plasma-treated SF/TO films had higher antibacterial activity compared to SF/TO films, as TO release increased due to plasma treatment resulting in about 6 log CFU/g reduction in *Salmonella typhimurium* counts in chicken and duck meats [34]. In addition, SF nanoparticle was also combined with natural antibacterial agents such as rosemary extracts [35], eugenol [36], and condensed tannin [37] to fabricate antimicrobial food packaging films. In several studies, SF was employed as a matrix and as a filler for developing composite films such as SF reinforced with cellulose nanocrystals and chlorogenic acid [38], poly(lactic acid)-poly(-caprolactone) [39], cellulose nanocrystals [40], poly(lactic acid) crystalline SF nanodiscs [41], silk nanofibril–polyethylene terephthalate [42], and carboxymethyl cellulose–graphene oxide [43]. Table 7.1 has illustrated several works on silk protein/fibroin–based active films for food packaging.

Table 7.1 Applications of silk protein/fibroin in the fabrication of active films for food packaging.

Polymers used	Film composition	Food product	TS (MPa)	EB (%)	Significance	Reference
Silk fibroin (SF)/ sodium alginate (SA)/ tannin (TN)	SF- 3.38% SA- 1.05% TN- 0.3%	Fresh-cut apple	0.99	91.38	The film showed higher scavenging ability and had high antibacterial activity against Staphylococcus aureus and Escherichia coli	[44]
Silk fibroin (SF)/ poly(ethylene oxide) (PEO)/thyme essential oil (TEO)	SF- 50% PEO- 50% TEO- 0.25 to 0.5%	Fresh chicken	12.50	17.06	SF/TEO nanofiber membrane treated with cold plasma exhibited good antibacterial properties in poultry meat against *Salmonella typhimurium* during storage	[34]
Silver nanoparticle (AgNP)/silk fibroin/ polyvinyl alcohol (PVA)	AgNPs- 2% SF- 2%, PVA solution	Food packaging system	1.26	15.4	The prepared film had high mechanical stability, as well as good antibacterial activity	[45]

(Continued)

Table 7.1 Applications of silk protein/fibroin in the fabrication of active films for food packaging. (*Continued*)

Polymers used	Film composition	Food product	TS (MPa)	EB (%)	Significance	Reference
Silk fibroin (SF)/ eugenol (EO)/ *Lycium barbarum* polysaccharide (LBP)	SF- 5, 15, 25, and 35 mg/ml EO- 13 mg/ml, LBP	Fresh pork	-	-	The obtained nanofibers exhibited robust antibacterial properties and had no adverse impact on the sensory qualities of pork	[36]
Silk fibroin (SF)/ pomegranate powder (PP)	SF:PP- 30:70 Glycerol- 20% w/w	Food packaging active pad	-	-	The pad demonstrated excellent water stability and exhibited both antioxidant and antimicrobial properties	[46]
Silk fibroin nanoparticle (SFNP)/poly(lactic acid) (PLA)/poly(ε-caprolactone) (PCL)	Various neat blends of PLA:PCL (90:10, 80:20, and 70:30) SFNP- 1%	Food packaging system	-	-	Incorporating 1% silk fibroin nanoparticles into a blend of PLA/PCL (70/30) resulted in improved barrier properties of the blend and compatibility between PLA and PCL	[39]

(*Continued*)

Table 7.1 Applications of silk protein/fibroin in the fabrication of active films for food packaging. (*Continued*)

Polymers used	Film composition	Food product	TS (MPa)	EB (%)	Significance	Reference
Silk fibroin (SF)/ polyvinyl alcohol (PVA)/rice starch (RS)	SF- 0.5 to 6% w/v PVA- 3% w/v RS- 0.1 to 6% w/v	Food packaging	28.46	9.47	The film exhibited high strength, flexibility, and transparency	[47]
Silk sericin (SS)/ chitosan (CS)/ carbon dots (CDs)	SS- IC CS- 1.5% (w/v) CDs- 20%	Lichi	3.41	54.70	The developed film delayed loss of nutrients during storage and maintained the quality of the litchi fruit up to 6 days of storage	[48]
Silk solution (SS)/ hydroxypropyl methyl cellulose (HPMC)/ n-octadecane (OD)/ hexadecyltrimethyl ammonium bromide (CTAB)	SS- 8 % HPMC- 5 % CTAB- 2 % OD- 0.2, 0.3, and 0.4g	Packaging material	-	-	The composite silk film has high water absorption capability that will be helpful in the transportation of temperature-sensitive products	[49]

*IC, indefinite concentration.

7.4.2 Silk Protein/Fibroin-Based Edible Coating

Fresh horticultural produce deteriorates rapidly because they remain metabolically active even after harvest. Coating of perishable foods, particularly fruits and vegetables with edible materials, is a recent approach to improve their postharvest shelf life and maintain quality for a longer duration. The coating forms a thin layer that provides a barrier against moisture and gas exchange, minimizes microbial contamination, and maintains organoleptic and sensory attributes [5]. SF-based coatings can effectively reduce respiration, prevent dehydration, and preserve the freshness and firmness of coated fruits like strawberries and bananas for longer periods. SF coatings with greater β sheet content had substantially reduced oxygen permeability and respiration rate resulting in significant improvement in maintaining fruit freshness for extended duration [13, 50]. Marelli *et al.* (2016) harnessed the fundamental characteristics of SF, including its polymorphic nature, adaptability, and hydrophobic properties, to create a protein-based suspension that spontaneously forms a layer on the surface of food items when dipped into it. Through this water-based postprocessing technique, they were able to manipulate the protein's polymorphism, which in turn allowed them to control the diffusion of gases, such as oxygen (O_2), carbon dioxide (CO_2), and water vapor, through the thin SF membranes. This control over gas diffusion is a crucial factor in preserving the freshness of food. The resulting SF film exhibited a 58% content of β sheet structures, a water vapor permeability rate of 6.49×10^{-11} g m^{-1} s^{-1} Pa^{-1}, an oxygen diffusion rate of 1.9×10^{-11} cm^2 s^{-1}, and extended shelf life of bananas by up to 7 days (Figure 7.2a) and strawberries by up to 14 days (Figure 7.2b) during storage at 22°C [50]. An edible coating of SF was also applied to *Solanum ferox*, a tropical fruit from Borneo, by dipping into

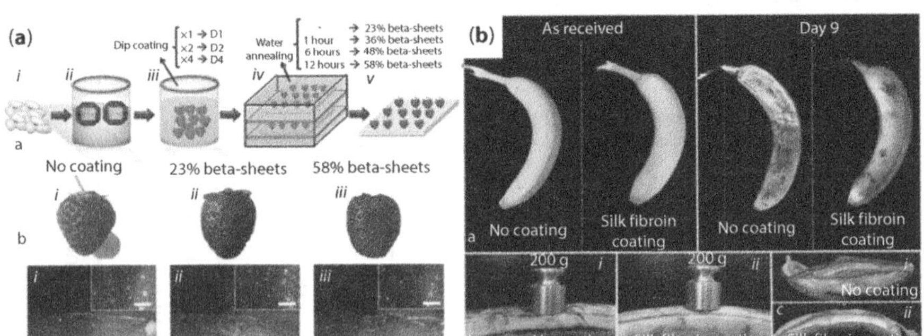

Figure 7.2 Effects of silk fibroin coating on freshly picked (a) strawberries and (b) bananas [50].

2% SF for 2 min, and the effectiveness of the biopolymer-based coating in retaining fruit firmness and reducing weight loss was compared with those achieved by carboxymethyl cellulose–coated and those in uncoated fruits. Weight loss and firmness were both 14.9% and 3.07 N, respectively, for fruit coated with SF, as opposed to 13.49% and 3.53 N, respectively, for that coated with carboxymethyl cellulose, and 28.11% and 4.3 N, respectively, for uncoated fruit [51]. In a separate investigation, SF derived from fibrous waste materials was used for coating strawberries, and the results showed a noticeable enhancement in the strawberries' visual appearance, and significantly prolonged shelf life of the strawberries [52]. Ghosh *et al.* (2021) developed an innovative edible coating from chitosan (CS) infused with silk nanodiscs (SND) for banana fruits that substantially extended the shelf life of bananas, preserving their original weight, optical attributes, firmness, and other desirable qualities for over 7 days at 25°C [53]. Ruggeri *et al.* (2020) aimed at enhancing the mechanical and gas barrier properties of edible coatings by combining SF with polyvinyl alcohol (PVOH). After evaluating various SF:PVOH weight ratios, a coating with a 1:1 ratio was found to exhibit the lowest water vapor permeability, which was chosen for coating fresh-cut apples stored at 4°C for 14 days. The coated samples experienced significantly reduced weight loss and color changes compared to the uncoated apples [54]. Table 7.2 has illustrated several works on silk protein/fibroin–based edible coating for food packaging.

7.5 Toxicological and Food Allergy Assessment of Silk Protein/Fibroin

The unique qualities of SF have generated research interest in its applications in food packaging and preservation to enhance the postharvest shelf life of perishable foods by reducing various metabolic activities [50, 58]. However, the use of SF in food necessitates an evaluation of its potential hazards to human health. Yigit *et al.* (2021) recently performed an in-depth toxicological analysis of SF on male and female Sprague-Dawley rats orally administering 500 mg/kg body weight/day for 14 days, and the results indicated high safety, showing no evidence of toxicity, mutagenicity, genotoxicity, or allergenicity. This study was further extended with daily doses of 0, 125, 250, or 500 mg/kg body weight for 28 days, and no significant effects were found in body weight growth, food intake, urine and serum analysis, hematological, behavior, and overall apparent health effects [59]. Silk-derived products like SF enzyme hydrolysates in human

Table 7.2 Applications of silk protein/fibroin in the fabrication of edible coating for food packaging.

Polymers used	Coating composition	Storage condition	Food product	Significance	Reference
Silk nano disc (SND)/ chitosan (CS)	CS- 1% w/v SND- 0.5, 1.0, and 1.5 wt %	T- 25°C SD- 7 days	Banana	The edible coating improved mechanical performance, wettability, and optical and thermal properties, as well as surface and morphology	[53]
Silk fibroins (SF)/ essential oil (EO)/ elicitor/modified nanocrystal chitin (MNCChit)	SF- 1% w/v EO- 0.01% v/v Elicitor- 8 mM (melatonin (MT) and phenylalanine (Phe)) Modified NCChit- 1% w/v	T- 5°C RH- 75% SD- 28 days	"Hass" avocado	Coating preserved the avocado quality until 30 days of cold storage	[55]
Silk fibroin (SF)/ polyvinyl alcohol (PVA)/ nanocurcumin (NC)/natural honey (NH)	SF- 5.95 % PVA- 3 % NC- 250 μg/ml (w.r.t. SF-PVA 1:1 volume) NH- 5%	Dried at room temperature for several days as per samples	Banana Apple Apple slice Zebrafish	Coating of banana (ripped and unripe) and apple (including slices) increased shelf life up to 8 to 9 days, up to 25 days, and up to 3 days, respectively.	[56]

(*Continued*)

Table 7.2 Applications of silk protein/fibroin in the fabrication of edible coating for food packaging. (*Contnued*)

Polymers used	Coating composition	Storage condition	Food product	Significance	Reference
Silk fibroin (SF)	SF- 1 wt %	T- 22°C RH- 38% SD- 14 days (strawberry) and 7 days (banana)	Strawberry Banana	The developed formulations extended the freshness of strawberries and bananas by reducing respiration and water loss and maintaining firmness	[50]
Silk sericin (SS)/ chitosan (CS)/aloe vera gel (AVG)	SS- 1.5 % CS- 1% AVG- 1% Glycerol- 1.5 %	T- 25°C RH- 70 % SD- 45 days	Tomato	The application of coatings on the tomato showed no deterioration signs, wrinkles (up to 21 days), and shrinkage in the tomatoes	[57]
Silk fibroin (SF)/ poly(vinyl alcohol) (PVOH)	SF- 6 % PVOH- 6 % Different ratios (SF:PVOH (1:0, 4:1, 2:1, 1:1, 1:2, 1:4, and 0:1))	T- 4°C SD- 14 days	Apple slices	The coated apple significantly lowered weight loss and browning index as compared to uncoated ones	[54]

*T, Temperature; RH, relative humidity; SD, storage days.

and animal cell cultures have shown antioxidant, anti-inflammatory, and neuroprotective properties and cognitive enhancement may be attributed to the presence of large fibroin contents that can protect the cells against free radicals and oxidative damage [60–63].

7.6 Conclusion and Future Perspective

SF protein generated from waste cocoons of *Bombyx mori* is one of the abundantly available biopolymers that can be used for the fabrication of various films and coatings for various applications in food packaging systems. Due to the abundance of availability, easy processability, and compatibility with numerous natural and synthetic components, silk proteins have been explored as an attractive alternative to synthetic polymers. Almost all protein engineering done so far has entailed altering ones that already exist, and engineered silk proteins by optimization of their amino acid sequence have shown to perform better than natural silk protein in applications, such as films and coatings to preserve perishables like fruits and vegetables by lowering gas permeability and by extending their postharvest shelf life. Silk proteins as structural biopolymers, including their nanostructures, may make it possible to create hierarchical materials or preservation matrices with better barrier properties ensuring more effective food preservation and safety. However, future studies should concentrate on closing the gap between research-level circumstances and requirements for large-scale applications, making use of improved manufacturing methods and processing parameters. Furthermore, it should also focus on preservation and storage strategies that are readily adaptable to the industrial setting. To ensure the safety of the environment and human health, more studies must be done on the short- and long-term toxicity of SF-based composites when consumed routinely in large quantities.

References

1. Jaiswal, K.K., Banerjee, I., M., V.P., Recent trends in the development and diversification of sericulture natural products for innovative and sustainable applications. *Bioresour. Technol. Rep.*, **13**, 100614, 2021.
2. Ssemugenze, B. *et al.*, Sericulture: Agro-based industry for sustainable socio-economic development: A review. *Int. J. Sci. Res. Publ.*, **11**, 9, 474–482, 2021.

3. Big, A., World silk production by country, Available from: https://www.atlasbig.com/en-in/countries-by-silk-production#:~:text=Worldwide%20609%2C332%20tonnes%20of%20silk,with%20161%2C127%20tonnes%20yearly%20production.
4. Kumar, S. *et al.*, Recent progress in pectin extraction and their applications in developing films and coatings for sustainable food packaging: A review. *Int. J. Biol. Macromol.*, **239**, 124281, 2023.
5. Tracey, C.T. *et al.*, Silk for post-harvest horticultural produce safety and quality control. *J. Food Eng.*, **337**, 111231, 2023.
6. Basumatary, I.B. *et al.*, Chitosan-based active coating for pineapple preservation: Evaluation of antimicrobial efficacy and shelf-life extension. *LWT*, **168**, 113940, 2022.
7. Basumatary, I.B. *et al.*, Edible films and coatings, in: *Biopolymer-Based Food Packaging*, pp. 445–475, 2022.
8. Kumar, S. *et al.*, Current progress in valorization of food processing waste and by-products for pectin extraction. *Int. J. Biol. Macromol.*, **239**, 124332, 2023.
9. Devi, L.S. *et al.*, Carnauba wax-based sustainable coatings for prolonging postharvest shelf-life of citrus fruits. *Sustain. Food Technol.*, **1**, 3, 415–425, 2023.
10. Majumder, R. *et al.*, Biopolymers, in: *Biopolymer-Based Food Packaging*, pp. 29–65, 2022.
11. Purkayastha, M.D. and Kumar, S., Protein-based films and coatings, in: *Biopolymer-Based Food Packaging*, pp. 178–224, 2022.
12. Koh, L.-D. *et al.*, Structures, mechanical properties and applications of silk fibroin materials. *Prog. Polym. Sci.*, **46**, 86–110, 2015.
13. Mondal, K. *et al.*, Sustainable nanostructured materials in food packaging, in: *Dynamics of Advanced Sustainable Nanomaterials and their Related Nanocomposites at the Bio-Nano Interface*, pp. 171–213, 2019.
14. Sun, H. and Marelli, B., Growing silk fibroin in advanced materials for food security. *MRS Commun.*, **11**, 1, 31–45, 2021.
15. Lujerdean, C. *et al.*, The contribution of silk fibroin in biomedical engineering. *Insects*, **13**, 3, 286, 2022.
16. Vidya, M., Rajagopal, S., Acierno, D., Silk fibroin: A promising tool for wound healing and skin regeneration. *Int. J. Polym. Sci.*, **2021**, 1–10, 2021.
17. Qi, Y. *et al.*, A review of structure construction of silk fibroin biomaterials from single structures to multi-level structures. *Int. J. Mol. Sci.*, **18**, 3, 237, 2017.
18. Manjunath, R.N., Kumar, A., Kumar, K.P.A., Utilisation of sericulture waste by employing possible approaches, in: *Contaminants in Agriculture*, M. Naeem, A.A. Ansari, S.S. Gill (Eds.), pp. 385–398, Springer, Springer Nature Switzerland AG, 2020.

19. Mala, P.H. and Chandrashekhar, S., Influence of application of seri-waste bio-digester liquid to mulberry on cocoon parameters of silkworm, Bombyx mori L. *Int. J. Curr. Microbiol. Appl. Sci.*, **9**, 4, 116–120, 2020.
20. Sharma, P. *et al.*, Potential use of sericultural by products: A review. *J. Pharm. Innov.*, **11**, 8, 1154–1158, 2020.
21. Wani, K.A. *et al.*, Bioconversion of seri waste to value-added products: Innovations in sericulture industry, in: *Innovative Waste Management Technologies for Sustainable Development*, R. Bhat, *et al.* (Eds.), IGI Global, Pennsylvania (USA), 2020122-133.
22. Yonesi, M. *et al.*, Silk fibroin: An ancient material for repairing the injured nervous system. *Pharmaceutics*, **13**, 3, 429, 2021.
23. Bucciarelli, A. *et al.*, A design of experiment rational optimization of the degumming process and its impact on the silk fibroin properties. *ACS Biomater. Sci. Eng.*, **7**, 4, 1374–1393, 2021.
24. Zhao, Z.-L. *et al.*, Using of hydrated lime water as a novel degumming agent of silk and sericin recycling from wastewater. *J. Clean. Prod.*, **172**, 2090–2096, 2017.
25. Wang, R. *et al.*, Degumming of raw silk via steam treatment. *J. Clean. Prod.*, **203**, 492–497, 2018.
26. Wang, H.Y., Wei, Z.G., Zhang, Y.Q., Dissolution and regeneration of silk from silkworm Bombyx mori in ionic liquids and its application to medical biomaterials. *Int. J. Biol. Macromol.*, **143**, 594–601, 2020.
27. Samie, M. *et al.*, Aqueous solution of a basic ionic liquid: A perspective solvent for extraction and regeneration of silk powder from bombyx mori silk cocoons. *J. Polym. Environ.*, **28**, 2, 657–667, 2019.
28. Anand, P., Pandey, J.P., Pandey, D.M., Study on cocoonase, sericin, and degumming of silk cocoon: Computational and experimental. *J. Genet. Eng. Biotechnol.*, **19**, 1, 32, 2021.
29. El-Sayed, H. *et al.*, Low temperature water-saving bio-degumming of natural silk using thermophilic protease. *Sustain. Chem. Pharm.*, **27**, 100681, 2022.
30. Sneha, Pandey, J.P., Pandey, D.M., Evaluating the role of trypsin in silk degumming: An in silico approach. *J. Biotechnol.*, **359**, 35–47, 2022.
31. Prakash, N.J. *et al.*, Enhancement of mechano-structural characteristics of silk fibroin using microwave assisted degumming. *Sustain. Chem. Pharm.*, **30**, 100902, 2022.
32. Wang, K. *et al.*, Review on fabrication and application of regenerated bombyx mori silk fibroin materials. *AUTEX Res. J.*, **23**, 2, 164–183, 2023.
33. Wang, H.Y., Zhang, Y.Q., Wei, Z.G., Excess acetone extraction in silk protein solution greatly accelerates the regeneration progress of silk fibroin for desalting and purification. *Int. J. Biol. Macromol.*, **146**, 588–595, 2020.
34. Lin, L., Liao, X., Cui, H., Cold plasma treated thyme essential oil/silk fibroin nanofibers against Salmonella Typhimurium in poultry meat. *Food Packag. Shelf Life*, **21**, 100337, 2019.

35. Hcini, K. *et al.*, Extraction and encapsulation of phenolic compounds of tunisian rosemary (*Rosmarinus officinalis* L.) extracts in silk fibroin nanoparticles. *Plants*, **10**, 11, 2312, 2021.
36. Lin, L. *et al.*, Eugenol/silk fibroin nanoparticles embedded Lycium barbarum polysaccharide nanofibers for active food packaging. *Food Packag. Shelf Life*, **32**, 100841, 2022.
37. Ma, M. *et al.*, Effects of addition of condensed tannin on the structure and properties of silk fibroin film. *Polym. Int.*, **66**, 1, 151–159, 2017.
38. Shuang, F.-F. *et al.*, Chlorogenic acid and cellulose nanocrystal–assisted crosslinking preparation of a silk-based film to extend the shelf life of strawberries. *LWT*, **172**, 114218, 2022.
39. Dadras Chomachayi, M. *et al.*, Biodegradable nanocomposites developed from PLA/PCL blends and silk fibroin nanoparticles: Study on the microstructure, thermal behavior, crystallinity and performance. *J. Polym. Environ.*, **28**, 4, 1252–1264, 2020.
40. Guidetti, G. *et al.*, Co-assembly of cellulose nanocrystals and silk fibroin into photonic cholesteric films. *Adv. Sustain. Syst.*, **5**, 6, 2000272, 2021.
41. Patwa, R., Kumar, A., Katiyar, V., Effect of silk nano-disc dispersion on mechanical, thermal, and barrier properties of poly(lactic acid) based bionanocomposites. *J. Appl. Polym. Sci.*, **135**, 38, 46671, 2018.
42. Sudsandee, S. *et al.*, Improving barrier performance of transparent polymeric film using silk nanofibril combine graphene oxide. *J. Taiwan Inst. Chem. Eng. J.*, **95**, 332–340, 2019.
43. Abdulkhani, A. *et al.*, Preparation and characterization of sodium carboxymethyl cellulose/silk fibroin/graphene oxide nanocomposite films. *Polym. Test.*, **52**, 218–224, 2016.
44. Zhang, X. *et al.*, Impacts of tannin inclusion on fabrication, characterization and antioxidant activity of sodium alginate-silk fibroin-tannin films and their application on fresh-cut apple packaging. *Food Biosci.*, **54**, 102897, 2023.
45. Tao, G. *et al.*, Biosynthesis and characterization of AgNPs-Silk/PVA film for potential packaging application. *Materials*, **10**, 6, 667, 2017.
46. Giannelli, M. *et al.*, Silk fibroin and pomegranate by-products to develop sustainable active pad for food packaging applications. *Foods*, **10**, 12, 2921, 2021.
47. Kuchaiyaphum, P. *et al.*, Composition optimization of polyvinyl alcohol/rice starch/silk fibroin-blended films for improving its eco-friendly packaging properties. *J. Appl. Polym. Sci.*, **129**, 5, 2614–2620, 2013.
48. Mei, S. *et al.*, Developing silk sericin-based and carbon dots reinforced bio-nanocomposite films and potential application to litchi fruit. *LWT*, **164**, 113630, 2022.
49. Zhao, L. *et al.*, Temperature buffering capacity of silk hydrogel: A useful packaging material. *Mater. Lett.*, **211**, 110–113, 2018.

50. Marelli, B. *et al.*, Silk fibroin as edible coating for perishable food preservation. *Sci. Rep.*, **6**, 25263, 2016.
51. Hamzah, M.H. *et al.*, Optimization of silk fibroin coating during storage using response surface methodology and its effect on the physicochemical properties of Solanum ferox (*S. ferox*). *J. Food Meas. Charact.*, **16**, 3385–3401, 2022.
52. Jaramillo-Quiceno, N. and Restrepo-Osorio, A., Water-annealing treatment for edible silk fibroin coatings from fibrous waste. *J. Appl. Polym. Sci.*, **137**, 13, 48505, 2019.
53. Ghosh, T. *et al.*, Silk nanodisc based edible chitosan nanocomposite coating for fresh produces: A candidate with superior thermal, hydrophobic, optical, mechanical and food properties. *Food Chem.*, **360**, 130048, 2021.
54. Ruggeri, E. *et al.*, A multilayered edible coating to extend produce shelf life. *ACS Sustain. Chem. Eng.*, **8**, 38, 14312–14321, 2020.
55. Funes, C.F. *et al.*, Active coatings based on oxidized chitin nanocrystals and silk fibroins for the control of anthracnose in 'Hass' avocados. *Int. J. Biol. Macromol.*, **253**, 126673, 2023.
56. Pratap Singh, D. and Packirisamy, G., Biopolymer based edible coating for enhancing the shelf life of horticulture products. *Food Chem.: Mol. Sci.*, **4**, 100085, 2022.
57. Tarangini, K., Kavi, P., Jagajjanani Rao, K., Application of sericin-based edible coating material for postharvest shelf-life extension and preservation of tomatoes. *eFood*, **3**, 5, e36, 2022.
58. Wang, K. *et al.*, Review on fabrication and application of regenerated bombyx mori silk fibroin materials. *AUTEX Res. J.*, **23**, 164–183, 2023.
59. Yigit, S. *et al.*, Toxicological assessment and food allergy of silk fibroin derived from Bombyx mori cocoons. *Food Chem. Toxicol.*, **151**, 112117, 2021.
60. Choi, B.-H. *et al.*, Phytochemicals and silk proteins in mature silkworm powders responsible for extended life expectancy and enhanced resistances to Parkinson's disease. *J. Asia Pac. Entomol.*, **20**, 4, 1425–1433, 2017.
61. Stohs, S.J. and Bucci, L.R., Effects of silk fibroin enzyme hydrolysates on memory and learning: A review. *Molecules*, **27**, 17, 104942, 2022.
62. Kang, Y.K. *et al.*, Effect of a fibroin enzymatic hydrolysate on memory improvement: A placebo-controlled, double-blind study. *Nutrients*, **10**, 2, 233, 2018.
63. Xu, Z. *et al.*, Neuroprotective effects of silk fibroin hydrolysate against Aβ25–35 induced cytotoxicity in SH-SY5Y cells and primary hippocampal neurons by regulating ROS inactivation of PP2A. *J. Funct. Foods*, **45**, 100–109, 2018.

8

Functional Agents from Agro-Waste for Active and Intelligent Food Packaging

Lokesh Kumar, Ram Kumar Deshmukh and Kirtiraj K. Gaikwad*

Department of Paper Technology, Indian Institute of Technology Roorkee, Roorkee, Uttarakhand, India

Abstract
Agricultural waste is the residue of various crops, fruits, and vegetables, having less economic value than the respective agricultural products. These waste byproducts can be converted into energy sources, fertilizers, additives, or materials by different value-added processes. Agricultural waste such as peels, pomace, and seeds of fruits and vegetables, along with corncob and wheat straw, contain several active compounds. Major active compounds are polyphenols which exhibit antioxidant and antimicrobial properties. Active and intelligent packaging systems can be developed by incorporating these active ingredients into packaging materials. Intelligent packaging systems include various indicators such as freshness, pH, and gas, which help in the real-time monitoring of packaged food products. This chapter aims to summarize different agricultural wastes, which are sources of polyphenols, antioxidants, antimicrobials, and biochemical indicators, and their uses as functional agents in biopolymer-based active and intelligent packaging systems.

***Keywords*:** Agro-waste, functional agents, polyphenols, antioxidants, antimicrobials, indicators, active packaging, intelligent packaging

8.1 Introduction

Agriculture activities generate a lot of waste in developing and developed countries. Agricultural wastes are the residue of the growing and processing of various agricultural products such as crops, vegetables, fruits, and

**Corresponding author*: kirtiraj.gaikwad@pt.iitr.ac.in

Santosh Kumar, Avik Mukherjee and Vimal Katiyar (eds.) Agro-Waste Derived Biopolymers and Biocomposites: Innovations and Sustainability in Food Packaging, (205–240) © 2024 Scrivener Publishing LLC

dairy products. These are considered to be non-product outputs of processing and production of agricultural produce, whose economic value is much less than the cost of collection, processing, and transportation. Still, they may contain compounds that can be beneficial for humans. Agro-waste can be in solid, liquid, or slurry form, and its composition will certainly depend on the type of agricultural activity [1]. Crop waste, such as corncob, wheat husk, peels of fruits and vegetables, food processing waste, and animal waste combined, is known as agro-waste. New technologies and processes are being explored to utilize such waste in the production of fuel, energy, and bioactive compounds including food preservatives. Such technologies will have twofold benefits: first, waste utilization will protect our environment, and second, additional income can be generated by producing value-added products from agro-waste. Estimating overall agricultural waste is rare, but they have a significant proportion of overall agricultural production. Agro-waste, mainly wheat straw, rice straw, and peels of fruits and vegetables, contains several active compounds such as polyphenols, flavonoids, and tannins. These active compounds have antioxidants and antimicrobial activities and can be extracted from agro-waste and utilized for food preservation. Additionally, these compounds can be incorporated in direct food packaging materials, known as active packaging that release active compounds in a controlled manner for food preservation applications [2].

The high amount of food waste threatens the long-term viability of the food supply chain. In the European Union alone, 89 million tonnes of food are wasted per year and are estimated to rise by 40% in the next four years. According to the International Food Policy Research Institute, one-third of the food produced worldwide is thrown away as food waste yearly. Food Wastage Footprint and Climate Change Report stated that about 25% of vegetables and 15% of fruits are being wasted at the end of the food production line [3]. These significant losses related to food waste in processing industries also affect the economy and environment. Recent innovations and research address problems related to food waste and its utilization in food, pharmaceutical, and packaging applications. Pomace, skin, seeds, and rind are underutilized agro-waste, which includes functional nutrients, phytochemicals, vitamins, enzymes, and oils. Incorporating fruit and vegetable waste extract and essential oil into food packaging materials will enhance the packaging film's functionality [4, 5].

Packaging is mainly intended to protect food products from the surrounding environment and ensures that consumers and industry's food safety requirements are satisfied simultaneously. Active packaging protects the packaged products, ensures food quality, improves product appearance, and prolongs shelf life. Active packaging incorporates active ingredients with functional properties that are released in a controlled manner into the food products or they may absorb moisture, gasses, flavor, etc., from the packaged product or in-package environment. Regulation 450/2009/EC and Regulation 1935/2004/EC mentioned that "articles and materials that maintain or improve the food condition, or enhance the food-shelf life" are active materials. Active packaging systems include antioxidant, antimicrobial, and antibrowning agents; moisture absorbers; and oxygen, ethylene, and carbon dioxide scavengers. Various essential oils are responsible for a specific aroma and flavor [6]. Agro-waste, such as peels, shells, and seeds of fruits and vegetables, contains a high concentration of bioactive materials like polyphenols, flavonoids, tocopherols, essential oils, and flavor compounds. The peels, seeds, and waste pomace, a byproduct of juice industries of various fruits such as citrus, pineapples, pomegranates, apples, and grapes, are the most common agro-waste used to inhibit lipid oxidation and growth of foodborne microorganisms. A study has evaluated the antioxidant properties of methanol-based coconut shell extract, and the results show that the extract has 90% DPPH radical scavenging activities, and food packaging films or coating incorporated with coconut shell extract can reduce packaged food's oxidation [7]. Methanolic extract of pineapple peel (PP) waste has 92% DPPH radical scavenging activities [8]. Several investigations have reviewed the antimicrobial properties of various essential oils for edible packaging applications [9–12].

Packaging protects the food product by separating it from the surrounding environment, and it communicates with consumers, providing product information. Food products are highly susceptible to deterioration and spoilage. According to the European Food Safety Authority, "Articles and materials that monitor the food condition or surrounding environment of food are known as intelligent packaging systems." Intelligent packaging monitors the product's condition and provides information to the consumers. An intelligent packaging system contains ingredients such as pigments or dyes that indicate changes in packaged food by changing their color [13]. These ingredients work as pH indicators, monitoring and tracing food products' freshness throughout the storage. pH indicators interact with food, the surrounding environment, and packaging material. The surrounding environment of food products changes the color of the indicator, indicating that the pH of food has changed, showing the level of food product deterioration. Thus, these pH indicators provide authentic information regarding the edibility and quality of food products, which may be different from the expiry date mentioned on the packaging. Synthetic dyes and pigments are a matter of concern for food safety, as they can migrate into food. These are being replaced by natural and biobased pH indicators extracted from the byproducts of vegetables and fruits. Researchers have developed natural pH indicators from potato peel waste by incorporating peel powder in sweet potato starch films. This film changes visible color when exposed to different pH environments [14]. The researchers have investigated the color-changing properties of the film made from corn starch incorporated with blueberry pomace. This film also works as a natural pH indicator, and visible change can be observed in different pH ranges [15].

Agro-wastes are unutilized byproducts of various crops and food processing industries. Utilizing peels, shells, waste pomace, and other byproducts to obtain active ingredients for food packaging applications is known as waste-to-wealth transformation. These active ingredients are food grade, as obtained from agricultural byproducts, will not alter the quality of food, and compromise consumer safety. Incorporating natural active agents in the packaging is a potential alternative to synthetic additives [16, 17].

8.2 Functional Agents in Active and Intelligent Packaging

The continuous demand for food products having high-quality and enhanced shelf life is changing the way food is produced, stored, and distributed. The packaging must fulfill all the essential functions by maintaining integrity, freshness, and food safety. Active and intelligent packaging systems are being developed to ensure consumer requirements and packaging performance [18]. Active packaging materials maintain or enhance the food condition by releasing themselves into the food or surrounding environment. Controlling the moisture transfer, reducing the oxidation rate, retarding microbial growth are well-explored areas of active food packaging. Intelligent packaging monitors food products' condition and then communicates with consumers about the quality of packaged food. Intelligent packaging examples include pH, biochemical, freshness, and time-temperature indicators. This section will discuss various agro-waste–based active ingredients containing polyphenols, antioxidants, antimicrobial, and biochemical indicators for food packaging applications, as depicted in Figure 8.1 [19].

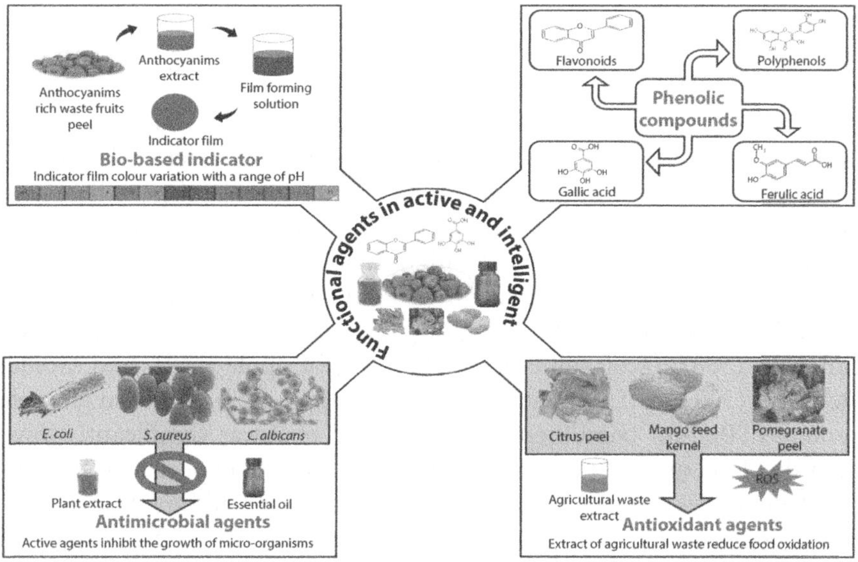

Figure 8.1 Schematic diagram representing biobased indicators, phenolic compounds, antimicrobials, and antioxidants extracted from agricultural waste as functional agents for active and intelligent packaging systems.

8.2.1 Polyphenolic Compounds from Agro-Waste

Functional additives derived from agro-waste are gaining much attention due to their therapeutic and health-promoting potential. Polyphenols have two categories: phenolic acid and flavonoids, the widely distributed bioactive molecules. Among these, phenolic acids are further divided into hydroxycinnamic and hydroxybenzoic acids. Flavonoids are classified into isoflavones, flavanols, flavonols, flavononse, and flavones. Agro-waste such as peel and pomace of fruits and vegetables are a rich flavonoid source. The amount of flavonoids depends on the type of waste species and exposure to sunlight. Each polyphenol is attributed to different biological activities, which depend on the structure and molecular range of the polyphenol. Research has shown that these polyphenols are nutritional compounds and effective against diseases such as hyperlipidemia, hyperglycemia, hypercholesterolemia, and cancer insurgence [20].

8.2.1.1 Polyphenolic Compounds from Apple Peel

Apple peels are considered waste byproducts of canned apple manufacture and apple sauce. Considering the research data, if any industry is processing 216 million pounds of apple, then 16 million pounds of waste apple peels can be generated from processing. These apple peels are valuable sources of phenolic acids and flavonoids. Waste apple peels contain more phenolic and flavonoids than apple flesh. The distribution and nature of phenolic compounds also differ for peel and flesh. A detailed study has been conducted regarding polyphenols and flavonoids of peel, flesh, and peel with flesh for Idared, Cortland, Rome Beauty, and Golden Delicious apples. Idared apple peels have the highest polyphenols, with a mean value of 589 mg of gallic acid equivalents (Eq.)/100 g of apple peel, followed by Rome Beauty apple peel (500 mg of gallic acid Eq./100 g of apple peel), Cortland (388 mg of gallic acid Eq./100 g of apple peel), and Golden Delicious (309 mg of gallic acid Eq./100 g of apple peel). The total phenolic content (TPC) for combined apple flesh and peel was significantly lower than the TPC of only peel. The value of TPC of apple peel with flesh for Cortland apples was observed to be 119 mg of gallic acid Eq./100 g of apple flesh with peel. Then, it increases to 120, 129, and 159 mg of gallic acid Eq./100 g of apple flesh with peel for Idared, Golden Delicious, and Rome Beauty apples, respectively. Apple peels have various polyphenols such as catechins, caffeic acid, chlorogenic acid, phloridizin, phloretin glycosides, procyanidins, and flavonoids like quercetin glycosides, which are not available in apple flesh. The peels of Rome Beauty have the highest amount of flavonoids (306 mg of catechin Eq./100 g of apple peel), followed by Idared apple peels (303 mg of catechin Eq./100 g of apple peel), Cortland apple peels (202 mg of catechin Eq./100 g of apple peel), and Golden Delicious apple peel (167 mg of catechin Eq./100 g of apple peel). The flavonoid contents of apple flesh with Peel were, in ascending order, 50, 55, 61, and 77 mg of catechin Eq./100 g of apple flesh with peel for Cortland, Idared, Golden Delicious, and Rome Beauty apples [21].

8.2.1.2 Polyphenolic Compounds from Citrus Fruit

The citrus fruit crop is one of the most abundantly produced worldwide that generates around 50% of waste in the form of citrus peel. These peels are a rich source of natural phenolic compounds, which are rare in other plants. The TPC of any plant material depends on the extraction process and solvent used, and methanol, ethanol, water, acetone, and diethyl ether are the most common solvents for extraction. Citrus peels are rich in hesperidin, narirutin, naringin, hydrocinnamic acid, and glycosylated flavones. Solvent-based extraction of orange peel contains caffeic acid (8.9 µg/g of dry orange peel), ferulic acid (578.8 µg/g of dry orange peel), naringin (32.7 µg/g of dry orange peel), hesperidin (904.5 µg/g of dry orange peel), and myricetin

(70.4 μg/g of dry orange peel) [22]. Lemon peels contain a wide variety of phenolic compounds relative to orange peels such as ellagic acid (611.4 μg/g of dry lemon peel), kaempferol (407.1 μg/g of dry lemon peel), quercetin (17.54 μg/g of dry lemon peel), as well as caffeic acid (114.2 μg/g of dry lemon peel), ferulic acid (644.3 μg/g of dry lemon peel), naringin (426 μg/g of dry lemon peel), hesperidin (1100 μg/g of dry lemon peel), and myricetin (165.4 μg/g of dry lemon peel) [23]. The conventional extraction process is solvent-based, but nonconventional methods such as microwave-assisted extraction, ultrasound extraction, and high-pressure extraction processes are also widely used for the high yield of phenolic contents. The conventional extraction process for *Citrus reticulata* peel containing polymethoxylated flavones yields 5.790 g/100 g of the dry weight of the peel, but ultrasound extraction yields 7.790 g/100 g of the dry weight of the peel. When the extraction method is changed from conventional to microwave-based, the yield for *Citrus unshiu* containing hesperidin increases from 1.88 to 1.94 g/100 g of the dry weight of peels [24].

8.2.1.3 Polyphenolic Compounds from Potato Peel

Potato peels are a byproduct and waste of agricultural produce, the peel extract contains a high amount of active phenolic and flavonoid agents. Caffeic acid, gallic acid, chlorogenic acid, and protocatechuic acid are the main phenolic compounds present in the aqueous extract of potato peel. The concentrations of phenolic acid before the freeze-drying process were 0.01 mg/g of potato peel extract of caffeic acid, 0.13 mg/g of potato peel extract of gallic acid, 0.30 mg/g of potato peel extract of chlorogenic acid, and 0.04 mg/g of potato peel extract of protocatechuic acid. The extract was analyzed after being freeze-dried at –60°C showing enhanced phenolic compounds' quantity. After freeze-drying, the concentrations of caffeic acid (0.02 mg/g of potato peel extract), gallic acid (4 mg/g of potato peel extract), chlorogenic acid (4.83 mg/g of potato peel extract), and protocatechuic acid (0.75 mg/g of potato peel extract) were observed [25]. As mentioned earlier, the amount of phenolic and flavonoid compounds also depends on solvents. Phenolic compounds extracted using methanol solvent have the highest yield with a TPC of 2.91 mg gallic acid equivalent (GAE)/g of potato peel waste dry weight. After that, ethanol has 101.5 mg GAE/g of DW, acetone has 45.37 mg GAE/g of DW, diethyl ether has 42.50 mg GAE/g of DW, and hexane has 16.83 mg GAE/g of DW of phenolic content. Like phenolic compounds, the methanol-based extract has the highest yield of flavonoid compounds (0.96 mg quercetin equivalent (QE)/g of the dry weight of potato peel waste) from potato peel waste. After that, ethanol has 0.81 mg QE/g of DW of potato peel waste, acetone has 0.65 mg QE/g of DW of potato peel waste, hexane has 0.56 mg QE/g of DW of potato peel waste, and diethyl ether has 0.72 mg QE/g of DW of potato peel waste of flavonoids compounds. These phenolic and flavonoid compounds are responsible for the antioxidant activities of potato peel waste extract [26].

8.2.1.4 Polyphenolic Compounds from Pineapple Peel

Pineapple fruits are mainly processed as juice, jam, and concentrate, and high production and demand for pineapple-processed items (mainly juice) lead to massive peel generation. These peels are rich in polyphenols and contain almost twice the polyphenols compared to the edible part of the fruit. High-performance liquid chromatography (HPLC) is a widely used method for quantifying and separating phenolic compounds from PPs. Ethanol, water, and ethyl acetate are solvents used to characterize polyphenols from the HPLC technique. The ethanol-based PP extract (PPE) contains gallic acid (0.147 μg/mg of dried PP), catechol (0.026 μg/mg of dried PP), caffeic acid (0.068 μg/mg of dried PP), syringic acid (1.0 *μg/* mg of dried PP), p-coumaric acid (0.53 μg/mg of dried PP), ferulic acid (0.158 μg/mg of dried PP), ellagic acid (1.82 μg/mg of dried PP), myricetin (0.0034 μg/mg of dried PP), cinnamic acid (0.229 μg/mg of dried PP), quercetin (4.18 μg/mg of dried PP), kaempferol (4.43 μg/mg of dried PP), and apigenin (4.9 μg/mg of dried PP). The water-based aqueous extract contains gallic acid (15.99 μg/mg of dried PP), catechol (0.026 μg/mg of dried PP), chlorogenic acid (0.035 μg/mg of dried PP), syringic acid (0.007 μg/mg of dried PP), p-coumaric acid (1.613 μg/mg of dried PP), ferulic acid (0.196 μg/mg of dried PP), ellagic acid (10.17 μg/mg of dried PP), cinnamic acid (2.145 μg/mg of dried PP), quercetin (5.08 μg/mg of dried PP), kaempferol (2.50 μg/mg of dried PP), and apigenin (1.50 μg/mg of dried PP). These polyphenols have antimicrobial, antioxidant, and anti-allergenic (physiological) properties [27].

8.2.1.5 Polyphenolic Compounds from Mango Kernel

Mango is a rich source of nutrition, has attractive sensory quality, and is an important tropical fruit worldwide. In 2017, the total global production of mango was 50.65 million tonnes. Mango processing and juice industries consume most of the mango and generate a high volume of waste as peel and kernel. Mango seed kernels are rich sources of phenolic compounds such as ellagic acid, gallates, and gallic acid. A study has shown that mango kernel contains 4.6 times higher phenolic compounds than mango pulp by using the HPLC technique. Results showed that mango kernel's ethanol-based extract had mangiferin (17.21 mg/g kernel), chlorogenic acid (6.88 mg/g kernel), myricetin (4.92 mg/g kernel), ferulic acid (4.49 mg/g kernel), quercetin (3.87 mg/g kernel), rutin (2.95 mg/g kernel), caffeic acid (2.04 mg/g kernel), ellagic acid (0.77 mg/g kernel), syringic acid (0.63 mg/g kernel), gallic acid (0.61 mg/g kernel), p-hydroxybenzoic acid (0.54 mg/g kernel), and apigenin (0.53 mg/g kernel) [28].

8.2.1.6 Polyphenolic Compounds from Grape

Grape is the most widely grown fruit crop worldwide with an annual production of more than 70 million tonnes. The primary consumer of the crop is the wine production industry, which utilizes almost 50% of the total production. These industries extract wine from grapes; the remaining grape pomace is considered agricultural waste, which is around 15 to 20% of the grape weight. Due to poor wine extraction techniques, the pomace residue contains many bioactive compounds and polyphenols [29]. The previous study has shown that aqueous extract of grape pomace contains gallic acid-glucoside (2.22 mg catechin equivalent (Eq.)/100 g pomace), gallic acid (2.71 mg catechin Eq./100 g pomace), procyanidin trimer C2 (16.54 mg catechin Eq./100 g pomace), procyanidin dimer (7.79 mg catechin Eq./100 g pomace), and malvidin (0.76 mg catechin Eq./100 g pomace) [30]. They have reported that gallic acid, gallic acid–3-glucopyranosides, gallic acid–4-glucopyranosides, trans-caftaric acid, trans- and cis-coutaric acid are phenolic acids; 3-glucuronide, 3-galactoside, kaempferol 3-glucoside, quercetin 3-glucoside, astilbin, eucryphia, and engeletin are flavonoids present in the grapefruit pomace. These polyphenols and flavonoids are directly associated with health benefits [31].

8.2.1.7 Polyphenolic Compounds from Pomegranate Peel

Pomegranate fruits are eaten fresh or processed for the wine and juice industries, and peels of the fruit are the byproduct. This peel constitutes around 40% of the weight of the fruits, and they contain a higher amount of active ingredients than pomegranate pulp. Previous study has shown that the pressurized water extraction (102.1-atm pressure) process yields more polyphenols and flavonoids than the solvent-based (methanol and water) extraction method. Pressurized water extraction has total phenolics (264 mg TAE/g of pomegranate peel), total flavonoids (13 mg CE/g of pomegranate peel), hydrolyzable tannins (262 mg TAE/g of pomegranate peel), and condensed tannins (9.5 mg CE/g of pomegranate peel); however, methanol-based extraction of pomegranate peel gives total phenolics (258 mg TAE/g of pomegranate peel), total flavonoids (18 mg CE/g of pomegranate peel), hydrolyzable tannins (260 mg TAE/g of pomegranate peel), and condensed tannins (9.2 mg CE/g of pomegranate peel). The water-based extraction of peel waste has the lowest yield of total phenolics (80 mg TAE/g of pomegranate peel), total flavonoids (6.2 mg CE/g of pomegranate peel), hydrolyzable tannins (82.6 mg TAE/g of pomegranate peel), and condensed tannins (3.7 mg CE/g of pomegranate peel) [32]. Another research has shown that pomegranate peel waste extracted by water/methanol (1:4 v/v) solvent mixture gives total phenols (216.9 mg GAE/g of peel waste), ellagic acid (0.117 mg/g of peel waste), and gallic acid (0.030 mg/g of peel waste) [33]. Nasr *et al.* (1996) investigated phenolic compounds in Lintong jingpitian pomegranate peel waste by HPLC technique and observed that gallic acid (2.59 mg/g of peel), punicalagin (64.98 mg/g of peel), catechin (12.65 mg/g of peel), chlorogenic acid (0.35 mg/g of peel), caffeic acid (0.02 mg/g of peel), epicatechin (0.94 mg/g of peel), rutin (0.36 mg/g of peel), and ellagic acid (2.83 mg/g of peel) were present in pomegranate peel [34].

8.2.1.8 Polyphenolic Compounds from Banana Peel

Banana peel is a byproduct of the banana processing industry, which contains bioactive phytochemicals such as carotenoids and polyphenols. The amount of polyphenols extracted from banana peels depends on the ripening stage of the fruit and the type of solvents. Sundaram *et al.* (2011) investigated the effect of solvents and the ripening stage on the yield of polyphenols. When the unripe banana peel was extracted using acetone/water (7:3), it showed the highest TPC of 1743 mg GAE/100 g compared to 1091 mg GAE/100 g in ripe banana peel, whereas for leaky ripe banana peel, it further reduces to 844 mg GAE/100 g. When the unripe, ripe, and leaky ripe banana peel was extracted using water solvent, 741, 371, and 155 mg GAE/100 g TPC were obtained, respectively, whereas with ethyl acetate, the yield was 992, 630, and 583 mg GAE/100 g, respectively [35]. Another research has shown that increasing the amount of ethanol as a solvent enhances the TPC for up to 50% ethanol, and TPC was observed to be maximum (31 mg GAE/g), while further increase in ethanol concentration reduces the yield of TPCs [36].

8.2.1.9 Polyphenolic Compounds from Corncob

Sweet corn is a widely cultivated whole grain in America and Asia. Consumption of these grains prevents chronic diseases such as Alzheimer's disease, obesity, and type II diabetes. Corn bran, corn residue, immature corn silks, and corncobs are various parts of corn. Corncobs are the waste of agricultural production and processing and are rich in phytochemicals responsible for immune-enhancing ability and antioxidant activities. Corncob contains various phytochemicals in free and bound forms. Gallic acid in corncob is mainly in free form (81 μg/100 g of fresh corncob) rather than bound form (28.3 μg/100 g of fresh corncob). Corncob also contains p-coumaric acid–free form (15.53 μg/100 g of fresh corncob) and bound form (917.5 μg/100 g of fresh corncob). 4-Hydroxycoumarin is a phytochemical only available in free form (32.63 μg/100 g of fresh corncob). Ferulic acid is majorly present in bound form (1121 μg/100 g of fresh corncob) compared to free form (43.24 μg/100 g of fresh corncob). Fresh corncob quercitrin (90.96 μg/100 g) is available in bound form only, but kaempferol is present only in free form (123.8 μg/100 g of fresh corncob) [37]. Another study has shown that the crop's location also affects the corncob's total polyphenolic content. Corncobs from different places (Turda 20A, Turda 200, Turda Star, Turda Favorit, Florencia, and Fundulea) were analyzed for their TPCs. Turda 200 contains the highest phenolic content (703.98 mg GAE/100 g dry weight) followed by Turda Star (545.80 mg GAE/100 g dry weight), Turda (20A 530.25 mg GAE/100 g dry weight), Turda Favorit (524.61 mg GAE/100 g dry weight), Florencia (487.00 mg GAE/100 g dry weight), and Fundulea (469.79 mg GAE/100 g dry weight) [38].

8.2.1.10 Polyphenolic Compounds from Wheat Straw

Worldwide, wheat production is more than 652 million tonnes annually. After chaff and grain have been separated from the wheat plant, the dry stalk of wheat as a byproduct is known as wheat straw, which contains phenolic acid, polyphenols, and flavonoids. Zhao *et al.* (2021) performed an HPLC analysis of wheat straw and extracted active compounds present as endo-phenolic compounds (IPC) and exo-phenolic compounds (EPC). Ethyl acetate soluble exo-phenolic compounds extract (EA-EPC) analysis showed that wheat straw has gallic acid (32.56 mg/g), ferulic acid (50.88 mg/g), epicatechin-3-gallate (132.69 mg/g), epigallocatechin-3-gallate (142.30 mg/g), naringin (126.01 mg/g), rutin (101.50 mg/g), phelligridin G (36.32 mg/g), inoscavin B (40.01 mg/g), and davallialectone (35.67 mg/g). N-butanol soluble exo-phenolic compounds extract (NB-EPC) analysis obtained gallic acid (52.46 mg/g), ferulic acid (29.85 mg/g), epicatechin-3-gallate (144.31 mg/g), epigallocatechin-3-gallate (110.55 mg/g), naringin (132.88 mg/g), rutin (132.37 mg/g), naringenin (0.21 mg/g), phelligridin G

(31.55 mg/g), inoscavin B (38.26 mg/g), and davallialectone (39.46 mg/g). Furthermore, endo-phenolic compounds extract has shown that wheat straw has gallic acid (93.41 mg/g), ferulic acid (62.44 mg/g), epicatechin-3-gallate (116.38 mg/g), epigallocatechin-3-gallate (124.72 mg/g), naringin (109.27 mg/g), rutin (91.66 mg/g), naringenin (1.23 mg/g), phelligridin G (39.35 mg/g), inoscavin B (19.21 mg/g), and davallialectone (21.90 mg/g) [39].

8.2.2 Antioxidants from Agro-Waste

Oxidation affects the nutritional and sensorial attributes of different food products, such as cut vegetables and fruits, meats, and fish. Incorporating active antioxidant materials in packaging reduces packaged food's oxidation rate, which is one of the main reasons for food spoilage. An active antioxidant packaging system has an advantage over directly adding antioxidant agents to the food, as active agents are released in a controlled manner when they are incorporated into the polymer matrix. The controlled release of antioxidant agents will maintain the concentrations of active agents for a much longer period of time, and it will extend the shelf life of food by reducing spoilage due to oxidation. Active antioxidant agents can be extracted from different agro-waste (peels, seeds, and pomace) in the form of powder, liquid, or essential oil. In order to prepare antioxidant films, these agents are mixed with a polymer matrix during the film-forming process. This section discusses various agro-waste materials containing antioxidant properties [40].

8.2.2.1 Antioxidant Compounds from Apple Peel

Apple is one of the rich antioxidant sources effective in preventing several chronic diseases. In canned apple manufacturing lines, apple peels are considered food processing waste with high antioxidant activities. Previous research has shown that apple peel contains higher antioxidant activities compared to apple flesh. A detailed study has been conducted on the antioxidant activities of peel, flesh, and peel with flesh for Idared, Cortland, Rome Beauty, and Golden Delicious apples. Among all these species, the Idared apple peel possessed the highest antioxidant activity of 312 μ of vitamin C equivalents (Eq.)/g of peel, followed by Rome Beauty with 228 μmol of vitamin C Eq./g of peel, Cortland with 159 μmol of vitamin C Eq./g of peel, and Golden Delicious with 111 μmol of vitamin C Eq./g of peel antioxidant activities. This research emphasizes that apple peel is best suitable for active packaging applications.

The antioxidant activities of Rome Beauty apple's flesh combined with peel were the highest (131 μmol of vitamin C Eq./g of flesh with peel), followed by Cortland (83 μmol of vitamin C Eq./g of peel), Idared (72 μmol of vitamin C Eq./g of peel) and Golden Delicious (67 μmol of vitamin C Eq./g of peel) [21].

8.2.2.2 *Antioxidant Compounds from Citrus Fruit*

Citrus peels and seeds are a large number of byproducts and waste produced by the food processing industry. This waste can be valorized into wealth as peels and seeds are rich sources of antioxidants. More than 170 antioxidants, such as phenolic compounds, terpenoids, pectin, mineral elements, and vitamins have been reported in the literature. Citrus peel is one of the significant sources of vitamin C, which shows free radical scavenging activities and breaks the oxidative chain reaction. Vitamin E, which has tocopherols and tocotrienols, also prevents lipid oxidation. Dried peel of candied orange has 5.60 mg/kg, tangerine peel has 4.50 mg/kg, and lemon peel has 11.40 mg/kg of vitamin E. Carotenoids are tetraterpenoids compounds, and they exhibit antioxidant activities by quenching oxygen radicals and eliminating them. Ponkan peel has 2.04 mg/g, wendun peel has 0.36 mg/g, and peiyou has 0.021 mg/g dried base of carotenoids [41].

8.2.2.3 *Antioxidant Compounds from Potato Peel*

Mainly, phenolic flavonoids, tannins, curcuminoids, terpenoids, and coumarins are responsible for the antioxidant activities of this agricultural waste. These active ingredients are extracted from potato peel and further used as antioxidant agents for food preservation. The water-based extract (WE) and acidified ethanol-based extract (AEE) of potato peel have shown antioxidant activities, and they were able to neutralize the DPPH, ABTS, and peroxyl radicals. The free radical scavenging capacity of AEE was significantly higher than WE for ABTS and DPPH methods. The antioxidant activity of AEE for ABTS was 1.40 mmol TE/g, for the DPPH was 0.38 mmol TE/g, and for the ORAC assay was 4.0 mmol TE/g. Water-based potato peel extract has antioxidant activities of 5.31 mmol TE/g, 0.51 mmol TE/g, and 0.10 mmol TE/g, respectively, for peroxyl radicals, ABTS, and DPPH methods. The yield of phenolic compounds from acidified ethanol-based potato peel extract was much higher than that from WE; higher phenolic compounds lead to higher antioxidant activities [25].

8.2.2.4 *Antioxidant Compounds from Pineapple Peel*

PPs are considered as waste of pineapple juice and jam processing industries, and they contain huge amounts of flavonoids and polyphenols responsible for antioxidant activities. Antioxidant activities of PPE can be measured by phosphorus-molybdenum and DPPH radical scavenging methods. PP has an inhibition concentration (IC_{50}) value of 1.13 mg/ml, higher than that of red grape pomace (0.039 mg/ml). One gram of PPE showed a total antioxidant capacity of 0.037 g ascorbic acid [42]. Other research has shown that fresh PP waste (PPW), extracted with 95% ethanol, has 0.6 mg/ml of antioxidant activity (IC_{50}), and dried PPW, extracted with 95% ethanol, has 1.2 mg/ml of antioxidant activity (IC_{50}) [43].

8.2.2.5 *Antioxidant Compounds from Mango Kernel*

Mango kernel shows antioxidant activities due to rich polyphenols and flavonoid contents. Research has been conducted to evaluate the antioxidant activities of mango kernel extract under various conditions. Mango kernel extraction was performed with water-ethanol (20/80 v/v) by varying amounts of kernel powder and sonication parameters. Sonication was performed at 20 kHz with 50% amplitude at varying times (10, 20, and 30 min). The kernel powder to solvent ratio was taken as 1/5, 1/10, and 1/20 g/ml. DPPH radical scavenging antioxidant activity of mango kernel extract with a solid-to-liquid ratio of 1/20 g/ml was the highest (2756 μmol Trolox equivalents (TE)/g). Comparing the effect of sonication time on the number of antioxidant activities, extract with 30 min of extraction time has the highest DPPH radical scavenging activities of 2226 μmol TE/g, whereas 10 min of extraction time has the lowest (1808 μmol TE/g) antioxidant activities [28].

8.2.2.6 *Antioxidant Compounds from Grape Pomace*

Niagara Rosada grape pomace has shown a DPPH antioxidant activity of 255 mg TEAC/kg, ABTS activity of 250 mg TEAC/kg, and FRAP activity of 109 mM Fe/kg, whereas Maximo grape pomace has a DPPH antioxidant activity of 343 mg TEAC/kg, ABTS activity of 401 mg TEAC/kg, and FRAP activity of 159 mM Fe/kg. However, Violeta grape pomace has shown the highest DPPH antioxidant activity of 358 mg TEAC/kg, ABTS activity of 470 mg TEAC/kg, and FRAP activity of 180 mM Fe/kg. Another research has shown that malbec grape pomace has ORAC assay antioxidant activities of 2756 μmol TE/g grape pomace extract [31, 44].

8.2.2.7 *Antioxidant Compounds from Pomegranate Peel*

Ganesh variety of Indian pomegranate peels was tested for DPPH and ABTS radical scavenging methods. Extraction of pomegranate peels with methanol has the highest DPPH antioxidant properties (around 80%), followed by 77% in the WE showed around, and then about 75% in ethanol/water (3:7), 70% in ethanol/water (7:3), and 67% in ethanolic extract. Ethanol/water (1:1) has shown significantly lower DPPH radical scavenging activities of around 62% [45]. Natanz peel has antioxidant activities of 45%, higher than its seed (34%) and juice (10%). Shahreza peel has shown antioxidant activities of 50%, more than its seed (26%) and juice (9%). Doorak peel has antioxidant activities of 58%, higher than its seed (54%) and juice (9%) [46]. These results show that pomegranate peel, a food processing waste, has higher potential benefits for human health [47, 48].

8.2.2.8 *Antioxidant Compounds from Banana Peel*

Banana peel, a food processing waste rich in polyphenols, also exhibits antioxidant properties. Sundaram *et al.* (2011) studied the effects of various solvents (acetone, water, chloroform, ethyl acetate, and hexane) on the antioxidant activities of unripe, ripe, and leaky banana peel [35]. Unripe banana peel has shown the highest DPPH radical scavenging properties irrespective of solvents, followed by ripe and leaky ripe banana peel. Unripe banana peel extracted with ethyl acetate has nearly 70% DPPH scavenging activity, followed by water (63%), acetone (61%), and chloroform (26%) extracts. Ripe banana peel, extracted with ethyl acetate, has nearly 40% DPPH scavenging activity, followed by acetone (35%), water (14%), and chloroform (11%) extracts. Leaky ripe banana peel, extracted with ethyl acetate, has nearly 23% DPPH scavenging activity, followed by acetone (22%), water (10%), and chloroform (8%) extracts. Another research has shown that extraction based on ethanol (75%) has a DPPH antioxidant activity of 80.1% and FRAP capacity of nearly 27% in the sonication technique. In the maceration technique, ethanol (75%) banana peel extract showed 77.78% DPPH scavenging activity and 25.3% FRAP capacity [36].

8.2.2.9 *Antioxidant Compounds from Corncob*

Various parts of sweet corn contain high polyphenolic agents and show antioxidant activities. Ioana *et al.* (2014) investigated the % DPPH radical scavenging capacity and FRAP of corncobs (Turda 20A, Turda 200, Turda

Star, Turda Favorit, Florencia, and Fundulea). Among all the corncobs, Fundulea has shown the highest DPPH activity of around 31%, and Turda 20 around 18%. Turda Favorit exhibited nearly 27%, followed by Turda 200 (26%), Florencia (24%), and Turda Star (22%). According to FRAP assay analysis, Turda 200 has shown the highest nearly 18 mmol TE/g reducing power, followed by Turda 20A (~16 mmol TE/g), Turda Star (14 mmol TE/g), and Turda Favorit (13 mmol TE/g). Florencia and Fundulea have shown almost equal FRAP capacity, around 12 mmol TE/g [38].

8.2.2.10 Antioxidant Compounds from Wheat Straw

LCCs extracted from wheat straw have intense antioxidant activities against reactive oxygen species. Water soluble LCC-leaf, LCC-stalk, and LCC-sheath were isolated from wheat straw by ball milling and further dissolution in 8% lithium chloride (LiCl)/dimethyl sulfoxide (DMSO) solvent. FRAP assay analysis was carried out by measuring the reducing capacity of ferric ion conversion. Increasing the concentration of LCC increases the absorbance at 700 nm. The 8 mg/ml concentration reaches the maximum for the leaf, stalk, and sheath. LCC- sheath has gained maximum absorbance (0.66), followed by LCC-stalk (0.52) and LCC-leaf (0.44). Similar to FRAP analysis, % absorbance in DPPH analysis also increases with concentrations of LCC. At 2 mg/ml concentration, LCC-sheath has shown the highest (74.91%) DPPH radical scavenging activities. After that, LCC-stalk has 74.55% scavenging activities, and LCC-leaf has the lowest (64.52%) scavenging activities. These results show that leaf, stalk, and sheath extracted from LCCs contain high antioxidant activities [49].

8.2.3 Antimicrobials Compounds from Agro-Waste

New technology regarding food preservation reduces food wastage and provides safe food products for consumption. Active antimicrobial packaging is emerging as a most promising technology, which controls food decay and retards the growth of microorganisms. Various potential active agents having intense antimicrobial activities are being extracted from agro-waste products such as seeds, pomace, and peel. Antimicrobial agents are incorporated in packaging material, and their release in a controlled manner to the packaged food products inhibits the growth of fungi and bacteria. In this section, we have discussed various antimicrobial agents obtained from different agro-wastes for food packaging applications [50–52].

8.2.3.1 Antimicrobials Compounds from Pomegranate Peel

Consumers, food safety authorities, and food industries are concerned about foodborne illnesses, and the use of synthetic preservatives in food is always a matter of concern for consumers. Few antibiotic-resistant pathogens like methicillin-resistant *Staphylococcus aureus* have increased attention on the search for plant-, herb-, and spice-based antimicrobial compounds. These herbs, spices, and plants contain several polyphenols, flavonoids, and tannins responsible for antimicrobial activities against a wide range of microorganisms. Methanolic extract (methanol and water, 4:1) of pomegranate peel was tested against 11 microorganisms by *in vitro* agar-diffusion method, and the results showed the highest zone of inhibition (ZOI) against *Listeria monocytogenes* ATCC (Z0I-20 mm) and the lowest for *Aspergillus niger* (ZOI-12 mm). Other than these microorganisms, the growth of *Yersinia enterocolitica* (Z0I-19 mm), *Pseudomonas aeruginosa* (Z0I-18 mm), *Candida utilis* (Z0I-18 mm), *Bacillus subtilis* (Z0I-17 mm), *Staphylococcus aureus* (MRSA) (Z0I-16 mm), *Escherichia coli* (Z0I-16 mm), *Klebsiella pneumoniae* (Z0I-16 mm), *Saccharomyces cerevisiae* (Z0I-14 mm), and *Staphylococcus aureus* ATCC 6538 (Z0I-13 mm) also has been inhibited by pomegranate peel extract. The methanol-water–based pomegranate peel extract contains tannins, flavonoids, glycosides, and total phenolic compounds (262.5 mg GAE/g) [53]. Another study has shown the effect of various solvents (methanol, ethanol, and water) on the growth of certain pathogenic strains. For *Staphylococcus aureus*, methanol-based pomegranate peel extract has shown the highest ZOI of 24.5 mm, followed by WE (ZOI-22.6 mm) and ethanol extract (ZOI-20.3 mm). The WE has the highest ZOI (ZOI-18.4 mm) for *Enterobacter aerogenes*; methanol extract has 18.2 mm, and ethanol-based extract has 17.3 mm. *Salmonella typhi* pathogen has minimum growth in the WE, and hence, the ZOI was maximum (ZOI-28 mm), methanol extract had a ZOI of 18.3 mm, and ethanol-based extract had a ZOI of 15.1 mm. *Klebsiella pneumoniae* pathogenic strain has the highest ZOI for ethanol-based extract (ZOI-12.4 mm), followed by methanol extract (ZOI-11.3 mm) and water extract (ZOI-11.2 mm) [45].

8.2.3.2 Antimicrobials Compounds from Grape Pomace

Grape is most widely cultivated for wine processing industries, and after processing, a large amount of grape residue in the form of pomace (mixture of seeds and skin) remains unutilized. The grape residue is rich in polyphenolic compounds and flavonoids and exhibits strong antioxidant, antiradical, anticarcinogenic, and antimicrobial activities. Grape pomace of Pinot Meunier (PM) and Pinot Noir (PN) varieties were extracted using methanol/water (1:1 v/v), ethanol/water (1:1 v/v), and acetone/water (1:1 v/v) solvents at 25°C and were tested against *Candida albicans* (fungi), *Staphylococcus aureus* (Gram-positive bacteria), and *Escherichia coli* (Gram-negative bacteria) [54].

8.2.3.3 Antimicrobials Compounds from Mango Kernel

Mango is one of the most popular fruits cultivated in the subtropical and tropical regions of the world. Mango seed kernel is a promising source of active polyphenols and shows antioxidant and antimicrobial activities. Kernel extract has shown remarkable antimicrobial activity against *S. aureus* (types B, C, D, E, TSST-1, and MRSA) with a MIC value of 50 ppm. *Clostridium botulinum* and *Cl. perfringens* strains have MIC values less than 50 ppm. Furthermore, *Yersinia enterocolitica* and *Campylobacter jejuni* have shown MIC values of 100 ppm, whereas *Bacillus cereus, Bacillus subtilis, Bacillus licheniformis*, and *Bacillus pumilus* have MIC values of 500 ppm [55]. Researchers have also investigated the antimicrobial activities of methanol-based (2:1 v/w) mango kernel extracts and reported the zones of inhibition for *E. coli* (18, 19, and 21 mm for 3000, 4500, and 6000 ppm) [56].

8.2.3.4 Antimicrobials Compounds from Citrus Fruit

Citrus peels are the byproduct of juice industries and are considered waste. However, the crude extract of citrus peel has limonoid and flavonoid compounds responsible for antiviral, antifungal, antibacterial, and anticancer activities [57]. Researchers have investigated the aqueous peel extract of *Citrus limetta* and *Citrus lemon* against several Gram-positive and Gram-negative bacteria. The zones of inhibition for *S. aureus* bacteria were 30 and 20 mm for *Citrus limon* and *Citrus limetta* extracts, respectively. For *M. catarrhalis*, the zones of inhibition were 30 and 20 mm, for *Citrus lemon* and *Citrus limetta* peel extracts, respectively [58]. Lemon peel extracted showed an average inhibition zone of 18.77 mm against *E. coli*, the highest among other peel extract samples [59]. Essential oil from a citrus plant has shown particular interest as an antimicrobial agent for food preservation. Previous research shows that lemon peel essential oil exhibits antimicrobial activities against *E. coli* and *S. aureus* microorganisms. Lemon peel oil (1000 μl/ml) has a strong ZOI, 8 mm for *S. aureus* and 10 mm for E. coli, as evaluated by the disc diffusion method [60].

8.2.3.5 Antimicrobials Compounds from Banana Peel

Bacillus cereus, Salmonella enteritidis, Escherichia coli, Bacillus subtilis, and *Staphylococcus aureus* bacteria were tested for yellow and green banana peel extracted with the help of ethyl acetate. Green peel extract has shown the highest antimicrobial activity against *Staphylococcus aureus* (12 mm), followed by *Bacillus cereus* (11 mm), *Bacillus subtilis* (10 mm), *Salmonella enteritidis* (10 mm), and *Escherichia coli* (9 mm). For yellow peel extract, *Staphylococcus aureus* and *Bacillus subtilis* have shown maximum inhibition zones of 3 mm for each. An inhibition zone of 2 mm was observed for both *Bacillus cereus* and *Escherichia coli,* whereas *Salmonella enteritidis* had the lowest ZOI (1 mm) [61]. In another study, the antimicrobial properties of banana peel were investigated against *Escherichia coli, Pseudomonas aeruginosa, Staphylococcus aureus,* and *Saccharomyces cerevisiae* by measuring disc diffusion assay [62].

8.2.3.6 Antimicrobials Compounds from Pineapple Peel

Researchers have evaluated the antimicrobial activities of PP (extracted using ethanol and water) against foodborne pathogens such as *Escherichia coli, Staphylococcus aureus, Klebsiella pneumonia,* and *Bacillus cereus.* For water-based peel extract, the highest ZOI of 14 mm was observed for *Bacillus cereus, Klebsiella pneumonia,* and *Staphylococcus aureus,* whereas for *Escherichia coli,* it was 12 mm. For ethanolic extract, *Bacillus cereus* had a maximum inhibition zone of 15 mm, followed by *Staphylococcus aureus, Escherichia coli,* and *Klebsiella pneumonia* with an inhibition zone of 14, 12.3, and 12 mm, respectively [63].

8.2.4 Biobased Indicators

Biobased indicators have emerged as a viable alternative for traditional synthetic materials and their application as indicators or sensors in food packaging applications. Natural dyes and extracts that can change color due to physiological variations in food materials could be used as indicators or sensors instead of synthetic ones. The pigments are usually integrated with and exhibit color variation in different conditions, such as low or high pH or CO_2 or O_2 concentration inside the package. Thus, natural pigments may be a less toxic and more environmentally friendly alternative to synthetic pigments [64]. At present, the European Union Policies make it compulsory for intelligent packaging applications, the material should be free from any toxins or allergens and thus the material should be derived from natural sources.

Freshness indicators in the intelligent packaging provide information about the quality status of the packed food in real-time monitoring. The indicators are usually based on the visual colorimetric change, which could be the reaction of a chemical compound or microbiological variation in the food product and react with the sensor informing changes in the product. The freshness indicator or sensor can be a label or visual sticker on the package [65]. Food degradation caused by microbial growth produces metabolites that can be identified by indicators present on the packaging. The unwanted metabolites such as nitrogen compounds, carbon dioxide, biogenic amine, or sulfur compounds present or formed in the packed food can be determined or identified. Meat freshness is lost due to microbial growth resulting in the formation of various substances such as acetic acid, n-butyrate, and biogenic amines. These primary and secondary metabolites can be detected by the indicator and send information to the producer and consumers for a better decision-making process [66].

A biobased intelligent film to monitor the food condition was developed with immobilized black rice bran anthocyanins (BRACNs) into a chitin nanocrystal/chitosan (CS) matrix, and significant color variation under visible and ultraviolet spectra has shown from red to grayish in the 2.0 to 12.0 pH range. The indicator was applied to monitor packaged fish and shrimp by visible color changes. The BRANCs with 3% effectively observed the color change in the fish and shrimp during storage. In particular, the color of BRACN solutions changed from red to pink, purple, and grayish green as the pH of the solution increased to 2.0 to 3.0, 4.0 to 5.0, 6.0 to 10.0, and 11.0 to 12.0, respectively [67]. A similar study was reported by Sun *et al.* (2020), where they developed a multifunctional nanocomposite with konjac glucomannan (KGM)/CS incorporated with ZnO nanoparticles. The anthocyanin was obtained from the mulberry extract, and the film added with anthocyanin was applied as an indicator in different pH conditions from 2.0 to 12.0, with color variation from red to blue, which can be differentiated with the naked eye [68]. A bilayer colorimetric film of gelatin (GN) added with ZnO as the upper layer (GN-ZnO) and gellan gum (GG) added with mulberry anthocyanin (MA) as the lower internal layer was fabricated with the solution casting method. The developed nanocomposite had good pH sensitivity due to the NH compound. This film was applied to detect the spoilage of the fish under the lower temperature of 4°C at 75 % RH [69]. The freshness indicator film was developed using starch, carbon dots, and anthocyanin to monitor the freshness of the pork meat at 25°C and showed the color variation from purple to green duration storage [70].

8.3 Active and Intelligent Agents in Biopolymer-Based Food Packaging

Active and intelligent packaging are new technologies in the world of food packaging. It has the potential to improve food traceability, safety, and quality significantly. Intelligent packaging is defined as the science and technology of introducing communication tools into food packaging systems to monitor changes in environmental conditions inside and outside the system and provide decision-making ability to producers and consumers. There are three major categories of intelligent packaging, which comprise sensors and indicators, as shown in Table 8.1. Active packaging involves creating and modifying the atmosphere so the condition inside the packaging will be favorable for food quality and safety. Active packaging system contains a compound that can emit or absorb the desirable and undesirable gasses, respectively, from the package headspace. Intelligent packaging and active packaging are not mutually exclusive, i.e., both can simultaneously work together to provide the utmost protection and convenience to the consumers and are thus called a smart packaging system.

Food quality indicators provide information about the freshness of the food commodity spoiled by microbial, chemical, or enzymatic degradation. The quality of the food may be visually analyzed with the naked eye or will be rated as per the color change in the label or indicator [88]. The food quality or freshness indicator is composed of the indicator component and the packaging matrix, as shown in Figure 8.2. The change in indicator is usually correlated with the chemical or biological degradation mechanism to provide the real-time value of the packed product. The food quality indicator can be of different types, such as carbon dioxide indicator, oxygen indicator, humidity indicator, pH indicator, moisture indicator, and specific chemical indicator.

8.3.1 Oxygen and Carbon Dioxide Indicators

Many foods produce undesirable changes under physiochemical variation when exposed to a high level of oxygen and carbon dioxide. The detector monitors the level of gasses inside the package and produces visual changes in the indicator labels. A conventional oxygen indicator was designed using natural compounds such as laccase, guaiacol, and cysteine. These components are physically separated by two compartments with a fragile barrier. The PET/LLDPE-based packaging material was used to develop an indicator that can provide an indication once the barrier is broken. Depending upon

Table 8.1 Intelligent food packaging system incorporated with different agro-waste–based active agents/indicators.

Packaging matrix	Agro-waste–based active agents/ indicators	Type of intelligent packaging	Application	References
Chitosan	Blueberry and blackberry pomace	pH indicator	Food packaging	[71]
Agar	Purple sweet potato extract	pH indicator	Pork packaging	[72]
Starch/polyvinyl alcohol	Roselle anthocyanin	Freshness indicator	Fish package monitoring	[73]
Chitosan/polyvinyl alcohol	Red cabbage extract	Time-temperature indicator	Milk packaging	[74]
gelatin/chitin nanocrystals	Black rice bran extract	Freshness indicator	Fish packaging	[75]
Cellulose paper–based	Ruellia simplex flowers dye extract	Time-temperature indicator	Shrimp packaging	[76]
Chitosan	Black corn extract	Time-temperature indicator	-	[77]
Polyethylene	Natural occurring dyes	CO_2 indicator	-	[78]
Polyvinyl/gelatin	Amaranthus leaf extract	pH indicator	Chicken packaging	[79]
Chitosan	Purple tomato anthocyanin extract	Colorimetric pH indicator	Milk and fish packaging	[80]
Gelatin	Dragon fruit skin	pH-sensitive skin	-	[81]
Krasch gum	Red cabbage extract	pH and NH_3 indicator	-	[82]
Carboxymethyl-cellulose/ starch	Purple sweet potato extract	NH_3 indicator	Fish packaging	[83]
Wheat gluten	Chlorophyll extract	Colorimetric indicator for oxidation	Sesame oil packaging	[84]
High-density polyethylene	Green and blue dye	Colorimetric pH indicator	Milk packaging	[85]
Polypropylene boxes	Soybean oil and tetradecane	Temperature indicator	Beef packaging	[86]
Polyethylene terephthalate bottle	Red cabbage extract	pH-sensitive	Meat packaging	[87]

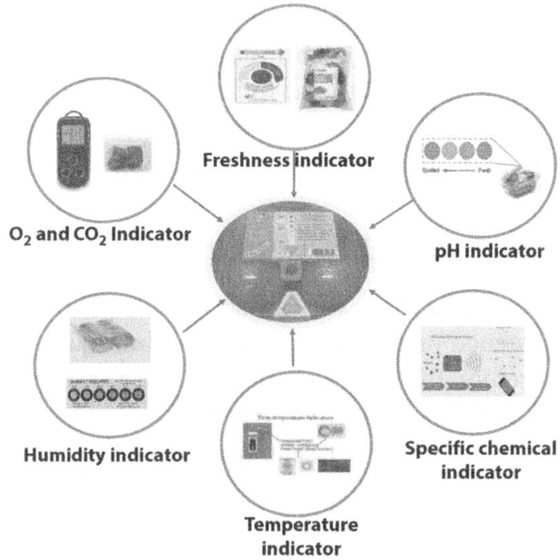

Figure 8.2 The development of different intelligent packaging in the food packaging application.

the concentration of the compound, the indicators exhibit different response times and colors [89]. The latest investigation reported that the researchers had developed a colorimetric-based oxygen indicator made of silver-loaded titanium dioxide nanotube, glycerol, methylene blue, and hydroxyethyl cellulose (HEC) that has the potential to accurately identify the oxygen concentration inside the package[90]. The indicator is composed of sodium alginate, sodium caseinate, and kappa κ-carrageenan–based sensors have been developed for the O_2 detection to monitor the O_2 level in the mozzarella cheese packaging, where they found that after 15 days of storage of mozzarella cheese under nitrogen storage, the significant color changes observed due to O_2 presence. The effectiveness of the indicator was observed at the O_2 concentration of 0.381%, and the developed indicator can potentially decrease food deterioration by informing the O_2 level inside the food package [91]. Sugarcane bagasse–based nanocellulose hydrogel was developed as a colorimetric freshness indicator to monitor the CO_2 content in packaged chicken breast. The hydrogel carries pH-responsive dyes (methyl red/bromothymol), where color changes according to pH variation. In this case, the color of the indicator transforms from green to red on the third day of storage, and the microbial population exceeds the acceptable limits for human consumption [92]. Another similar study for a CO_2-sensitive detector has been developed with black goji anthocyanin (BGA) extract to monitor the freshness of the breast chicken meat using the metal complexation method. The detector was fabricated using hydroxymethyl cellulose hydrogel with 5% BGA extract, and the result showed that the indicator changes color from red to purple and then blue after 3 days of storage due to microbial load up to 7 CFU/g [93].

8.3.2 Moisture or Humidity Indicator

Foods like bakery products are very sensitive to moisture or humidity; the products get moistened or dry due to improper humidity conditions inside the package. Therefore, the ability to monitor the moisture content of food packaging in real time is beneficial. A variety of moisture monitoring options are now available, including the measurement of variation in dielectric properties, capacitance, and crystal resonant frequency [94]. Because proteins' dielectric characteristics are humidity-sensitive, they are frequently used in the development of intelligent packaging systems for food packaging. Gelatin-based nanocomposite film with ZnO nanoparticles was prepared *via* solution casting for relative humidity monitoring. In different environments, the gelatin-based film detected relative humidity with a high sensitivity of 99.47% for the developed nanocomposite material at 25°C [95]. Another study used the eco-friendly material wheat gluten protein to create a composite material for measuring relative humidity. The dielectric permittivity and dielectric loss range between 5.01 0.04 and 9.79 0.06 and 0.39 0.01 and 1.48 0.02, respectively. The dielectric property of the packaging film is too sensitive to relative humidity, and it fluctuates significantly due to a comparable exponential increase in the relative humidity of the wheat gluten–based composite. Wheat gluten's dielectric permittivity and dielectric loss on relative humidity make it suitable for relative humidity monitoring in food [96].

8.3.3 pH Indicators

Visual pH sensors are typically made up of two major components: a solid substrate and a pH-sensitive dye, and they are divided into three major categories. The first type consists of liquid-based dye immersion, where the physical absorption occurs onto the solid support; the second type is based on the dye covalently linked to hydrophilic supports like cellulose; and the third type is physically entrapped into the polymer matrixes [97]. Generally, there are different pH-sensitive synthetic dyes like bromothymol blue, chlorophenol, methyl red, and xylenol that have been utilized to develop the pH indicator for food packaging applications. However, owing to the potential toxicity of these synthetic pigments, it is challenging to meet consumers' expectations about food safety, and thus, natural dyes obtained from plants are appealing alternatives for use as indicators in food packaging materials [98].

CS-based smart films containing blueberry and blackberry pomace extract were developed, which change color from rose to green with blueberry and from red to dark violet with blackberry due to changes in pH from 2 to 10 [99]. Another similar work was reported where the pH-sensitive color-changing film was fabricated with gelatin and carrageenan as a base matrix for food packaging applications. The root part of the gromwell plant (*Lithospermum erythrorhizon*) was utilized to extract the solution for pH sensitivity in the film. The developed intelligent film was successfully utilized to monitor the freshness of the milk that exhibited color-changing phenomena between pH 2 and 12 [100].

8.3.4 Temperature Indicator

The temperature indicator works on the temperature gradient in the product or packed product. This type of indicator is integrated with a digital display unit and is generally environmentally friendly and cost-effective. Exposure to abnormal temperature ranges above or below the critical temperature of storage for a long time can cause alteration in the organoleptic and textural changes in the foodstuffs packed, and irreversible color change occurs in the film or strip indicator [101]. The time-temperature indicator was developed using the anthocyanin extracted from the *Brassica oleracea* var. capitata (red cabbage) incorporated in the PVA/CS matrix for monitoring pasteurized milk [102]. Another time-temperature indicator consists of tryptic soy agar with glycerol fabricated for the monitoring of beef meat [103].

8.3.5 Specific Chemical Indicator

There is a variety of harmful chemical compounds produced during food processing and storage, including nitrogen-containing compounds, biogenic amines, biotoxins, and sulfur-containing chemicals. Thus, identifying the presence or absence of these compounds can indicate food quality and spoilage. Food such as meat and seafood can produce volatile compounds like trimethylamine (TMA), dimethylamine (DMA), and ammonia during storage. At present, there are three sensors for the detection and monitoring of these chemicals in the package that are being used, namely, a pH-sensitive colorimetric gas sensor, a semiconductor gas sensor, and a nanoparticle plasma sensor. The mechanism of these sensors is based on the pH alteration in the system due to the amine's alkaline nature, which led to a change in the dye color [104]. Hydrogen sulfide is the major volatile compound released during the deterioration of high-protein-containing food such as meat, eggs, and milk. Hydrogen sulfide generates a pronounced odor that most consumers do not find unacceptable. Electrochemical and metal nanoparticle plasma sensors have recently been used for the detection and monitoring of hydrogen sulfide [105]. The other toxic substances can also be detected and monitored through intelligent packaging systems like organophosphate pesticides and aflatoxin. A simple colorimetric paper–based sensor was developed using a degradable nanoenzyme, MnOH nanowire, and 3,3', 5,5'-tetramethylbenzidine (TMB) as a color-based indicator to detect rapidly and accurately values of organophosphate pesticide and acetylcholinesterase [106].

8.4 Conclusion and Perspective

Polyphenols, antioxidants, and antimicrobial agents derived from agro-waste enhance the functionality of food packaging. Smart packaging has a significant impact on food quality and control because it allows consumers to directly interpret the freshness and safety of the food packaged within it. Active packaging refers to the incorporation of additives into packaging systems with the goal of maintaining or extending the quality and shelf life of fresh vegetable or livestock products, whereas intelligent packaging systems monitor the condition of packaged foods to provide information about the quality of the packaged food during transport and storage. The investigations based on the development of active and intelligent packaging demonstrated high sensitivity to pH and ammonia, with differences in the color of the films indicating potential use as a visual indicator. The use of pigments derived from food waste also contributes to the recovery of waste food materials. These materials are suitable for use as food packaging due to their easy availability and nontoxicity. These wrapping films also have the added benefit of attracting customers with their vibrant colors. Smart packaging systems containing naturally derived pigments from food and food waste thus provide a simple, easy-to-produce, safe, and cost-effective solution for continuous monitoring of fresh perishable food with real-time commercialization potential. Biobased safe, value-added, green materials have promising properties such as biodegradability, renewability, lower production costs, waste reduction, and reduced environmental impact. As a result, future research should concentrate on treatments or additives that improve the thermal resistance, barrier properties, and mechanical properties of related biomaterials.

References

1. Obi, F., Ugwuishiwu, B., Nwakaire, J., Agricultural waste concept, generation, utilization and management. *Niger. J. Technol.*, 35, 957, 2016, https://doi. org/10.4314/njt.v35i4.34.
2. Kauldhar, B.S. and Yadav, S.K., Turning waste to wealth: A direct process for recovery of nano-silica and lignin from paddy straw agro-waste. *J. Clean. Prod.*, 194, 158–166, 2018, https://doi.org/10.1016/j.jclepro.2018.05.136.
3. Basri, M.S.M., Shah, N.N.A.K., Sulaiman, A. *et al.*, Progress in the valorization of fruit and vegetable wastes: Active packaging, biocomposites, by-products, and innovative technologies used for bioactive compound extraction. *Polym. (Basel)*, 13, 3503, 2021, https://doi.org/10.3390/polym13203503.
4. Baig, M.B., Al-Zahrani, K.H., Schneider, F. *et al.*, Food waste posing a serious threat to sustainability in the Kingdom of Saudi Arabia – A systematic review. *Saudi J. Biol. Sci.*, 26, 1743–1752, 2019, https://doi.org/10.1016/J.SJBS.2018.06.004.
5. Deshmukh, R.K., Kumar, P., Tanwar, R., Gaikwad, K.K., Pectin-polyvinylpyrrolidone based antimicrobial and antioxidant nanocomposite film impregnated with Titania nanoparticles and Bael shell extract. *Food Bioprocess Techmol.*, 15, 12, 1–15, 2022.
6. Kumar, L., Deshmukh, R.K., Gaikwad, K.K., Antimicrobial packaging film from cactus (Cylindropuntia fulgida) mucilage and gelatine. *Int. J. Biol. Macromol.*, 215, 596–605, 2022, https://doi.org/10.1016/J.IJBIOMAC.2022.06.162.
7. Tanwar, R., Gupta, V., Kumar, P. *et al.*, Development and characterization of PVA-starch incorporated with coconut shell extract and sepiolite clay as an antioxidant film for active food packaging applications. *Int. J. Biol. Macromol.*, 185, 451–461, 2021, https://doi.org/10.1016/j.ijbiomac.2021.06.179.
8. Kumar, P., Tanwar, R., Gupta, V. *et al.*, Pineapple peel extract incorporated poly(vinyl alcohol)-corn starch film for active food packaging: Preparation, characterization and antioxidant activity. *Int. J. Biol. Macromol.*, 187, 223–231, 2021, https://doi.org/10.1016/J.IJBIOMAC.2021.07.136.
9. Gaikwad, K.K., Singh, S., Ajji, A., Moisture absorbers for food packaging applications. *Environ. Chem. Lett.*, 17, 609–628, 2019, https://doi.org/10.1007/S10311-018-0810-Z/TABLES/4.
10. Carpena, M., Nuñez-estevez, B., Soria-lopez, A. *et al.*, Essential oils and their application on active packaging systems : A review. *Resources*, 10, 1, 1–20, 2021.
11. Deshmukh, R.K. and Gaikwad, K.K., Natural antimicrobial and antioxidant compounds for active food packaging applications. *Biomass Convers. Biorefin.*, 1, 1–22, 2022, https://doi.org/10.1007/S13399-022-02623-W.
12. Kumar, L., Ramakanth, D., Akhila, K., Gaikwad, K.K., Edible films and coatings for food packaging applications: A review. *Environ. Chem. Lett.*, 20, 1, 875–900, 2021, https://doi.org/10.1007/S10311-021-01339-Z.

13. Kumar, J., Akhila, K., Gaikwad, K.K., Recent developments in intelligent packaging systems for food processing industry: A review. *J. Food Process. Technol.*, 12, 1–9, 2021, https://doi.org/10.35248/2157-7110.21.12.895.
14. Sohany, M., Tawakkal, I.S.M.A., Ariffin, S.H. *et al.*, Characterization of anthocyanin associated purple sweet potato starch and peel-based pH indicator films. *Foods*, 10, 2005, 2021, https://doi.org/10.3390/ FOODS10092005.
15. Luchese, C.L., Sperotto, N., Spada, J.C., Tessaro, I.C., Effect of blueberry agro-industrial waste addition to corn starch-based films for the production of a pH-indicator film. *Int. J. Biol. Macromol.*, 104, 11–18, 2017, https://doi.org/10.1016/J.IJBIOMAC.2017.05.149.
16. Akhila, K., Ramakanth, D., Gaikwad, K.K., Development of novel gallic acid- and cellulose acetate-coated paper as pH-responsive oxygen indicator for intelligent food packaging. *J. Coat. Technol. Res.*, 19, 5, 1–14, 2022, https://doi. org/10.1007/S11998-022-00622-0/FIGURES/10.
17. Müller, P. and Schmid, M., Intelligent packaging in the food sector: A brief overview. *Foods*, 8, 16, 2019, https://doi.org/10.3390/foods8010016.
18. Kruijf, D.D., Beest, V.V., Rijk, R. *et al.*, Active and intelligent packaging:Applications and regulatory aspects. *Food Addit. Contam.*, 19, 144–162, 2010, https://doi.org/10.1080/02652030110072722.
19. Siró, I., Active and intelligent packaging of food, in: *Progress in Food Preservation*, pp. 23–48, 2012, https://doi.org/10.1002/9781119962045.CH2.
20. Abbas, M., Saeed, F., Anjum, F.M. *et al.*, Natural polyphenols: An overview. *Int. J. Food Prop.*, 20, 1689–1699, 2017, https://doi.org/10.1080/10942912.2016.1220393.
21. Wolfe, K., Wu, X., Liu, R.H., Antioxidant activity of apple peels. *J. Agric. Food Chem.*, 51, 609–614, 2003, https://doi.org/10.1021/jf020782a.
22. Ayala, M. and Leticia, A., Determination of naringin and hesperidin in citrus fruit by high-performance liquid chromatography. The antioxidant potential of citrus fruit, *Acta Chromatogr.*, 17, 108, 2012.
23. Larrauri, J.A., Rupérez, P., Bravo, L., Saura-Calixto, F., High dietary fibre powders from orange and lime peels: Associated polyphenols and antioxidant capacity. *Food Res. Int.*, 29, 757–762, 1996, https://doi.org/10.1016/S0963-9969(97)00003-3.
24. M'hiri, N., Ioannou, I., Ghoul, M., Boudhrioua, N.M., Extraction methods of citrus peel phenolic compounds. *Food Rev. Int.*, 30, 265–290, 2014.
25. Silva-Beltrán, N.P., Chaidez-Quiroz, C., López-Cuevas, O. *et al.*, Phenolic compounds of potato peel extracts: Their antioxidant activity and protection against human enteric viruses. *J. Microbiol. Biotechnol.*, 27, 234–241, 2017, https://doi.org/10.4014/jmb.1606.06007.
26. Mohdaly, A.A.A., Sarhan, M.A., Smetanska, I., Mahmoud, A., Antioxidant properties of various solvent extracts of potato peel, sugar beet pulp and sesame cake. *J. Sci. Food Agric.*, 90, 218–226, 2010, https://doi.org/10.1002/jsfa.3796.

27. Lubaina, A.S., Renjith, P.R., Roshni, A.S., Identification and quantification of polyphenols from pineapple peel by high performance liquid chromatography analysis. *Adv. Zool. Bot.*, 8, 431–438, 2020, https://doi.org/10.13189/azb.2020.080507.
28. Gómez-Maldonado, D., Lobato-Calleros, C., Aguirre-Mandujano, E. *et al.*, Antifungal activity of mango kernel polyphenols on mango fruit infected by anthracnose. *LWT*, 126, 109337, 2020, https://doi.org/10.1016/j.lwt.2020.109337.
29. Bordiga, M., Travaglia, F., Locatelli, M., Valorisation of grape pomace: An approach that is increasingly reaching its maturity – a review. *Int. J. Food Sci. Technol.*, 54, 933–942, 2019, https://doi.org/10.1111/ijfs.14118.
30. Lu, Y. and Foo, L.Y., The polyphenol constituents of grape pomace. *Food Chem.*, 65, 1–8, 1999.
31. Monteiro, G.C., Minatel, I.O., Junior, A.P. *et al.*, Bioactive compounds and antioxidant capacity of grape pomace flours. *LWT*, 135, 110053, 2021, https://doi.org/10.1016/j.lwt.2020.110053.
32. Çam, M. and Hişil, Y., Pressurised water extraction of polyphenols from pomegranate peels. *Food Chem.*, 123, 878–885, 2010, https://doi.org/10.1016/j.foodchem.2010.05.011.
33. Li, J., He, X., Li, M. *et al.*, Chemical fingerprint and quantitative analysis for quality control of polyphenols extracted from pomegranate peel by HPLC. *Food Chem.*, 176, 7–11, 2015, https://doi.org/10.1016/j.foodchem.2014.12.040.
34. Ben Nasr, C., Ayed, N., Metche, M., Quantitative determination of the polyphenolic content of pomegranate peel. *Z. Lebensm. Unters. Forsch.*, 203, 374–378, 1996, https://doi.org/10.1007/BF01231077.
35. Sundaram, S., Anjum, S., Dwivedi, P., Rai, G.K., Antioxidant activity and protective effect of banana peel against oxidative hemolysis of human erythrocyte at different stages of ripening. *Appl. Biochem. Biotechnol.*, 164, 1192–1206, 2011, https://doi.org/10.1007/s12010-011-9205-3.
36. Chaudhry, F., Ahmad, M.L., Hayat, Z. *et al.*, Extraction and evaluation of the antimicrobial activity of polyphenols from banana peels employing different extraction techniques, *Separations*, 9, 7, 165, 2022.
37. Yang, T., Guang Hu, J., Yu, Y. *et al.*, Comparison of phenolics, flavonoids, and cellular antioxidant activities in ear sections of sweet corn (Zea mays L. saccharata Sturt). *J. Food Process. Preserv.*, 43, 1–9, 2019, https://doi.org/10.1111/jfpp.13855.
38. Ioana, V.S., Alin, T., Mariana, M. *et al.*, Preliminary study regarding to the total polyphenols and antioxidant capacity of yellow maize corncobs. *Analele Universității din Oradea, Fascicula Protecția Mediului*, XXIII, 179–184, 2014.
39. Zhao, W., Huang, P., Zhu, Z. *et al.*, Production of phenolic compounds and antioxidant activity via bioconversion of wheat straw by Inonotus obliquus under submerged fermentation with the aid of a surfactant. *J. Sci. Food Agric.*, 101, 1021–1029, 2021, https://doi.org/10.1002/jsfa.10710.

40. Gómez-Estaca, J., López-de-Dicastillo, C., Hernández-Muñoz, P. *et al.*, Advances in antioxidant active food packaging. *Trends Food Sci. Technol.*, 35, 42–51, 2014, https://doi.org/10.1016/j.tifs.2013.10.008.
41. Zou, Z., Xi, W., Hu, Y. *et al.*, Antioxidant activity of Citrus fruits. *Food Chem.*, 196, 885–896, 2016, https://doi.org/10.1016/j.foodchem.2015.09.072.
42. Li, T., Shen, P., Liu, W. *et al.*, Major polyphenolics in pineapple peels and their antioxidant interactions. *Int. J. Food Prop.*, 17, 1805–1817, 2014, https://doi.org/10.1080/10942912.2012.732168.
43. Saraswaty, V., Risdian, C., Primadona, I. *et al.*, Pineapple peel wastes as a potential source of antioxidant compounds. *IOP Conf. Ser. Earth Environ. Sci.*, 60, 012013, 2017, https://doi.org/10.1088/1755-1315/60/1/012013.
44. Antoniolli, A., Fontana, A.R., Piccoli, P., Bottini, R., Characterization of polyphenols and evaluation of antioxidant capacity in grape pomace of the cv. Malbec. *Food Chem.*, 178, 172–178, 2015, https://doi.org/10.1016/j.foodchem.2015.01.082.
45. Malviya, S., Arvind, Jha, A., Hettiarachchy, N., Antioxidant and antibacterial potential of pomegranate peel extracts. *J. Food Sci. Technol.*, 51, 4132–4137, 2014, https://doi.org/10.1007/s13197-013-0956-4.
46. Derakhshan, Z., Ferrante, M., Tadi, M. *et al.*, Antioxidant activity and total phenolic content of ethanolic extract of pomegranate peels, juice and seeds. *Food Chem. Toxicol.*, 114, 108–111, 2018, https://doi.org/10.1016/j.fct.2018.02.023.
47. Ain, H. B. U., Tufail, T., Bashir, S., Ijaz, N., Hussain, M., Ikram, A., ... & Saewan, S. A., Nutritional importance and industrial uses of pomegranate peel: A critical review. *Food Science & Nutrition*, 11, 6, 2589–2598, 2023. https://doi.org/10.1002/fsn3.3320
48. Malviya, S., Arvind, Jha, A., Hettiarachchy, N., Antioxidant and antibacterial potential of pomegranate peel extracts. *J. Food Sci. Technol.*, 51, 4132–4137, 2014, https://doi.org/10.1007/s13197-013-0956-4.
49. Xie, D., Gan, T., Su, C. *et al.*, Structural characterization and antioxidant activity of water-soluble lignin-carbohydrate complexes (LCCs) isolated from wheat straw. *Int. J. Biol. Macromol.*, 161, 315–324, 2020, https://doi.org/10.1016/j.ijbiomac.2020.06.049.
50. Krepker, M., Shemesh, R., Danin Poleg, Y. *et al.*, Active food packaging films with synergistic antimicrobial activity. *Food Control*, 76, 117–126, 2017, https://doi.org/10.1016/j.foodcont.2017.01.014.
51. Kumar, P., Kumar, L., Tanwar, R. *et al.*, Active edible coating based on guar gum with mint extract and antibrowning agents for ber (Ziziphus mauritiana) fruits preservation. *J. Food Meas. Charact.*, 17, 1, 1–14, 2022.
52. Rout, S., Tambe, S., Deshmukh, R.K. *et al.*, Recent trends in the application of essential oils: The next generation of food preservation and food packaging. *Trends Food Sci. Technol.*, 129, 421–439, 2022. https://doi.org/10.1016/j.tifs.2022.10.012

53. Al-Zoreky, N.S., Antimicrobial activity of pomegranate (Punica granatum L.) fruit peels. *Int. J. Food Microbiol.*, 134, 244–248, 2009, https://doi.org/10.1016/j.ijfoodmicro.2009.07.002.
54. Cheng, V.J., Bekhit, A.E.D.A., McConnell, M. *et al.*, Effect of extraction solvent, waste fraction and grape variety on the antimicrobial and antioxidant activities of extracts from wine residue from cool climate. *Food Chem.*, 134, 474–482, 2012, https://doi.org/10.1016/j.foodchem.2012.02.103.
55. Kabuki, T., Nakajima, H., Arai, M. *et al.*, Characterization of novel antimicrobial compounds from mango (Mangifera indica L.) kernel seeds. *Food Chem.*, 71, 61–66, 2000, https://doi.org/10.1016/S0308-8146(00)00126-6.
56. Abdalla, A.E.M., Darwish, S.M., Ayad, E.H.E., El-Hamahmy, R.M., Egyptian mango by-product 2: Antioxidant and antimicrobial activities of extract and oil from mango seed kernel. *Food Chem.*, 103, 1141–1152, 2007, https://doi.org/10.1016/j.foodchem.2006.10.026.
57. Yu, J., Wang, L., Walzem, R.L. *et al.*, Antioxidant activity of citrus limonoids, flavonoids, and coumarins. *J. Agric. Food Chem.*, 53, 2009–2014, 2005, https://doi.org/10.1021/JF0484632/ASSET/IMAGES/LARGE/JF0484632F00003.JPEG.
58. Kadhim Hindi, N.K. and Ghani Chabuck, Z.A., Antimicrobial activity of different aqueous lemon extracts. *J. Appl. Pharm. Sci.*, 3, 74–78, 2013, https://doi.org/10.7324/JAPS.2013.3611.
59. Ali, J., Das, B., Saikia, T., Antimicrobial activity of lemon peel (Citrus Limon) extract. *Int. J. Curr. Pharm. Res.*, 9, 79, 2017, https://doi.org/10.22159/ijcpr.2017v9i4.20962.
60. Roy, S.D., Bania, R., Chakraborty, J. *et al.*, Pharmacognostic, phytochemical, physicochemical property and antimicrobial activity studies of lemon peel oil. *J. Nat. Product Plant Resour.*, 2, 431–435, 2012.
61. Mokbel, M.S. and Hashinaga, F., Antibacterial and antioxidant activities of banana (Musa, AAA cv. Cavendish) fruits peel. *Am. J. Biochem. Biotechnol.*, 1, 125–131, 2005, https://doi.org/10.3844/AJBBSP.2005.125.131.
62. Chaudhry, F., Ahmad, M.L., Hayat, Z. *et al.*, Extraction and evaluation of the antimicrobial activity of polyphenols from banana peels employing different extraction techniques, *Separations*, 9, 7, 165, 2022.
63. Dabesor, A., Asowata-Ayodele, A.M., Umoiette, P., Phytochemical compositions and antimicrobial activities of *Ananas comosus* peel (M.) and *Cocos nucifera* kernel (L.) on selected food borne pathogens. *Am. J. Plant Sci.*, 2, 73–76, 2017, https://doi.org/10.11648/j.ajpb.20170202.15.
64. Alizadeh-Sani, M., Mohammadian, E., Rhim, J.W., Jafari, S.M., pH-sensitive (halochromic) smart packaging films based on natural food colorants for the monitoring of food quality and safety. *Trends Food Sci. Technol.*, 105, 93–144, 2020, https://doi.org/10.1016/J.TIFS.2020.08.014.
65. Kuswandi, B., Freshness sensors for food packaging, in: *Reference Module in Food Science*, Elsevier, University of Jember, Jember, Indonesia, 2017, https://doi.org/10.1016/B978-0-08-100596-5.21876-3.

66. Poyatos-Racionero, E., Ros-Lis, J.V., Vivancos, J.L., Martínez-Máñez, R., Recent advances on intelligent packaging as tools to reduce food waste. *J. Clean. Prod.*, 172, 3398–3409, 2018, https://doi.org/10.1016/J.JCLEPRO.2017.11.075.
67. Wu, C., Sun, J., Zheng, P. *et al.*, Preparation of an intelligent film based on chitosan/oxidized chitin nanocrystals incorporating black rice bran anthocyanins for seafood spoilage monitoring. *Carbohydr. Polym.*, 222, 115006, 2019, https://doi.org/10.1016/J.CARBPOL.2019.115006.
68. Sun, J., Jiang, H., Wu, H. *et al.*, Multifunctional bionanocomposite films based on konjac glucomannan/chitosan with nano-ZnO and mulberry anthocyanin extract for active food packaging. *Food Hydrocoll.*, 107, 105942, 2020, https://doi.org/10.1016/J.FOODHYD.2020.105942.
69. Yang, Z., Zhai, X., Zou, X. *et al.*, Bilayer pH-sensitive colorimetric films with light-blocking ability and electrochemical writing property: Application in monitoring crucian spoilage in smart packaging. *Food Chem.*, 336, 127634, 2021, https://doi.org/10.1016/J.FOODCHEM.2020.127634.
70. Koshy, R.R., Koshy, J.T., Mary, S.K. *et al.*, Preparation of pH sensitive film based on starch/carbon nano dots incorporating anthocyanin for monitoring spoilage of pork. *Food Control*, 126, 108039, 2021, https://doi.org/10.1016/J.FOODCONT.2021.108039.
71. Kurek, M., Garofulić, I.E., Bakić, M.T. *et al.*, Development and evaluation of a novel antioxidant and pH indicator film based on chitosan and food waste sources of antioxidants. *Food Hydrocoll.*, 84, 238–246, 2018, https://doi. org/10.1016/j.foodhyd.2018.05.050.
72. Choi, I., Lee, J.Y., Lacroix, M., Han, J., Intelligent pH indicator film composed of agar/potato starch and anthocyanin extracts from purple sweet potato. *Food Chem.*, 218, 122–128, 2017, https://doi.org/10.1016/j.foodchem.2016.09.050.
73. Zhai, X., Shi, J., Zou, X. *et al.*, Novel colorimetric films based on starch/polyvinyl alcohol incorporated with roselle anthocyanins for fish freshness monitoring. *Food Hydrocoll.*, 69, 308–317, 2017, https://doi.org/10.1016/j.foodhyd.2017.02.014.
74. Pereira, V.A., de Arruda, I.N.Q., Stefani, R., Active chitosan/PVA films with anthocyanins from Brassica oleraceae (red cabbage) as time–temperature indicators for application in intelligent food packaging. *Food Hydrocoll.*, 43, 180–188, 2015, https://doi.org/10.1016/j.foodhyd.2014.05.014.
75. Ge, Y., Li, Y., Bai, Y. *et al.*, Intelligent gelatin/oxidized chitin nanocrystals nanocomposite films containing black rice bran anthocyanins for fish freshness monitorings. *Int. J. Biol. Macromol.*, 155, 1296–1306, 2020, https://doi.org/10.1016/j.ijbiomac.2019.11.101.
76. Listyarini, A., Sholihah, W., Imawan, C., A paper-based colorimetric indicator label using natural dye for monitoring shrimp spoilage. *IOP Conf. Ser. Mater. Sci. Eng.*, 367, 012045, 2018, https://doi.org/10.1088/1757-899X/367/1/012045.

77. Rachmelia, D. and Imawan, C., Time temperature indicator label using black corn extract and chitosan matrix. *J. Phys.: Conf. Ser.*, IOP Publishing, 1120, 012041, 2018.
78. Saliu, F. and della Pergola, R., Carbon dioxide colorimetric indicators for food packaging application: Applicability of anthocyanin and poly-lysine mixtures. *Sens. Actuators B Chem.*, 258, 1117–1124, 2018, https://doi.org/10.1016/j.snb.2017.12.007.
79. Kanatt, S.R., Development of active/intelligent food packaging film containing Amaranthus leaf extract for shelf life extension of chicken/fish during chilled storage. *Food Packag. Shelf Life*, 24, 100506, 2020, https://doi.org/10.1016/j. fpsl.2020.100506.
80. Li, Y., Wu, K., Wang, B., Li, X., Colorimetric indicator based on purple tomato anthocyanins and chitosan for application in intelligent packaging. *Int. J. Biol. Macromol.*, 174, 370–376, 2021, https://doi.org/10.1016/j.ijbiomac.2021.01.182.
81. Azlim, N.A., Mohammadi Nafchi, A., Oladzadabbasabadi, N. *et al.*, Fabrication and characterization of a pH-sensitive intelligent film incorporating dragon fruit skin extract. *Food Sci. Nutr.*, 10, 597–608, 2022.
82. Liang, T., Sun, G., Cao, L. *et al.*, A pH and NH3 sensing intelligent film based on Artemisia sphaerocephala Krasch. gum and red cabbage anthocyanins anchored by carboxymethyl cellulose sodium added as a host complex. *Food Hydrocoll.*, 87, 858–868, 2019, https://doi.org/10.1016/j.foodhyd.2018.08.028.
83. Jiang, G., Hou, X., Zeng, X. *et al.*, Preparation and characterization of indicator films from carboxymethyl-cellulose/starch and purple sweet potato (Ipomoea batatas (L.) lam) anthocyanins for monitoring fish freshness. *Int. J. Biol. Macromol.*, 143, 359–372, 2020, https://doi.org/10.1016/j.ijbiomac.2019.12.024.
84. Chavoshizadeh, S., Pirsa, S., Mohtarami, F., Sesame oil oxidation control by active and smart packaging system using wheat gluten/chlorophyll film to increase shelf life and detecting expiration date. *Eur. J. Lipid Sci. Technol.*, 122, 1900385, 2020.
85. Romero, A., Sharp, J.L., Dawson, P.L. *et al.*, Evaluation of two intelligent packaging prototypes with a pH indicator to determine spoilage of cow milk. *Food Packag. Shelf Life*, 30, 100720, 2021, https://doi.org/10.1016/j.fpsl.2021.100720.
86. Singh, S., Gaikwad, K.K., Lee, M., Lee, Y.S., Temperature sensitive smart packaging for monitoring the shelf life of fresh beef. *J. Food Eng.*, 234, 41–49, 2018, https://doi.org/10.1016/j.jfoodeng.2018.04.014.
87. Shukla, V., Kandeepan, G., Vishnuraj, M.R., Soni, A., Anthocyanins based indicator sensor for intelligent packaging application. *Agric. Res.*, 5, 205–209, 2016, https://doi.org/10.1007/s40003-016-0211-0.

88. Kuswandi, B., Maryska, C., Jayus, *et al.*, Real time on-package freshness indicator for guavas packaging. *J. Food Meas. Charact.*, 7, 29–39, 2013, https://doi.org/10.1007/s11694-013-9136-5.
89. Won, K., Jang, N.Y., Jeon, J., A natural component-based oxygen indicator with in-pack activation for intelligent food packaging. *J. Agric. Food Chem.*, 64, 9675–9679, 2016, https://doi.org/10.1021/acs.jafc.6b04172.
90. Wen, J., Huang, S., Jia, L. *et al.*, Visible colorimetric oxygen indicator based on Ag-loaded TiO2 nanotubes for quick response and real-time monitoring of the integrity of modified atmosphere packaging. *Adv. Mater. Technol.*, 4, 1900121, 2019.
91. Deshwal, G.K., Panjagari, N.R., Singh, A.K., Alam, T., Performance evaluation of a biopolymer-based in-package UV activated colorimetric oxygen indicator with modified atmosphere packaged mozzarella cheese. *J. Packag. Technol. Res.*, 5, 51–57, 2021, https://doi.org/10.1007/s41783-021-00108-1.
92. Lu, P., Yang, Y., Liu, R. *et al.*, Preparation of sugarcane bagasse nano-cellulose hydrogel as a colourimetric freshness indicator for intelligent food packaging. *Carbohydr. Polym.*, 249, 116831, 2020, https://doi.org/10.1016/j.carbpol.2020.116831.
93. Choi, I., Choi, H., Lee, J.-S., Han, J., Novel color stability and colorimetry-enhanced intelligent CO2 indicators by metal complexation of anthocyanins for monitoring chicken freshness. *Food Chem.*, 404, 134534, 2023, https://doi. org/10.1016/j.foodchem.2022.134534.
94. Gaikwad, K.K., Singh, S., Ajji, A., Moisture absorbers for food packaging applications. *Environ. Chem. Lett.*, 17, 609–628, 2019.
95. Pereira, P.F.M., Picciani, P.H.S., Calado, V.M.A., Tonon, R.V., Gelatin-based nanobiocomposite films as sensitive layers for monitoring relative humidity in food packaging. *Food Bioprocess Technol.*, 13, 1063–1073, 2020, https://doi.org/10.1007/s11947-020-02462-5.
96. Bibi, F., Guillaume, C., Vena, A. *et al.*, Wheat gluten, a bio-polymer layer to monitor relative humidity in food packaging: Electric and dielectric characterization. *Sens. Actuators A Phys.*, 247, 355–367, 2016, https://doi.org/10.1016/j. sna.2016.06.017.
97. Prietto, L., Mirapalhete, T.C., Pinto, V.Z. *et al.*, pH-sensitive films containing anthocyanins extracted from black bean seed coat and red cabbage. *LWT*, 80, 492–500, 2017, https://doi.org/10.1016/j.lwt.2017.03.006.
98. Balbinot-Alfaro, E., Craveiro, D.V., Lima, K.O. *et al.*, Intelligent packaging with pH indicator potential. *Food Eng. Rev.*, 11, 235–244, 2019, https://doi.org/10.1007/s12393-019-09198-9.
99. Kurek, M., Garofulić, I.E., Bakić, M.T. *et al.*, Development and evaluation of a novel antioxidant and pH indicator film based on chitosan and food waste sources of antioxidants. *Food Hydrocoll.*, 84, 238–246, 2018, https://doi.org/10.1016/j.foodhyd.2018.05.050.
100. Roy, S. and Rhim, J.-W., Preparation of gelatin/carrageenan-based color-indicator film integrated with Shikonin and Propolis for smart food

packaging applications. *ACS Appl. Bio Mater.*, 4, 770–779, 2021, https://doi.org/10.1021/acsabm.0c01353.

101. Maciel, V.B.V., Yoshida, C.M.P., Franco, T.T., Development of a prototype of a colourimetric temperature indicator for monitoring food quality. *J. Food Eng.*, 111, 21–27, 2012, https://doi.org/10.1016/j.jfoodeng.2012.01.037.
102. Pereira, V.A., de Arruda, I.N.Q., Stefani, R., Active chitosan/PVA films with anthocyanins from Brassica oleraceae (Red Cabbage) as time–temperature indicators for application in intelligent food packaging. *Food Hydrocoll.*, 43, 180–188, 2015, https://doi.org/10.1016/j.foodhyd.2014.05.014.
103. Mataragas, M., Bikouli, V.C., Korre, M. *et al.*, Development of a microbial Time Temperature Indicator for monitoring the shelf life of meat. *Innov. Food Sci. Emerg.*, 52, 89–99, 2019, https://doi.org/10.1016/j.ifset.2018.11.003.
104. Tseng, S.-Y., Li, S.-Y., Yi, S.-Y. *et al.*, Food quality monitor: Paper-based plasmonic sensors prepared through reversal nanoimprinting for rapid detection of biogenic amine odorants. *ACS Appl. Mater. Interfaces*, 9, 17306–17316, 2017, https://doi.org/10.1021/acsami.7b00115.
105. Koskela, J., Sarfraz, J., Ihalainen, P. *et al.*, Monitoring the quality of raw poultry by detecting hydrogen sulfide with printed sensors. *Sens. Actuators B Chem.*, 218, 89–96, 2015.
106. Huang, L., Sun, D.-W., Pu, H. *et al.*, A colorimetric paper sensor based on the domino reaction of acetylcholinesterase and degradable γ-MnOOH nanozyme for sensitive detection of organophosphorus pesticides. *Sens. Actuators B Chem.*, 290, 573–580, 2019, https://doi.org/10.1016/j.snb.2019.04.020.

9

Starch from Agro-Waste for Food Packaging Applications

Shashank T. Mhaske*, Jyoti Darsan Mohanty and Pavan Y. Borse

Department of Polymer and Surface Engineering, Institute of Chemical Technology, Mumbai, Maharashtra, India

Abstract

The development of biodegradable packaging from renewable sources has been encouraged by studies into environmental challenges. Reduced usage of non-biodegradable and nonrenewable materials in the packaging sector is possible by using natural polymers like starch. A naturally occurring biopolymer found in various vital commodities like rice, potatoes, wheat, maize, cereals, roots, and tubers is known as starch. The principal ingredient in many foods produced for human consumption is starch, and research into how it interacts with other components like lipids and water is crucial to understanding human nutrition. Starch is a possible raw material for manufacturing food-grade coatings and packaging film. Starch's essential qualities, including crystallinity, packaging density, flexibility, heat stability, processability, barrier properties, and water absorption capacity, are greatly influenced by its main structural component ratio of amylose and amylopectin. Because starch is more readily available and less expensive than other biopolymers, its potential uses are more and more economically feasible. The modification of starch through different chemical, physiochemical, and physical modifications produces starch with enhanced qualities for various applications. The starch can be modified with ester and ether functionality and cross-linked with diether and diester through a chemical route. Similarly, the physiochemical modification process includes gelatinization, retrogradation, and pasting. The thermophysical method includes plasticizing starch with glycerol and water to alter its structure and disrupt the intermolecular bonding force. This chapter summarizes the sources, modifications, properties, and broader applications of starch in the food packaging sector. Also, industrial and agricultural items require

**Corresponding author*: stmhaske@gmail.com; st.mhaske@ictmumbai.edu.in

Santosh Kumar, Avik Mukherjee and Vimal Katiyar (eds.) Agro-Waste Derived Biopolymers and Biocomposites: Innovations and Sustainability in Food Packaging, (241–266) © 2024 Scrivener Publishing LLC

packaging with excellent performance and long-term property preservation. The starch can be used to create a high-performance bioplastic as an alternative to synthetic and biopolymers due to its high market demand, easy availability, and economic viability.

***Keywords*:** Agro-waste, food packaging, seed starch, tuber starch, renewable films, biodegradable

9.1 Introduction

Starch or amylum, a polymeric carbohydrate, comprises many glucose units connected by glycosidic linkages. All green plants synthesize starch, a white and granular organic molecule, as a form of energy storage, which is insoluble in cold water, alcohol, or other solvents. Starch is a soft and tasteless powder having the empirical formula $(C_6H_{10}O_5)_n$. A starch polysaccharide comprises glucose monomers linked together by α-1,4 and α-1,6 glycosidic links. The linear polymer amylose is the most basic type of starch, and amylopectin is the branched type. Starch is synthesized in the green leaves of plants from extra glucose produced during photosynthesis as a reserve food. The cassava plant's roots, potato's tuber, stem pith, seeds, and other storage organs contain granulated starch chloroplasts. Starch is traditionally extracted from small sources like grains, tubers, and cassava, and it is used to improve the functional qualities of numerous food and nonfood formulations. Utilizing unusual starch sources as a raw material for commercial applications may offer an alternative to traditional starches while lowering production costs. Due to their unique physicochemical and functional properties, starch from nontraditional sources generates much interest. To have tangible viability at an industrial scale, a significant study of nonconventional sources and their attributes must be well understood [1–3]. Most of the commercial starch is obtained from corn but is also extracted from wheat, tapioca, and potato. Commercial starch is obtained by crushing or grinding tubers or seeds, combining the pulp with water, and then removing any leftover contaminants before drying the paste. Starch is used in brewing and as a thickener in baked foods and confectioneries. Starch is added throughout the paper production process to make it stronger for making corrugated paperboard, paper bags, and boxes.

The specific properties of starch are given below:

- It is a nonreducing carbohydrate with no taste and odor.
- Strong hydrogen bonding makes it immiscible in most organic solvents and cold water.

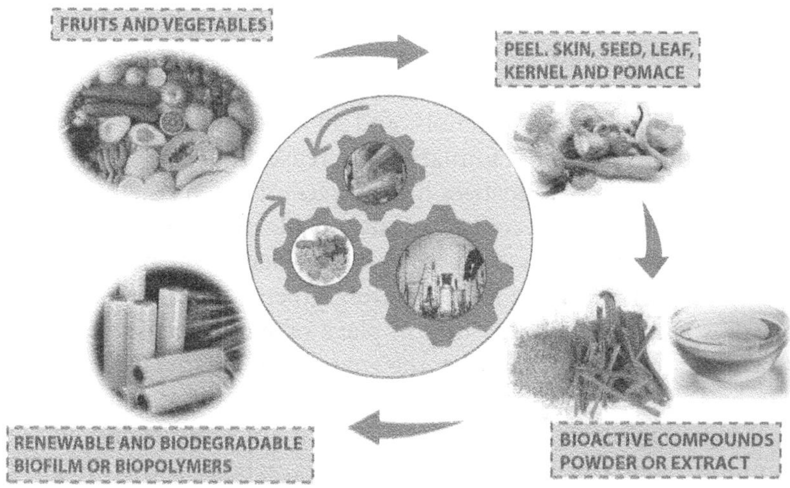

Figure 9.1 Utilization of agro-waste for the production of starch-based food packaging film.

- The molecular weight of two components of starch, i.e., amylose and amylopectin, are 10^5 and 10^7 g/mol, respectively.
- The density of starch is 18 g/cm^3.
- The melting point is 256°C to 258°C.
- The gelatinization temperature is between 55°C and 60°C.

Animals, plants, algae, and other living beings such as microorganisms produce natural biopolymers. The most prevalent polysaccharide within these biopolymers, starch, has drawn a lot of interest as a biodegradable material. From plant sources including potato, wheat, cassava, and maize, starch differs in its physicochemical and functional properties. Owing to its biological compatibility, complete decay without leaving behind unwanted residues, low cost, abundant accessibility, renewability, and thermoplastic nature, starch is well known for its usage in food processing and packaging industries as presented in Figure 9.1. It is possible to convert starch into thermoplastic materials using a variety of industrial processes, including injection molding and extrusion to create porous or dense materials [4, 5].

9.2 Starch from Agro-Waste

Rice, wheat, maize, barley, cassava, and potato are the major plant sources of starch. The byproducts and residues of the starch processing can considerably harm the environment if they are not adequately handled.

Therefore, researchers and producers are increasingly interested in using starch wastes as they are abundant, affordable, and sustainable sources of recovering bioactive chemicals such as peptides, oryzanol, xylose, lignin, carotenoids, quinones, polyphenols, and alkaloids. Utilizing these wastes will reduce the risk of contamination, increase the value of agricultural products, and facilitate industries like nutraceuticals, medicines, and cosmetics, particularly in the context of sustainable development. To address the limitation of polymers derived from petroleum, scientists have been searching for other sources for generating greener polymers. Using thermoplastic starch (TPS)–based film in place of synthetic plastics is one of the few solutions that could lessen the environmental impact [6].

Starch is an inexpensive, plentiful, renewable, sustainable, palatable, and biodegradable polymer, sources of which are roots, stalks, and seeds of common crops like rice, potatoes, tapioca, wheat, and corn. In these crops, starch is regarded as the primary form of carbohydrate storage [7]. In addition to its amazing qualities, such as effective reinforcement, it may be modified or blended with other polymers to improve their processability, physicochemical, mechanical, barrier, and natural biodegradability [8, 9]. The structure and properties of raw starch and the processing techniques employed significantly impact the microstructures, mechanical characteristics, and nutritional attributes of materials based on starch. Growing trends favor the creation of goods generated from biodegradable starches [10]. Several companies depend on food crops to supply these demands. Because it could result in the overexploitation of food crops, which would have an impact on food security, this is not a sustainable approach. Therefore, one of the notable answers will be to use sustainable alternatives, such as food that is made from agro-waste [11].

9.2.1 Tuber Wastes

Starchy roots and tubers are the subterranean stems, roots, rhizomes, corms, and tubers of plants that contain edible starch. Potatoes and yams are tubers, but taro and cocoyams are obtained from corms, underground stems, and swelling hypocotyls. Root and tuber crops are typically found in tropical climates and are the second most staple energy sources after cereals [12]. The principal substrate for starch extraction has been historically described as wheat; however, potatoes quickly became a useful source for starch production after being introduced into European agriculture [13, 14]. Production of potato starch appears to be more straightforward than wet milling of corn and wheat [15]. Potato tissue may be pulverized using relatively light mechanical forces, such as rasping, and because potato cells are often not densely filled with starch granules, potato tuber structure and composition are favorable for starch accessibility [16]. For the best yield, the potato tissue must be completely ruptured and all cells must be opened [17, 18].

Cassava (*Manihot esculenta*), often known as manioc, tapioca, or yuca, is one of the most important food crops in the humid tropics that can grow in soil having poor nutrient availability and can withstand drought [19]. The characteristics that set tapioca starch apart from other starches are low residual material content (fat, protein, and ash), lower amylose content, and high amylose and amylopectin molecular weights. Due to partial removal, the trace quantity of phosphorus in tapioca starch is not as tightly linked as it is in potato starch as a phosphate ester. It also contains extremely low amounts of protein and lipids compared to the starches found in cereals [20, 21]. Cassava starch contains 17 to 20% amylose, and the amylose content does not vary significantly, unlike rice (0 to 40% amylose content) and corn (0 to 70% amylose content). The beta-amylolysis limit of cassava starch, which is lower than that of corn, potato, rice, and wheat starches, reveals that the amylose molecules are not entirely unbranched like those of other starches, and thus, cassava amylose has higher molecular weight [22]. Legume starch contains 30 to 40% amylose and 60 to 70% amylopectin; however, the majority of other dietary starches only contain 25 to 30% amylose and 70 to 75% amylopectin [23].

9.2.2 Seed Wastes

The need for starch in a variety of industries, such as food, medicine, paper, glue, and textiles, has increased interest in finding additional sources in other plants or fruits. The utilization of starch derived from several plant sources, including rice, wheat, corn, and potatoes, has been studied extensively which is shown in Figure 9.2. Yet, there is not much research on starch extraction from fruit byproducts such as seeds [24]. The popular tropical fruit known as the mango (*Mangifera indica*) is grown and cultivated in many tropical areas of the world, including Malaysia, Thailand, and India. Mango seeds are typically thrown away as rubbish after the fruits are consumed. As a result, this kind of agricultural waste could be used to create added-value products because it is rich in starch and is readily available. Thus, mango seeds are used to extract starch using various procedures [25]. Other than mango seeds, avocado seeds are also used to extract starch as they contain a quite high amount (80.1%) of starch [26].

Researchers have focused a lot of emphasis on using jackfruit (*Artocarpus heterophyllus*) seed as a source of starch flour. Jackfruit is widely grown in

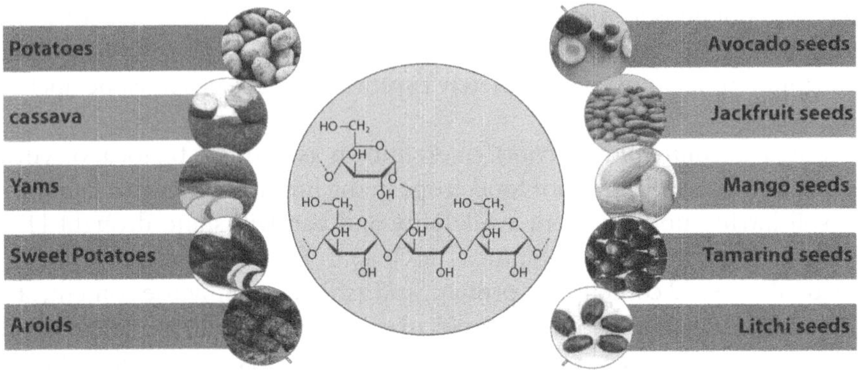

Figure 9.2 Different seeds and tuber waste for extracting starch.

Southeast Asian countries. Jackfruit seed is regarded as a low-cost source of carbohydrates because it makes up 8 to 15% of the weight of the fruit and often includes 60 to 80% of the dry content. Jackfruit starch offers a lot of promise for use in a variety of applications due to its distinct structure and functional characteristics [27]. Jackfruit seeds are a substantial source of starch (22%) and dietary fiber (3.19%) [28]. Tamarind seed starch contains between 67 and 70% total starch, which is regarded as inedible. However, this is a cheap source and can be used for a wide range of industrial applications. Thus, the starch extracted from tamarind seeds is also used to develop biodegradable packaging materials. To produce tamarind seed powder, the seeds were first collected, washed, and then further roasted, decorticated, and ground. There are 27.55 wt % amylose and 72.45 wt % amylopectin in defatted tamarind seed starch [29].

Litchi (*Litchi chinensis*) seeds are the primary waste products produced when the litchi fruit is processed into beverages, jellies, wine, and squash. These seeds are rich in complex polysaccharides, with starch accounting for 40% of them. This starch exhibits various functional and structural traits because of its low amylose and high amylopectin concentration. In addition to solubility, swelling power (SP), and water-holding capacity, these various functional food features also include micrometric properties (Hausner's ratio, Carr's index, and angle of repose). These qualities guarantee their use as a thickening agent, edible packaging, a useful component, and an excipient in pharmaceutical products and targeted drug delivery systems in nonfood products. Utilizing the starch from litchi seeds can help the business more commercially viable for processing the fruits while also helping to lessen waste production and environmental damage [30].

9.3 Modifications in Starch for Food Packaging

One of the potential and most promising materials for biodegradable food packaging is starch; however, it has many drawbacks in terms of processing and final properties. Therefore, structural modification in starch is needed to overcome such hurdles. Several modification techniques change starch properties to suit today's stringent technical requirements. Starch modification aims to address one or more of the problems, increasing its adaptability and meeting market demand. Thus, numerous chemical, physical, enzymatic, or combined methods are being used to modify starch as shown in Figure 9.3, which provide significant value addition and allow the creation of various manufactured products for food packaging applications. Even though physical modification has a clear advantage over the other two, from the perspectives of customers and the environment; however, chemical modifications such as oxidation, esterification, and etherification are typically chosen. The researchers claim that native starches' (NSs') hydrophilic behavior often diminishes their dimensional stability and mechanical properties. As a result, the acetyl, succinyl, and carboxymethyl hydroxypropyl groups demonstrate esterification and etherification procedures that are necessary to change NSs for hydrophobicity by replacing the starch's hydroxyl groups with big hydrophobic groups. The hydrophobicity of NSs can also be increased chemically by cross-linking, grafting, and condensing procedures. Hydrophobicity is needed to decrease the influence of water or moisture on films made from starch or TPS. NSs' strong

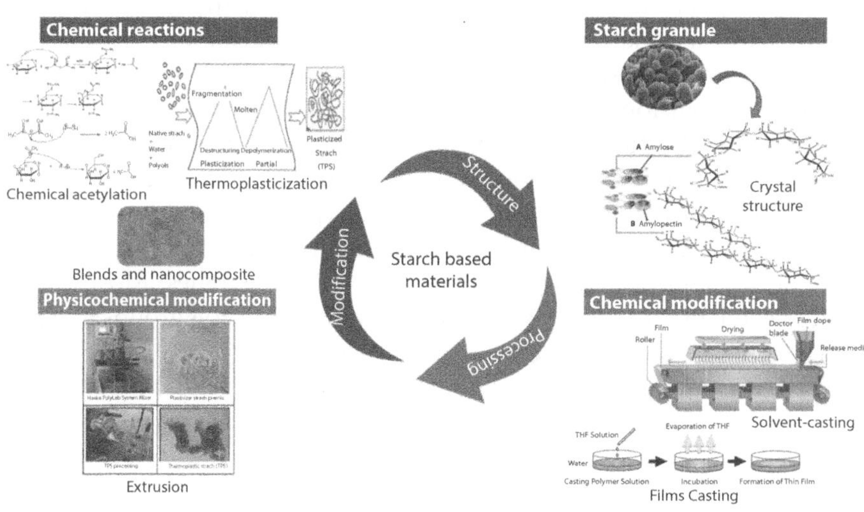

Figure 9.3 Modification of starch for packaging film [56].

hydrophilic characteristics severely restrict their industrial applications. These changes are intended to bring about desired changes in the starch structure, attend to the selected properties, and improve the processability. The starch can be modified through different methods to improve its thermal stability, processability, and physical properties. The common practice of modification is physical blending and chemical grafting. The modification method generally depends on applying starch, whether to use it in food products or food packaging. The food packaging sector is one of the many areas where modified starches with improved hydrophobicity can be applied. These large hydrophobic groups added to starch molecules prevent or diminish the retrogradation or recrystallization phenomena. The functional groups were altered chemically, but the starches also acquired new chemical, physical, and functional characteristics. However, the resultant starch's structural and organizational integrity could be more balanced. Therefore, the improved starch hydrophobicity would facilitate its use in the application of food packaging [31–35].

9.3.1 Chemical Modification

9.3.1.1 Acetylation

Acetylation of starch is a chemical modification method in which the hydroxyl group of an anhydrous glucose unit converts into an acetyl group. Starch reacts with acetic anhydride in an alkaline media usually in NaOH as a catalyst because it is recoverable and reusable. Acetylated starch is also known as starch acetate. Several factors influenced the molar substitution of starch acetylation, which include the type of catalysts, the concentration of acetic anhydride, the origin of the starch, reaction time, and the pH of the suspension. According to the research, starch acetate can be used for various applications, including surface sizing, biodegradable film, and paper tape with gum. Additionally, they claimed that acetylation reduced the relative crystallinity, viscosity, pasting temperature, and upsurge SP of barley starch. Additionally, it has been noted that acetylated starches are more shear- and heat-resistant than unacetylated starches. Most importantly, acetylation causes hydrophobicity by adding acetyl groups and partially depolymerizing amylopectin and amylose molecules.

Different amounts of acetic acid were used in relation to the starch weight to obtain different levels of acetylation, ranging from 61.5 g of acetic acid anhydride to 100 g of starch [36]. The efficacy of both the acetylation process and its dependence on the concentration of starch milk, reaction pH, temperature, and reaction time was carefully examined [37]. The enthalpy of gelatinization (H) in the context of gelatinization falls with acetylation [38]. When jack bean starch was acetylated, its gelatinization value dropped from 0.75 to 0.56 J g^{-1}. According to a study by Wojeicchowski *et al.* [39], the values of H and onset-temperature (To) fall in the case of acetylated common beans. According to the study, the starch's ability to paste decreased as a result of the acetylation process. Native Bambara starch was shown to have a greater pasting temperature, hot paste viscosity, peak viscosity, and cold paste viscosity than its acetylated counterpart. Common bean starch that has been acetylated has the minimum breakdown viscosity.

9.3.1.2 Acid Chloride Modification

The modification of starch is done with various chloride and fatty acid chloride (FAC) such as benzoyl, acetyl octanoyl, myristoyl, and stearoyl chlorides. Starch's flexibility and emulsifying abilities are improved by FAC modification. The incorporation of acid chloride leads to the destruction of the crystal structure of amylopectin. A unique modified starch was created by esterifying waxy maize starch with several FACs, such as octanoyl, myristoyl, and stearoyl chlorides, in a NaOH solution reaction media. As the concentration of FACs increased, a rising trend in the DS and a declining trend in yield and reaction efficiency (RE) were seen. The lowest DS (0.027, 0.017, and 0.012 for octanoyl, myristoyl, and stearoyl chlorides, respectively) and best yield and RE were obtained at the concentration of 0.1 ml of FAC/g of starch. According to the dynamic vapor sorption (DVS) results, increasing the DS resulted in an increase in hydrophobicity. The O/W emulsions made with 0.1 ml of FACs/g of starch had the best emulsifying abilities. According to XRD findings, the modified waxy maize starch lost its A-type crystalline structure and became amorphous. Therefore, as compared to conventional procedures, the esterification of starch with FACs in the absence of organic solvent might be recommended as a viable strategy to enhance the physicochemical and emulsifying characteristics of starch for food applications [40]. With the use of acyl chlorides (octanoyl, lauroyl, and palmitoyl chloride) in aqueous media, hydrophobic alteration of potato and maize starch was effectively accomplished with high yields. In comparison to potato starch, maize starch that has been grafted with acyl chlorides has an excellent yield. Using the dialysis approach, hydrophobically modified starch might create nanoparticles. The laser diffraction particle size analyzer analysis validated the generation and size of hydrophobically modified starch nanoparticles, and their particle size ranged from 360 to 500 nm. According to research on the relationship between dialysis time and nanoparticle size, particle size decreases as dialysis time increases. Hydrophobically modified starch nanoparticles aggregated following freeze-drying, according to SEM analyses of starch and grafted starch nanoparticles [41].

9.3.1.3 Octenyl Succinic Anhydride (OSA) Modification

Amphiphilic modifications of starch are usually carried out to increase its hydrophobicity, which is essential for food packaging applications. A hydrophobic and hydrophilic dual-functional group is introduced through OSA modification for its use in many food systems as a stabilizer and emulsifier. In order to enhance the starch coatings' flexibility and barrier properties, ethylene-modified poly(vinyl alcohol) and OSA were mixed with hydroxypropylated maize and potato starches. To increase the starch's hydrophobicity, starch nanoparticles were created by the self-assembly of short-chain amylose that was debranched from cooked taro starch and altered by OSA. By esterifying NS with succinic anhydride, succinylation is accomplished. Under alkaline circumstances, starch succinylation occurs, adding hydrophilic, negatively charged succinyl groups that give the starch its hydrophilicity. As a result, starch succinate offers a number of desirable qualities, including increased consistency, thickening, decreased gelatinization temperature, and improved stability during cold storage [42]. According to studies, succinylation improves the native mung bean starch's ability to expand and the SP grows over time as the temperature rises. As the temperature rose, succinylation's influence on solubility was observed to be diminishing. Following succinylation, the natural mung bean starch's propensity to absorb water rises. The decrease of starch retrogradation caused by the addition of a succinyl group increases the water-binding and -holding ability of the material. On succinylation, the natural mung bean starch's ability to absorb oil likewise declines [43].

9.3.1.4 Hydropropylation

In hydroxypropylation (HPT), starch and propylene oxide are used to create starch ether. The hydroxypropyl group produced during this modification causes starch's hydrogen bonds to be broken, weakening the structure of modified starch and increasing its accessibility to water [44]. Compared to NS, HPT-modified starches showed reduced temperatures of gelatinization and simplified the film fabrication from hydroxypropylated high amylose starches (HASs). The packaging films made from hydroxypropylated high amylose maize starch were more flexible and had ideal TS and water vapor permeability (WVP) [45]. Combining HAS with HPT has excellent potential for generating better films. For use in film applications, hydroxypropylated starch has the benefits and desirable flexibility, transparency, and processability. HPT proved to have the opposite impact on starch films as high amylose concentration did, increasing elongation at break, WVP, and film solubility but decreasing tensile strength [46]. Therefore, it was concluded that high amylose and HPT may significantly enhance the film properties of typical starch film. To boost the amylose content of regular starch, debranching was frequently done [47, 48]. However, there has not been much research on dual modification, which combines debranching and HPT, to enhance the characteristics of starch films.

9.3.1.5 Oxidation of Starch

Oxidation aims to create more functional groups, namely, carbonyl and carboxyl in the modified oxidized starch (ODS), enhancing its functionality and reactivity compared to NS. It has been proven that ODS is whiter and less likely to retrograde than NS. Sodium hypochlorite is the oxidant of choice for creating ODSs, and other oxidants include hydrogen peroxide (H_2O_2), per-acetic acid (CH_3CO_3H), potassium permanganate ($KMnO_4$), chromic acid (H_2CrO_4), and nitrogen dioxide (NO_2). In starch oxidation, the hydroxyl groups are initially changed into carbonyl groups and later into carboxyl groups by oxidation. Accordingly, the amount of carbonyl and carboxyl groups in the ODS, which mostly occur at the hydroxyl groups of the C2, C3, and C6 positions of anhydrous glucose units (AGUs), determines the degree of oxidative substitution. Oxidation causes a reduction in the starch's relative crystallinity, viscosity, SP, and pasting temperature. Oxidation increases the distances between chains and inhibits amylose molecules from realigning, lessening the retrogradation. Compared to NSs, ODSs generate more translucent and flexible films. ODSs' film-forming and binding capabilities in the food industry are correlated with their higher stability, lower viscosity, and increased clarity. The characteristics of starch are significantly impacted by oxidation processes. Researchers discovered an increase in water absorption after oxidizing the starch of Bambara groundnut (Voandzeia subterranean) [49]. The natural starch form had a water absorption capacity of 2.00 0.10 g g^{-1}, whereas the oxidized form had a water absorption capacity of 2.40 0.01 g g^{-1}. The oxidized jack bean starch demonstrated the highest water absorption capacity, with modified starch having a considerably greater water absorption capacity than native jack bean starch [38]. In one study, researchers used gaseous ozone to oxidize carbohydrates derived from pigeon peas, lima beans, and jack beans. The SP of ozone-ODSs was greater than that of their original forms. Pigeon pea and lima bean solubility both increased with ozonation. The jack bean, however, showed decreased solubility after 15 minutes of ozone formation [50].

9.3.1.6 *Cross-Linking of Starch*

The cross-linking of starch is done to improve the mechanical and thermal properties and to reduce swelling as well as solubility. The cross-linked starch shows more hydrophobic behavior as compared to the NS. Cross-linking agents significantly improve the mechanical and barrier properties of most polymeric materials including starch. For instance, cross-linking of sodium trimetaphosphate (STMP) and sodium tripolyphosphate (STPP) with wheat starch resulted in increased tensile strength (TS) and reduction in elongation at break (EAB) [51, 52]. The NS's functional characteristics are impacted by cross-linking. Due to the strengthening of the starch granules' structure, which restricts the starch chain's mobility in the amorphous region and lowers its capacity to absorb water, cross-linked starch has poor solubility [53]. Additionally, cross-linking reduces the ability of legume starch to bind water [54]. The swelling capacity and solubility of the modified starch decrease as the degree of cross-linking increases in comparison to the native form, according to a group of researchers who studied the cross-linking of the kudzu starch (a vine of the leguminous plant) by treating it with sodium trimetaphosphate at various times, temperatures, and concentrations. The structural reorganization of amylose and amylopectin may cause a decrease in starch's ability to swell, and the fact that the cross-linking bonds are stronger than the hydrogen bonds may explain a drop in solubility. The outcome is that the starch granules are difficult to break down [55].

9.3.2 Physical Modification of Starch to Thermoplastic Starch (TPS)

Owing to its low mechanical strength and high permeability to water vapor, starch exhibits limited film-forming applications in its natural state. The physical modification of starch is done by adding a plasticizer to the starch. With the combined effects of heat and shear, a suitable quantity of plasticizer is used to create the deformable thermoplastic material known as TPS. During gelatinization, starch granules must be broken apart to form TPS. During this procedure, amylose molecules dissolve, and amylopectin crystallites melt, forming a gel. Strong hydrogen bonds (intermolecular) between starch chains restrict the processing of gelatinized starch after drying. To facilitate the processes of extrusion or injection molding, plasticizers are added. Chain mobility is increased by the addition of a plasticizer because intermolecular tensions along polymer chains are reduced resulting in the film being more flexible and less brittle which prevents shrinkage. The most effective starch plasticizers are water and glycerol. Other plasticizers that work well with TPS include ethane-1,2-diol, sorbitol, glucose, sucrose, fructose, urea, maltodextrin, and amino acids. TPS films are still inferior to petroleum-based films despite having good barrier and mechanical characteristics. The critical properties of TPS are dependent on its glass transition temperature (Tg). Tg of TPS is generally observed between 75°C and 100°C. Above the Tg, starch loses stability and swells irreversibly. This issue needs to be addressed while processing starch into film where the tolerance is minimal. In the extrusion process, starch is transformed using thermal and mechanical energy. Three key events are observed at different structural levels: partial depolymerization of the starch polymers, hydrogen bond breaking between starch molecules, which results in the loss of crystallinity, and fragmentation of starch granules.

9.4 Starch-Based Composite, Nanocomposite, and Hybrid Films

Products made from TPS are hydrophilic and water-sensitive, which would result in less favorable mechanical and physical qualities. These properties of starch can be modified using a variety of techniques, including combining two polymers to form composite materials, incorporating nano-sized fillers into the starch to form nanocomposite, and combining more than two polymers and/or nanofillers to form hybrid materials.

9.4.1 Starch-Based Blends

TPS can be blended with various synthetic polymers as well as biopolymers to impart biodegradability. Extensive research has been done on commercializing starch-based blends for packaging applications. Compared to NS, adding functionalized tapioca starch to PBS polymeric blends exhibited superior thermal stability and strength [57]. PLA-starch (70/30 or 80/20) with epoxidized soybean oil, maleic anhydride, castor oil, and hexamethylene diisocyanate offered desirable mechanical properties and could possibly substitute for nondegradable petroleum-derived polymers owing to their functional characteristics, polymer availability, sufficient food contact properties, and competitive cost. However, compared to NS film, PLA-starch multilayer films showed enhanced mechanical resistance and excellent water vapor and gas barrier capability. The addition of hydrophilic starch and hydrophobic PLA layer develops a synergistic blend for packaging applications [58].

Lactic acid from sugar and beets is converted into PLA, an aliphatic semicrystalline polyester. PLA is a biopolymer that breaks down through hydrolysis, making it suitable for use in food and pharmaceutical areas because of its biocompatible qualities. However, compared to traditional polymers, PLA has poorer thermal stability, is more costly, and is brittle resulting in its limited use. In order to lower the price of PLA and enhance its biodegradability, starch is combined with PLA. However, due to low barrier and mechanical qualities as well as distinct polarity in the two polymers, their compatibility is constrained. A well-known aliphatic polyester with outstanding flexibility, high elongation, and good processability is polybutylene adipate terepthalate (PBAT). A combination of sodium benzoate, potassium sorbate, PBAT, and TPS was utilized to increase the shelf life of packaged rice noodles [56]. After seven reprocessing cycles, a ternary mix of PP/PBAT/TPS blends was compared to the polypropylene matrix for degradation. Due to the minimal degradation temperature, they discovered that the PBAT/TPS phase degraded before the PP matrix, and the recycling process increased the elasticity modulus and yield strength despite lowering impact strength.

The O_2 and aroma barrier properties of polyvinyl alcohol make it an ideal material for use in a variety of food packaging applications. It is also odorless and nontoxic. Through possible plasticizing and cross-linking agents, the mechanical properties of PVOH can be improved. Because starch and other polymers are less expensive than polyvinyl alcohol (PVOH), combining PVOH with inexpensive starch might be an option to reduce the cost of mixes overall while improving the qualities of the starch. Given its polar nature, PVOH (65 to 70%) reported interacting with cassava starch in a way that had a physical bonding strength comparable to PVA's [59]. Additionally, PVOH/corn starch films had excellent thermal stability than PVOH alone. To increase TS (1.8 MPa) and EAB (113%) and avoid surface cracks, extruded corn starch might be blended with glycerol and then with PVOH.

Due to their resistance to oxidation and photodegradation, polyolefin polymers, such as polyethylene (PE) are very stable. To hasten the environment's mechanical breakdown of polyolefins, it is advised that they be blended with polymers made of carbohydrates. TPS has been combined with several types of PE, such as LDPE and LLDPE, HDPE, and LLDPE, in the past for enhanced packaging applications for food. In an effort to make TPS/PE mixtures with distinctive characteristics, they employed a combined twin-screw/single screw extrusion process to create PE/TPS (55:45) blends by employing glycerol (36%) as a plasticizer while maintaining the degree of elongation (94%) at break and modulus (76%) of PE. Without any alterations, this mixture had identical ductility and modulus properties as the original PE at very high TPS concentrations. LDPE/LLDPE (70:30) was mixed with TPS (5 to 20 wt%) and PE-grafted maleic anhydride (PE-g-MA) at 3 wt% in a twin-screw extruder before being blown, and the result showed that the LLDPE had a more dramatic influence on starch crystallization than LDPE.

9.4.2 Starch-Based Composite and Nanocomposite

Citric acid and gelatin were incorporated with starch through solution blending to fabricate biocomposite film with improved function properties, and the results showed that adding citric acid up to 5% in the starch films considerably reduced the amount of free hydroxyl groups that exist in the starch matrix through cross-linking. Modified starch films were stiffer and more transparent than neat starch films. The solubility, swelling index, moisture content, water vapor, migration rate, and moisture rate were all improved by adding high gelatin content in starch [60]. Natural starches have weak mechanical qualities, less prolonged stability, and high water sensitivity; their direct application is constrained [61]. The incorporation of different nanofillers, such as layers, fibers, and particles into the starch matrix to create nanocomposites, has drawn a lot of interest [60], thereby improving the mechanical, thermal, moisture-resistant, and oxygen barrier of nanocomposite materials. Additionally, nanocomposites have many advantages over conventional composites, including the ability to achieve desired properties with only a small amount of fillers (less than 15%). Due to their size being considerably smaller than the wavelength of light and the fact that they do not scatter light, nanofillers can also boost the optical transparency of the composite [62]. A stiff molecular network will form as a result of strong interface adhesion and good dispersion of the nanofillers inside the starch matrix, which can increase the variety of characteristics of the nanocomposites [40]. Low water barrier and tensile qualities were enhanced by TPS/nanocellulose, as were mechanical properties [63].

A relatively small layer of silicate can be included in the starch matrix to enhance the biopolymer's mechanical and barrier properties. Montmorillonite (MMT), which is known for its environmental friendliness, low cost, and processability, is the most used layered silicate in starch-based nanocomposite [64]. As compared to NS films, the nanocomposite films showed remarkable gains in mechanical characteristics and a decrease in WVP [46, 47, 65, 66]. Additionally, studies on MMT and potato starch nanocomposites showed a significant increase in modulus and tensile strength [67–69].

Starch and talc nanoparticle–based composite film were fabricated and used for manufacturing packaging film/bag. Talc concentrations above 3% w/w considerably raised TPS films' stiffness without influencing their elongation at break. These findings suggest that talc nanoparticle–reinforced TPS sheets could create more durable and robust packaging without sacrificing flexibility [70]. Polyester and starch blend at a ratio of 70/30 with little addition of nanoclay MMT about 1.5 to 5 wt% produces a film with a more exfoliated structure and greater tensile properties [71].

Cellulose nanocrystal (5 and 30 wt%) made from flax fibers was employed to fortify with plasticized starch (PS) matrix and create nanocomposite materials using the solution casting process. Cellulose nanocrystals as filler had high adherence in the interfacial region and demonstrated good dispersion inside the PS matrix and decreased flexibility of starch molecular chains [72]. A modified potato starch matrix was employed with sorbitol and water as plasticizers and 5% cellulose nanowhisker, and the result showed an improvement in tensile and yield strength by 90 MPa (24%) and 100 MPa (27%), respectively [73]. The effects of MMT nanoclay and different thyme essential oil concentrations on the characteristics of sweet potato starch–based nanocomposite were also investigated, and the results showed improvements in tensile strength, Young modulus, elongation at break, WVP, and reduced film solubility [74].

9.5 Food Packaging Applications

In recent decades, the usage of synthetic plastics for food packaging has increased at a pace of 5% per year, and synthetic plastics are the second most used material for food packaging [75]. There have been a lot of investigations on the use of starch-based polymers for food packaging applications due to their abundant availability, affordable pricing, nontoxicity, biodegradability, and compostability and more are given in Figures 9.4 and 9.5 [56]. Nevertheless, a TPS can be created by adding plasticizers, applying heat, high pressure, and mechanical shearing force while starch is being processed. By breaking down hydrogen bonds, plasticizers lower the intermolecular force resulting in more flexible starch, which has significantly lower Tg and Tm [76, 77]. The main focus of the development of starch-based products is due to their biodegradability, availability of a variety of raw resources, and reasonable price [78–80]. Compared to synthetic polymers, the blended material deteriorates rapidly and has better mechanical properties [81, 82]. There are several benefits of using biodegradable materials based on starch in food packaging such as reduced petroleum product

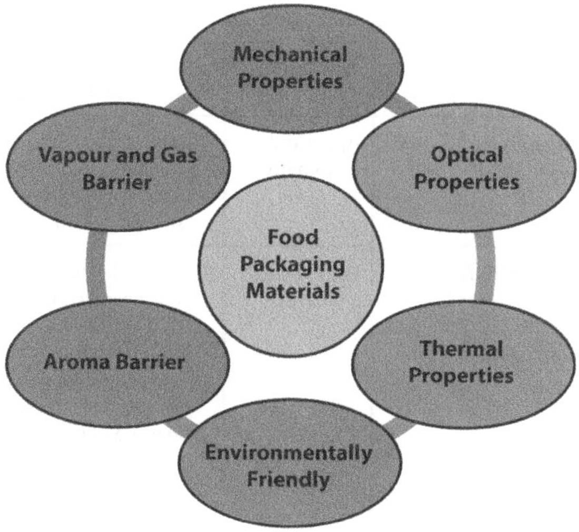

Figure 9.4 Desire properties of starch for food packaging.

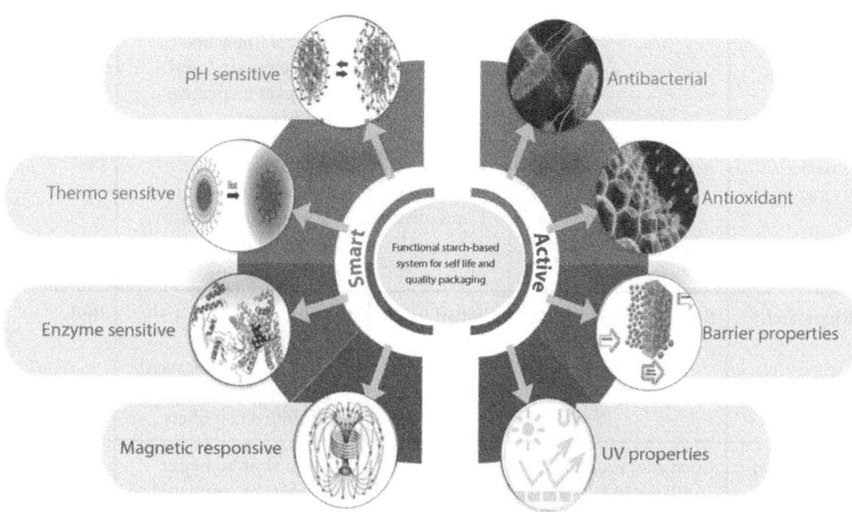

Figure 9.5 Starch for food packaging applications [56].

consumption, environmental friendliness, and d safe food packaging [83]. Biodegradable materials have limited use due to their high manufacturing costs and subpar mechanical and barrier qualities when compared to standard synthetic packaging materials [84]. In recent years, several researchers worked on the utilization of agro-waste–derived starches for food packaging applications, which are summarized in Table 9.1.

Table 9.1 Starch from agro-waste and their applications in food packaging.

Source	Composition of packaging material	Application on food	Effects on food quality and shelf life	References
Wheat starch	Starch/glycerol (70/30%)	Apple and tomato	Vitamin C content remains almost constant over the storage period	[85]
Potato starch	Starch (67%)/glycerol (30%), MMT nanoclay (3%), and thyme (0, 2, 4, and 6%)	Baby spinach leave	The lower bacterial activity was observed for a neat starch film packed with baby spinach leaves, and no bacterial activity for 4 and 6% loading of thyme	[86]
Cassava starch	PLAs/starch (90/10%), modified with glycidyl methacrylate (80/20%)	Capsicum	Retained the quality attributes of the packaged capsicum up to 12 and 24 days at 25°C and 8°C, respectively, as compared to the control unpacked capsicum *viz* 4 and 9 days at respective temperatures	[87]
Cassava starch	Acylated cassava starch/glycerol (70/30%) and pea protein (5, 10, and 20%)	Vegetable oil (soybean and olive)	10 and 20% protein loading showed good dimensional stability and a higher barrier to lipid oxidation of oil	[88]
Rice starch	Rice starch/glycerol (70/30%) and oregano essential oil (4.5%)	Chilled fish fillet	The film containing 4.5% (m/v) OEO, 30% glycerol, and 3% starch increased the shelf life of chilled fish fillets	[89]
Corn starch	Corn starch/chitosan/ Plasticizer (glycerol/sorbitol)/ antimicrobial agent (potassium sorbate)	Bread	The film showed better properties compared to neat LDPE and LLDPE film	[90]
Corn starch	Corn Starch/glycerol (90/10%), gelatin and guabiroba pulp	Olive oil	The addition of guabiroba pulp into a blend of starch/glycerol/gelatin improved the shelf life of olive oil	[91]

9.6 Conclusion and Perspectives

Starch-based packaging materials have a lot of potential as a sustainable and greener alternative to synthetic plastic packaging in the food sector. These materials lack their synthetic equivalent's optical, barrier, and mechanical qualities, and their general manufacturing for food applications is limited. Starch has been physically and chemically modified and blended with other biopolymers, plasticizers, and nanofillers to mitigate these limitations. Although starch-based packaging materials are biodegradable, chemical additives may reduce their biodegradability. Starch-based biodegradable materials may play a significant role in developing sustainable food packaging materials in the future because they use less energy and are less harmful to the ecosystem than petroleum-based synthetic packaging materials. Further research is needed to scale up and commercialize starch-based packaging materials for food applications.

References

1. Bangar, S.P., Whiteside, W.S., Ashogbon, A.O., Kumar, M., Recent advances in thermoplastic starches for food packaging: A review. *Food Packag. Shelf Life*, **30**, July, 100743, 2021.
2. Fang, J. and F., P., The use of starch and its derivatives as biopolymer sources of packaging materials. *J. Food Agric. Environ.*, 1, 82–84, August 2003.
3. Khan, B., Bilal Khan Niazi, M., Samin, G., Jahan, Z., Thermoplastic starch: A possible biodegradable food packaging material-A review. *J. Food Process Eng.*, **40**, 3, e12447, 2017.
4. Tang, X.Z., Kumar, P., Alavi, S., Sandeep, K.P., Recent advances in biopolymers and biopolymer-based nanocomposites for food packaging materials. *Crit. Rev. Food Sci. Nutr.*, **52**, 5, 426–442, 2012.
5. Shlush, E. and Davidovich-Pinhas, M., Bioplastics for food packaging. *Trends Food Sci. Technol.*, **125**, 03, 66–80, 2022.
6. Zhang, Y., Rempel, C., Liu, Q., Thermoplastic starch processing and characteristics—A review. *Crit. Rev. Food Sci. Nutr.*, **54**, 10, 1353–1370, 2014.
7. Sanyang, M.L., Ilyas, R.A., Sapuan, S.M., Jumaidin, R., Sugar palm starch-based composites for packaging applications, in: *Bionanocomposites for Packaging Applications*, pp. 125–147, International Publishing, Springer, Cham, 2018.
8. Atikah, M.S.N., Ilyas, R.A., Sapuan, S.M., Ishak, M.R., Zainudin, E.S., Ibrahim, R., Atiqah, A., Ansari, M.N.M., Jumaidin, R., Degradation and physical properties of sugar palm starch/sugar palm nanofibrillated cellulose bionanocomposite. *Polimery*, **64**, 10, 680–689, 2019.

9. Ilyas, R.A., Sapuan, S.M., Ishak, M.R., Zainudin, E.S., Development and characterization of sugar palm nanocrystalline cellulose reinforced sugar palm starch bionanocomposites. *Carbohydr. Polym.*, **202**, May, 186–202, 2018.
10. Wei, M., Andersson, R., Xie, G., Salehi, S., Boström, D., Xiong, S., Properties of cassava stem starch being a new starch resource. *Starch - Stärke*, **70**, 5–6, 1700125, 2018.
11. Liu, H., Xie, F., Yu, L., Chen, L., Li, L., Thermal processing of starch-based polymers. *Prog. Polym. Sci.*, **34**, 12, 1348–1368, 2009.
12. Chandrasekara, A. and Josheph Kumar, T., Roots and tuber crops as functional foods: A review on phytochemical constituents and their potential health benefits. *Int. J. Food Sci.*, **2016**, 1–15, 2016.
13. Grommers, H.E. and van der Krogt, D.A., Potato starch, in: *Starch*, pp. 511–539, Elsevier, Amsterdam, The Netherlands, 2009.
14. Yang, L., Zhang, H., Huang, B., Hao, S., Li, S., Li, P., Yu, H., Studying the role of potato powder on the physicochemical properties and dough characteristics of wheat flour. *Gels*, **9**, 2, 73, 2023.
15. Wang, F.C., Chung, D.S., Seib, P.A., Kim, Y.S., Optimum steeping process for wet milling of Sorghum. *Cereal Chem. J.*, **77**, 4, 478–483, 2000.
16. Pilling, E. and Smith, A.M., Growth ring formation in the starch granules of potato tubers. *Plant Physiol.*, **132**, 1, 365–371, 2003.
17. Schmidt, J.M., Damgaard, H., Greve-Poulsen, M., Larsen, L.B., Hammershøj, M., Foam and emulsion properties of potato protein isolate and purified fractions. *Food Hydrocoll.*, **74**, 367–378, 2018.
18. Ayoub, A.S. and Rizvi, S.S.H., An overview on the technology of cross-linking of starch for nonfood applications. *J. Plast. Film Sheeting*, **25**, 1, 25–45, 2009.
19. Shewry, P.R., Tuber storage proteins. *Ann. Bot.*, **91**, 7, 755–769, 2003.
20. Singh, V. and Zakiuddin Ali, S., Estimation of phosphorus in native and modified starches. Improvement in the molybdovanadophosphoric acid method. *Starch - Stärke*, **39**, 8, 277–279, 1987.
21. Guan, Y., Zhao, G., Thaiudom, S., Evaluation of the physico-chemical properties of potato starch-based foods and their interactions with milk protein and soybean oil. *Food Chem. X*, **16**, October, 100495, 2022.
22. Waterschoot, J., Gomand, S.V., Fierens, E., Delcour, J.A., Production, structure, physicochemical and functional properties of maize, cassava, wheat, potato and rice starches. *Starch - Stärke*, **67**, 1–2, 14–29, 2015.
23. Kaur, M. and Sandhu, K.S., Sweet potato flour and starch, in: *Tropical Roots and Tubers*, pp. 479–506, John Wiley & Sons, Ltd., Chichester, UK, 2016.
24. Tetlow, I.J., Starch biosynthesis in developing seeds. *Seed Sci. Res.*, **21**, 1, 5–32, 2011.
25. Shahrim, N.A., Sarifuddin, N., Ismail, H., Extraction and characterization of starch from mango seeds. *J. Phys. Conf. Ser.*, **1082**, 1, 012019, 2018.
26. Maryam, Kasim, A., Santosa, Utilization starch of avocado seed (Persea Americana Mill.) as a raw material for dextrin. *J. Food Sci. Eng.*, **6**, 1, 32–37, 2016.

27. Suzihaque, M.U., Zaki, N.A.M., Alwi, H., Ibrahim, U.K., Abd Karim, S.F., Anuar, N.K., Jackfruit seed as an alternative replacement for starch flour. *Mater. Today Proc.*, **63**, S451–S455, 2022.
28. Swami, S.B., Thakor, N.J., Haldankar, P.M., Kalse, S.B., Jackfruit and its many functional components as related to human health: A review. *Compr. Rev. Food Sci. Food Saf.*, **11**, 6, 565–576, 2012.
29. Sudharsan, K., Chandra Mohan, C., Azhagu Saravana Babu, P., Archana, G., Sabina, K., Sivarajan, M., Sukumar, M., Production and characterization of cellulose reinforced starch (CRT) films. *Int. J. Biol. Macromol.*, **83**, 385–395, 2016.
30. Bangar, S.P., Kumar, M., Whiteside, W.S., Tomar, M., Kennedy, J.F., Litchi (Litchi chinensis) seed starch: Structure, properties, and applications - A review. *Carbohydr. Polym. Technol. Appl.*, **2**, February, 100080, 2021.
31. Marichelvam, M.K., Jawaid, M., Asim, M., Corn and rice starch-based bio-plastics as alternative packaging materials. *Fibers*, **7**, 4, 1–14, 2019.
32. Kanatt, S.R., Irradiation as a tool for modifying tapioca starch and development of an active food packaging film with irradiated starch. *Radiat. Phys. Chem.*, **173**, March, 108873, 2020.
33. Lauer, M.K. and Smith, R.C., Recent advances in starch-based films toward food packaging applications: Physicochemical, mechanical, and functional properties. *Compr. Rev. Food Sci. Food Saf.*, **19**, 6, 3031–3083, 2020.
34. Medina V, O.J., Pardo C, O.H., Ortiz, C.A., Modified arracacha starch films characterization and its potential utilization as food packaging. *Vitae*, **19**, 2, 186–196, 2012.
35. Gutiérrez, T.J. and Valencia, G.A., Reactive extrusion-processed native and phosphated starch-based food packaging films governed by the hierarchical structure. *Int. J. Biol. Macromol.*, **172**, 439–451, 2021.
36. Nuñez-Santiago, C., Garcia-Suarez, F.J.L., Roman-Gutierrez, A.D., Bello-Pérez, L.A., Effect of reagent type on the acetylation of barley and maize starches. *Starch - Stärke*, **62**, 9, 489–497, 2010.
37. Han, F., Liu, M., Gong, H., Lü, S., Ni, B., Zhang, B., Synthesis, characterization and functional properties of low substituted acetylated corn starch. *Int. J. Biol. Macromol.*, **50**, 4, 1026–1034, 2012.
38. Lawal, O.S. and Adebowale, K.O., Physicochemical characteristics and thermal properties of chemically modified jack bean (Canavalia ensiformis) starch. *Carbohydr. Polym.*, **60**, 3, 331–341, 2005.
39. Wojeicchowski, J.P., de Siqueira, G.L., de, A., Lacerda, L.G., Schnitzler, E., Demiate, I.M., Physicochemical, structural and thermal properties of oxidized, acetylated and dual-modified common bean (Phaseolus vulgaris L.) starch. *Food Sci. Technol.*, **38**, 2, 318–327, 2018.
40. Mirzaaghaei, M., Nasirpour, A., Keramat, J., Goli, S.A.H., Dinari, M., Desobry, S., Durand, A., Chemical modification of waxy maize starch by esterification with saturated fatty acid chlorides: Synthesis, physicochemical and emulsifying properties. *Food Chem.*, **393**, November 2021, 133293, 2022.

41. Namazi, H., Fathi, F., Dadkhah, A., Hydrophobically modified starch using long-chain fatty acids for preparation of nanosized starch particles. *Sci. Iran.*, **18**, 3 C, 439–445, 2011.
42. Mehboob, S., Ali, T.M., Alam, F., Hasnain, A., Dual modification of native white sorghum (Sorghum bicolor) starch via acid hydrolysis and succinylation. *LWT - Food Sci. Technol.*, **64**, 1, 459–467, 2015.
43. García-Gurrola, A., Rincón, S., Escobar-Puentes, A.A., Zepeda, A., Pérez-Robles, J.F., Martínez-Bustos, F., Synthesis and succinylation of starch nanoparticles by means of a single step using sonochemical energy. *Ultrason. Sonochem.*, **56**, September 2018, 458–465, 2019.
44. Kim, H.R., Hermansson, A.-M., Eriksson, C.E., Structural characteristics of hydroxypropyl potato starch granules depending on their molar substitution. *Starch - Stärke*, **44**, 3, 111–116, 1992.
45. Fu, Z., Zhang, L., Ren, M.-H., BeMiller, J.N., Developments in hydroxypropylation of starch: A review. *Starch - Stärke*, **71**, 1–2, 1800167, 2019.
46. Kim, H.-Y., Jane, J., Lamsal, B., Hydroxypropylation improves film properties of high amylose corn starch. *Ind. Crops Prod.*, **95**, 175–183, 2017.
47. Miao, M., Jiang, B., Zhang, T., Effect of pullulanase debranching and recrystallization on structure and digestibility of waxy maize starch. *Carbohydr. Polym.*, **76**, 2, 214–221, 2009.
48. Luo, Z., Zou, J., Chen, H., Cheng, W., Fu, X., Xiao, Z., Synthesis and characterization of amylose–zinc inclusion complexes. *Carbohydr. Polym.*, **137**, 314–320, 2016.
49. Adebowale, K.O., Adeniyi Afolabi, T., Lawal, O.S., Isolation, chemical modification and physicochemical characterisation of Bambarra groundnut (Voandzeia subterranean) starch and flour. *Food Chem.*, **78**, 3, 305–311, 2002.
50. Oladebeye, A.O., Oshodi, A.A., Amoo, I.A., Karim, A.A., Gaseous ozonation of Pigeon Pea, Lima Bean, and Jack Bean starches: Functional, thermal, and molecular properties. *Starch - Stärke*, **70**, 11–12, 1700367, 2018.
51. Sun, S., Liu, P., Ji, N., Hou, H., Dong, H., Effects of various cross-linking agents on the physicochemical properties of starch/PHA composite films produced by extrusion blowing. *Food Hydrocoll.*, **77**, 964–975, 2018.
52. Chandak, A., Dhull, S.B., Punia Bangar, S., Rusu, A.V., Effects of cross-linking on physicochemical and film properties of lotus (Nelumbo nucifera G.) Seed Starch. *Foods*, **11**, 19, 3069, 2022.
53. Ashogbon, A.O., Akintayo, E.T., Oladebeye, A.O., Oluwafemi, A.D., Akinsola, A.F., Imanah, O.E., Developments in the isolation, composition, and physicochemical properties of legume starches. *Crit. Rev. Food Sci. Nutr.*, **61**, 17, 2938–2959, 2021.
54. Vanier, N.L., da Rosa Zavareze, E., Pinto, V.Z., Klein, B., Botelho, F.T., Dias, A.R.G., Elias, M.C., Physicochemical, crystallinity, pasting and morphological properties of bean starch oxidised by different concentrations of sodium hypochlorite. *Food Chem.*, **131**, 4, 1255–1262, 2012.

55. Chen, B., Dang, L., Zhang, X., Fang, W., Hou, M., Liu, T., Wang, Z., Physicochemical properties and micro-structural characteristics in starch from kudzu root as affected by cross-linking. *Food Chem.*, **219**, 93–101, 2017.
56. García-Guzmán, L., Cabrera-Barjas, G., Soria-Hernández, C.G., Castaño, J., Guadarrama-Lezama, A.Y., Rodríguez Llamazares, S., Progress in starch-based materials for food packaging applications. *Polysaccharides*, **3**, 1, 136–177, 2022.
57. Ayu, R., Khalina, A., Harmaen, A., Zaman, K., Jawaid, M., Lee, C., Effect of modified tapioca starch on mechanical, thermal, and morphological properties of PBS blends for food packaging. *Polym. (Basel)*, **10**, 11, 1187, 2018.
58. Muller, J., González-Martínez, C., Chiralt, A., Combination of poly(lactic) acid and starch for biodegradable food packaging. *Mater. (Basel)*, **10**, 8, 952, 2017.
59. Ge, C., Lansing, B., Lewis, C.L., Thermoplastic starch and poly(vinyl alcohol) blends centered barrier film for food packaging applications. *Food Packag. Shelf Life*, **27**, December 2020, 100610, 2021.
60. Zakaria, N.H., Muhammad, N., Abdullah, M.M.A.B., Potential of starch nanocomposites for biomedical applications. *IOP Conf. Ser. Mater. Sci. Eng.*, **209**, 1, 012087, 2017.
61. Sanyang, M., Sapuan, S., Jawaid, M., Ishak, M., Sahari, J., Effect of plasticizer type and concentration on tensile, thermal and barrier properties of biodegradable films based on sugar palm (Arenga pinnata) starch. *Polym. (Basel)*, **7**, 6, 1106–1124, 2015.
62. Bajer, D., Kaczmarek, H., Bajer, K., The structure and properties of different types of starch exposed to UV radiation: A comparative study. *Carbohydr. Polym.*, **98**, 1, 477–482, 2013.
63. Nazrin, A., Sapuan, S.M., Zuhri, M.Y.M., Ilyas, R.A., Syafiq, R., Sherwani, S.F.K., Nanocellulose reinforced Thermoplastic Starch (TPS), Polylactic Acid (PLA), and Polybutylene Succinate (PBS) for food packaging applications. *Front. Chem.*, **8**, 1–12, 2020.
64. Zhou, M. and Xu, D., Starch-MMT composite films: Effects of bio-inspired modification on MMT. *Starch - Stärke*, **67**, 5–6, 470–477, 2015.
65. Monteiro, M.K.S., Oliveira, V.R.L., Santos, F.K.G., Barros Neto, E.L., Leite, R.H.L., Aroucha, E.M.M., Silva, R.R., Silva, K.N.O., Incorporation of bentonite clay in cassava starch films for the reduction of water vapor permeability. *Food Res. Int.*, **105**, December 2017, 637–644, 2018.
66. Romo-Uribe, A., Reyes-Mayer, A., Calixto-Rodriguez, M., Sarmiento-Bustos, E., Elastomeric biodegradable starch/bentonite nanocomposites. Structure-thermo-mechanical correlation and degradation behavior. *Carbohydr. Polym.*, **304**, September 2022, 120514, 2023.
67. Avella, M., De Vlieger, J.J., Errico, M.E., Fischer, S., Vacca, P., Volpe, M.G., Biodegradable starch/clay nanocomposite films for food packaging applications. *Food Chem.*, **93**, 3, 467–474, 2005.

68. Tang, X., Alavi, S., Herald, T.J., Barrier and mechanical properties of starch-clay nanocomposite films. *Cereal Chem.*, **85**, 3, 433–439, 2008.
69. Tang, X., Alavi, S., Herald, T.J., Effects of plasticizers on the structure and properties of starch-clay nanocomposite films. *Carbohydr. Polym.*, **74**, 3, 552–558, 2008.
70. Castillo, L., López, O., López, C., Zaritzky, N., García, M.A., Barbosa, S., Villar, M., Thermoplastic starch films reinforced with talc nanoparticles. *Carbohydr. Polym.*, **95**, 2, 664–674, 2013.
71. McGlashan, S.A. and Halley, P.J., Preparation and characterisation of biodegradable starch-based nanocomposite materials. *Polym. Int.*, **52**, 11, 1767–1773, 2003.
72. Cao, X., Chen, Y., Chang, P.R., Muir, A.D., Falk, G., Starch-based nanocomposites reinforced with flax cellulose nanocrystals. *Express Polym. Lett.*, **2**, 7, 502–510, 2008.
73. Kvien, I., Sugiyama, J., Votrubec, M., Oksman, K., Characterization of starch based nanocomposites. *J. Mater. Sci.*, **42**, 19, 8163–8171, 2007.
74. Issa, A.T., Schimmel, K.A., Worku, M., Shahbazi, A., Ibrahim, S.A., Tahergorabi, R., Sweet potato starch-based nanocomposites:Development, characterization, and biodegradability. *Starch - Stärke*, **70**, 7–8, 1700273, 2018.
75. Sadeghizadeh-Yazdi, J., Habibi, M., Kamali, A.A., Banaei, M., Aplicación de películas a base de almidón comestibles y biodegradables en elenvasado de alimentos: Revisión sistemática y metanálisis. *Investig. Actual en Nutr. y Cienc. Aliment.*, **7**, 3, 624–637, 2019.
76. Glenn, G.M., Orts, W., Wood, D.F., Glenn, G.M., Orts, W., Imam, S., Chiou, B., Wood, D.F., Applications agriculture applications, in: *Starch Plast. Packag. Agric. Appl*, pp. 526–536, 2017.
77. Żołek-Tryznowska, Z. and Kałuża, A., The influence of starch origin on the properties of starch films: Packaging performance. *Mater. (Basel)*, **14**, 5, 1–11, 2021.
78. Liu, D., Zhao, P., Chen, J., Yan, Y., Wu, Z., Recent advances and applications in starch for intelligent active food packaging: A review. *Foods*, **11**, 18, 2879, 2022.
79. Abioye, A.A. and Obuekwe, C.C., Investigation of the biodegradation of low-density polyethylene-starch Bi-polymer blends. *Results Eng.*, **5**, November 2019, 100090, 2020.
80. Dammak, M., Fourati, Y., Tarrés, Q., Delgado-Aguilar, M., Mutjé, P., Boufi, S., Blends of PBAT with plasticized starch for packaging applications: Mechanical properties, rheological behaviour and biodegradability. *Ind. Crops Prod.*, **144**, 1–8, May 2019, 2020.
81. Castaño, J., Rodríguez-Llamazares, S., Sepúlveda, E., Giraldo, D., Bouza, R., Pozo, C., Morphological and structural changes of starch during processing by melt blending. *Starch/Staerke*, **69**, 9–10, 1–9, 2017.

82. Hu, H., Xu, A., Zhang, D., Zhou, W., Peng, S., Zhao, X., High-toughness poly(lactic acid)/starch blends prepared through reactive blending plasticization and compatibilization. *Molecules*, **25**, 24, 1–16, 2020.
83. Liu, P., Li, Y., Shang, X., Xie, F., Starch–zinc complex and its reinforcement effect on starch-based materials. *Carbohydr. Polym.*, **206**, July 2018, 528–538, 2019.
84. Kuz, P. and Ateş, M., Starch-based bioplastic materials for packaging industry. *J. Sustain. Constr. Mater. Technol.*, **5**, 1, 399–406, 2020.
85. Saiful, Helwati, H., Saleha, S., Iqbalsyah, T.M., Development of bioplastic from wheat Janeng starch for food packaging. *IOP Conf. Ser. Mater. Sci. Eng.*, **523**, 1, 1–6, 2019.
86. Tahergorabi, R., Sweet potato starch/clay nanocomposite film: New material for emerging biodegradable food packaging. *MOJ Food Process. Technol.*, **3**, 3, 313–315, 2016.
87. Mangaraj, S., Ranjan, R., Ajay, T., Development and characterization of PLA and Cassava based novel biodegradable film used for food packaging application. *J. Food Process. Preserv.*, **2050**, September 2021, 1–14, 2022.
88. Huntrakul, K., Yoksan, R., Sane, A., Harnkarnsujarit, N., Effects of pea protein on properties of cassava starch edible films produced by blown-film extrusion for oil packaging. *Food Packag. Shelf Life*, **24**, August 2019, 100480, 2020.
89. Martins, P.C. and Bagatini, D.C., Oregano essential oil addition in rice starch films and its effects on the chilled fish storage. *J. Food Sci. Technol.*, **58**, April, 1562–1573, 2021.
90. Jha, P., International Journal of Biological Macromolecules effect of plasticizer and antimicrobial agents on functional properties of bionano-composite films based on corn starch-chitosan for food packaging applications. *Int. J. Biol. Macromol.*, **160**, 571–582, 2020.
91. Miriam, N., Camila, A., Cristiani, R., Paula, A., Yamashita, F., Tormen, L., Matta, F., Ignacio, J., Canhadas, L., Corn starch and gelatin-based films added with guabiroba pulp for application in food packaging. *Food Packag. Shelf Life*, **19**, July 2018, 140–146, 2019.

10

Chitosan from Agro-Waste for Food Packaging Applications

Mekala Pavani, Sushil Kumar Singh and Poonam Singha*

Department of Food Process Engineering, National Institute of Technology (NIT) Rourkela, Odisha, India

Abstract

The use of synthetic plastics should be avoided due to their non-renewable, non-compostable, and non-biodegradable nature. The utilization of alternative environmentally friendly biopolymers like chitosan has increased. Chitin and its derivative, i.e., chitosan is the second most plentiful polysaccharides after cellulose, which has gained predominant value due to their unique functional and biological properties, wide availability, nontoxicity, biocompatibility, biodegradability, and lower cost. It is mainly obtained from diverse sources, including marine wastes, fungi, insects, and agricultural wastes. Marine wastes are the common source of chitosan, but their use is decreasing due to seasonal availability and the generation of toxic effluents. Therefore, there is a requirement for substitute sources to fabricate sustainable and economically viable options for chitosan production. This chapter mainly focuses on chitin or chitosan from agricultural waste, their production process from agricultural waste, and their functional properties. Furthermore, detailed information about various methods to fabricate films and coatings and their applications in preserving meat, fish, fruits, and vegetables are also elaborated.

Keywords: Agricultural waste, chitosan, chitin, coating, packaging film, perishable foods

**Corresponding author*: poonamsingha2652@gmail.com; singhap@nitrkl.ac.in

Santosh Kumar, Avik Mukherjee and Vimal Katiyar (eds.) Agro-Waste Derived Biopolymers and Biocomposites: Innovations and Sustainability in Food Packaging, (267–294) © 2024 Scrivener Publishing LLC

List of Abbreviations

ABTS - 2,2′-azino-bis (3-ethylbenzothiazoline-6-sulfonic acid
CA - Cashew apple
CAJ - Cashew apple juice
CFU - Colony forming unit
CO -NH - Amide group
CSL - Corn steep liquor
CW - Cheese whey
DA - Degrees of acetylation
DD - Degree of deacetylation
DNA - Deoxyribonucleic acid
DPPH - 2,2-diphenylpicrylhydrazyl
MW - Molecular weight
-OH - Hydroxyl groups
PPJ - Papaya peel juice
RNA - Ribonucleic acid
ROS - Reactive oxygen species

10.1 Introduction

Packaging of food protects it from physical, chemical, and biological hazards along the food chain and extends food products' shelf life by maintaining quality [1]. For the last half-century, petroleum-based polymers including polyester, ethylene vinyl alcohol, and polypropylene have been widely utilized materials for food packaging due to their heat seal ability, higher mechanical, and barrier properties at lower cost [2]. However, extensive use of petroleum-based synthetic plastic materials could negatively affect the environment and human health due to their nonrenewable, non-compostable, and non-biodegradable properties [3]. In this context, recently, food researchers and industries have focused on developing eco-friendly biopolymers to reduce environmental challenges as well as protect various food products from deterioration caused by external conditions. In addition, these biopolymers show better antioxidant, antimicrobial, and barrier characteristics, which makes them appropriate material for the development of food packaging films/coatings to protect various food products [4, 5]. In general, polysaccharides, proteins, lipids, essential oils, and plant extracts are commonly used raw materials to develop food packaging films. Among all, polysaccharides like chitosan, cellulose, agar, pectin, sodium alginate, gelatin, polylactic acid, carrageenan, and starch are predominant and common base materials because of their better film-forming properties, being eco-friendly, nontoxicity, and lower cost [1, 6].

Chitosan is a naturally derived cationic polysaccharide consisting of randomly lined β-(1–4)-2-amino-D-glucose and N-acetyl-D-glucosamine units. It is derived from chitin by alkaline deacetylation at higher temperatures [7]. Traditionally, chitosan is mainly obtained from shrimp, crab, prawn, krill, and fish waste. However, chitosan extracted from these sources is restricted due to seasonal availability and scarcity in certain countries. Additionally, the extraction process can lead to environmental pollution, as it generates a large amount of acidic and alkaline effluents. Furthermore, the production of chitosan requires high temperatures and a costlier process, which can affect its properties and quality [8]. In this context, alternative sources of chitosan such as fungi, insects, and mushrooms is gaining popularity in pharmaceutical, biomedical, and food industries due to their better biocompatibility, progressive biodegradability, nontoxicity, immunomodulatory, antimicrobial, antioxidant, and anticancer properties [9, 10]. Therefore, this chapter aims to investigate various agricultural waste sources for the extraction of chitin or chitosan and their functional properties. Furthermore, detailed information about various methods used for the fabrication of chitosan-based films/coating and their applications for shelf-life enhancement of perishable foods such as meat, fish, fruits, and vegetables is discussed.

10.2 Sources of Chitosan

Chitosan can be produced from various sources including marine (shrimp, crab, lobster, krill, coral, crayfish, jellyfish, barnacles, and prawn shells), insects (brachiopods, beetles, ants, scorpions, and cockroaches), microorganisms (bacteria and fungus), algae (green algae, brown algae, and diatoms), and agricultural waste (corn sweep liquor, date waste, wheat hydrolysates, wheat starch, and apple waste.) as shown in Figure 10.1 [11]. The important sources of chitin and chitosan and various research studies conducted on them will be explained in the subsections below.

10.2.1 Agro-Waste

Agricultural waste is also an important source of chitosan. Various researchers have identified industrial byproducts, namely, cassava wastewater, steep corn liquor, and molasses, as some of the sources of chitosan, which can be produced using various fungal strains. Chitin and chitosan

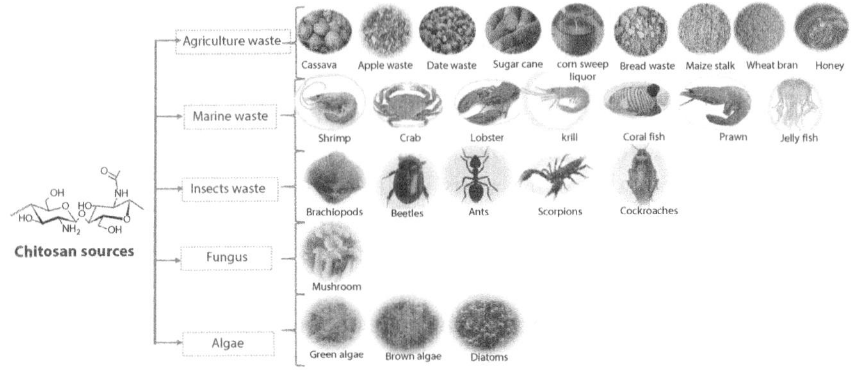

Figure 10.1 Various sources of chitosan.

were produced from molasses and corn steep liquor using *Rhizopus arrhizus* and *Cunninghamella elegans*, and their antimicrobial and physicochemical properties were also evaluated [12]. The highest yields of chitin and chitosan of 83.20 and 49.31 mg/g, respectively, were obtained using *Rhizopus arrhizus* strain, and antimicrobial study of prepared chitosan showed reduced microbial growth against *Escherichia coli, Salmonella enterica, Staphylococcus aureus, Listeria monocytogenes, Pseudomonas aeruginosa*, and *Yersinia enterocolitica.* In another study, chitosan and chitin extracted from yam bean medium using *Mucor circinelloides* (UCP 050) and investigated their physicochemical and functional properties, and the results showed the highest level of chitin and chitosan of 500 and 64 mg/g, respectively [13]. Likewise, chitosan was extracted from Mucorales fungi using papaya peel juice (PPJ) and corn steep liquor (CSL) as substrates and reported that chitosan produced from all the nine fungi ranged from 5.02 to 15.63 mg/g, whereas highest chitosan (37.25 mg/g) was produced using *C. elegans* medium containing PPJ and CSL of 42.5% and 9.43%, respectively [14]. Similarly, fungal chitosan is produced from cashew apple juice and cheese whey using Mucorales fungi, and the amount of chitosan cultivated in the CAJ-CW medium ranged from 27.58 (*Mucor hiemalis* UCP 1309) to 65.40 mg/g (*C. elegans* UCP 1306) [15]. However, the highest chitosan production (65.40 mg/g) was obtained by *C. elegans* with a medium containing CAJ and CW of 40 and 30%, respectively. Chitosan has been also produced by *Rhizopus oryzae* using whey as a substrate alongwith plant growth hormones, viz., gibberellic acid, indole-3-acetic acid, and indole-3-butyric acid [16]. It has been reported that more than 50% of chitosan was produced from 1 L of whey containing 0.1 mg/L of gibberellic acids. Studies related to extraction of chitosan from agricultural waste are listed in Table 10.1.

Table 10.1 Production of chitosan from agro-waste and their yields.

Spices and genes type	Cultivar	Yield	Reference
Cunninghamella elegans	Corn steep liquid	8.8 g/100 ml	[17]
Mucorales	Cashew apple juice and cheese whey	27.58–65.40 mg/g	[15]
Mucorales fungi	Corn steep liquor and papaya peel juice	5.02–15.63 mg/g	[14]
Mucor circinelloides	Corn steep liquor	7 g/ml	[18]
Zygomycete and *Rhizopus delemar*	Bread waste	36.9 and 19.7%	[19]
Syncephalastrum racemosum	Sugar cane substrates	50 mg/g	[20]
Mucor hiemalis	Wheat and starch hydrolysates	37.3–46.7%	[21]
M. indicus CCUG 22424	Wheat straw	0.68 g/g	[22]
Rhizopus oryzae	Corn straw	60.1–207.1%	[23]
Rhizopus oryzae PAS 17	Soil samples	14.45%	[24]
Rhizopus oryzae TISTR3189	Potato, peel and substandard (low-quality) chips	10.8 g/kg	[25]
Aspergillus niger Tiegh	Potato dextrose agar	7.0% w/w	[26]
Rhizopus oryzae NRRL1526	Potato dextrose broth	13.43% w/w	[27]
Trichoderma koningii	Wheat bran	4.84 IU/gds	[28]
Aspergillus terreus	Apple waste extract	140.9 mg g^{-1}	[29]
Cunninghamella elegans UCP/ WFCC 0542	Corn steep liquor and cassava wastewater	57.82 mg/g	[30]

(*Continued*)

Table 10.1 Production of chitosan from agro-waste and their yields. (*Continued*)

Spices and genes type	Cultivar	Yield	Reference
Mucor subtilissimus UCP 1262 and *Lichtheimia hyalospora UCP 1266*	Corn steep liquor and cassava wastewater	63.18 mg/g	[31]
Rhizopus arrhizus UCP 402	Corn steep liquor and honey	29.3 mg/g	[32]
Cunninghamella elegans and *Rhizopus arrhizus strains*	Corn steep liquor and molasses	20.30 and 17.80%	[12]

10.2.2 Sources Other Than Agro-Waste

10.2.2.1 Terrestrial Insects

Insects are the secondary source of chitosan, which has been used in various fields including agriculture, medicine, food, and animal science due to their biological activities. Chitosan can be extracted from wings and cuticles of insects that contain protein (30 to 45%), fat (25 to 40%), and chitin (10 to 15%). The yield of chitosan mainly depends on the extraction method, extraction time, life cycle, and various external factors [33, 34]. The extraction of chitosan from insects has various advantages including higher yield and lower mineral content and requires mild extraction conditions. Around 900,000 species of insects are available on Earth, some of which including beetle larvae, beetles, crickets, corpses of bumblebees, grasshoppers, silkworm (*Bombyx mori*) pupae, and honeybees are identified as sources of chitosan [35, 36]. For example, from *Hermetia illucens* and *Tenebrio molitor*, approximately 5% (50g/kg) of chitosan can be extracted, whereas a lesser amount has been extracted from house flies (11.9 g/kg), Tebo worms (11.1 g/kg), black soldier flies (21 g/kg), and Turkestan cockroaches (6.7 g/kg) [37]. Torres-Castillo *et al.* (2015) extracted both chitin (11.8%) and chitosan (58.8%) from *Pterophylla beltrani Bolivar*, which showed antifungal activity against *Metarhizium anisopliae (Metschnikoff) Sorokin* [38]. In another study, chitosan obtained from *Brachystola magna and Tenebrio molitor* was compared with commercial chitosan to evaluate their mechanical properties, and the results showed that the insect-extracted chitosan-based films had lower elongation modulus and tensile strength, and higher water vapor permeability and solubility than the commercial chitosan-based films [39].

10.2.2.2 Microbial Sources

Fungus is the second-highest source of chitosan after insects. Nowadays, fungal chitosan has been gaining popularity in various industrial applications including agriculture, food, and biomedical due to its multiple advantages such as immunomodulatory, biodegradability, antioxidant, anticancer, non-toxicity, and antimicrobial activities [9]. Also, chitosan extracted from these sources has higher purity with better physicochemical properties and helps to lower the risk of various diseases compared to animal-derived chitosan. Furthermore, fungal chitosan can be produced using a mild acid treatment that shows less effect on polysaccharide structure and generates less amount of waste [40]. The physicochemical properties of fungal chitosan can be differentiated from other sources in terms of its viscosity, degree of acetylation (DA), charge distribution, and molecular weight (MW). For example, chitosan extracted from fungal mycelia shows low to medium MW ($1 - 10^4$ to 12×10^4 Da); however, chitosan obtained from crustaceans' has higher MW (about 1.5×10^6 Da). Furthermore, fungal chitosan does not contain any heavy metals including copper and nickel.

The different fungal spices including *Zygomycetes*, *Chytridiomycetes*, *Ascomycetes*, and *Basidiomycetes* have been used to produce chitosan [11]. Among all, zygomycetes contain a higher amount of both chitin and chitosan compared to other spices. According to Poverenov *et al.* (2018), underutilized portion of mushroom showed a higher yield of chitosan (176 mg/g) compared to it's caps (105 mg/g) [41]. In another study, chitins such as A, B, and C were obtained from shiitake stipes by alkaline treatment and their yields were 283.2, 367.2, and 250.8 mg/g, respectively [42]. They also extracted chitosan from shiitake stipes by alkaline N-deacetylation of crude chitin, and the yields were 69.3 to 83.4 and 218.4 to 240.3 mg/g, respectively. Table 10.1 shows the various studies on chitosan extraction from fungal strains. The algae are also considered a source of chitosan, and chitin and collagen are mainly found in coralline algae's structural tissues [43].

10.2.2.3 Marine Sources

Marine-based foods are an important source of protein, and the non edible parts and external shells are around half the total weight, which are considered waste or byproducts that are good sources of chitosan. Marine sources including crab, krill, prawn, shrimp, crayfish, and lobster are the most predominant and commercial sources of chitin, which are deacetylated into chitosan. For example, Hajji *et al.* (2014) studied the structural difference between chitosan and chitin obtained from shrimp waste, cuttlefish bones, and crab shells and reported that shrimp waste yielded higher chitin (more than 20%) compared to other sources without any residual salt and protein [44]. Chitosan has also been extracted from crab shells using the chemical method, and the yield was 21%, in which microwave heating reduced the extraction time from 6 to 10 hours to 10 to 15 min to achieve the same degree of deacetylation (DD) and higher MW chitosan [45]. In another study, chitosan was extracted from shrimp, fish shells, and fish scales, and the obtained chitosan had 78, 75, and 70% DD, respectively [46]. Among all, shrimp shells were considered the best choice for extraction of chitosan due to their better physicochemical properties like high DD, higher MW, and better solubility compared to other sources. Chitosan has various applications in pharmacology, food industries (coating, films, preservatives, antioxidants, and antimicrobial agents), medicine, and cosmetology [36].

10.3 Chitosan Extraction

Chitosan extraction from agricultural waste involves various steps including pretreatment, alkaline-based cleaning, acid and base treatments, and chitin extraction, and deacetylation into chitosan. Chitosan is found in two forms: covalently bonded beta-glucan and free amino glucoside. First, beta-glucan and chitin accumulate separately in cell walls followed by the formation of interpolymer linkage [11] leading to the formation of a complex rigid network in the cell wall, which affects the extraction of the glucan and chitosan. Therefore, strong alkali treatment is needed to break down the chitin-glucan linkage. Furthermore, high-quality chitosan can be extracted using enzymatic digestion using glucanase, amylase, and chitinase [47]. Pretreatment is crucial to remove impure materials affecting the extracted chitosan quality. Most agricultural waste is a wet mass, so it must be cleaned, sorted, and dried before processing, and then powdered. Prior to extraction, the powder samples are treated with an alkaline solution to break the chitin-glucan linkage. A similar step is also carried out to produce chitin or chitosan from marine waste. According to Berger *et al.* (2014), various process conditions like high temperature (80°C to 121°C), short contact times (15 min to 3 hours), high alkali concentration

(1 to 4 M), and low substrate loading (1:30 to 1:40 w/v) are considered for extraction of chitin from agriculture residue [12]. Chitosan is produced from the fungal source using a two-step extraction process such as base and acid treatment. After alkaline treatment, the remaining cell walls containing proteins and chitosan are considered alkaline-insoluble materials. The insoluble materials are treated with acids including acetic or lactic acid and hydrochloric acid for 24 hours. The extraction of chitosan using fungus does not involve deproteinization and demineralization processes due to a lack of protein compounds [11, 47]. The fungal chitosan is washed around four times using distilled water followed by freeze-drying. Then, the dried samples were kept in the desiccator until constant weight was reached.

10.4 Chitosan and Its Functional Properties

Chitin is the second most abundant source of polysaccharides on Earth after cellulose. It comprises three functional groups: i) primary, ii) secondary -OH groups at C-3 and C-6 positions, and iii) tertiary CO-NH group at C-2 [48]. Chitosan can be differentiated based on the presence of the N-acetyl amine group at the C-2 position, whereas chitin contains an amino group at C-2 in the backbone (Figure 10.2a and 10.2b). According to Azmana *et al.* (2021), a small variation in MW, polydispersity, and DD affects the various physicochemical properties like solubility, hydrophilic/hydrophobic properties, crystallinity, and cell response [5]. Chitosan is used as a natural biopolymer in the food packaging industry due to its

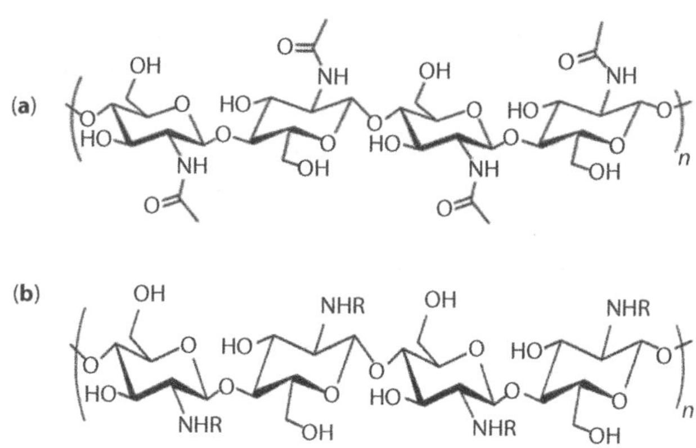

Figure 10.2 Molecular structure of (a) chitin and (b) chitosan.

physicochemical, barrier, and functional properties. The functional properties like antimicrobial, antioxidant, film-forming properties, and solubility of the fungal chitosan are mentioned in the sections below.

10.4.1 Antimicrobial Activity

Food packaging materials should have the capacity to delay microbial growth, enhance the shelf life of food products, and ensure the safety of the products. The chitosan obtained from agricultural waste can be used as an antimicrobial compound due to its superior action against various microorganisms including fungi, yeast, and Gram-negative and Gram-positive bacteria [48, 49]. Various research studies have been conducted on the antimicrobial properties of chitosan and a few established mechanisms have been reported such as i) electrostatic interaction between the amide groups of chitosan and anionic charges on the microbial surface results in leakage of the cell components and then ultimately causing the cell death, ii) chitosan binds with metals and inhibit different metabolic enzymes of the microbial cell decreasing their growth, and iii) MW also plays an important role on the microbial activity. For example, high MW chitosan forms an impermeable layer on the surface of microbial cell walls resulting in blockage of nutrients entry through the cell wall, whereas low MW chitosan enters cytosol and binds with cell's DNA resulting in inhibition of RNA and protein synthesis leading to cell death [50, 51]. Chitosan's antimicrobial properties mainly depend on the chitosan source and their concentration, DD, DA, cationic nature, period of exposure, MW, pathogen type, and pH of the medium. Lower pH chitosan shows that more antimicrobial activity compared to the higher pH chitosan could be due to chitosan chains becoming more protonated at a higher acidic medium than in a basic medium [52]. Few investigations have focused on the antimicrobial properties of chitosan obtained from numerous agricultural wastes. The antimicrobial properties of chitosan obtained from mushroom waste were studied by Poverenov *et al.* (2018) and they found that chitosan inhibited both *Escherichia coli* bacteria and *Saccharomyces cerevisiae* growth at a concentration of 1 and 2%, respectively [41]. In another study, chitosan obtained from fungal and crustacean were compared and reported that chitosan obtained from both sources was nontoxic and showed better antimicrobial properties against tested pathogens. Therefore, these can be used as food ingredients [53]. Similarly, the antimicrobial properties of fungal chitosan obtained from date waste were compared with potassium sorbate, and the results showed fungal chitosan has better antimicrobial properties than potassium sorbate [54].

10.4.2 Antioxidant Properties

Antioxidant activity is defined as the compound's ability to delay or block another compound's oxidation [2]. The antioxidant activity of chitosan is mainly due to the presence of the -OH and CO-NH groups in the backbone of chitosan that helps to remove the excessive reactive oxygen species (ROS) [55–57]. This ability of the chitosan has driven an interest in the development of chitosan-based food packaging films [57]. The antioxidant property of films/coatings is analyzed using DPPH activity, total phenolic content, and ABTS radical scavenging. In previous years, synthetic compounds like butylated hydroxytoluene (BHT) and butylated hydroxy anisole (BHA) have been used as antioxidants for various applications. However, nowadays application of these compounds is restricted due to their toxic properties and carcinogenicity, and the use of antioxidants obtained from various natural sources has been increasing. Chitosan is one among them, which has better antioxidant activity mainly due to -OH and other functional groups in its backbone. The antioxidant activity of the chitosan mainly depends on their MW, and it has been observed that the DPPH activity of the low MW chitosan is higher compared to the higher MW chitosan.

Numerous researchers investigated on antioxidant properties of chitosan extracted from agricultural waste, for example, antioxidant properties of fungal chitosan obtained from shiitake stipes were reported, and both fungal chitosan B and C showed better antioxidant properties especially scavenging activity against hydroxy radicals, antioxidant activity, and chelating ability on ferrous ions [58]. In another study, antioxidant activity and stability of liposomes modified with bio-surfactants loaded with betulinic acid using fungal chitosan obtained from *Agaricus bisporus* (Lange) *Sing. Chaidam.* They found that fungal chitosan's antioxidant activity was more than normal [59]. Likewise, the antimicrobial and antitumor properties of the chitosan obtained from shiitake stipes and crab shells were compared, and the result showed that chitosan obtained from both sources had excellent antimicrobial properties against eight species of Gram-negative and -positive bacteria with inhibition zones of 11.4 to 26.8 mm at 0.5 mg/ml; however, shiitake stipes derived chitosan showed better antimicrobial and antitumor properties compared to crab-based chitosan [60].

10.4.3 Film-Forming Ability

Chitosan has been used as a packaging material to enhance the shelf life and prevent the contamination of food products due to their unique properties and advantages [61]. However, they show poor mechanical and barrier properties because of their hydrophilic nature, which restricts their applications in food packaging. Therefore, numerous strategies have been used to enhance its mechanical and barrier properties, for example, the blending of chitosan with nanoparticles such as clays and metal oxides. Soni *et al.* (2016) demonstrated that nanocomposite films based on chitosan and TEMPO-oxidized cellulose nanofibers had improved mechanical strength due to the strong interaction of chitosan with oxidized cellulose nanofibers [62].

10.4.4 Solubility

The solubility of chitosan is an important property, which mainly depends on temperature, MW, ionic strength, particle size, pH of the solution, origin, alkali concentration, DD, and ionic interaction with chitosan [37, 63]. In general, chitin is insoluble in both organic and inorganic solvents due to the presence of inter- and intramolecular hydrogen bonds. Therefore, there is a need for deep eutectic solvents, ionic liquids, and alkali/urea aqueous systems to solubilize chitin. Chitosan is an important derivate of chitin, which undergoes structural modification due to the presence of amine groups [64]. These amine groups in the chitosan formed helps to solubilize the chitosan in the aqueous phase particularly at acidic pH (below 6) due to the fully protonated nature of amide groups. After a pH of 6, the solubility of the chitosan starts decreasing due to the loss of the positive charges and de-protonation of the amino groups resulting in the precipitation of the chitosan. Higher DD leads to the formation of protonated amino acids, which facilitates the dissolving of the chitosan, and thus increases solubility [65, 66].

10.5 Chitosan-Based Composites and Nanocomposites

10.5.1 Coating Formulations

The coating of food products can be performed using various methods including dipping, spreading, and spraying. The selection of an appropriate coating method for food products mainly depends on the overall cost, intended effect, and product characteristics. In spray-based coating, the solution is applied on the product surface using an aerosol sprayer, whereas in the dipping process, the product is submerged in the prepared solution containing preservatives and plasticizers [55, 67].

Then, the excess solution is drained away followed by drying the products at desirable temperature and humidity conditions. Among all the coating methods, dipping is considered a simpler and more effective method. Chitosan has been gaining importance as the coating material due to its excellent antimicrobial and antioxidant characteristics [49, 68]. For example, Poverenov *et al.* (2018) applied fungal chitosan coating on the surface of fresh-cut melon to investigate various quality parameters during storage, and the results showed that fungal chitosan coating significantly improved firmness, whereas the microbial count and off-flavors were reduced [41]. In another study, the fungal chitosan nanoparticle–based coating was applied on table grapes and reported that the coating delayed the ripening process, resulting in reduced soluble solids, sugar content, weight loss, and the preservation of titratable acidity and sensory characteristics [69]. Similarly, de Oliveira *et al.* (2014) applied a chitosan-based coating obtained from *Mucor circinelloides* to evaluate the quality of the grapes during storage, and the results showed that the chitosan coating controlled pathogens and maintained the grapes' quality during storage [18]. Chitosan derived from *Cunninghamella elegans* grown in the CSL is used as a coating to evaluate table grapes' physicochemical and sensory properties [17].

10.5.2 Packaging Films

Chitosan is used in the development of packaging films due to its antimicrobial properties, film forming, and superior biocompatibility. These films can be divided into two categories, i.e., edible and nonedible, and in most cases, edible films are used on the surface of the food product. Chitosan-based packaging films are fabricated mainly by solution casting, electrospinning, and extrusion (Figure 10.3). In solution casting, chitosan is dissolved in suitable solvents particularly acidified water followed by pouring of the prepared mixture on a flat surface, and then drying is performed to evaporate the solvent. The developed film's final property mainly depends on MW of chitosan, solvent pH, type of acid, and DD.

The extrusion technique has been used to produce packaging films as an alternative to casting methods due to its lower energy consumption and faster processing time. Furthermore, extrusion processing can enhance the mechanical and thermal properties of the film [70]. The extrusion process involves various steps, i.e., feed formulation is formed by mixing the different ingredients, and then formed blends are fed into an extruder to form the pellets followed by drying using a hot air oven [71–81]. Then, dried pellets are made into sheets using a secondary extruder. Last, blowing

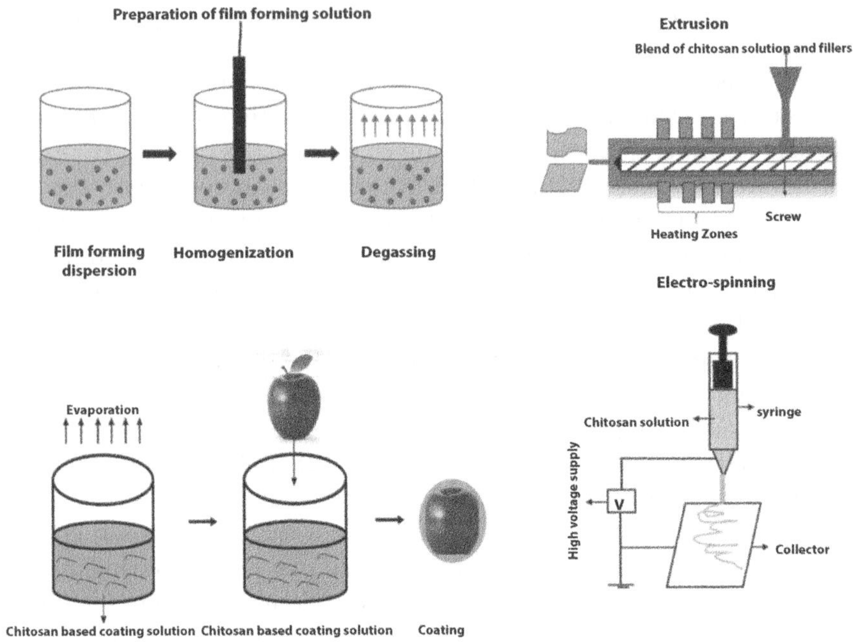

Figure 10.3 Film fabrication techniques.

mixed resins into a film using a blown film extruder [48]. Even though the extrusion process has been used as a promising process for film fabrication, there are no studies on the application of extrusion for developing chitosan-based films.

The electrospinning technique has been gaining importance in food packaging and coating due to its lower cost, versatility, simplicity, higher surface area to volume, higher porosity, and production in a short time [82]. It mainly works based on the high voltage electric field, which produces chitosan-based mats containing nanostructures or nanofibers [83]. The electrospinning process mainly consists of four main parts stainless steel capillary or needle, high voltage electric source, and a ground collector either a flat plate or orienting drum [84, 85]. In this process, chitosan or a combination of other polymers is blended, filled into the syringe, and loaded at a specific rate using the syringe pump followed by the fabrication of the nanofibers.

Yu *et al.* (2013) studied the methyl cellulose/hydroxypropyl starch nanocomposite film reinforced with chitosan nanoparticles encapsulating cinnamon essential oil and found that nanocomposites exhibited better mechanical, barrier, antioxidant, and antimicrobial properties [86]. Similarly, chitosan plasticized with glycerol and nano-reinforced by both CNC and nano-keratin has been reported, and the study showed that the addition of nanocrystals exhibited better thermomechanical and barrier properties compared to starch-based packaging material [87]. In another study, chitosan composite films were supplemented with copolymerized (with polyacrylonitrile monomers) cellulose nanofibers and diatomite nanocomposite at different concentrations. According to their results, the addition of cellulose nanofibers enhanced the physicochemical properties of the films [88].

10.6 Food Packaging Applications

10.6.1 Fish and Meat Products

Fish and meat products are nutritionally denser mainly containing higher protein (various amino acids), omega-3 fatty acids, and trace elements (magnesium, potassium, iron, selenium, and vitamins) [89]. The high nutritional content can cause microbial spoilage/contamination that negatively affects the shelf life and quality of the food and may cause foodborne illness. Numerous researchers have used high- and low-temperature processing/preservation techniques to control microbial growth and oxidative spoilage. The use of synthetic preservatives impacts the quality by reducing the nutritional value and their long-term consumption can adversely affect consumers' health [90]. In this sense, researchers have recently focused on developing biopolymer-based packaging films and coating materials for extending the shelf life of perishable foods [91]. In this regard, natural polysaccharides like chitosan extracted from agricultural waste have gained significant relevance because of their antimicrobial, film-forming, antifungal, biodegradable, and biocompatible properties [61]. Few studies have been conducted on the application of fungal chitosan-based films/coating to enhance the shelf life of chicken, pork, beef, and fish shown in Table 10.2. The fungal chitosan incorporated with pomegranate extract was coated on the Nile tilapia fillets and found that fillet coating reduced the microbial count during storage [92]. In another study, chitosan from date waste was applied on the meat surface as the preservative compared to the artificial potassium sorbate, and the results showed that fungal chitosan had higher antimicrobial activity compared to potassium sorbate [54].

Table 10.2 Chitosan and chitosan-based films/coatings on the shelf life and quality of fish and meat products.

Composition of chitosan films or coatings	Fish and meat products	Effects on quality and shelf life of fish and meat products	Reference
Portulaca oleracea extract (POE)/chitosan (CS) based films	Chilled meat	CS/POE-based films showed better shelf life and preservation effect	[93]
Chitosan/polyvinyl alcohol/Jacaranda cuspidifolia anthocyanin	Fish	Composite film maintained the freshness of the fish at room temperature	[94]
PLA/chitosan/caseinate/rosemary essential oil coating	Fresh minced chicken breast	Composite coating reduced meat oxidation during the storage conditions	[95]
Carboxymethyl chitosan (CMCS)/pomegranate peel extract (PPE)	Fish flesh	CMCS-PPE coating helped to inhibit oxidation and texture softening and reduced microbial growth	[96]
Chitosan/epigallocatechin gallate coating	Bighead carp fillets	Chitosan-EGCG coating retained both the flavor and quality of fillets	[97]
Gelatine/chitosan/gallic acid/clove oil coating	Fresh salmon fillet	Composite coating preserved the salmon fillets and extended shelf life for 5 days	[98]
Fish myofibrillar protein/chitosan/rosemary extract composite edible films	Grass carp fillets	FMP/CS/RE composite film preserved the quality as well as maintained lipid oxidation	[99]
Chitosan coating	Pastirma (traditional dry-cured meat product)	Chitosan coating reduced microbial plate count by 1-log cycle/g and maintained meat quality	[100]
Chitosan/wampee seed essential oil composite film	Golden pompano fillets	Edible chitosan-WSEO composite film preserved and extended the shelf life of golden pompano fillets at refrigerated conditions	[101]
Tannic acid (TA)/grafted chitosan (g-CH) coating	Fresh pork slices	TA-g-CH coatings extended the shelf life of pork slices by delaying protein and lipid oxidation as well as suppressing microbial proliferation	[102]

(Continued)

Table 10.2 Chitosan and chitosan-based films/coatings on the shelf life and quality of fish and meat products. (*Continued*)

Composition of chitosan films or coatings	Fish and meat products	Effects on quality and shelf life of fish and meat products	Reference
Chitosan/free or nano-encapsulated *Satureja* plant essential oil coating	Lamb meat	Composite coating retarded chemical spoilage and microbial growth	[103]
Almond gum/polyvinyl alcohol/chitosan/ thyme essential oil nanoemulsion composite films	Chicken breast fillets	Composite film containing TEO nanoemulsion enhanced shelf life and protected breast fillets from UV light	[104]
Chitosan (CHI)/gelatin (GEL)/nisin (NIS)/ grape seed extract (GSE)	Fresh pork	CHI-GEL-GSE–based composite coating preserved pork, however, incorporation of NIS did not show any effect on pork	[105]
Nanochitosan/free or nanoencapsulated essential oil of Golpar	Rainbow trout fish	NCS-NEO coating enhanced the shelf life of the rainbow trout fish	[106]

10.6.2 Fruits and Vegetables

Consumption of fruits and vegetables has been increasing due to lifestyle changes and increasing preference for healthy foods [107]. However, these commodities are highly perishable, and their high moisture content and continuous respiration reduce their quality and shelf life [108, 109]. Furthermore, the complex process of food ripening results in various physiological and biochemical changes including changes in sugar content, chlorophyll degradation, respiration activity, ethylene production, softening, change in the level of an aromatic compound, and breakdown of the cell wall through enzymatic means [48, 110]. Various traditional preservation techniques including waxing, freezing, drying, canning, and vacuum packaging have been used for fruits and vegetables. Biobased coating and packaging films are effective in maintaining shelf life and microbial safety while meeting the demand for nutritious, fresh, and healthy products. Composite films and coatings can act as a barrier against carbon dioxide, atmosphere gases, humidity, and water vapor, and thus help to enhance both the quality and shelf life of packaged/coated food products [107]. Few studies have been conducted to develop chitosan-based films to apply on various fresh produce including fruits and vegetables shown in Table 10.3.

Table 10.3 Chitosan and chitosan-based films/coatings on the shelf life and quality of fruits and vegetables.

Composition of chitosan films/coatings	Fruits and vegetables	Effects of films/coatings	Reference
Essential oil from Assam lemon peels/chitosan	Grape	Chitosan coating containing 1% EO improved the shelf life of the grapes for 15 days at room temperature	[111]
Guar gum/chitosan-based composite edible coating	Kinnow fruits	GG-TAM–based composite coating extended the shelf life of kinnow fruit up to 25 days without damaging quality at ambient temperature	[112]
Imine-chitosan films	Fresh-cut pineapple	Composite film extended microbial shelf life and slowed down the enzymatic browning.	[113]
Chitosan/zinc oxide, calcium chloride, nanoclay/polyethylene glycol/thyme oil	Sweet cherry	The composite film maintained postharvest quality and increased the shelf life of cherry	[114]
Chitosan/carboxymethyl gellan gum/mustard essential oil coating	Mango	Composite film maintained quality and appearance for 20 days	[115]
Chitosan/casein/*Origanum vulgare L.* essential oil coating	Cherry tomato	Coating inhibited fungal growth for up to 28 days and delayed the spoilage for long-term storage	[116]
Chitosan/natural kaolinite clay (KC)/*Ficus carica* leaves extract (FLE) film	Fresh-cut apple slices	Chitosan/KC/FLE-based film maintained the quality in terms of freshness, moisture, and phenolic content of apple slices	[117]
Water-soluble chitosan (WSC)/tannic acid (TA)	Passion fruit	Coating maintained a shelf life of up to 7 days and preserved the quality of passion fruit	[118]
Chitosan/montmorillonite nanocomposites	"Hom Thong" banana fruit	The chitosan-based film containing 2% montmorillonite maintained the postharvest quality of the banana fruit	[119]

(Continued)

Table 10.3 Chitosan and chitosan-based films/coatings on the shelf life and quality of fruits and vegetables. (*Continued*)

Composition of chitosan films/coatings	Fruits and vegetables	Effects of films/coatings	Reference
Chitosan/natamycin	*Prunus salicina* Lindl. *cv.* "Shazikongxinli" fruit	Chitosan-natamycin better maintained postharvest quality compared to other treatments	[120]
Chitosan/titanium dioxide nanocomposite	Mango fruit	Composite coating maintained the quality of fruit at 13°C without affecting nutritional value	[121]
Chitosan/chia mucilage/ levan-based composite coatings	Sweet cherry	The shelf life of cherry was maintained for up to 31 days	[122]

10.7 Conclusion and Future Perspectives

There is considerable interest in developing biodegradable, eco-friendly, and cost-effective biopolymer-based packaging films to avoid and reduce the risk associated with the use of petroleum-based synthetic plastics. Chitosan is the predominant and abundantly available biopolymer on the planet that has various advantages like renewability, biodegradability, non-toxicity, film-forming ability, and biocompatibility, making it suitable for various food packaging applications. Chitosan is mainly obtained from marine sources and insects, but due to their seasonal availability, use of toxic solvents, and generation of harmful effluents, it is avoided nowadays. Chitosan production from alternative sources like agricultural waste is gaining predominant attention due to its low cost and easy availability. Agricultural waste–based chitosan could be an alternative source for developing packaging films and coatings for food packaging applications.

References

1. Almeida, A.d.R. *et al.*, Production of chitosan and poly (vinyl alcohol) films functionalized with hop extract (*Humulus lupulu L. var. Cascade*) for food packaging application. *Food Packag. Shelf Life*, **32**, 100833, 2022.
2. Flórez, M. *et al.*, Chitosan for food packaging: Recent advances in active and intelligent films. *Food Hydrocoll.*, **124**, 107328, 2022.

3. Zhang, L. *et al.*, Modulating physicochemical, antimicrobial and release properties of chitosan/zein bilayer films with curcumin/nisin-loaded pectin nanoparticles. *Food Hydrocoll.*, **133**, 107955, 2022.
4. Purkayastha, M.D. and Kumar, S., Protein-based films and coatings, in: *Biopolymer-Based Food Packaging*, pp. 178–224, 2022.
5. Azmana, M. *et al.*, A review on chitosan and chitosan-based bionanocomposites: Promising material for combatting global issues and its applications. *Int. J. Biol. Macromol.*, **185**, 832–848, 2021.
6. Alves, Z. *et al.*, Design of heat sealable starch-chitosan bioplastics reinforced with reduced graphene oxide for active food packaging. *Carbohydr. Polym.*, **291**, 119517, 2022.
7. Cheng, S., Li, F., Mei, X., Structure, mechanical and physical properties of hordein/chitosan composite films. *LWT*, **163**, 113596, 2022.
8. Cai, J. *et al.*, Enzymatic preparation of chitosan from the waste Aspergillus niger mycelium of citric acid production plant. *Carbohydr. Polym.*, **64**, 2, 151–157, 2006.
9. Hamidi, M. *et al.*, Fungal exopolysaccharides: Properties, sources, modifications, and biomedical applications. *Carbohydr. Polym.*, **284**, 119152, 2022.
10. Liu, L. *et al.*, Antimicrobial and antioxidant AIE chitosan-based films incorporating a Pickering emulsion of lemon myrtle (*Backhousia citriodora*) essential oil. *Food Hydrocoll.*, **133**, 107971, 2022.
11. Huq, T. *et al.*, Sources, production and commercial applications of fungal chitosan: A review. *J. Bioresour. Bioprod.*, **7**, 2, 85–98, 2022.
12. Berger, L. *et al.*, Green conversion of agroindustrial wastes into chitin and chitosan by *Rhizopus arrhizus* and *Cunninghamella elegans* strains. *Int. J. Mol. Sci.*, **15**, 5, 9082–9102, 2014.
13. Fai, A.E.C. *et al.*, Physico-chemical characteristics and functional properties of chitin and chitosan produced by Mucor circinelloides using yam bean as substrate. *Molecules*, **16**, 8, 7143–7154, 2011.
14. Ramos Berger, L.R. *et al.*, Chitosan produced from Mucorales fungi using agroindustrial by-products and its efficacy to inhibit Colletotrichum species. *Int. J. Biol. Macromol.*, **108**, 635–641, 2018.
15. Berger, L.R.R. *et al.*, Agroindustrial waste as ecofriendly and low-cost alternative to production of chitosan from Mucorales fungi and antagonist effect against Fusarium solani (Mart.) Sacco and Scytalidium lignicola Pesante. *Int. J. Biol. Macromol.*, **161**, 101–108, 2020.
16. Chatterjee, S. *et al.*, Enhancement of growth and chitosan production by Rhizopus oryzae in whey medium by plant growth hormones. *Int. J. Biol. Macromol.*, **42**, 2, 120–126, 2008.
17. de Oliveira, C.E.V. *et al.*, Effects of chitosan from Cunninghamella elegans on virulence of post-harvest pathogenic fungi in table grapes (*Vitis labrusca L.*). *Int. J. Food Microbiol.* 3er45rrrrrrrrrrfdcx, **171**, 54–61, 2014.

18. de Oliveira, C.E.V. *et al.*, Effects of post-harvest treatment using chitosan from Mucor circinelloides on fungal pathogenicity and quality of table grapes during storage. *Food Microbiol.*, **44**, 211–219, 2014.
19. Svensson, S.E. *et al.*, Turning food waste to antibacterial and biocompatible fungal chitin/chitosan monofilaments. *Int. J. Biol. Macromol.*, **209**, 618–630, 2022.
20. Amorim, R.V.S. *et al.*, Chitosan from syncephalastrum racemosum using sugar cane substrates as inexpensive carbon sources. *Food Biotechnol.*, **20**, 1, 43–53, 2006.
21. Heidary Vinche, M. *et al.*, Ethanol and chitosan production from wheat hydrolysate by Mucor hiemalis. *J. Chem. Technol. Biotechnol.*, **88**, 2, 255–260, 2013.
22. Asachi, R. and Karimi, K., Enhanced ethanol and chitosan production from wheat straw by Mucor indicus with minimal nutrient consumption. *Process Biochem.*, **48**, 10, 1524–1531, 2013.
23. Tai, C. *et al.*, Chitosan production from hemicellulose hydrolysate of corn straw: Impact of degradation products on Rhizopus oryzae growth and chitosan fermentation. *Lett. Appl. Microbiol.*, **51**, 3, 278–284, 2010.
24. Tasar, O.C., Erdal, S., Taskin, M., Chitosan production by psychrotolerant Rhizopus oryzae in non-sterile open fermentation conditions. *Int. J. Biol. Macromol.*, **89**, 428–433, 2016.
25. Kleekayai, T. and Suntornsuk, W., Production and characterization of chitosan obtained from Rhizopus oryzae grown on potato chip processing waste. *World J. Microbiol. Biotechnol.*, **27**, 5, 1145–1154, 2010.
26. Abdel-Gawad, K.M. *et al.*, Technology optimization of chitosan production from Aspergillus niger biomass and its functional activities. *Food Hydrocoll.*, **63**, 593–601, 2017.
27. Sebastian, J. *et al.*, Microwave-assisted extraction of chitosan from Rhizopus oryzae NRRL 1526 biomass. *Carbohydr. Polym.*, **219**, 431–440, 2019.
28. da Silva, L.C.A. *et al.*, Optimization of chitosanase production by trichoderma koningii sp. Under solid-state fermentation. *Food Bioprocess Technol.*, **5**, 5, 1564–1572, 2010.
29. Habibi, A. *et al.*, Key parameters optimization of chitosan production from Aspergillus terreus using apple waste extract as sole carbon source. *Bioprocess Biosyst. Eng.*, **44**, 2, 283–295, 2020.
30. Berger, L. *et al.*, Effect of corn steep liquor (CSL) and cassava wastewater (CW) on chitin and chitosan production by cunninghamella elegans and their physicochemical characteristics and cytotoxicity. *Molecules*, **19**, 3, 2771–2792, 2014.
31. de Souza, A.F. *et al.*, Biotechnological strategies for chitosan production by mucoralean strains and dimorphism using renewable substrates. *Int. J. Mol. Sci.*, **21**, 12, 4286, 2020.
32. Cardoso, A. *et al.*, Microbial enhance of chitosan production by Rhizopus arrhizus using agroindustrial substrates. *Molecules*, **17**, 5, 4904–4914, 2012.

33. Ma, J. *et al.*, Terrestrial insects as a promising source of chitosan and recent developments in its application for various industries. *Food Chem.*, **373**, 131407, 2022.
34. Kaya, M. *et al.*, Changes in physicochemical properties of chitin at developmental stages (larvae, pupa and adult) of *Vespa crabro* (wasp). *Carbohydr. Polym.*, **145**, 64–70, 2016.
35. Zainol Abidin, N.A. *et al.*, The potential of insects as alternative sources of chitin: An overview on the chemical method of extraction from various sources. *Int. J. Mol. Sci.*, **21**, 14, 4978, 2020.
36. Joseph, S.M. *et al.*, A review on source-specific chemistry, functionality, and applications of chitin and chitosan. *Carbohydr. Polym. Technol. Appl.*, **2**, 100036, 2021.
37. Mohan, K. *et al.*, Recent insights into the extraction, characterization, and bioactivities of chitin and chitosan from insects. *Trends Food Sci. Technol.*, **105**, 17–42, 2020.
38. Torres-Castillo, J.A. *et al.*, Evaluation of biochemical components from Pterophylla beltrani (Bolivar & Bolivar) (Orthoptera: Tettigoniidae): A forest pest from Northeastern Mexico. *Southwest. Entomol.*, **40**, 4, 741–751, 2015.
39. Saenz-Mendoza, A.I. *et al.*, Characterization of insect chitosan films from *Tenebrio molitor* and *Brachystola magna and* its comparison with commercial chitosan of different molecular weights. *Int. J. Biol. Macromol.*, **160**, 953–963, 2020.
40. Bierhalz, A.C.K., Westin, C.B., Moraes, Â.M., Comparison of the properties of membranes produced with alginate and chitosan from mushroom and from shrimp. *Int. J. Biol. Macromol.*, **91**, 496–504, 2016.
41. Poverenov, E. *et al.*, Potential of chitosan from mushroom waste to enhance quality and storability of fresh-cut melons. *Food Chem.*, **268**, 233–241, 2018.
42. Yen, M.-T. and Mau, J.-L., Selected physical properties of chitin prepared from shiitake stipes. *LWT - Food Sci. Technol.*, **40**, 3, 558–563, 2007.
43. Rahman, M.A. and Halfar, J., First evidence of chitin in calcified coralline algae: New insights into the calcification process of Clathromorphum compactum. *Sci. Rep.*, **4**, 1, 6162, 2014.
44. Hajji, S. *et al.*, Structural differences between chitin and chitosan extracted from three different marine sources. *Int. J. Biol. Macromol.*, **65**, 298–306, 2014.
45. Sreeja, S.J. *et al.*, Production of chitin and conversion into chitosan from crab (*Scylla tranquebarica*) shells and evaluation of its antioxidant activities. *Biomass Convers. Biorefin.*, 1–7, 2023.
46. Kumari, S. *et al.*, Physicochemical properties and characterization of chitosan synthesized from fish scales, crab and shrimp shells. *Int. J. Biol. Macromol.*, **104**, 1697–1705, 2017.
47. Li, B. *et al.*, Purification of chitosan by using sol-gel immobilized pepsin deproteinization. *Carbohydr. Polym.*, **88**, 1, 206–212, 2012.

48. Kumar, S., Mukherjee, A., Dutta, J., Chitosan based nanocomposite films and coatings: Emerging antimicrobial food packaging alternatives. *Trends Food Sci. Technol.*, **97**, 196–209, 2020.
49. Basumatary, I.B., Mukherjee, A., Kumar, S., Chitosan-based composite films containing eugenol nanoemulsion, ZnO nanoparticles and Aloe vera gel for active food packaging. *Int. J. Biol. Macromol.*, **242**, 124826, 2023.
50. Hosseinnejad, M. and Jafari, S.M., Evaluation of different factors affecting antimicrobial properties of chitosan. *Int. J. Biol. Macromol.*, **85**, 467–475, 2016.
51. Yu, D. *et al.*, Chitosan modified with bio-extract as an antibacterial coating with UV filtering feature. *Int. J. Biol. Macromol.*, **230**, 123145, 2023.
52. Abd El-Hack, M.E. *et al.*, Antimicrobial and antioxidant properties of chitosan and its derivatives and their applications: A review. *Int. J. Biol. Macromol.*, **164**, 2726–2744, 2020.
53. Darwesh, O.M. *et al.*, Bio-evaluation of crustacean and fungal nano-chitosan for applying as food ingredient. *Toxicol. Rep.*, **5**, 348–356, 2018.
54. Tayel, A.A. *et al.*, Production of fungal chitosan from date wastes and its application as a biopreservative for minced meat. *Int. J. Biol. Macromol.*, **69**, 471–475, 2014.
55. Qu, B. and Luo, Y., A review on the preparation and characterization of chitosan-clay nanocomposite films and coatings for food packaging applications. *Carbohydr. Polym. Technol. Appl.*, **2**, 100102, 2021.
56. Hou, T. *et al.*, Essential oils and its antibacterial, antifungal and anti-oxidant activity applications: A review. *Food Biosci.*, **47**, 101716, 2022.
57. Oladzadabbasabadi, N. *et al.*, Recent advances in extraction, modification, and application of chitosan in packaging industry. *Carbohydr. Polym.*, **277**, 118876, 2022.
58. Yen, M.-T. *et al.*, Antioxidant properties of fungal chitosan from shiitake stipes. *LWT - Food Sci. Technol.*, **40**, 2, 255–261, 2007.
59. Wu, J. *et al.*, Fungal chitosan from *Agaricus bisporus* (Lange) Sing. Chaidam increased the stability and antioxidant activity of liposomes modified with biosurfactants and loading betulinic acid. *Int. J. Biol. Macromol.*, **123**, 291–299, 2019.
60. Chien, R.-C., Yen, M.-T., Mau, J.-L., Antimicrobial and antitumor activities of chitosan from shiitake stipes, compared to commercial chitosan from crab shells. *Carbohydr. Polym.*, **138**, 259–264, 2016.
61. Zhang, X. *et al.*, Emerging chitosan-essential oil films and coatings for food preservation - A review of advances and applications. *Carbohydr. Polym.*, **273**, 118616, 2021.
62. Soni, B. *et al.*, Transparent bionanocomposite films based on chitosan and TEMPO-oxidized cellulose nanofibers with enhanced mechanical and barrier properties. *Carbohydr. Polym.*, **151**, 779–789, 2016.

63. Priyadarshi, R. and Rhim, J.-W., Chitosan-based biodegradable functional films for food packaging applications. *Innov. Food Sci. Emerg.*, **62**, 102346, 2020.
64. Khayrova, A., Lopatin, S., Varlamov, V., Obtaining chitin, chitosan and their melanin complexes from insects. *Int. J. Biol. Macromol.*, **167**, 1319–1328, 2021.
65. Moran, H.B.T. *et al.*, Immunomodulatory properties of chitosan polymers. *Biomaterials*, **184**, 1–9, 2018.
66. Zargar, V., Asghari, M., Dashti, A., A review on chitin and chitosan polymers: Structure, chemistry, solubility, derivatives, and applications. *ChemBioEng Rev.*, **2**, 3, 204–226, 2015.
67. Basumatary, I.B. *et al.*, Chitosan-based active coating for pineapple preservation: Evaluation of antimicrobial efficacy and shelf-life extension. *LWT*, **168**, 113940, 2022.
68. Majumder, R. *et al.*, Biopolymers, in: *Biopolymer-Based Food Packaging*, pp. 29–65, 2022.
69. Castelo Branco Melo, N.F. *et al.*, Effects of fungal chitosan nanoparticles as eco-friendly edible coatings on the quality of postharvest table grapes. *Postharvest Biol. Technol.*, **139**, 56–66, 2018.
70. Wang, W. *et al.*, High performance extrusion blown starch/polyvinyl alcohol/clay nanocomposite films. *Food Hydrocolloids*, **79**, 534–543, 2018.
71. Singha, P., Muthukumarappan, K., Krishnan, P., Influence of processing conditions on apparent viscosity and system parameters during extrusion of distiller's dried grains-based snacks. *Food Sci. Nutr.*, **6**, 101–110, 2018.
72. Singha, P. and Muthukumarappan, K., Effects of processing conditions on the system parameters during single screw extrusion of blend containing apple pomace. *J. Food Process Eng.*, **40, 4**, 72. 40, 2017.
73. Singha, P. *et al.*, Physicochemical and nutritional properties of extrudates from food grade distiller's dried grains, garbanzo flour, and corn grits. *Food Sci. Nutr.*, **6**, 7, 1914–1926, 2018.
74. Singha, P., Singh, S.K., Muthukumarappan, K., Textural and structural characterization of extrudates containing blends of apple pomace, defatted soy flour and corn grits. *J. Food Process Eng.*, 42, e13046, 2019.
75. Singh, S.K., Singha, P., Muthukumarappan, K., Modeling and optimizing the effect of extrusion processing parameters on nutritional properties of soy white flakes-based extrudates using response surface methodology. *Anim. Feed Sci. Technol.*, **254**, 114197, 2019.
76. Singha, P. and Muthukumarappan, K., Single screw extrusion of apple pomace-enriched blends: Extrudate characteristics and determination of optimum processing conditions. *Food Sci. Technol. Int.*, **24**, 5, 447–462, 2018.
77. Singh, S.K. and Muthukumarappan, K., A viscosity model for soy white flakes-based aquafeed dough in a single screw extruder. *J. Food Process Eng.*, **40**, 2, 1–7, 2017.

78. Singh, S.K. and Muthukumarappan, K., Effect of different extrusion processing parameters on physical properties of soy white flakes and high protein distillers dried grains-based extruded aquafeeds. *J. Food Res.*, **3**, 6, 107–123, 2014.
79. Singh, S.K. and Muthukumarappan, K., Rheological characterization and cfd simulation of soy white flakes based dough in a single screw extruder. *J. Food Process Eng.*, **40**, 2, 1–10, 2017.
80. Singh, S.K. and Muthukumarappan, K., Single screw extrusion processing of soy white flakes based catla feed. *J. Food Res.*, **4**, 1, 1–9, 2015.
81. Singh, S.K. and Muthukumarappan, K., Effect of feed moisture, extrusion temperature and screw speed on properties of soy white flakes based aquafeed: A response surface analysis. *J. Sci. Food Agric.*, **96**, 6, 2220–9, 2016.
82. Liu, F. *et al.*, Preparation and antibacterial properties of ε-polylysine-containing gelatin/chitosan nanofiber films. *Int. J. Biol. Macromol.*, **164**, 3376–3387, 2020.
83. Kasaai, M.R., Bio-nano-composites containing at least two components, chitosan and zein, for food packaging applications: A review of the nanocomposites in comparison with the conventional counterparts. *Carbohydr. Polym.*, **280**, 119027, 2022.
84. Hemmati, F. *et al.*, Electrospun antimicrobial materials: Advanced packaging materials for food applications. *Trends Food Sci. Technol.*, **111**, 520–533, 2021.
85. Maftoonazad, N. and Ramaswamy, H., Novel techniques in food processing:Bionanocomposites. *Curr. Opin. Food Sci.*, **23**, 49–56, 2018.
86. Yu, X. *et al.*, A hydroxypropyl methylcellulose/hydroxypropyl starch nano-composite film reinforced with chitosan nanoparticles encapsulating cinnamon essential oil: Preparation and characterization. *Int. J. Biol. Macromol.*, 242, 124605, 2023.
87. Khanzada, B., Mirza, B., Ullah, A., Chitosan based bio-nanocomposites packaging films with unique mechanical and barrier properties. *Food Packag. Shelf Life*, **35**, 101016, 2023.
88. Mujtaba, M. *et al.*, Understanding the effects of copolymerized cellulose nanofibers and diatomite nanocomposite on blend chitosan films. *Carbohydr. Polym.*, **271**, 118424, 2021.
89. Zhou, L., Zhang, W., Wang, J., Recent advances in the study of modified cellulose in meat products: Modification method of cellulose, meat quality improvement and safety concern. *Trends Food Sci. Technol.*, **122**, 140–156, 2022.
90. Sun, Y. *et al.*, Double-layer indicator films aided by BP-ANN-enabled freshness detection on packaged meat products. *Food Packag. Shelf Life*, **31**, 100808, 2022.
91. Yuan, D. *et al.*, A novel composite edible film fabricated by incorporating W/O/W emulsion into a chitosan film to improve the protection of fresh fish meat. *Food Chem.*, **385**, 132647, 2022.

92. Ban, Z. *et al.*, Efficient production of fungal chitosan utilizing an advanced freeze-thawing method; quality and activity studies. *Food Hydrocoll.*, **81**, 380–388, 2018.
93. Fan, X. *et al.*, Incorporating *Portulaca oleracea* extract endows the chitosan-starch film with antioxidant capacity for chilled meat preservation. *Food Chem.:X*, **18**, 100662, 2023.
94. Amaregouda, Y., Kamanna, K., Gasti, T., Fabrication of intelligent/active films based on chitosan/polyvinyl alcohol matrices containing Jacaranda cuspidifolia anthocyanin for real-time monitoring of fish freshness. *Int. J. Biol. Macromol.*, **218**, 799–815, 2022.
95. Fiore, A. *et al.*, Active packaging based on PLA and chitosan-caseinate enriched rosemary essential oil coating for fresh minced chicken breast application. *Food Packag. Shelf Life*, **29**, 100708, 2021.
96. Yu, D. *et al.*, Multifunctional bioactive coatings based on water-soluble chitosan with pomegranate peel extract for fish flesh preservation. *Food Chem.*, **374**, 131619, 2022.
97. Dai, W. *et al.*, Effect of chitosan-epigallocatechin gallate coating on volatile flavor compounds retention in bighead carp (*Aristichthys nobilis*) fillets during chilled storage. *LWT*, **169**, 114027, 2022.
98. Xiong, Y. *et al.*, Incorporation of salmon bone gelatine with chitosan, gallic acid and clove oil as edible coating for the cold storage of fresh salmon fillet. *Food Control*, **125**, 107994, 2021.
99. Du, H. *et al.*, Development and characterization of fish myofibrillar protein/chitosan/rosemary extract composite edible films and the improvement of lipid oxidation stability during the grass carp fillets storage. *Int. J. Biol. Macromol.*, **184**, 463–475, 2021.
100. Abdallah, M.R.S. *et al.*, Improving the sensory, physicochemical and microbiological quality of pastirma (A traditional dry cured meat product) using chitosan coating. *LWT*, **86**, 247–253, 2017.
101. Wang, S. *et al.*, Chitosan-wampee seed essential oil composite film combined with cold plasma for refrigerated storage with modified atmosphere packaging: A promising technology for quality preservation of golden pompano fillets. *Int. J. Biol. Macromol.*, **224**, 1266–1275, 2023.
102. Zhang, H. *et al.*, Effect of tannic acid-grafted chitosan coating on the quality of fresh pork slices during cold storage. *Meat Sci.*, **188**, 108779, 2022.
103. Pabast, M. *et al.*, Effects of chitosan coatings incorporating with free or nano-encapsulated Satureja plant essential oil on quality characteristics of lamb meat. *Food Control*, **91**, 185–192, 2018.
104. Mirsharifi, S.M. *et al.*, Production, characterization, and antimicrobial activity of almond gum/polyvinyl alcohol/chitosan composite films containing thyme essential oil nanoemulsion for extending the shelf-life of chicken breast fillets. *Int. J. Biol. Macromol.*, **227**, 405–415, 2023.

105. Xiong, Y. *et al.*, Incorporating nisin and grape seed extract in chitosan-gelatine edible coating and its effect on cold storage of fresh pork. *Food Control*, **110**, 107018, 2020.
106. Abdollahzadeh, M. *et al.*, Effects of nano-chitosan coatings incorporating with free/nano-encapsulated essential oil of Golpar (*Heracleum persicum L.*) on quality characteristics and safety of rainbow trout (*Oncorhynchus mykiss*). *Int. J. Food Microbiol.*, **385**, 109996, 2023.
107. Kerch, G., Chitosan films and coatings prevent losses of fresh fruit nutritional quality: A review. *Trends Food Sci. Technol.*, **46**, 2, 159–166, 2015.
108. Barbhuiya, R.I., Singha, P., Singh, S.K., Pomelo wastes, in: *Handbook of Fruit Wastes and By-Products. Chemistry, Processing Technology, and Utilization*, K. Muzaffar, S. Ahmad Sofi, S.A. Mir (Eds.), pp. 19–38, CRC Press, Boca Raton, 2022.
109. Barbhuiya, R.I., Singh, S.K., Singha, P., Mangosteen wastes, in: *Handbook of Fruit Wastes and By-Products. Chemistry, Processing Technology, and Utilization*, K. Muzaffar, S. Ahmad Sofi, S.A. Mir (Eds.), pp. 113–124, CRC Press, Boca Raton, 2022.
110. Rios, D.A.d.S. *et al.*, Food coating using vegetable sources: Importance and industrial potential, gaps of knowledge, current application, and future trends. *Appl. Food Res.*, **2**, 1, 100073, 2022.
111. Chetia, J. *et al.*, Extraction of essential oil from Assam lemon peels and its incorporation in chitosan based coating for maintaining grape quality. *Sustain. Chem. Pharm.*, **32**, 101034, 2023.
112. Bhan, C. *et al.*, Guar gum and chitosan-based composite edible coating extends the shelf life and preserves the bioactive compounds in stored Kinnow fruits. *Int. J. Biol. Macromol.*, **222**, 2922–2935, 2022.
113. Heras-Mozos, R., Gavara, R., Hernández-Muñoz, P., Responsive packaging based on imine-chitosan films for extending the shelf-life of refrigerated fresh-cut pineapple. *Food Hydrocoll.*, **133**, 107968, 2022.
114. Zehra, A. *et al.*, Preparation of a biodegradable chitosan packaging film based on zinc oxide, calcium chloride, nano clay and poly ethylene glycol incorporated with thyme oil for shelf-life prolongation of sweet cherry. *Int. J. Biol. Macromol.*, **217**, 572–582, 2022.
115. Yang, Z. *et al.*, Amphiphilic chitosan/carboxymethyl gellan gum composite films enriched with mustard essential oil for mango preservation. *Carbohydr. Polym.*, **300**, 120290, 2023.
116. Roshandel-hesari, N. *et al.*, Investigation of physicochemical properties, antimicrobial and antioxidant activity of edible films based on chitosan/casein containing *Origanum vulgare L.* essential oil and its effect on quality maintenance of cherry tomato. *Food Chem.*, **396**, 133650, 2022.
117. El Mouzahim, M. *et al.*, Food packaging composite film based on chitosan, natural kaolinite clay, and Ficus. carica leaves extract for fresh-cut apple slices preservation. *Int. J. Biol. Macromol.*, **233**, 123430, 2023.

118. Zhou, Y. *et al.*, A multifunctional chitosan-derived conformal coating for the preservation of passion fruit. *LWT*, **163**, 113584, 2022.
119. Wantat, A., Seraypheap, K., Rojsitthisak, P., Effect of chitosan coatings supplemented with chitosan-montmorillonite nanocomposites on postharvest quality of 'Hom Thong' banana fruit. *Food Chem.*, **374**, 131731, 2022.
120. Li, G. *et al.*, Interaction between chitosan and natamycin maintain freshness quality and enzymatic activity of post-harvest *Prunus salicina Lindl. cv.* 'Shazikongxinli' fruit. *Sustain. Chem. Pharm.*, **27**, 100704, 2022.
121. Xing, Y. *et al.*, Effect of chitosan/Nano-TiO2 composite coatings on the postharvest quality and physicochemical characteristics of mango fruits. *Sci. Hortic.*, **263**, 109135, 2020.
122. Mujtaba, M. *et al.*, Understanding the effects of chitosan, chia mucilage, levan based composite coatings on the shelf life of sweet cherry. *Food Chem.*, **416**, 135816, 2023.

11

Biodegradable Synthetic Poly(Lactic Acid) (PLA) for Food Packaging Application

Shikha Sharma, Tabli Ghosh, Neha Mulchandani and Vimal Katiyar*

Department of Chemical Engineering, Indian Institute of Technology (IIT) Guwahati, Guwahati, Assam, India

Abstract

Poly(lactic acid) (PLA) is a microbial polysaccharide that has gained popularity as a biodegradable polymer and an effective alternative to synthetic plastics. In the context of increasing consumers' concerns about the harmful impacts of synthetic plastic packaging on their health and the environment, researchers have been developing biopolymer-based packaging material that can replace this synthetic plastic packaging. This chapter is a synoptic discussion on the synthesis and types of PLA and PLA-derived biodegradable polymers, their properties, improvements in mechanical, physicochemical, and barrier properties, and food packaging applications. Moreover, composites of PLA with synthetic plastic polymers such as polyethylene, polypropylene, polyvinyl alcohol, and polyethylene terephthalate, and with other biopolymers such as chitosan, cellulose, and starch, have also been explored and evaluated. Additionally, stereocomplex (SC) PLA, blends of poly(L-lactic acid) and poly(D-lactic acid), and their improved mechanical and barrier properties have been elaborated. The incorporation of fillers such as graphene oxide, montmorillonite, cellulose nanocrystals, and other active agents has been shown to improve the functional properties of biopolymer-based composite packaging materials. Finally, the chapter summarizes applications of PLA-based composite materials in the packaging of perishable foods like fruits, meat, vegetables, and processed foods like cheese, cake, and others.

Keywords: Biodegradable polymer, bioplastic, PLA, food packaging, lactic acid

**Corresponding author*: vkatiyar@iitg.ac.in

Santosh Kumar, Avik Mukherjee and Vimal Katiyar (eds.) Agro-Waste Derived Biopolymers and Biocomposites: Innovations and Sustainability in Food Packaging, (295–318) © 2024 Scrivener Publishing LLC

11.1 Introduction

Due to the desired properties of plastics such as lightweight, transparency, flexibility, low cost, durability, mechanical strength, easy handling, and others, they penetrated the human lifestyle. Plastics are widely used throughout the world in various stages of life such as from kitchen to aerospace, packaging to textiles to 3D printing, automobile, healthcare, and electronics, almost everywhere. Approximately, more than 300 million tons of plastic is produced every year worldwide, as shown in Figure 11.1a, among which 9 million tonnes are produced in India every year, and forecasted that the production may cross 400 million tons by 2050 worldwide [1, 2]. It is notified by different agencies that approximately 60% of the produced plastics are discarded as waste after use [3]. These plastic wastes are being reused or recycled but mainly they go for either landfill or incineration. As a result, it increases the carbon footprint and thus increases global warming, produces environmental pollutants, and makes us dependent on petroleum. An article published in *The Guardian* in January 2016 written by Ellen MacArthur forecasted that the weight of plastics as waste present in the ocean will overshoot the weight of entire fish present in the ocean by 2050 [4]. Due to these issues, there is an urgent requirement to have an alternative such as biobased plastics, which should be sustainable and environment friendly.

Furthermore, if we compare the per capita consumption of plastics with developed nations, Indian people consume only 9.7 kg/person of plastics in comparison to the United States of America (USA), which is 109 kg/person. The current plastic consumption in India is approximately 8 million tonnes, which has increased to 20 million tonnes in 2020. The demand breakup of various types of plastics is given in Figure 11.1b, in which, polypropylene (PP) and polyethylene (PE) combined have 67% share and 79% share, respectively, in the flexible packaging combined with PE terephthalate (PET) [5]. These data give evidence showing the massive growth potential of plastic industries in India. On the other hand, the comparable properties of biomass-based plastics have the potential to replace these conventional plastics.

The technical definition of biobased plastics is that these are biobased polymers derived from biomass or are synthesized by polymerization of monomers derived from biomass and are processed and shaped into finished products [6]. Several biobased polymers, such as starch, cellulose, protein-based polymers, poly(3hydroxybutyrate) (PHB), poly(lactic acid) (PLA), bio-PP, bio-PE, bio-PET, polybutylene succinate (PBS),

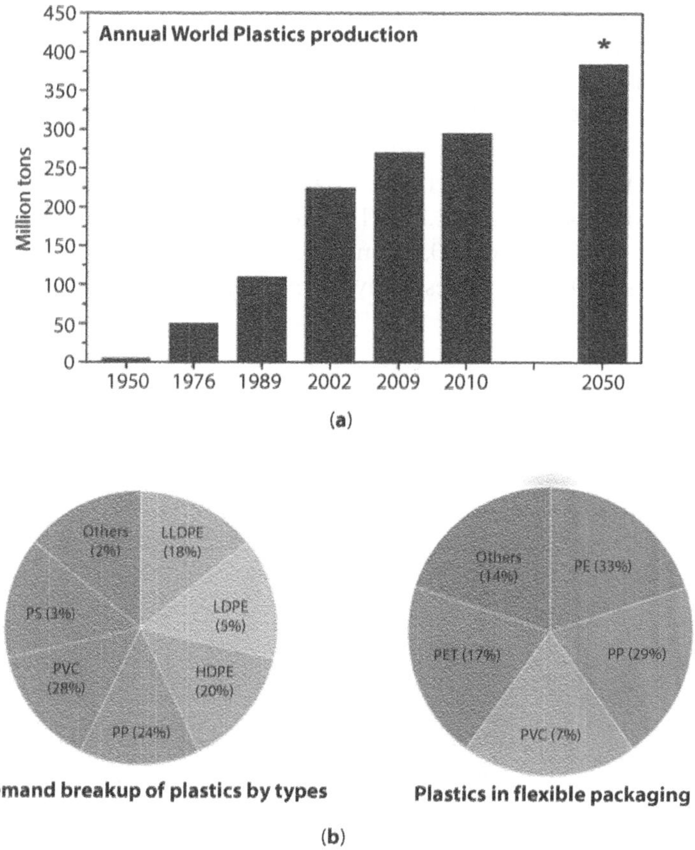

Figure 11.1 (a) Trend for annual plastic production in the world (Source: Plastic Europe Market Research Group) and (b) demand breakup of plastics by type and share in flexible packaging.

polyhydroxyalkanoates (PHA), and polyglycolic acid (PGA), are available in the market [7–9]. These polymers have different biomass content, which is responsible for their biodegradability. A market study on biobased polymers in the world done by Nova Institute for Ecology and Innovation, Germany, reported that, in 2011, approximately 3.5 million tonnes of biobased polymers were produced by 247 companies worldwide, which was equivalent to 1.5% of the total production of all type of polymers, and have increased to 12 million tonnes in 2020 [10]. It is also reported that most of the world's companies will invest their capital in Asia and South America due to its better accessibility for feedstock and favorable political framework among which Asia will contribute ~55% share. Among 247 companies, 25 companies are producing more than 180,000 tonnes per year of PLA at 30 different locations worldwide, and this has increased to 800,000 tonnes per year in 2020 (Figure 11.2a) [11]. Another study reported the production capacities of biodegradable plastics including PLA, starch-based biopolymers, PHA, PBS, and polybutylene adipate terephthalate (PBAT), as well as those of non-biodegradable plastics such as PE, polyamides (PA), PET, polytrimethylene terephthalate (PTT), and PP from the year 2017 to 2024 (Figure 11.2b). The global productions of PLA and PET were about 400,000 and 170,000 metric tonnes, respectively, which were approximately 19 and 8 %, respectively, of the global production in the year

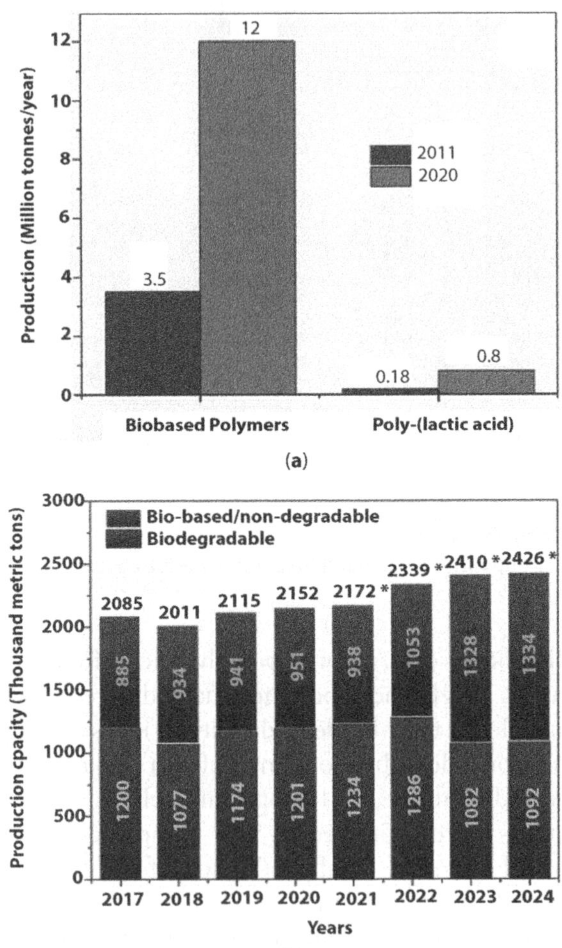

Figure 11.2 Global production (a) of biobased polymers and PLA and (b) capacities of bioplastics (Source: Statistica, Production capacities of bioplastics worldwide by material type).

2020 [12]. Moreover, PLA is expected to account for 19.5% of the global production of biobased/nonbiodegradable plastics and is expected to rise further to 2.43 million metric tonnes by 2024. The observed data indicates that the production of PLA in developing countries like India is going to enhance drastically in the near future. As per the literature available, PLA is one of the most attractive alternatives, as a substitute to synthetic plastics. Moreover, PLA as a biobased polymer can fill the gap for stringent food packaging and beverage industries. In adverse, other biopolymers such as chitosan, gums, and to a certain extent, starch and cellulose are hydrophilic in nature, which hinders their usability in packaging application.

Furthermore, PLA is a biodegradable, bioabsorbable, nontoxic, biocompostable polymer, produced through polymerization of monomer lactic acid, which is derived from plants such as sugar cane, cassava, and potato, and after its service life, can be decomposed into nontoxic content by natural and soil composting. PLA has 100% biobased content, and its properties can be tuned through various routes according to the necessity. The general life cycle for the PLA is shown in Figure 11.3, where it is synthesized from the polymerization of either lactide or lactic acid, which is produced by fermentation of carbohydrate-based biomaterials (glucose, sucrose, or galactose), and at last, degrades in an environment-friendly way. The current work discusses its synthesis, properties, characterization, composting, and application as a packaging material specifically for food products in the subsequent sections.

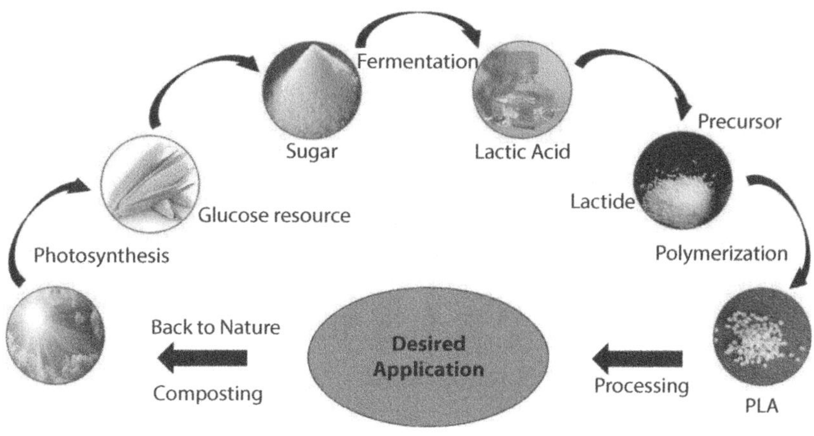

Figure 11.3 The life cycle of PLA.

11.2 Synthesis of PLA

PLA belongs to the α-hydroxy acid family. It is derived from lactic acid bacteria fermentation, and numerous renewable sources such as corn starch, sugarcane, cassava roots, and potatoes can be valorized for commercial manufacturing of PLA. The isolation of lactic acid, a monomeric unit of PLA, from sour milk was reported by Carl Wilhelm Scheele, a Swedish chemist, in the 18th century [13]. Lactic acid is chiral and forms two enantiomeric homopolymers, i.e., poly(L-lactic acid) (PLLA; laevorotatory) and poly(D-lactic acid, dextrorotatory) (PDLA) acid, and the mixture of these two is called racemic lactic acid, as shown in Figure 11.4a [14]. Compared to PDLA, PLLA biodegradation takes longer, and the latter has better biocompatibility. Also, PDLA possesses relatively lesser strength

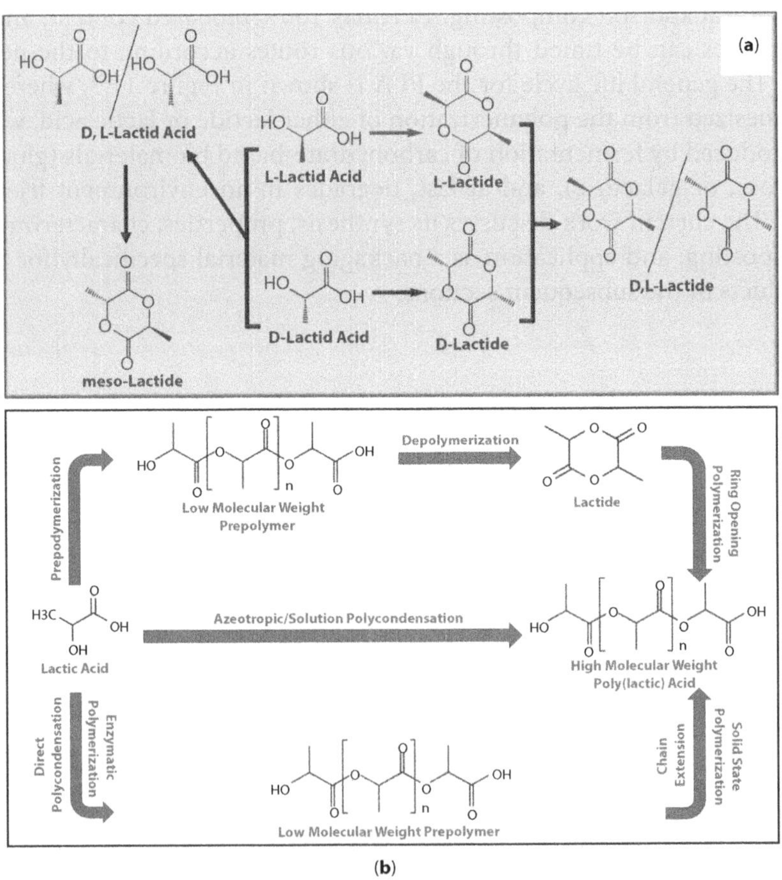

Figure 11.4 (a) Isomers of lactic acid and corresponding lactide and (b) synthesis of PLA [15].

and modulus compared to PLLA. Both of these polymers are crystalline in nature, however, a racemic mixture of D- and L-lactic acid forms amorphous poly(DL-lactic acid) (PDLLA) upon polymerization [16, 17].

PLA can be synthesized from lactic acid (2-hydroxypropanoic acid) by polycondensation, which was first discovered by Theophile-Jules Pelouze (Figure 11.4b). Lactide (cyclic lactone derived from lactic acid) is another precursor that can be used to manufacture PLA by ring-opening polymerization (ROP) process, as developed by Wallace Hume Carothers in 1932 [18]. The process of synthesis of lactide was patented by Reinhold Gruter in 1913 [19]. This method manufactures high molecular weight PLA at a commercial scale. The precursor lactide production involves the three most common methods (Figure 11.5). The first method involves a two-step synthesis of lactide, during which lactic acid is polycondensed to form oligomer lactic acid prepolymer, and then depolymerized into lactide dimer ring at high temperature and low pressure. The second method is gas-phase synthesis of lactide, where vaporized lactic acid is reacted over a catalyst bed. The third includes the elimination of water from an aqueous lactic acid solution at the time of ring closing in a condensation reaction, and clearly, it is a one-step liquid-phase process [11]. The ROP of

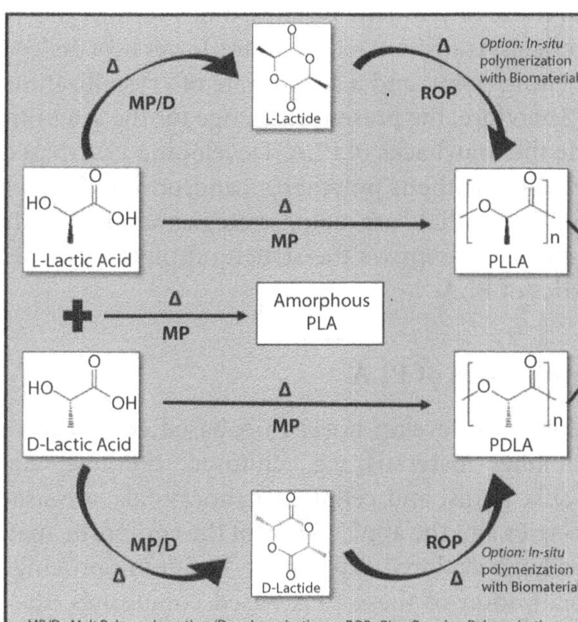

Figure 11.5 Different PLA production pathways depending upon prospective applications in various sectors.

lactide is achieved by several mechanisms that include the coordination-insertion mechanism, activated monomer, activated chain mechanisms, and nucleophilic activation, which are carried out at the end of the polymer chain. Additionally, the catalytic system of ROP is classified as a cationic catalyst, metal catalyst, and anionic catalyst. A greener method for metal-based catalysts is enzymatic ROP (eROP) of lactides using lipases and esterases, which can fetch additional advantages, e.g., better physical and chemical properties of the synthesized PLA [11]. PLA can also be produced by the azeotropic dehydrative condensation polymerization of the lactic acid in diphenyl ether (diphenyl oxide) solvent using tin powder as a catalyst [15]. Furthermore, solid-state polycondensation or chain extenders are used to increase the molecular weight of PLA. Clearly, after manufacturing PLA by a combination of fermentation and polymerization, the biopolymer can be further modified using different approaches depending upon the intended applications of the PLA.

11.3 Properties of PLA

Even though the thermal, mechanical, and barrier properties of PLA are comparable to those of synthetic plastics, it has various shortcomings, e.g., relatively lower glass transition temperature, lower heat deflection temperature, lesser melting point, and a slower rate of crystallization, which limit its uses [20]. Therefore, the present challenge for the scientific community is to overcome the drawbacks of PLA. Developing PLA-based composites, blending PLA with synthetic polymer(s), and/or other biopolymer(s), and synthesis of SC PLA (sPLA) are major strategies adopted by the researchers and manufacturers to improve the structural physicochemical and functional properties of PLA.

11.3.1 Composites of PLA

The opportunities to develop novel PLA-based composites are plentiful, among which filler materials, e.g., chitosan, cellulose, clays, graphene oxide (GO), silk, gums, and cellulose nanocrystals, are used to improve relevant properties and the applicability of the composite materials in food packaging has been explored extensively. Besides improving the potential for commercialization of these PLA-based composites, the strategy also enables waste/byproduct valorization, which promotes sustainable ecological balance, economic feasibility, and business profitability. Food packaging applications demand effective gas and moisture barrier properties of PLA/

PLA-based composites. Layer by layer (LbL), other blending techniques, and PLA surface treatment(s) can achieve such improvements rendering PLA composites to be more useful in food packaging applications by prolonging the shelf life of the packaged food products. Aulin *et al.* (2013) used the LbL method to prepare multilayer PLA composites, in which polyethylenimine (PEI) and nanofibrillated cellulose (NFC) or carboxymethyl cellulose (CMC) layers were applied on the surface of PLA and reported that the developed composite films coated with 50 bilayers of PEI/NFC and PEI/CMC had oxygen permeance (~10 $cm^3/m^2 \cdot day \cdot atm$) lower (by more than an order of magnitude) than the uncoated PLA (151.5 $cm^3/m^2 \cdot day \cdot atm$) at 23°C and 50% RH [21]. The LbL deposition method also resulted in less moisture-sensitive PLA composite films that showed more than 80% transparency to visible light [21]. The LbL method employed by Laufer *et al.* (2012) showed a four-fold reduction in oxygen permeability leading to retarded oxidative degradation of the packaged food, i.e., enhanced shelf life of the food [22]. These researchers sequentially dipped PLA film in chitosan and montmorillonite (MMT) clay solutions followed by drying, which formed the desired coatings on the biopolymer film. Thirty such coatings reduced the oxygen permeability of PLA from 177.2×10^{-16} $cm^3.cm.cm^{-2}.s^{-1}.Pa^{-1}$ to less than 0.03×10^{-16} $cm^3.cm.cm^{-2}.s^{-1}.Pa^{-1}$, which could be due to the formation of a nanolayer wall of the clay making the composite film significantly more impermeable to oxygen.

The gas or water vapor barrier properties of polymers can also be improved by surface treatments, i.e., UV or ozone treatment, treatment at elevated temperature, exposing the film to plasma, and cross-linking by hyperthermal hydrogen. Hyperthermal hydrogen-induced cross-linking causes selective breakage of C-H or Si-O bonds without using any chemical reagents and without affecting other atomic bonds. The breakage of the C-H bond generates carbon radicals, which facilitates the cross-linking by hyperthermal hydrogen, which occurs because of exposure of the composite film to plasma. Du *et al.* (2015) reduced the water vapor permeability of PLA film using this crosslinking technology, which resulted in cross-link formation between carbon radicals on the PLA film surface [23]. The cross-linked surface layers serve as a barrier, which reduces the water vapor permeability of PLA from 26.74 to 22.96 $gm/m^2 \cdot day$ and improves its hydrophobicity. Blending of polymers is another approach that can significantly improve the barrier properties. Armentano *et al.* (2015) have found that a 15% addition of PHB into the PLA matrix increases its crystallinity due to enhanced crystal packing, thereby reducing the oxygen permeability by ~35% [24]. Along with the gas barrier properties, PLA also lacks mechanical and thermomechanical characteristics compared to

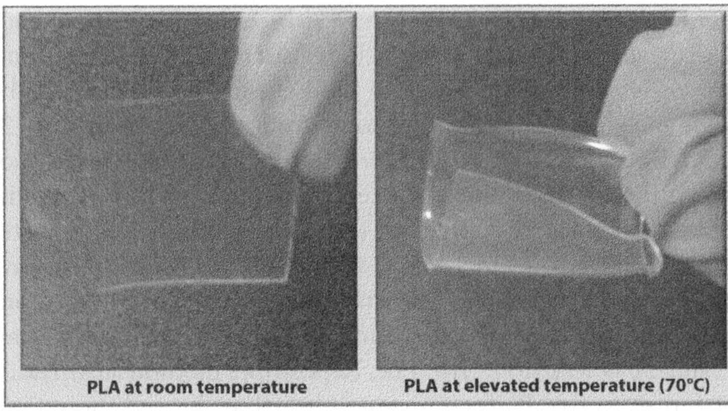

Figure 11.6 Mechanical stability of PLA at elevated temperature.

synthetic plastics, which can be enhanced by the development of biocomposites by fabricating polymer blends or by other methods. It is proven that the development of biocomposite affects different methods for making PLA which is comparatively mechanically strong (Figure 11.6). A new type of crystal is formed by mixing the two PLA enantiomers, i.e., PLLA and PDLA together, and they are called SC crystals [25]. It has been reported that the melting temperature of sPLA was 50°C higher than both the pure PLA enantiomers.

11.3.2 Stereocomplex of PLA

Stereocomplexation is a special type of interaction between two PLA enantiomer chains, and the resulting crystals of sPLA show improved physicochemical, mechanical, and barrier properties. The pure PLA enantiomers possess 10_3 helical configurations, while sPLA crystals have 3_1 helical configurations that result in the densely packing of the polymer chains leading to improvements in various properties of the sPLA. Xu *et al.* (2015) reported that during solution mixing for GO-based film fabrication, SC nanospheres were formed, and they surrounded the GO sheets, which acted as locations for nucleation resulting in the development of spherulite [26]. The GO matrix drove a steady increase in the concentration of SC, and the homocrystal formation was suppressed. The composite films showed reduced oxygen permeability from 4.869×10^{-15} $cm^3.cm.cm^{-2}.s^{-1}.Pa^{-1}$ to 0.2553×10^{-15} $cm^3.cm.cm^{-2}.s^{-1}.Pa^{-1}$ due to the addition of 0.5 wt % GO, and annealing of the composite film further reduced it to 0.06264×10^{-15} $cm^3.cm.cm^{-2}.s^{-1}.Pa^{-1}$. The GO matrix was also responsible for the formation of an impermeable barrier against gas diffusion and for improving the SC crystallinity in the composite film, which in turn significantly reduced oxygen permeability.

The production of PLA grafted fillers by *in situ* polymerization, i.e., lactide polymerization in the presence of the fillers, is also an effective technique that improves the dispersion of the filler material(s) into the polymeric matrix. The homogenous dispersion of the fillers is achieved through interaction between the matrix and the filler molecules during the grafting of PLA chains on the surface of the filler. Re *et al.* (2014) have reported both stereocomplexation and the use of filler, i.e., a dual approach, which significantly contributes to the mitigation of shortcomings in the properties of the polymer matrix [27]. They first prepared and purified the PLLA grafted clay (Cloisite 30B), and then melt blended with PDLA to get the homogeneous dispersion of grafted clay followed by the formation of SC, which leads to a 46% reduction in the oxygen permeance of the film. Clearly, the structural modifications through these chemical changes, as elaborated in these studies, can significantly enhance the structural and barrier properties of PLA making it closely comparable with that of synthetic plastics.

11.3.2.1 Stereocomplex PLA

As defined by Slager *et al.* (2003), SC of a polymer occurs when two complementary stereoregular polymers interact with each other through a stereoselective interlocking that converts them into a new composite, the physical properties of which are different from those of the parent polymers [28]. Tsuji *et al.* (2005) also have defined SC somewhat similarly and stated that polymers having different tacticity engage in stereoselective association between themselves and it prevails over same tacticity interaction can be termed as stereocomplexation or formation of SC that happens between isotactic and syndiotactic, i.e., optically active, opposite enantiomers of a polymer [29]. Precipitation, melt blending, solution casting, and stereoblock of the two PLA enantiomers (PLLA and PDLA) are some of the commonly used methods developed for the formation of SC in PLA leading to a crystalline arrangement of the two PLA enantiomer chains. The primary drawback of the method is that along with the sPLA, some homo crystals of relatively low molecular weight (<100 kDa) PLA are also formed, which can negatively impact the mechanical properties of the fabricated biopolymer-based films, the improvement of which is possible through several techniques, as elaborated by researchers. Ikada *et al.* (1987) carried out solution mixing of PLLA and PDLA for the first time, i.e., PLLA and PLDA were dissolved separately in methylene chloride followed by mixing with each other and precipitation in methanol that led to the formation of sPLA [25]. The synthesized sPLA had about 50°C higher melting point, compared to that of the normal PLA, which could be due to Van der Waals forces, e.g., dipole-dipole interactions between the two enantiomeric helical chains in solution, in which molecular motion is high enough to form the SC.

Characterization of the sPLA is quite complicated, and analytical techniques such as nuclear magnetic resonance (NMR), differential scanning calorimetry (DSC), and X-ray diffraction (XRD) are effective in examining the complex polymer. sPLA is significantly less soluble than each of the individual PLA enantiomers in solvents that can dissolve PLA, for example, dissolution in HFIP can be used to confirm the formation of sPLA. Due to the different crystallite formations in sPLA, it is easy to analyze. When the XRD technique was used to determine the existence and stability of α, β, and γ polymorphs of pure PLA enantiomers, α was found to be the most stable among them, crystallizing two 10_3 helices in the pseudo-orthorhombic unit cell having dimensions a = 1.07 nm, b = 0.595 nm, and c = 2.78 nm. On the other hand, chain segments of L and D-lactide form 3_1 helical conformations by parallel packing into dimensions of a = b = 0.916 nm, c = 0.870 nm, and α = β = γ = 109.2° which is a triclinic unit cell. The XRD patterns (Figure 11.7a) of sPLA have peaks at 2θ values of 11.9°, 20.9°, and 23.9° corresponding to the (110), (300/030), and (220) planes of the SC crystals, respectively, whereas (110/200) and (203) planes of the α homocrystals of PLA are represented by the peaks at 16.7° and 19°, respectively [30]. The DSC analysis shown in Figure 11.7b clearly demonstrates that sPLA has a greater melting point than that of pure PLA. In terms of the production and marketing of the PLA, several companies have put industrial-level plants for production. Well-known, large-scale PLA manufacturers are Corbian Purac, Sulzer Technology, and Nature Works LLC (Ingeo as trade name), however, only a few of them are working to develop the technology for commercial manufacturing of sPLA.

11.3.2.2 Additives to Enhance Stereocomplexation of PLA

Researchers have explored the incorporation of a third component (other than the two PLA enantiomers) to improve the stereocomplexation process and to enhance sPLA crystals in the finished product. Li *et al.* (2009) added poly-(4-vinyl phenol) (PVPh) in the mixture of the two PLA enantiomers PLLA and PDLA and reported accelerated crystallization as well as reduction in the size of spherulites in the SC. The addition of low PVPh content in the blend also accelerated the ability of nonisothermal

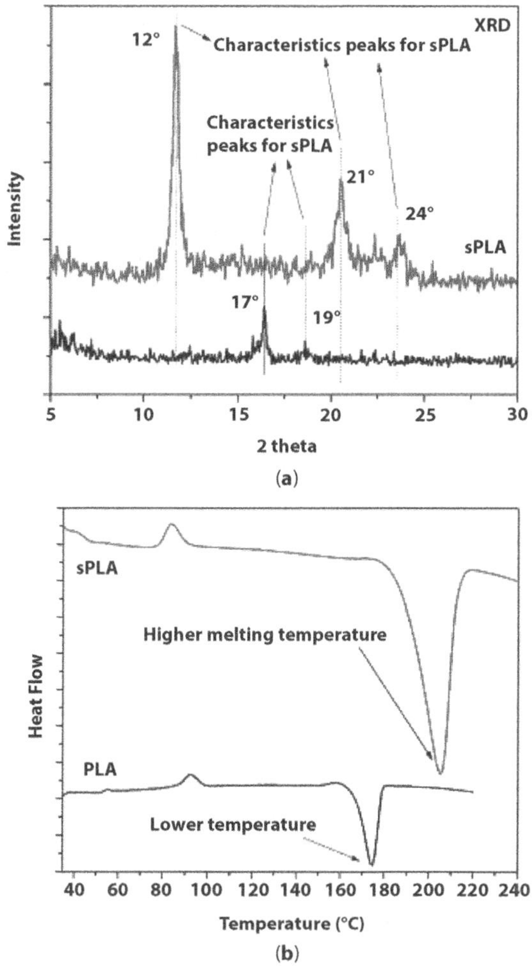

Figure 11.7 PLA and stereocomplex PLA; (a) XRD patterns and (b) DSC thermogram.

SC crystallization [31]. Chang and Woo (2011) reported that the addition of PHB reduced sPLA crystallization [32], while Sun and He (2012) found that the incorporation of GO as nanofillers, the crystallization of sPLA and GO, resulted in increased crystallinity, and hence higher SC crystallites contents in the composite [33]. It is also found that the increased amount of GO can block and hinder crystal growth by retarding polymer chain mobility. Quan *et al.* (2012) used direct melt mixing of carbon nanotubes in the sPLA reaction blend and found that the crystallinity of the composite increased to 62% from 43% with 1% carbon nanotubes due to the nucleation effect of CNT [34]. Layered silicate nanocomposites with sPLA were also evaluated (Furuhashi *et al.*, 2013) [35]. Another study conducted by Bao *et al.* (2014) found that increasing low molecular weight poly-(ethylene glycol) (PEG content showed increased SC crystal formation. The same study also reported decreased glass transition temperature (t_g) of the sPLA from 62°C to 37°C, as PEG had a plasticizing effect on the composite [36]. Purnama and co-workers (2014) synthesized the SC poly-(lactide) containing cellulose nanowhiskers by supercritical fluid technology and, found that the 100% SC crystallite was formed which had a melting point of more than 200°C [37]. Zinc phenylphosphonate (PPZn) was used as a nucleation agent in the stereocomplexation of PLA, and the results showed that PPZn nucleation not only enhanced the crystallization rate but also selectively promoted the formation of SC crystallites [38].

sPLA has shown tremendous potential as an effective packaging material for food/beverages consumed at elevated temperatures (hot tea cups) and for any other liquid food as well as other engineering applications. Bor *et al.* (2014) reported that upon exposure or coming in contact with different food items, sPLA showed a slower rate of release of oligomeric chains and/or cyclic lactic acid compared to that shown by the pure PLA enantiomers [39]. Moreover, they reported that the higher crystalline nature of sPLA compared to relatively amorphous PLA enantiomers led to increased migration resistance of cyclic lactic acid from the composite packaging into food stimulants such as water and 10% ethanol solution during microwave heating. From the above discussion, the PLA is the future of the plastic industries, and enhancement in the properties of PLA through stereocomplexation may be the tool for the development of polymers for different types of packaging and engineering applications.

11.4 Food Packaging Applications of PLA

Thanks to its biodegradability and other important functionalities, PLA is a potential candidate to be greatly used in packaging industries, especially food packaging applications, e.g., biodegradable and compostable disposable cups that are biodegradable, beverages mugs that are compostable, films, coatings for perishables, as well as for coating of other food packaging such as paper coating, fiber coating, and laminates. Besides food packaging applications, PLA can be effectively applied in agriculture and electronics industries, including 3D printing. Thus, PLA has immense promise as an alternative that can effectively replace synthetic plastics. The recent studies on applications of PLA for food packaging are presented in Table 11.1.

Table 11.1 Application of PLA for food packaging.

Food product	Packaging materials	Packaging method	Storage condition	Outcome	Reference
Shiitake mushroom	PLA and cinnamaldehyde essential oil (CEO) coating on modified cellulose fiber (MC, source: water bamboo leaf sheath) sheet	Sealed with developed film	4°C for 30 days	The developed sheet had high tensile strength, the water contact angle was 132.12° and decreased water absorption rate; cinnamaldehyde imparted antibacterial activity; the Browning index was lowest, L* value and firmness were maintained, and psychrophilic, mesophilic, yeast, and molds count were significantly lower in shiitake mushrooms for 30 days stored in 5% CEO in PLA/MC box than PE and PE with CEO box	[40]
Cut beef tenderloin	PLA, cinnamaldehyde, and butterfly pea flower extract (BPA) electrospun nanofiber film	Plastic boxes having 1 cm × 1 cm developed the smart film in the center	Incubator (25°C, 75% RH) for 2 days refrigerator (4°C) for 8 days	Box with smart film showed maintained total volatile basic nitrogen (TVB-N) compared to the control; PLA-BPA 6 wt %/CIN film increased shelf-life to 6 days at 4°C and indicated freshness, sub-freshness, and spoilage levels clearly by color change	[41]

(*Continued*)

Table 11.1 Application of PLA for food packaging. (*Continued*)

Food product	Packaging materials	Packaging method	Storage condition	Outcome	Reference
Sunflower oil	Atmospheric cold plasma pre-functionalized PLA laminate (2 PLA film layers)	Heat-sealed pouch of 20 × 10 cm	35°C for 64 days at 50 % RH	The peroxide value of oil in the nonactivated film exceeded the threshold after 25 days while in the activated film it reached after 37 days; activated oil retained a more shiny yellow color; lipid oxidation kinetics reduced in the activated sample	[42]
	Atmospheric cold plasma pre-functionalized PLA laminate (2 PLA film layers) and activated with ascorbic acid/sodium alginate gel between 2 PLA layers				
Genovese pesto sauce	Atmospheric cold plasma pre-functionalized PLA laminate (2 PLA film layers)		25°C for 41 days and 45°C for 34 days at 50% RH	Peroxide value at 25°C exceeds the threshold after the 27th day in the activated pouch but in the nonactivated one after the 10th-day peroxide value crossed the limit and at 45°C, it is almost similar for both samples; textural properties were maintained in pesto samples stored at 25°C; color and water activity decrement was drastic at 45°C than 25°C; mesophilic, yeast and spore-forming bacteria were slightly low in activated pouch than nonactivated one	
	Atmospheric cold plasma pre-functionalized PLA laminate (2 PLA film layers) and activated with ascorbic acid/sodium alginate gel between 2 PLA layers				

(*Continued*)

Table 11.1 Application of PLA for food packaging. (*Continued*)

Food product	Packaging materials	Packaging method	Storage condition	Outcome	Reference
Shrimp	Cast sheet extrusion (single screw extruder) of PBAT/PLA with lime peel oil (LPO), and ginger oil (GO)	Three shrimps in a UV light sterilized and sealed bag film	4°C for 6 days	PBAT 30%/PLA 70% with GO effectively inhibited the formation of melanosis, while LPO retarded the growth of microorganisms	[43]
Cotton cupcakes	PLA thermoformed trays incorporated with cinnamaldehyde (CIN) and carvacrol (CAR)	The thermoformed tray was sterilized by UV and heat-sealed with aluminum foil laminated polyethylene film	25°C for 5 days	CAR had stronger inhibition of molds than CIN	[44]
Tuna meat cut pieces	PLA and betel leaf ethanolic extract (BLEE) (5 wt % and 10 wt %)	Fabricated pouch from film	4°C 7 days	Composite films preserve tuna from spoilage including lipid oxidation. The blending of 5 wt % BLEE had higher efficiency than 10 wt % extract. pH, trimethyl amine, thiobarbituric acid reactive substances, and peroxide values were lower in wrapped tuna than in the control.	[45]

(*Continued*)

Table 11.1 Application of PLA for food packaging. (*Continued*)

Food product	Packaging materials	Packaging method	Storage condition	Outcome	Reference
Almonds	Film 1: PLA with PEG [poly(ethylene glycol)] and 4% lemon extract (LE) Film 2: PLA and 6% lemon extract	Almonds (without shell) was treated with hot water and then ground and vacuum-packaged	60 days at 40°C	Regarding the film's ability to delay lipid oxidation, PLA/PEG and PLA/PEG/4LE showed a lower value of malonaldehyde (MDA) equivalents in almonds for 30 days; PLA and PLA/6LE had low p-Anisidine value until the 15th day, and overall hydrogen peroxide value was lower of PLA/6LE as compared to PLA for 7 days	[46]
Beef meat	PLA and lemon extract (6%)	Vacuum-packaged	11 days at 4°C	The developed coating inhibited total microbial count and antioxidant activity	
Steamed glutinous rice	Single screw extruded PLA/PBS with galangal essential oils (GEO)	PLA/PBS/GEO films placed on Polystyrene trays and then packed into PA/LDPE bags and sealed	25°C for 10 days	The composite film prevented fungal growth for up to 3 days by blending with GEO	[47]

(*Continued*)

Table 11.1 Application of PLA for food packaging. (*Continued*)

Food product	Packaging materials	Packaging method	Storage condition	Outcome	Reference
Cheese cubes	PLA/PBAT, TiO_2, and cinnamon essential oil	UV sterilized and sealed in the film	12-day storage period at 4°C	The composite film with 7 % TiO_2 showed the lowest weight loss and antibacterial activity	[48]
Red grapes	Extruded PLA/PBS film with bis[2-(2-butoxyethoxy) ethyl] adipate (BXA) and ADR-4368	Packed in sealed bags (the top was hot sealed after the grapes were packed) with PLA and PLA-based films (size: 35 cm × 20 cm, thickness: 25 ± 5 μm)	Light-proof sample room for 12 days at a temperature of 25°C and 78 % RH	80 PLA/20 PBS with BXA and ADR-4368 showed the best preservation results; Reactive oxygen species metabolic balance was maintained and respiration was effectively inhibited; lowest loss of weight (8.64% on the 12th day), and color was plump with bright color at the end of storage.	[49]

PE: Polyethylene, PBS: polybutylene succinate, PBAT: polybutylene adipate terephthalate

11.5 Conclusion and Future Prospects

Synthetic plastics, particularly single-use plastics, profusely applied in food packaging have significant harmful impacts on consumers' and environmental health. PLA and its composites PLA have been extensively studied as an attractive alternative to synthetic plastics for food packaging applications. PLA composites and blends, e.g., sPLA, have shown improved physicochemical, thermomechanical, and barrier properties, making them even better food packaging materials. However, there are still considerable research and development opportunities for the improvement of PLA-based active, biodegradable food packaging systems, their commercial applications through scale-up studies, and potential cost-reduction in manufacturing of the composite packaging materials. Sustainable food packaging being a global issue, PLA and PLA-derived biodegradable materials and other microbial biopolymers are attractive alternatives to develop such environment-friendly food packaging systems that can have relevant contributions toward a circular economy.

References

1. A report on plastic industry, in: *2nd National Conference Plastic Packaging-The Sustainable Choice*, TATA Strategic Management Group, New Delhi, 2016.
2. Fornasiero, P. *et al.*, *Renewable resources and renewable energy: A global challenge*, S. Edition, p. 504, CRC Press Taylor & Francis Group, Boca Raton, Florida, USA, 2011.
3. Geyer, R., Jambeck, J.R., Law, K.L., Production, use, and fate of all plastics ever made. *Sci. Adv.*, **3**, 7, e1700782, 2017.
4. Wearden, G., More plastic than fish in the sea by 2050, says Ellen MacArthur, Davos, 2016, [Web] 2016 [cited 2016 20 december].
5. *Potential of plastics industry in Northern Indian with special focus on plasticulture and food processing, a report on plastics industry*, Federation of Indian Chamber of Commerce and Industry, New Delhi, 2014.
6. Pilla, S., Engineering applications of bioplastics and biocomposites — An overview, in: *Handbook of Bioplastics and Biocomposites Engineering Applications*, pp. 1–15, John Wiley & Sons, Inc., Beverly, Massachusetts, USA, 2011.
7. Kumar, S. *et al.*, An overview of natural biopolymers in food packaging, in: *Biopolymer-Based Food Packaging*, pp. 1–28, 2022.
8. Majumder, R. *et al.*, Biopolymers, in: *Biopolymer-Based Food Packaging*, pp. 29–65, 2022.

9. Naser, A.Z., Deiab, I., Darras, B.M., Poly(lactic acid) (PLA) and polyhydroxyalkanoates (PHAs), green alternatives to petroleum-based plastics: A review. *RSC Adv.*, **11**, 28, 17151–17196, 2021.
10. Mirabal, A.S., Scholz, L., Carus, M., *Market study on bio-based polymers in the world capacities, production and applications: Status quo and trends towards 2020*, Nova Institute for Ecology and Innovation GmbH, Hürth, Rhein-Erft-Kreis, Germany, 2013.
11. Balla, E., Daniilidis, V., Karlioti, G., Kalamas, T., Stefanidou, M., Bikiaris, N.D., Vlachopoulos, A., Koumentakou, I., Bikiaris, D.N., Poly(lactic acid): A versatile biobased polymer for the future with multifunctional properties applications. *Polymers*, 13, 11, 1822, 2021.
12. Ali, S.S. *et al.*, Bioplastic production in terms of life cycle assessment: A state-of-the-art review. *Environ. Sci. Ecotechnology*, **15**, 100254, 2023.
13. Ghaffar, T. *et al.*, Recent trends in lactic acid biotechnology: A brief review on production to purification. *J. Radiat. Res. Appl. Sci.*, **7**, 2, 222–229, 2014.
14. Miller, C. *et al.*, 3.17 - Industrial production of lactic acid A2, in: *Comprehensive Biotechnology*, Second Edition, M. Moo-Young, (Ed.), pp. 179–188, Academic Press, Burlington, 2011.
15. Desai, H., Mehta, T., Shah, N., Azeotropic dehydrative (solution) polycondensation of lactic acid to polylactic acid (PLA): A in-depth review of an overlooked method for manufacturing PLA. *Polym.-Plast. Technol. Mater.*, 62, 11, 1394–1402, 2023, https://doi.org/10.1080/25740881.2023.2209644.
16. Srivatsan, T.S., Biomaterials: A nano approach, *Mater. Manuf. Process.*, 29, 1510–1511, 2014.
17. Ghalia, M.A. and Dahman, Y., Biodegradable poly(lactic acid)-based scaffolds: synthesis and biomedical applications. *J. Polym. Res.*, 24, 74, 2017.
18. Moya-Lopez, C. *et al.*, Polylactide perspectives in biomedicine: From novel synthesis to the application performance. *Pharmaceutics*, **14**, 1673, 2022.
19. Mulchandani, N., Gupta, A., Katiyar, V., Polylactic acid based hydrogels and its renewable characters: Tissue engineering applications, in: *Cellulose-Based Superabsorbent Hydrogels*, M.I.H. Mondal (Ed.), pp. 1–24, Springer International Publishing, Cham, 2018.
20. Nagarajan, V., Mohanty, A.K., Misra, M., Perspective on polylactic acid (PLA) based sustainable materials for durable applications: Focus on toughness and heat resistance. *ACS Sustain. Chem. Eng.*, 4, 6, 2899–2916, 2016.
21. Aulin, C. *et al.*, Transparent nanocellulosic multilayer thin films on polylactic acid with tunable gas barrier properties. *ACS Appl. Mater. Interfaces*, **5**, 15, 7352–7359, 2013.
22. Laufer, G. *et al.*, Clay–chitosan nanobrick walls: Completely renewable gas barrier and flame-retardant nanocoatings. *ACS Appl. Mater. Interfaces*, 4, 3, 1643–1649, 2012.
23. Du, W. *et al.*, Cross-linking poly(lactic acid) film surface by neutral hyper-thermal hydrogen molecule bombardment. *J. Agric. Food Chem.*, **63**, 49, 10604–10610, 2015.

24. Armentano, I. *et al.*, Bio-based PLA_PHB plasticized blend films: Processing and structural characterization. *LWT - Food Sci. Technol.*, **64**, 2, 980–988, 2015.
25. Ikada, Y. *et al.*, Stereocomplex formation between enantiomeric poly(lactides). *Macromolecules*, **20**, 4, 904–906, 1987.
26. Xu, H. *et al.*, Thermostable and impermeable "nano-barrier walls" constructed by poly(lactic acid) stereocomplex crystal decorated graphene oxide nanosheets. *Macromolecules*, 48, 7, 2127–2137, 2015.
27. Re, G.L. *et al.*, Stereocomplexed PLA nanocomposites: From *in situ* polymerization to materials properties. *Eur. Polym. J.*, **54**, 138–150, 2014.
28. Slager, J. and Domb, A.J., Biopolymer stereocomplexes. *Adv. Drug Deliv. Rev.*, **55**, 4, 549–583, 2003.
29. Tsuji, H., Poly(lactide) stereocomplexes: Formation, structure, properties, degradation, and applications. *Macromol. Biosci.*, 5, 7, 569–597, 2005.
30. Tsuji, H., Poly(lactide) stereocomplexes: Formation, structure, properties, degradation, and applications. *Macromol. Biosci.*, 7, 12, 1299–1299, 2007.
31. Li, Q., S.-H., Woo, E.M., Kinetic analysis on effect of poly(4-vinyl phenol) on complex-forming blends of poly(L-lactide) and poly(D-lactide). *Polym. J.*, 41, 5, 374–382, 2009.
32. Chang, L. and Woo, E.M., Effects of molten poly(3-hydroxybutyrate) on crystalline morphology in stereocomplex of poly(L-lactic acid) with poly(D-lactic acid). *Polymer*, **52**, 1, 68–76, 2011.
33. Sun, Y. and He, C., Synthesis and stereocomplex crystallization of poly-(-lactide)–graphene oxide nanocomposites. *ACS Macro Lett.*, **1**, 6, 709–713, 2012.
34. Quan, H. *et al.*, The electrical properties and crystallization of stereocomplex poly(lactic acid) filled with carbon nanotubes. *Polymer*, **53**, 20, 4547–4552, 2012.
35. Furuhashi, Y. *et al.*, Preparation and some properties of stereocomplex-type poly(lactic acid)/layered silicate nanocomposites. *J. Appl. Polym. Sci.*, 127, 3, 1615–1622, 2013.
36. Bao, R.-Y. *et al.*, Enhanced formation of stereocomplex crystallites of high molecular weight poly(l-lactide)/poly(d-lactide) blends from melt by using poly(ethylene glycol). *ACS Sustain. Chem. Eng.*, 2, 10, 2301–2309, 2014.
37. Purnama, P. and Kim, S.H., Bio-based composite of stereocomplex polylactide and cellulose nanowhiskers. *Polym. Degrad. Stab.*, 109, 430–435, 2014.
38. Han, L. *et al.*, Stereocomplex crystallization of high-molecular-weight poly(l-lactic acid)/poly(d-lactic acid) racemic blends promoted by a selective nucleator. *Polymer*, **63**, 144–153, 2015.
39. Bor, Y., Alin, J., Hakkarainen, M., Polylactide stereocomplexation leads to reduced migration during microwave heating in contact with food simulants. *J. Food Eng.*, **134**, 1–4, 2014.

40. Wang, W. *et al.*, Development of bio-based PLA/cellulose antibacterial packaging and its application for the storage of shiitake mushroom. *Food Chem.*, **429**, 136905, 2023.
41. Liu, X. *et al.*, A polylactide based multifunctional hydrophobic film for tracking evaluation and maintaining beef freshness by an electrospinning technique. *Food Chem.*, **428**, 136784, 2023.
42. Glicerina, V. *et al.*, Influence of an innovative, biodegradable active packaging on the quality of sunflower oil and "pesto" sauce during storage. *Appl. Food Res.*, **3**, 2, 100313, 2023.
43. Laorenza, Y. and Harnkarnsujarit, N., Ginger oil and lime peel oil loaded PBAT/PLA via cast-extrusion as shrimp active packaging: Microbial and melanosis inhibition. *Food Packag. Shelf Life*, **38**, 101116, 2023.
44. Gonon, H. *et al.*, PLA thermoformed trays incorporated with cinnamaldehyde and carvacrol as active biodegradable bakery packaging. *Food Packag. Shelf Life*, **38**, 101123, 2023.
45. Radhalakshmi, V., Raman, M., Joy, M.R., Development of active packaging film based on poly (lactic acid) incorporated with Piper betel leaf ethanolic extract and its application in the shelf-life extension of tuna meat. *Int. J. Biol. Macromol.*, **246**, 125751, 2023.
46. Andrade, M.A. *et al.*, LDPE and PLA active food packaging incorporated with lemon by-products extract: Preparation, characterization and effectiveness to delay lipid oxidation in almonds and beef meat. *Foods*, **12**, 2450, 2023, DOI:10.3390/foods12132450.
47. Wongphan, P. *et al.*, Lesser galangal (*Alpinia officinarum Hance*) essential oil incorporated biodegradable PLA/PBS films as shelf-life extension packaging of cooked rice. *Food Packag. Shelf Life*, **37**, 101077, 2023.
48. Sharma, S. *et al.*, Active film packaging based on bio-nanocomposite TiO2 and cinnamon essential oil for enhanced preservation of cheese quality. *Food Chem.*, **405**, 134798, 2023.
49. Hu, W. *et al.*, Antioxidant mechanism of a continuous microenvironment regulation system for prolonging the shelf-life of red grapes. *J. Food Sci.*, **88**, 6, 2496–2511, 2023.

12

Pectin from Agro-Waste Residues for Food Packaging

Kona Mondal and Vimal Katiyar*

Department of Chemical Engineering, Indian Institute of Technology Guwahati, Guwahati, Assam, India

Abstract

Pectin is an environment-friendly heteropolysaccharide that is primarily derived from various agricultural sources and their byproducts. Pectin is generally recognized as safe (GRAS) and serves a variety of functions in the food industry, including emulsifier, hydrocolloid, encapsulant, gelling agent, stabilizing agent, and thickening agent. Pectin is made up of D-galacturonic acids connected by α-1,4-glycosidic bond, typically found in the middle lamella of plant cell walls. Because its backbone has many hydroxyl and carboxyl groups, pectin has a unique structure that makes it easy to change and work with a wide range of bioactive substances. Extracting pectin from the cell walls of agro-waste residues is a critical unit operation; however, rising demand has revealed that the traditional method of extracting pectin has some flaws, particularly in terms of efficiency and product consistency. Pectin is a suitable polymeric matrix for the creation of active food packaging films and coatings. This chapter discusses the various forms of agricultural waste that contain pectin as well as the techniques for extracting it and using it as edible coatings and films in food packaging.

Keywords: Pectin, extraction, agro-waste, food packaging, edible film, active coating

**Corresponding author*: vkatiyar@iitg.ac.in

Santosh Kumar, Avik Mukherjee and Vimal Katiyar (eds.) Agro-Waste Derived Biopolymers and Biocomposites: Innovations and Sustainability in Food Packaging, (319–348) © 2024 Scrivener Publishing LLC

12.1 Introduction

Packaging plays an essential role in the preservation, protection, and presentation of food products, and it also serves as a marketing tool providing various information to consumers about the packaged product [1]. However, conventional packaging materials, such as synthetic plastics and metals, have several drawbacks, including nonbiodegradability, nonrenewability, and potential toxicity to human health and the environment [2]. Furthermore, it has been estimated that less than 5% of all synthetic plastics are recycled worldwide. This led to a 25-fold increase in the production of new plastic materials, which quickly accumulate in the environment [3]. According to the EPA (2011), the increased production of synthetic plastic in the United States alone resulted in 31 million tonnes of waste in 2010 [4]. This amount accounted for 12.4% of all municipal solid waste, also known as trash or garbage. Food wraps contribute millions tonnes of waste to landfills each year, which has a negative impact on the environment. Therefore, there is a growing interest in developing sustainable and eco-friendly packaging materials [5]. In addition, over the past 20 years, consumers' interest in issues like health, nutrition, food safety, and the environment has grown, and this has resulted in a rapid increase in sustainable packaging research. One potential source of eco-friendly packaging materials is agro-waste residues, which are generated in large quantities worldwide. Agro-waste residues are byproducts of agricultural activities and include various plant parts, such as stems, leaves, shells, and seeds, that are not used for human consumption. These residues can be a rich source of various biopolymers, and the overview of these biopolymers has been depicted in Figure 12.1. Among these biopolymers, pectin is one of the important and valuable polysaccharides, which has numerous applications in the food industry, including as a gelling agent, thickener, and stabilizer.

Pectin is a type of biopolymer that is also a complex polysaccharide found in the cell walls of plant parts, such as fruits, vegetables, and even some grains [1]. The term "pectic acid" was historically given to the substance by the French scientist Henri Braconnot, who first isolated pectin from vegetables in 1825 [6, 7]. Pectin has unique physicochemical properties, such as the ability to form gels in the presence of sugar and acid, which make it an ideal ingredient for various food products, including jams, jellies, and fruit preserves. In addition, pectin has excellent film-forming and barrier properties, which make it a promising material for food packaging applications. In this context, pectin-based edible films and coatings have been developed using various techniques and applied to various food products to prolong their shelf life (Figure 12.2). The demand

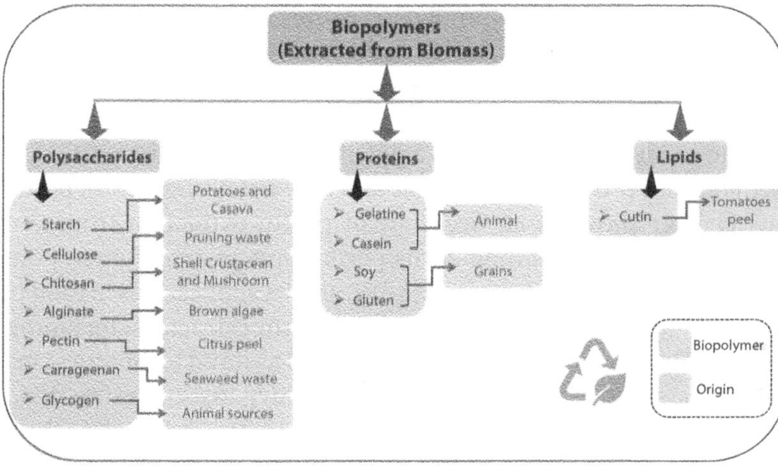

Figure 12.1 Types of biopolymers obtained from renewable sources.

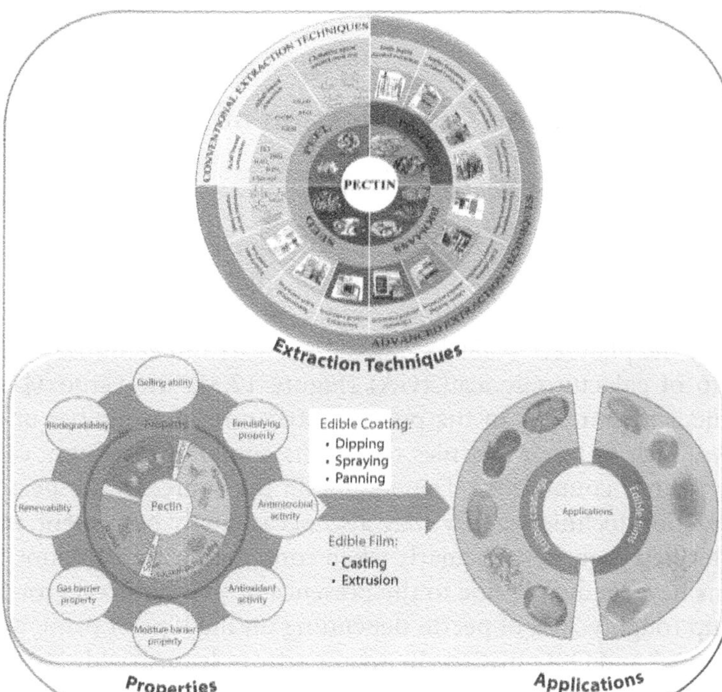

Figure 12.2 Various extraction techniques [2] and properties of pectin and its application on food as edible packaging material using various techniques [1].

for pectin is increasing, and its production mainly relies on the extraction from citrus peels and apple pomace, which are limited resources and subject to price fluctuations. Therefore, there is a need to explore alternative sources of pectin, and other agro-waste residues could be a viable option. Pectin extraction from agro-waste residues not only offers a potential solution to the waste disposal problem but also provides a sustainable and cost-effective source of pectin.

This chapter provides an overview of pectin extraction from various agro-waste residues, including citrus peels, apple pomace, grape pomace, carrot pomace, and sugar beet pulp (SBP), and its applications in food packaging [1, 2]. The conventional methods of pectin extraction, including hot acid extraction, and alkaline extraction, as well as novel techniques, such as ultrasound-assisted extraction (UAE), microwave-assisted extraction, and enzymatic extraction, are discussed (Figure 12.2). The physicochemical properties of pectin extracted from different sources, and its potential as a material for food packaging, including its film-forming and barrier properties are also reviewed. The exploration of alternative sources of pectin not only provides a sustainable and cost-effective solution but also offers a promising material that can contribute to reducing the environmental impact of packaging on human health and the planet.

12.2 Structure and Classification of Pectin

12.2.1 Structure of Pectin

The cell walls of plants and agricultural waste contain pectin, a complex heteropolysaccharide with a wide range of applications. It is often found in the same place, where cellulose, lignin, or polyphenols are also present inside the plant cell walls [8, 9]. The majority of pectin molecules are made up of galacturonic acid (GA) (Figure 12.3). The carboxyl groups of uronic acid can exist in the polymer structure in a number of different ways, such as free molecules or as salts containing sodium, calcium, or other minor counterions. Depending on the source of the pectin and the extraction technique, they may also occasionally be present as naturally esterified groups, particularly when combined with methanol. Pectin solutions have acidic pH due to the presence of free carboxyl groups. GA makes up roughly 70% of pectin depending on the type of plant, and it is linked at O-1 and O-4 positions in every pectic polysaccharide [10]. Pectin has a linear anionic backbone, with "hairy regions" that have nonionic side chains and "smooth regions" that do not have any side chains [11]. This colloidal heteropolysaccharide contains homogalacturonan (HG),

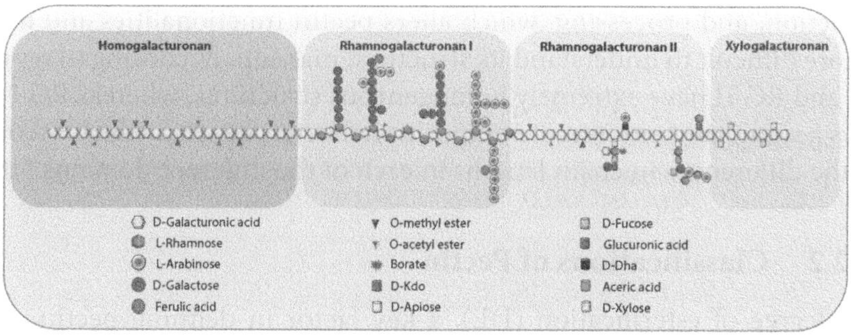

Figure 12.3 Schematic of structural representation of pectin containing different domains as adapted from [12].

rhamnogalacturonan (RG-I), and rhamnogalacturonan (RG-II), which are structurally distinct regions or domains (Figure 12.3).

HG is a linear polymer of α-1,4-linked GA that makes up about 65% of pectin and is partially methyl-esterified at C-6 and O-acetylated in positions 2 and 3 [13]. The presence of different distinguishable structural domains can influence the properties of pectin (Figure 12.3). HG can turn into gels when calcium is present, depending on the amount and pattern of methyl esterification, and this is why it is the most studied pectin [14]. A neutral side chain is joined to the O-4 position and, occasionally, the O-3 position of the repeating GA and rhamnose (Rha) disaccharide that makes up the backbone of RG-I, which constitutes 20 to 35% of pectin. According to the source of the plant as well as the extraction methods used, 20 to 80% of the Rha residues have C-4 substitutions [15]. Similar to HG, RG-I is susceptible to methylation and acetylation [16]. The minor sugars glucuronic acid, fucose, methyl-esterified glucuronic acid, apiose, 2-O-methylxylose, and 2-O-methylfucose are all found in RG-II, which make up 2 to 10% of pectin. The HG backbone of RG-II is very branchy and has complex side chains. Among pectin molecules, RG-II is thought to be the most conserved domain [16]. Due to their branching patterns, xylogalacturonan (XG) and other branched regions like RG-I and RG-II are known as the "hairy regions," whereas HG is frequently referred to as the "smooth region" due to its linear structure [17]. Most experts agree that pectic polysaccharides have strong covalent bonds and have a high degree of cross-links, and thus, HG, RG-I, and RG-II can be separated using harsh chemical processes or digestion by enzymes that break down pectin. Additionally, it has been suggested that other substances, like GA and apiogalacturonan (AP), may take the place of GA units in some regions of the pectin chain [8]. Pectin structures can change during plant storage, extraction, and processing, which alters pectin functionalities and makes it more difficult to understand its structural makeup. According to reports, HG and RG II have extremely homogeneous structures, whereas RG I displays a significant amount of compositional heterogeneity, which accounts for the differences in chain lengths in each of the different domains [18].

12.2.2 Classifications of Pectin

The degree of esterification (DE), a key factor in defining pectin applications, is the proportion of carboxyl groups in the pectin structure that have undergone esterification. DE is frequently used to categorize pectin, as the emulsifying, texturizing, and gelling properties depend on the DE of pectin. Esters with long hydrocarbon chains are hydrophobic, and as the DE increases, they become less soluble in water. Additionally, as DE increases, the rate of gelation also gets better, resulting in faster gelling of pectin [19]. Also, the molecular weight of pectin and the number as well as the type of neutral sugars have a major effect on their rheological properties [20]. The food industry uses high-methoxyl pectin (HMP), which has DE levels above 50%, most frequently for its thickening and gelling properties. According to reports, HMP is highly acid-sensitive and requires a high concentration of sugars to form gel [21]. Due to hydrogen bonds and hydrophobic interactions between the pectin chains, HMP gels at low pH levels and high concentrations of soluble solids. Gels are produced by the cross-linking of HG fragments, which results in the formation of three-dimensional crystalline networks capable of stabilizing solutes [22]. Over the past few decades, there has been a great deal of research done on the intricate formation of HMP gels that is explained by the glass transition hypothesis. Different processes, such as de-esterification, self-aggregation, and entanglement, led to HMP gel formation under alkaline conditions. Electronic attraction makes Na^+ or K^+ bind to the free carboxyl groups on HMP. This makes it easier for HMP molecules to move toward each other which helps in gel network formation [23]. Jiang *et al.* (2020) reported the emulsifying properties of HMP in systems with only water and ethanol and proposed that ethanol promotes pectin aggregation and weakens electrostatic attraction [24]. The water-ethanol mixtures lost electrostatic repulsion when HMP was added, and the helix shape was broken. Thus, pectin is more likely to stick to the oil-water interface because of close-packed lipophilic conformation. The mixture produced an emulsion that showed good stability after the addition of 21% ethanol.

Low methoxyl pectin (LMP) typically results from the de-esterification of HMP when the DE of esterification falls below 50%. LMP can be produced from HMP using a variety of substances, including concentrated acids, pectin methyl esterase enzymes, and alkalis like sodium hydroxide or ammonia [20]. Because LMP does not require an adequate amount of sucrose to form a gel, the food industry frequently uses it to create jams with low sugar content. It is less sensitive to acidic environments and needs Ca^{2+} ions to form gels. LMP gels are obtained as a result of the creation of calcium bonds between the two carboxyl groups of adjacent chains. Han *et al.* (2017) recently investigated how the concentrations of pectin, soluble solids, calcium, and pH affect the gel strength of LMP gels. They also developed various hypotheses regarding the formation of pectin gels in light of their rheological characteristics [25]. They found that calcium bridges formed at dissociated carboxyl groups when the calcium concentration was high and the pH was close to the isoelectric point (pH 3.50), and this made the gel stronger. Because neutral sugars like hydroxyl groups help to stabilize the gel and create hydrogen bonds that aid in immobilizing free water, the presence of sucrose also increases gel strength. However, Yang *et al.* (2018) investigated whether LMP gels could form under an alkaline environment and proposed that LMP can produce comparatively stable gels with NaOH acting as a pH regulator in the range of 3.5 to 9.5 [26]. Furthermore, they explored how the calcium ion concentration affected the thermal and structural properties of the LMP gels and concluded that the ion decreased the thermal stability and crystallinity of the LMP.

12.3 Agro-Waste as Sources of Pectin

Due to the high demand for pectin, abundant research has been conducted on how to get pectin from agro-waste biomass. Pectin is primarily obtained from orange and apple pomace, which are byproducts of the juice [8]. Citrus peel has a high pectin content (25 to 35% on a dry basis), whereas apple pomace, sugar beet, and sunflower contain 10 to 15%, 10 to 20%, and 15 to 25% of pectin on a dry basis, respectively [27]. Table 12.1 provides an overview of the various sources of pectin from agro-waste residue. Citrus fruit peel is the primary industrial source of pectin due to its favorable characteristics and high extraction yield. The most widely used technique for removing pectin from orange peels is hydrothermal extraction, which is carried out at 75°C to 95°C, and for long periods of time (60 to 300 min). Pectin is extracted using hydrothermal methods in acidic environments, with water as the solvent. An acidic environment helps to reduce the presence of other substances like polyphenols due to the hydrophilic nature of pectin, enhanced extraction yields, and maintained pectin quality [8].

Table 12.1 Various agro-waste sources for the extraction of pectin using different techniques.

Source (agro-waste)	Extraction technique	Pectin content (%)	References
Citrus peel	Acid	25.5	[29]
Citric waste	Acid	78	[30]
Orange peel	Acid	7.3–52.9	[31]
Orange peel	Water	2.2	[32]
Orange peel	Alcohol precipitation	4–13.8	[33]
Lemon peel	Acid	0.8–8	[34]
Lime peel and pulp	Microwave-assisted	8–17.9	[35]
Dragon fruit peel	Ultrasound-assisted	1.89–7.65	[36]
Banana (stem, leaf, and peel)	Ultrasound-assisted	4–13.8	[33]
Cocoa peel	Microwave-assisted	42.3	[37]
Tomato waste	Ultrasound-assisted	15.1–35.7	[38]
Mango peel	Acid	20.8	[39]
Jackfruit wastes	Acid	38.42	[40]
Passion fruit peels	Acid	2.25–14.6	[41]
Pumpkin peels	Soxhlet extraction	6.8–7.7	[42]
Watermelon rind	Acid and enzymatic extraction	65	[43]
Mangosteen rind	Chemical	23.5	[44]
Durian rind	Acid	2.1–10.25	[45]
Apple pomace	Acid	12.9–20.9	[46]

Citrus byproduct extraction can be sped up using other strategies, for example, using a microwave, the extraction time from lime and pomelo peels was cut down to 2 to 5 min, using high microwave powers (700 to 1100 W). Pectin made from leftover orange peel was also obtained using the hydrodynamic cavitation (HC) method, in which a relatively lower amount of solvent was required (2.86 mL/g of dry waste), but the process took a long time (~270 min) [28]. This high-quality polysaccharide can also be made from the peel of a tomato, cocoa pod, banana, mango, or eggplant [8]. In recent years, research on tropical fruits to obtain HMP has also taken place, for instance, intriguing sources of pectin have been suggested, including the peels of jackfruit, durian, and passion fruit. In a study, a 10-min UAE was used at 450 W, and a water-to-dry sample ratio of 20 mL/g on the rind of the passion fruit. The results showed that HGs were the main constituent of the obtained pectin, and due to the high level of methylation, the passion fruit pectin could be used as a good quality gelling agent. In order to extract pectin in an environment-friendly manner, hybrid techniques have been used such as combining microwave assisted extraction (MAE) and UAE. Pectin has been extracted from sisal waste using a combination of enzymatic and ultrasonic processes because enzymatic treatment breaks the bonds between cellulose and xyloglucans in the sisal cell wall, and ultrasonic treatment breaks up the sisal structure to make it easier to get the pectin out [8]. Even though microwave or ultrasonic extraction equipment costs more upfront than traditional extraction equipment, it makes more profit in the long run because it uses less energy, takes less time, and needs fewer expensive reagents.

12.4 Techniques for Pectin Extraction from Agro-Waste

Pectin extraction is a physical and chemical process that involves a number of hydrolysis steps that sequentially remove pectin macromolecules from plant tissues. Several process variables, like pH, temperature, and time, control the continuity and extraction process length, when dissolved in bulk solvents. In conventional methods for extraction of pectin from fruits and plant tissues, a hot acid wash containing acetic, citric, lactic, hydrochloric, nitric, phosphoric, and sulfuric acids as well as other mineral and organic acids is frequently used. Marić *et al.* (2018) have already provided a thorough overview of traditional extraction methods along with variables affecting pectin recovery [13]. Therefore, this chapter has talked about a few new techniques for getting pectin out of the cell walls of agro-waste materials. Standardizing cutting-edge pectin extraction methods that are also good for the environment is important for chemical and biochemical processing industries. The solvent is not required in these methods, they use less energy and save time. These cutting-edge ways to get pectin extraction use enzymes, microwaves, pressure, subcritical water, ultrasound waves

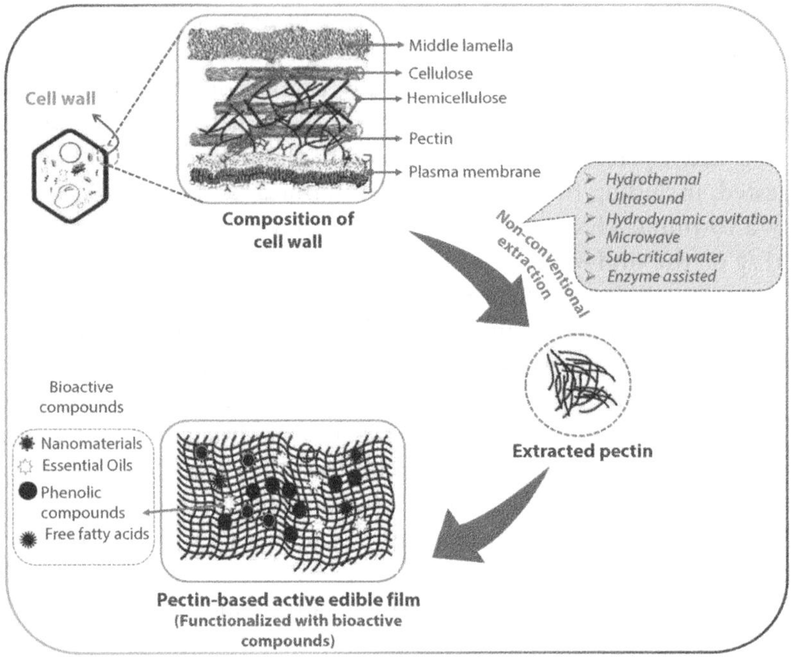

Figure 12.4 Schematic representation of plant cell wall consisting of pectin and other materials and extraction of pectin by nonconventional techniques followed by functionalization for commodity applications as adapted from [47].

and combined with microwaves, and enzymes. Figure 12.4 schematically illustrates the extraction of pectin using nonconventional techniques followed by commodity applications.

12.4.1 Hydrothermal Extraction

Pectin is a complex heteropolysaccharide that is found in the cell walls of plants and makes up most of the skeleton of many fruits and vegetables. In recent years, there has been more research on pectin-derived oligosaccharides made by hydrothermal treatment because it has been suggested that they are emerging prebiotics that may be good for human health [48]. Most of the time, different components of biomass are removed through hydrolysis and depolymerization using chemicals or environment-friendly solvents. Hydrothermal pretreatment has attracted enough attention because of its benefits of being simple to use, requiring only moderate amounts of energy, finishing quickly, being affordable, and causing little equipment corrosion [26]. Hydrothermal treatment only uses liquid water, steam, and heat. This means that biomass with a high moisture content can be used without having to go through expensive drying processes. This method can handle corrosion and hard reactions because it works in less harsh conditions without corrosive chemicals and does not need expensive materials or unusual designs. During the hydrothermal process, when water is heated above its boiling point, it acts like an organic solvent, which makes it easier to dissolve organic compounds in water [48]. In this process, water serves not only as a solvent but also as a reactant or catalyst. The hydronium ions that are made when water is ionized cause polysaccharides to break apart during autohydrolysis, by cleaving acetyl groups and selectively hydrolyzing heterocyclic ether bonds. The hydronium ion created by the acetyl group can be used as a catalyst later in the reaction to quicken the autohydrolysis process. As acetyl groups are crucial components of the hydrolysis process, this method performs better on materials with a high-acetyl content, such as hardwoods, and the majority of agricultural waste. Numerous studies investigated the effects of process variables and reaction mechanisms on the physicochemical characteristics of the pretreated biomass, and the chemical makeup of the liquid fraction [48]. Hydrothermal pretreatment under favorable operating conditions results in the selective dissolution of pectin or hemicelluloses, with the dissolved products primarily consisting of oligosaccharides, and the solid fraction retaining most of the cellulose and lignin [49].

12.4.2 Ultrasound-Assisted Extraction

The term "ultrasound" describes sound waves that are too high for humans to hear (16 Hz to 16 kHz) and have frequencies higher than 20 kHz [12]. Cycles of expansion and compression, which push molecules together and apart, respectively, occur during the transmission of ultrasound. When the applied negative pressure is greater than the liquid partial pressure, cavities develop in liquid media before collapsing. The term "cavitation" refers to the process by which bubbles emerge, expand, and then burst. Wang *et al.* (2018) say that phytochemicals are extracted when sound waves create cavitation bubbles close to the tissue [50]. These bubbles weaken the cell walls and help the solvent get into the cells leading to the release of cell contents. Pectin has been extracted using UAE from *Opuntia ficusindica* and *Artocarpus heterophyllus* fruit peels, tomato waste, orange peels, and industrial *Musa balbisiana* waste, among other sources [12]. The yield of pectin from the combined enzymatic and ultrasonic method is 31.1%, which is 1.5 to 3.5 times higher than that of separate enzymatic extraction (9.4%) or acid extraction (5.4%) because ultrasound treatment breaks up the cellulose network [12]. In addition to increasing yields, sonication has been shown to affect pectin structure and bioactive properties. Also, the shock wave could cause swelling and watering, which caused the plant cell walls to get bigger pores, and the tissue to change its shape. As a result, mass transfer is enhanced and inner cell material diffusion is encouraged. Compared to conventional heating methods, using ultrasound for extraction has a number of benefits, including greater operator safety, higher yield, reduced solvent usage, shorter treatment times, and lower energy consumption [13]. Due to its ability to improve extraction processes, ultrasound has been used to get pectin out of different plant foods, wastes, and byproducts.

Pomegranate (*Punica granatum*) processing generates a significant amount of byproducts including peel (78%) and seeds (22%). Pomegranate byproducts have been focused on pectin extraction and are good sources of antioxidants [51]. An ultrasound machine with a 130 W maximum power output was used at 20 kHz to extract pectin, and the authors concluded that the highest pectin extraction yield was obtained with a solid-liquid ratio of 1:18 g/mL and at processing parameters of 62°C for 29 min. Xu *et al.* (2014) looked at the effects of conventional heating extraction (CHE) and ultrasound-assisted heating extraction (UAHE) on the grapefruit peel's pectin yield and swelling behavior [52]. Using CHE as the processing method, the maximum pectin yield was 23.44% after 72.26 min, whereas the best results were found in the case of UAHE (25.9 min), which yielded 26.74% pectin. Additionally, the UAHE approach had a higher yield (12.35%) at a temperature during the extraction procedure that was 20°C lower and a treatment time that was 39.53% shorter, opening the door for the UAHE approach to have industrial potential.

Both the food industry and researchers are interested in tomato waste and byproducts, especially the seeds, skin, and pulp because they contain vital essential compounds, especially pectin. Grassino *et al.* (2016), for example, extracted pectin from tomato waste using two methods: conventional extraction (CE) and ultrasonographic extraction (UZE) [38]. UZE was carried out using an ultrasonic bath with a dual frequency (37 to 80 kHz), among which the lower frequency (37 kHz) was used throughout the entire research in order to produce stronger agitation. The CE was run in two stages, i.e., reflux was used in the first step, which lasted for 24 hours, and fresh solvent was used in the second step. The UAE took place for 15, 30, 45, 60, and 90 min. The results showed that the CE treatment at 60°C produced more pectin than the UAE treatment, but the UAE treatment took less time (15 min) to extract the pectin. Additionally, pectin from the UAE was of higher quality in terms of anhydrouronic acid, methoxy, and calcium pectate contents, as well as esterification levels. After comparing the results, it was found that the pectin yield was roughly the same for both methods at 80°C, but with a noticeably shorter treatment time in UAE (CE for 36 hours and UAE for 15 min). Pectin extraction in the UAE has been found to be beneficial and offers an environment-friendly method at a lower temperature (than CE), and with significant energy savings.

The plant material obtained from the sugar industry is another source of great interest. During the production of sugar, enormous amounts of SBP are produced as a byproduct. Pectin was extracted from SBP with the help of subcritical water and ultrasound [13]. RSM design was used to figure out how to get the best yield of pectin from the process parameters, in which UAE was used at a frequency of 25 kHz, and the parameters of the extraction process that had the highest yield of pectin (24.63%) were extraction time (30.49 min), extraction pressure (10.70 MPa), liquid/solid ratio (44.03), and extraction temperature (120.72°C). Galacturonic acid (GalA) and arabinose (Ara) were found in 59.12 and 21.66% of the extracted pectin, respectively. Furthermore, UAE extracts pectin with high purity and little DM. However, because the intensity of ultrasound waves decreases with increasing distance from the emitter, the ultrasound waves that eventually reach the dispersed sample are not uniform, leading to poor pectin uniformity and batch variation [12].

12.4.3 Hydrodynamic Cavitation

Cavitation is a phenomenon that can be achieved by changes in pressure caused by mechanical, especially hydrodynamic, and acoustic methods. Pressure changes in HC are brought by mechanical waves and high-frequency sonic waves (>20 kHz) in acoustic cavitation (AC), also known as ultrasound cavitation. In both cases, pressure differences are produced by a series of compression cycles, which repeatedly cause bubbles to form and burst. The formation of vapor-filled bubbles or cavities within a bulk liquid can occur when the applied pressure is lower than the liquid vapor pressure. These cavities will explode when a higher pressure is applied, creating localized shock waves that produce turbulent (nonlaminar) flow that extremely disrupts nearby surfaces, particles, or fibers. The size and strength of the pressure waves as well as the characteristics of the bulk fluid all affect how strong cavitation is. The properties of solvent and solute, such as viscosity, surface tension, and particle volume fraction, all have an impact on waves, bubbles, and how stable they are. Reactive species can alter the chemical behavior of macromolecules, in addition to dissolving and wearing down particles. Cavitation, shear, thermal, and chemical effects can be challenging to distinguish from one another [53]. HC has been used as a sterilant and wastewater treatment agent in various industrial settings. Sun *et al.* (2020) have given a clear explanation of how HC works and how it can be used to treat waste [54]. HC can also be used on a microliter scale, such as in microfluidic devices to make radical-induced fluorescence. Microorganisms have been differentially inactivated using a variety of HC setups, and therefore, HC has been used at the end of fermentation to sterilize liquid food. Ashokkumar *et al.* (2015) have suggested that HC could be used instead of AC in situations where a lot of food needs to be processed [55]. Meneguzzo *et al.* (2019) used a circular Venturi-shaped reactor and processing times longer than 120 min without temperature control to extract low DM pectin, polyphenols, and terpenes from orange peel [28]. Katariya *et al.* (2020) processed orange juice with HC for 5 to 25 min to pasteurize it to preserve the bioactive components while inhibiting the activity of enzymes like pectin methyl esterase and peroxidase [56].

12.4.4 Microwave-Assisted Extraction

The science and technology of food have greatly benefited from the use of microwaves, and in the industrial setting, microwaves can be used for food drying, enzyme inactivation, microorganism inhibition, and extraction of phytochemicals [12]. MAE is used as an adjunct technique in combination with a chemical solvent to extract bioactive substances such as pectin, polyphenols, and essential oils from agro and food residues. The MAE method uses less solvent and is more efficient. Pectin quality is the same throughout the solvent because the temperature is uniform, and there is no temperature gradient, as in the case of traditional methods. Yoo *et al.* (2012) found that extracting pumpkin powder with a microwave at 120°C for 3 min doubled the amount of pectin yield without changing its quality [57]. According to Bagherian *et al.* (2011), the yield, GalA content, and DE of the extracted pectin all increased, as microwave power and heating time increased [58]. Ideal conditions for making pistachio green hull pectin are microwave power of 700 W, an irradiation time of 165 s, and a pH value of 1.5 [12]. This results in a high yield of pistachio green hull pectin (18.13%) having a higher percentage of HG (64%) and lower DE (12.1 %). The de-esterification of GA chains is likely the cause of the reduction of DE under extreme conditions (low pH, high microwave power, and long irradiation times) [12].

12.4.5 Subcritical Water Extraction

Subcritical water is hot water that is kept in a liquid state at a critical temperature between the boiling point of water at 100°C and the critical point of water at 374°C, at a critical pressure of between 1 and 22.1 MPa. When this water is used to extract bioactive compounds, the process is called subcritical water extraction (SWE), superheated water extraction (SHWE), or pressurized hot water extraction (PHWE). Surface tension, dielectric constant, and water viscosity all decrease as temperature rises. This increased temperature increases diffusion, vapor pressure, and mass transfer while decreasing surface tension and viscosity. Water has a dielectric constant of 80 at a temperature of 25°C. It is interesting to note that the dielectric constant decreases to 25 at a higher temperature of 250°C and 25 bar pressure. As a result, water begins to resemble some properties of organic solvents, which can dissolve a wide range of low to medium-polarity compounds [47]. One of the intriguing features of SWE is its capacity to modify water dielectric constant by adjusting pressure and temperature [59]. Additionally, convection and diffusion both contribute to mass transfer in SWE. Subcritical water could get in the way of interactions between solutes and the matrix by lowering the activation energy needed for desorption. High pressure can aid in the extraction of pectin by applying a strong force to the water molecules to allow them to enter the plant matrix pores, which is not possible under normal pressure. SWE was used to extract pectin from fresh sunflower heads, and a pectin yield of 6.57 ± 0.60% was optimized at 8 bar, 120°C for 20 min, and a liquid-solid ratio of 7 mL/g [47]. The SWE technique was used to remove Berg fruit pectin (*Campomanesia xanthocarpa*), and a yield of 5.70 wt % was optimized at 150 bar and 120°C [60]. Water is a cheap, readily available, nonflammable, nontoxic, environmentally friendly solvent that SWE uses for extraction [61]. SWE equipment is straightforward and affordable, allowing for quick and high-throughput pectin extraction, however, some heat-sensitive compounds may be thermally degraded by SWE because it uses elevated temperatures for extraction [47].

12.4.6 Enzyme-Assisted Extraction

Finding alternative methods for solvent conservation is now essential due to the increased pressure on the food industries to develop sustainable extraction strategies. Enzymes speed up hydrolysis in a very selective way, and they can reduce the amount of solvent needed and yield high-quality pectin. Enzymes that hydrolyze pectin structure and disassemble or deconstruct plant cell walls to increase permeability are the two categories of enzymes.

Some of the enzymes frequently used for pectin extraction include xylanase, protease, cellulase, hemicellulose, polygalacturonase, α-amylase, and pectinesterase [12]. Variables such as enzyme concentration, enzyme type, reaction time, pH, temperature, and plant source affect how well the extraction process works. Liew *et al.* (2016) used the commercial enzyme Cellucast to extract pectin from peels of passion fruit [62]. The best conditions for pectin extraction were a Cellucast concentration of 1.67% (w/w) at 61°C. However, this pectin was found to be of higher quality and contained more HMP. The pectin yield was 7.12%, comparable to that of CE. In another study laminex, a *Penicillium funiculosm* product was used to extract 23% of the pectin from lime peel, with the process being optimized at pH 3.5, 4 hours, and 50°C, in which about 82% of the extracted pectin was methylated [47]. Conventional methods require a lot of neutralization of the effluent, but enzyme-assisted extraction cuts down this step by using enzyme adsorption. The method also helps to improve the yield and quality of pectin extracted from agro-waste while consuming less energy and solvent. The high cost of enzymes is one of the main disadvantages of this extraction technique, which can cause the cost of extraction to increase significantly when more raw materials are used.

12.5 Food Packaging Applications of Pectin-Based Films and Coatings

The Food and Agriculture Organization has estimated that each year, about 1.3 billion tonnes, or roughly one-third of the total amount, of foods produced globally is wasted [63]. The general consensus is that significant amounts of food waste are due to food deterioration, which occurs because of a variety of circumstances, including the environment and food's individual characteristics. Therefore, a better solution is required to keep the food products safe with maintained quality throughout their storage life. Packaging is important for keeping food safe and keeping it from getting ruined. It acts as a barrier between food and the outside world, which keeps food away from interacting with microorganisms, moisture, gasses, etc., that make it spoil. In the food industry, conventional synthetic plastics–based packaging is frequently used; however, their widespread use generates a tremendous amount of "white pollution" and plastic waste. Worldwide, only 3% of plastic is recycled annually, and in the United States, 32 million tonnes of plastic is wasted yearly [64]. Biodegradable packaging, which is made of eco-friendly, nontoxic, and degradable materials, has gained enormous attention for food packaging applications as a result of the growing awareness of environmental protection and consumers' health. Active packaging made of biodegradable materials is currently the best substitute for conventional packaging, which can meet consumer demands for healthier food and better protection of the environment. Active packaging has made the field of packaging more advanced by putting more attention on how food preservation and packaging work together instead of just focusing on boundary functions. Due to the increasing demand for biopolymer-based packaging, numerous studies have been conducted [63]. Pectin, a biopolymer, comes under the GRAS category, which is not only one of the best materials for making films but is also cheap, nontoxic, and poses film-forming properties.

12.5.1 Pectin-Based Active Films

Pectin-based active films are gaining attention in the food packaging industry and can be used to wrap foods. Pectin-based films can be incorporated with active compounds such as antimicrobial agents, antioxidants, and flavorings, which can be released gradually over time, providing additional functionality to the packaging material. For example, antimicrobial agents can help inhibit the growth of spoilage and pathogenic microorganisms, and antioxidants can help prevent oxidative deterioration of the food, thus extending the shelf life of the wrapped perishable foods [65]. Casting, spraying, and extrusion are the three primary techniques used to prepare film.

12.5.1.1 Film Fabrication by Casting

Casting is the most common way to make a film, especially at the lab or pilot scales [66]. The casting process consists of three steps: dissolving, casting, and drying. First, a solution for making a film (film-forming solution) is made by mixing pectin, plasticizers, and other ingredients using a suitable solvent. After the film-forming solution has been prepared, it is poured/cast into a mold in such a way that air bubbles are not formed. Lastly, drying of the casted film is done to evaporate the solvent. Most of the time, drying temperatures range from 25 to 50°C, and drying time ranges from 10 hours to 2 days. Notably, the drying condition influences various characteristics of the film. The effects of temperature and humidity on film formation were examined, and the results showed that the film's mechanical properties were closely related to the temperature at which it was dried and that it leaked more water vapor when the relative humidity was higher [67]. The solution casting method is low-cost and can be carried out without any specialized equipment. It has some shortcomings too, for example, its lengthy drying time, which might make it challenging to use in large-scale commercial production.

12.5.1.2 Film Fabrication by Spraying

The following steps are involved in the spraying process: firstly, a film-forming solution is prepared by dissolving all the ingredients in the suitable solvent followed by homogenization and then sprayed on a smooth surface, and finally, the solvent is evaporated by drying that could take place concurrently with or following the spraying process. A pectin-chitosan film was fabricated using the spray technique by Kamboj *et al.* (2015) [68]. The resulting film had good mechanical and thermal stability, as well as physicochemical characteristics because of strong molecular interactions between the film components due to spraying. Spraying parameters like temperature, speed, and others have a significant impact on the process of spraying as well as the characteristics of the resulting films. Due to its high efficiency and quick speed, the spray method is a good way to mix and homogenize materials that form a film. Spraying is also hard to use for mass-scale production because it is difficult to control uniform film thickness.

12.5.1.3 Film Fabrication by Extrusion

Extrusion is different from both casting and spraying because it is a dry process that leaves less water in the film matrix and has the potential to change the properties and structure of film materials. The extrusion process is divided into three zones: the feeding zone, where the polymers are treated by rotating screws with increasing pressure, mixed, and moved to the next zone; the kneading zone, which uses mechanical help and air to compress and soften the ingredients; and the heating zone, where heating and softening will happen when the temperature is above the glass transition temperature (Tg). Extrusion is usually used to make multilayer films, but it is not used to make active films with only one layer. There are two categories of factors that affect the extrusion process, i.e., the properties of film-forming polymers and the extruder operating condition [63]. The properties of polymers that are necessary for the extrusion of films include their thermoplasticity, glass transition, and gelatinization. Conditions during extrusion, like screw speed, feed speed, extrusion temperature, extrusion pressure, and die configuration, can have a big impact on the physicochemical properties of the film. Extrusion of films at different screw speeds was studied, and it was found that the surface of a film made at 80 rpm was the most uniform compared to those made at 40 and

120 rpm [65]. Extrusion is a more efficient process that requires less time and energy and can be used for industrial mass production, unlike casting and spraying. Also, there is no need for solvents, and concern with solubility, which is a big problem with the wet methods. However, the high cost of specialized equipment for extrusion limits its widespread application, but strict requirements for specific polymers that can handle high temperatures and low humidity compel the use of extrusion.

12.5.2 Pectin-Based Active Coatings

Active coating made from pectin can be used to keep perishable food fresh due to its antimicrobial, antioxidant, and barrier properties. Kumar *et al.* (2020) found that oregano essential oil, cinnamon essential oil, nisin, and titanium dioxide nanoparticles all have strong antibacterial properties and have been used to improve the shelf life of coated foods [47]. Pectin can neutralize free radicals due to its hydroxyl or carboxyl groups that can form complexes with metal ions and prevent the formation of free radicals [69]. When carboxyl groups were turned into esters, the conformation of pectin changed, which helped to eliminate free radicals [63]. Ogutu and Mu (2017) found that the antioxidant capacity goes up as the molecular weight goes down [70]. Yan *et al.* (2016) discovered that pectin that had undergone ultrasound treatment contained more Gal-A, had a lower molecular weight, and displayed stronger antioxidant activity [71]. Barrier qualities are another crucial component of active packaging, along with antimicrobial and antioxidant properties that help create boundaries and limit compound losses. Pectin's tight network structure makes it a good barrier against oxygen and carbon dioxide; however, it has inferior water vapor barrier properties. In this context, bio-fillers such as curcumin, clove bud essential oil, and other substances can be added to pectin-based coatings to improve their barrier properties. Pectin-based coatings can be used to keep vegetables and fruits fresh, as well as poultry, fish, and dairy products.

12.5.2.1 Preservation of Fruits and Vegetables

Fruits and vegetables are perishable in nature and spoil very quickly when handled, transported, or stored due to several factors such as tissue damage, sweating, breathing, insect infestation, and microbial growth. To preserve the quality of fruits and vegetables, pectin-based active coating has been used to slow down the natural processes and keep food away from the environment. Freshly cut apples, potatoes, and carrots that were wrapped in an edible film made of whey protein and pectin lost less weight after harvesting than samples coated with whey protein alone [72]. In comparison to the control group (carrots packaged directly in PVC without being coated), pectin-coated carrots accumulated fewer lignin precursors and flavorless flavonoid compounds, leading to more appetizing organoleptic characteristics [73]. Pectin, whey protein, and transglutaminase were combined to create a coating that was placed on apples, tomatoes, and carrots for consumption, and after 10 days, there were no changes in the hardness and chewiness of the coated samples [63]. However, the texture index of uncoated products dropped by 35 to 60% at the end of storage. Pectin mixed with calcium chloride to make a coating was applied on freshly cut persimmons, and the results showed lower a* values signifying the pectin-based coating reduced browning in the coated persimmons. [74]. Ascorbic acid (AA)–based pectin coating on apricots preserved more vitamin C (46.5%), total phenolics (110 mg GA/100 g), and antioxidant capacity [63]. As a result of the synergistic effects of pectin and AA, oxygen was scavenged and consumed, and PPO-catalyzed reactions were inhibited. Furthermore, the higher oxygen barrier qualities of pectin-based packaging contributed to its use at a lower degree of oxidation [75]. Losing water is unfavorable for keeping fruits and vegetables fresh because it leads to a change in their physiology and quality. Sganzerla *et al.* (2020) developed a pectin-active coating that included feijoa peel flour and starch, extending the shelf life of apples to 15 days [76]. During storage, the coated apples lost less water than the uncoated ones due to the hydrophobic surface of the coating [77]. Pectin-based coating added with essential oil/citrus extract reduced *L. monocytogenes* in coated carrots (1.59 log CFU/g) than the carrots coated with commercial antimicrobial (4.9 log CFU/g) on the 15th day of storage [78].

12.5.2.2 Preservation of Meat, Poultry, and Aquatic Animal Products

Protein denaturation, lipid oxidation, and microbial growth all contribute to the spoilage of muscle foods such as poultry and aquatic animal products. Now, active packaging made of pectin is often used to keep poultry and aquatic products fresh by restricting their deterioration. Pectin-based coating incorporated with eugenol was developed by Wagle *et al.* (2019) for coating chicken wingettes [79]. In comparison to uncoated samples, the active coating significantly decreased the amount of *Campylobacter jejuni* in the coated chicken wingettes by 6.76%. Moreover, the pectin-based edible coating is also used to maintain quality and improve the shelf life of muscle food during storage by maintaining moisture content, water activity, pH, and texture of the food. Several studies have reported the effectiveness of pectin-based edible coating on muscle food, which are depicted in Table 12.2.

Table 12.2 Pectin-based edible coating applications on muscle foods.

Coating material	Food applications	Storage condition	Effectiveness of coating	Reference
Pectin and green tea leaf extract	Pork patty	Refrigerated storage (10°C) for 14 days	Restricted lipid oxidation, improved radical scavenging activity, and inhibited microbial count	[80]
Pectin, oregano essential oil, and resveratrol	Fresh pork loin	Refrigerated storage (4°C) for 15 days, under high oxygen-modified atmosphere packaging	Extended the shelf life, minimized the pH and color change, retarded lipid and protein oxidation, maintained meat tenderness, and inhibited microbial growth	[81]
Pectin, essential oils, and/or extracts of *Thymus vulgaris* (thyme) and *Thymbra spicata* (thymbra)	Sliced bolognas	Cold storage (4°C)	Prolonged shelf life, effective antimicrobial activity against Listeria monocytogenes, Staphylococcus aureus, and Salmonella typhimurium	[82]
Pectin and clove essential oil	Bream (*Megalobrama amblycephala*) fillets	Refrigeration over a period of 15 days	Extended the shelf life of bream filets by at least 15 days, reduced the extent of lipid oxidation, and superior antimicrobial activity	[83]
Pectin, chitosan, and tarragon essential oil	Mackerel (*Scomberomorus commerson*)	Refrigerated storage for 16 days	Improved physicochemical properties and shelf life of fish	[84]

(*Continued*)

Table 12.2 Pectin-based edible coating applications on muscle foods. (*Continued*)

Coating material	Food applications	Storage condition	Effectiveness of coating	Reference
Pectin and gallic acid (GA)	Fresh Japanese sea bass (*Lateolabrax japonicas*) fillets	Refrigerated storage (4°C) for 20 days	Lowered total volatile basic nitrogen, lipid oxidation, and total sulfhydryls, but higher Ca2+-ATPase activity during the storage period	[85]
Pectin, oregano essential oil, and ginger essential oil	Large yellow croaker (*Pseudosciaena crocea*)	Iced storage	Prolonged shelf life extended by 20 to 27 days, delayed increment of pH, thiobarbituric acid, and total volatile basic nitrogen, reduced microbial growth	[86]
Pectin, caproic acid, caprylic acid, linalool, and cuminaldehyde	Chicken eggs	Refrigerated storage (4°C) for 21 days	Effectively reduced Salmonella Heidelberg count	[87]
Pectin, extract of citrus bioflavonoids, and glucono-δ-lactone	Chicken cuts	Refrigerated storage (0, 5, and 10°C)	Prolonged shelf life and improved commercial value of broiler chicken cuts	[88]
Pectin and glycerol	Eggs	Refrigerated storage (5°C), and ambient storage (25°C),	Effectively maintained egg quality during 35 days of storage and extended shelf life of egg at ambient storage	[89]

12.6 Conclusion and Perspective

The utilization of agro-waste as a source of pectin extraction presents significant advantages, such as minimizing environmental pollution, reducing landfill waste, and utilizing renewable resources. Agro-waste, such as citrus peels, apple pomace, and other fruits and vegetables residues, contains a considerable amount of pectin, which can be efficiently extracted using various conventional and novel extraction methods. Among novel techniques, supercritical water extraction, ultrasound, microwave, and ultrasonic microwave-assisted extraction are superior and eco-friendly. This not only adds value to the waste but also contributes to the overall sustainability of the food industry. The application of pectin as an edible coating and film offers numerous benefits for food preservation. Pectin coatings act as a barrier against moisture loss, oxygen exposure, and microbial growth, thereby slowing down the deterioration processes in fruits, vegetables, muscle foods, and various food products. The film-forming properties of pectin provide an additional protective layer, enhancing the physical and chemical stability of the coated items. Additionally, pectin coatings can act as carriers for bioactive compounds, such as antimicrobials, antioxidants, and flavor enhancers, further enhancing the preservation and sensory properties of the coated products. Moreover, the extraction of pectin from agro-waste and its application as an edible coating and film holds immense potential for extending the shelf life of fruits, vegetables, meat, fish, and poultry products. The use of pectin-based coatings and films also aligns with consumer preferences for natural and sustainable food preservation methods. With increasing awareness and demand for clean-label and eco-friendly products, pectin-based coatings provide an attractive solution that meets these expectations.

References

1. Kumar, S., Reddy, A.R.L., Basumatary, I.B., Nayak, A., Dutta, D., Konwar, J., Mukherjee, A., Recent progress in pectin extraction and their applications in developing films and coatings for sustainable food packaging: A review. *Int. J. Biol. Macromol.*, 239, 124281, 2023.
2. Kumar, S., Konwar, J., Purkayastha, M.D., Kalita, S., Mukherjee, A., Dutta, J., Current progress in valorization of food processing waste and by-products for pectin extraction. *Int. J. Biol. Macromol.*, 239, 124332, 2023.

3. Sutherland, W.J., Clout, M., Côté, I.M., Daszak, P., Depledge, M.H., Fellman, L. *et al.*, A horizon scan of global conservation issues for. *Trends Ecol. Evol.*, 25, 1, 1–7, 2010.
4. Mellinas, C., Ramos, M., Jiménez, A., Garrigós, M.C., Recent trends in the use of pectin from agro-waste residues as a natural-based biopolymer for food packaging applications. *Materials*, 13, 3, 673, 2020.
5. Azeredo, H.M.C., Mattoso, L.H.C., Wood, D., Williams, T.G., Avena-Bustillos, R.J., McHugh, T.H., Nanocomposite edible films from mango puree reinforced with cellulose nanofibers. *J. Food Sci.*, 74, 5, N31–N35, 2009.
6. Muzzarelli, R.A.A., Boudrant, J., Meyer, D., Manno, N., DeMarchis, M., Paoletti, M.G., Current views on fungal chitin/chitosan, human chitinases, food preservation, glucans, pectins and inulin: A tribute to Henri Braconnot, precursor of the carbohydrate polymers science, on the chitin bicentennial. *Carbohydr. Polym.*, 87, 2, 995–1012, 2012.
7. Yanakieva, I., Kussovski, V., Kratchanova, M., Isolation, characterization and modification of citrus pectins. *J. Biosci. Biotechnol.*, 1, 3, 223–233, 2012.
8. Mellinas, C., Ramos, M., Jiménez, A., Garrigós, M.C., Recent trends in the use of pectin from agro-waste residues as a natural-based biopolymer for food packaging applications. *Materials*, 13, 3, 673, 2020.
9. Noreen, A., Nazli, Z.H., Akram, J., Rasul, I., Mansha, A., Yaqoob, N., Iqbal, R., Tabasum, S., Zuber, M., Zia, K.M., Pectins functionalized biomaterials; a new viable approach for biomedical applications: A review. *Int. J. Biol. Macromol.*, 101, 254–272, 2017.
10. Mohnen, D., Pectin structure and biosynthesis. *Curr. Opin. Plant Biol.*, 11, 266–277, 2008.
11. Naqash, F., Masoodi, F.A., Rather, S.A., Wani, S.M., Gani, A., Emerging concepts in the nutraceutical and functional properties of pectin—A review. *Carbohydr. Polym.*, 168, 227–239, 2017.
12. Mao, G., Wu, D., Wei, C., Tao, W., Ye, X., Linhardt, R.J., Chen, S., Reconsidering conventional and innovative methods for pectin extraction from fruit and vegetable waste: Targeting rhamnogalacturonan I. *Trends Food Sci. Technol.*, 94, 65–78, 2019.
13. Marić, M., Grassino, A.N., Zhu, Z., Barba, F.J., Brnčić, M., Brnčić, S.R., An overview of the traditional and innovative approaches for pectin extraction from plant food wastes and by-products: Ultrasound-, microwaves-, and enzyme-assisted extraction. *Trends Food Sci. Technol.*, 76, 28–37, 2018.
14. Celus, M., Kyomugasho, C., Van Loey, A.M., Grauwet, T., Hendrickx, M.E., Influence of pectin structural properties on interactions with divalent cations and its associated functionalities. *Compr. Rev. Food Sci. Food Saf.*, 17, 6, 1576–1594, 2018.
15. Kaya, M., Sousa, A.G., Crepeau, M.J., Sorensen, S.O., Ralet, M.C., Characterization of citrus pectin samples extracted under different conditions: Influence of acid type and pH of extraction. *Ann. Bot.*, 114, 6, 1319–1326, 2014.

16. Sun, L., Ropartz, D., Cui, L., Shi, H., Ralet, M.C., Zhou, Y., Structural characterization of rhamnogalacturonan domains from Panax ginseng C. A. Meyer. *Carbohydr. Polym.*, 203, 119–127, 2019.
17. Pfaltzgraff, L.A., Bruyn, M.D., Cooper, E.C., Budarin, V., Clark, J.H., Food waste biomass: A resource for high-value chemicals. *Green Chem.*, 15, 2, 307–314, 2013.
18. Voragen, A.G.J., Coenen, G.J., Verhoef, R.P., Schols, H.A., Pectin, a versatile polysaccharide present in plant cell walls. *Struct. Chem.*, 20, 263–275, 2009.
19. Sañudo Barajas, J.A., Ayón, M., Velez, R., Verdugo-Perales, M., Lagarda, J., Allende, R., Pectins: From the gelling properties to the biological activity, pp. 203–224, Nova Publishers, Hauppauge, NY, USA, 2014.
20. Morales-Contreras, B.E., Rosas-Flores, W., Contreras-Esquivel, J.C., Wicker, L., Morales-Castro, J., Pectin from Husk Tomato (Physalis ixocarpa Brot.): Rheological behavior at different extraction conditions. *Carbohydr. Polym.*, 179, 282–289, 2018.
21. Giacomazza, D., Bulone, D., San Biagio, P.L., Marino, R., Lapasin, R., The role of sucrose concentration in self-assembly kinetics of high methoxyl pectin. *Int. J. Biol. Macromol.*, 112, 1183–1190, 2018.
22. do Nascimento, G.E., Simas-Tosin, F.F., Iacomini, M., Gorin, P.A.J., Cordeiro, L.M.C., Rheological behavior of high methoxyl pectin from the pulp of tamarillo fruit (Solanum betaceum). *Carbohydr. Polym.*, 139, 125–130, 2016.
23. Wang, H., Wan, L., Chen, D., Guo, X., Liu, F., Pan, S., Unexpected gelation behavior of citrus pectin induced by monovalent cations under alkaline conditions. *Carbohydr. Polym.*, 212, 51–58, 2019.
24. Jiang, W., Qi, J.-R., Huang, Y., Zhang, Y., Yang, X.-Q., Emulsifying properties of high methoxyl pectins in binary systems of water-ethanol. *Carbohydr. Polym.*, 229, 115420, 2020.
25. Han, W., Meng, Y., Hu, C., Dong, G., Qu, Y., Deng, H., Guo, Y., Mathematical model of Ca2+ concentration, pH, pectin concentration and soluble solids (sucrose) on the gelation of low methoxyl pectin. *Food Hydrocoll.*, 66, 37–48, 2017.
26. Yang, X., Nisar, T., Liang, D., Hou, Y., Sun, L., Guo, Y., Low methoxyl pectin gelation under alkaline conditions and its rheological properties: Using NaOH as a pH regulator. *Food Hydrocoll.*, 79, 560–571, 2018.
27. Abdul Khalil, H.P.S., Chong, E.W.N., Owolabi, F.A.T., Asniza, M., Tye, Y.Y., Rizal, S., Nurul Fazita, M.R., Mohamad Haafiz, M.K., Nurmiati, Z., Paridah, M.T., Enhancement of basic properties of polysaccharide-based composites with organic and inorganic fillers: A review. *J. Appl. Polym. Sci.*, 136, 47251, 1–21, 2019.
28. Meneguzzo, F., Brunetti, C., Fidalgo, A., Ciriminna, R., Delisi, R., Albanese, L., Zabini, F., Gori, A., dos Santos Nascimento, L.B., De Carlo, A. *et al.*, Real-scale integral valorization of waste orange peel via hydrodynamic cavitation. *Processes*, 7, 581, 2019.

29. Panchami, P.S. and Gunasekaran, S., Extraction and characterization of pectin from fruit waste. *Int. J. Curr. Microbiol. Appl. Sci.*, 6, 8, 943–948, 2017.
30. Gama, B., Silva, C.D.F., Da Silva, L.O., Abud, A.D.S., Extraction and characterization of pectin from citric waste. *Chem. Eng. Trans.*, 44, 259–264, 2015.
31. Tiwari, A.K., Saha, S.N., Yadav, V.P., Upadhyay, U.K., Katiyar, D., Mishra, T., Extraction and characterization of pectin from orange peels. *Int. J. Biotechnol. Biochem.*, 13, 1, 39–47, 2017.
32. Liu, Y., Shi, J., Langrish, T.A.G., Water-based extraction of pectin from flavedo and albedo of orange peels. *Chem. Eng. J.*, 120, 3, 203–209, 2006.
33. Bhavya, D.K. and Suraksha, R., Value added products from agriculture: Extraction of pectin from agro waste product Musa Acuminata and citrus fruit. *Food Ind. Mirobiol.*, 3, 6, 2015.
34. Bagde, P.P., Dhenge, S., Bhivgade, S., Extraction of pectin from orange peel and lemon peel. *Int. J. Eng. Technol. Sci. Res.*, 4, 3, 1–7, 2017.
35. Fishman, M.L., Chau, H.K., Hoagland, P.D., Hotchkiss, A.T., Microwave-assisted extraction of lime pectin. *Food Hydrocoll.*, 20, 8, 1170–1177, 2006.
36. Abang Zaidel, D.N., Md Rashid, J., Hamidon, N.H., Mohd Kassim, A.S., Extraction and characterisation of pectin from dragon fruit (Hylocereus polyrhizus) peels. *Chem. Eng. Trans.*, 56, 805–810, 2017.
37. Sarah, M., Hanum, F., Rizky, M., Hisham, M.F., Microwave-assisted extraction of pectin from cocoa peel. *IOP Conf. Ser.: Earth Environ. Sci.*, 122, 1, 012079, 2018. IOP Publishing.
38. Grassino, A.N., Brnčić, M., Vikić-Topić, D., Roca, S., Dent, M., Brnčić, S.R., Ultrasound assisted extraction and characterization of pectin from tomato waste. *Food Chem.*, 198, 93–100, 2016.
39. Sudhakar, D.V. and Maini, S.B., Isolation and characterization of mango peel pectins. *J. Food Process. Preserv.*, 24, 3, 209–227, 2000.
40. Sundarraj, A.A., Vasudevan, R.T., Sriramulu, G., Optimized extraction and characterization of pectin from jackfruit (Artocarpus integer) wastes using response surface methodology. *Int. J. Biol. Macromol.*, 106, 698–703, 2018.
41. Liew, S.Q., Chin, N.L., Yusof, Y.A., Extraction and characterization of pectin from passion fruit peels. *Agric. Agric. Sci. Proc.*, 2, 231–236, 2014.
42. Hamed, A., Elkhedir, A., Mustafa, S., Effect of Soxhlet method extraction on characterization of pectin of pumpkin peels. *J. Exp. Food Chem.*, 3, 1, 1–3, 2017.
43. Petkowicz, C.L.O., Vriesmann, L.C., Williams, P.A., Pectins from food waste: Extraction, characterization and properties of watermelon rind pectin. *Food Hydrocoll.*, 65, 57–67, 2017.
44. Gan, C.Y. and Latiff, A.A., Extraction of antioxidant pectic-polysaccharide from mangosteen (Garcinia mangostana) rind: Optimization using response surface methodology. *Carbohydr. Polym.*, 83, 2, 600–607, 2011.
45. Wai, W.W., Alkarkhi, A.F., Easa, A.M., Effect of extraction conditions on yield and degree of esterification of durian rind pectin: An experimental design. *Food Bioprod. Process.*, 88, 2-3, 209–214, 2010.

46. Canteri-Schemin, M.H., Fertonani, H.C.R., Waszczynskyj, N., Wosiacki, G., Extraction of pectin from apple pomace. *Braz. Arch. Biol. Technol.*, 48, 259–266, 2005.
47. Kumar, M., Tomar, M., Saurabh, V., Mahajan, T., Punia, S., del Mar Contreras, M., Kennedy, J.F., Emerging trends in pectin extraction and its antimicrobial functionalization using natural bioactives for application in food packaging. *Trends Food Sci. Technol.*, 105, 223–237, 2020.
48. Yue, P., Hu, Y., Tian, R., Bian, J., Peng, F., Hydrothermal pretreatment for the production of oligosaccharides: A review. *Bioresour. Technol.*, 343, 126075, 2022.
49. Marcondes, W.F., Milagres, A.M.F., Arantes, V., Co-production of xylooligosaccharides, xylose and cellulose nanofibrils from sugarcane bagasse. *J. Biotechnol.*, 321, 35–47, 2020.
50. Wang, W., Chen, W., Zou, M., Lv, R., Wang, D., Hou, F. *et al.*, Applications of power ultrasound in oriented modification and degradation of pectin: A review. *J. Food Eng.*, 234, 98–107, 2018.
51. Moorthy, I.G., Maran, J.P., Surya, S.M., Naganyashree, S., Shivamathi, C.S., Response surface optimization of ultrasound assisted extraction of pectin from pomegranate peel. *Int. J. Biol. Macromol.*, 72, 1323–1328, 2015.
52. Xu, Y., Zhang, L., Bailina, Y., Ge, Z., Ding, T., Ye, X. *et al.*, Effects of ultrasound and/or heating on the extraction of pectin from grapefruit peel. *J. Food Eng.*, 126, 72–81, 2014.
53. Chu, J., Metcalfe, P., Linford, H.V., Zhao, S., Goycoolea, F.M., Chen, S., Orfila, C., Short-time acoustic and hydrodynamic cavitation improves dispersibility and functionality of pectin-rich biopolymers from citrus waste. *J. Clean. Prod.*, 330, 129789, 2022.
54. Sun, X., Liu, J., Ji, L., Wang, G., Zhao, S., Yoon, J.Y., Chen, S., A review on hydrodynamic cavitation disinfection: The current state of knowledge. *Sci. Total Environ.*, 737, 139606, 2020.
55. Ashokkumar, M., Applications of ultrasound in food and bioprocessing. *Ultrason. Sonochem.*, 25, 17–23, 2015.
56. Katariya, P., Arya, S.S., Pandit, A.B., Novel, non-thermal hydrodynamic cavitation of orange juice: Effects on physical properties and stability of bioactive compounds. *Innovative Food Sci. Emerg. Technol.*, 62, 102364, 2020.
57. Yoo, S.H., Lee, B.H., Lee, H., Lee, S., Bae, I.Y., Lee, H.G. *et al.*, Structural characteristics of pumpkin pectin extracted by microwave heating. *J. Food Sci.*, 77, 11, C1169–C1173, 2012.
58. Bagherian, H., Ashtiani, F.Z., Fouladitajar, A., Mohtashamy, M., Comparisons between conventional, microwave- and ultrasound-assisted methods for extraction of pectin from grapefruit. *Chem. Eng. Process. Process Intensif.*, 50, 11–12, 1237–1243, 2011.
59. Gbashi, S., Adebo, O.A., Piater, L., Madala, N.E., Njobeh, P.B., Subcritical water extraction of biological materials. *Sep. Purif. Rev.*, 46, 1, 21–34, 2017.

60. Dias, I.P., Barbieri, S.F., Fetzer, D.E.L., Corazza, M.L., Silveira, J.L.M., Effects of pressurized hot water extraction on the yield and chemical characterization of pectins from Campomanesia xanthocarpa Berg fruits. *Int. J. Biol. Macromol.*, 146, 431–443, 2020.
61. Nova, M.V., Nothnagel, L., Thurn, M., Travassos, P.B., Herculano, L.S., Bittencourt, P.R.S. *et al.*, Development study of pectin/Surelease® solid microparticles for the delivery of L-alanyl-L-glutamine dipeptide. *Food Hydrocoll.*, 89, 921–932, 2019.
62. Liew, S.Q., Ngoh, G.C., Yusoff, R., Teoh, W.H., Sequential ultrasound-microwave assisted acid extraction (UMAE) of pectin from pomelo peels. *Int. J. Biol. Macromol.*, 93, Pt A, 426–435, 2016.
63. Huang, J., Hu, Z., Hu, L., Li, G., Yao, Q., Hu, Y., Pectin-based active packaging: A critical review on preparation, physical properties and novel application in food preservation. *Trends Food Sci. Technol.*, 118, 167–178, 2021.
64. Mohamed, S.A.A., Ei-Sakhaway, M., Ei-Sakhawy, M.A.M., Polysaccharides, protein and lipid-based natural edible films in food packaging: A review. *Carbohydr. Polym.*, 238, 116178, 2020.
65. Suhag, R., Kumar, N., Petkoska, A.T., Upadhyay, A., film formation and deposition methods of edible coating on food products: A review. *Food Res. Int.*, 136, 109582, 2020.
66. Chen, Y., Xu, L.L., Wang, Y.J., Chen, Z.Q., Zhang, M. *et al.*, Characterization and functional properties of a pectin/tara gum based edible film with ellagitannins from the unripe fruits of Rubus chingii Hu. *Food Chem.*, 325, 126964, 2020.
67. Ortiz, C.M., Moraes, J.O., Vicente, A.R., Laurindo, J.B., Mauri, A.N., Scale-up of the production of soy (Glycine max L.) protein films using tape casting: Formulation of film-forming suspension and drying conditions. *Food Hydrocoll.*, 66, 110–117, 2017.
68. Kamboj, S., Singh, K., Tiwary, A.K., Rana, V., Optimization of microwave assisted Maillard reaction to fabricate and evaluate corn fiber gum-chitosan IPN films. *Food Hydrocoll.*, 44, 260–276, 2015.
69. Wang, H., Fei, S.Y., Wang, Y.F., Zan, L.S., Zhu, J., Comparative study on the self-assembly of pectin and alginate molecules regulated by calcium ions investigated by atomic force microscopy. *Carbohydr. Polym.*, 231, 115673, 2020.
70. Ogutu, F.O. and Mu, T.H., Ultrasonic degradation of sweet potato pectin and its antioxidant activity. *Ultrason. Sonochem.*, 38, 726–734, 2017.
71. Yan, J.K., Wang, Y.Y., Ma, H.L., Wang, Z.B., Ultrasonic effects on the degradation kinetics, preliminary characterization and antioxidant activities of polysaccharides from Phellinus linteus mycelia. *Ultrason. Sonochem.*, 29, 251–257, 2016.
72. Marquez, G.R., Pierro, P.D., Mariniello, L., Esposito, M., Giosafatto, C.V.L. *et al.*, Fresh-cut fruit and vegetable coatings by transglutaminase-cross-linked whey protein/pectin edible films. *Lebensm. Wiss. Technol.*, 75, 124–130, 2017.

73. Ranjitha, K., Rao, D.V.S., Shivashankara, K.S., Oberoi, H.S., Roy, T.K. *et al.*, Shelf-life extension and quality retention in fresh-cut carrots coated with pectin. *Innovative Food Sci. Emerg. Technol.*, 42, 91–100, 2017.
74. Sanchis, E., Gonzalez, S., Ghidelli, C., Sheth, C.C., Mateos, M. *et al.*, Browning inhibition and microbial control in fresh-cut persimmon (Diospyros kaki Thunb. cv. Rojo Brillante) by apple pectin-based edible coatings. *Postharvest Biol. Technol.*, 112, 186–193, 2016.
75. Vayghan, R.S., Peighambardoust, S.H., Hesari, J., Peressini, D., Effects of osmotic dehydration (with and without sonication) and pectin-based coating pretreatments on functional properties and color of hot-air dried apricot cubes. *Food Chem.*, 311, 125978, 2020.
76. Sganzerla, W.G., Rosa, G.B., Ferreira, A.L.A., Rosa, C.G., Beling, P.C. *et al.*, Bioactive food packaging based on starch, citric pectin and functionalized with Acca sellowiana waste by-product: Characterization and application in the postharvest conservation of apple. *Int. J. Biol. Macromol.*, 147, 295–303, 2020.
77. Sucheta, D.S., Chaturvedi, K., Sharma, N., Yadav, S.K., Composite edible coatings from commercial pectin, corn flour and beetroot powder minimize post-harvest decay, reduces ripening and improves sensory liking of tomatoes. *Int. J. Biol. Macromol.*, 133, 284–293, 2019.
78. Fadhel, Y.B., Maherani, B., Manus, J., Salmieri, S., Lacroix, M., Physicochemical and microbiological characterization of pectin-based gelled emulsions coating applied on pre-cut carrots. *Food Hydrocoll.*, 101, 105573, 2020.
79. Wagle, B.R., Shrestha, S., Arsi, K., Upadhyaya, I., Donoghue, A.M. *et al.*, Pectinor chitosan coating fortified with eugenol reduces Campylobacter jejunion chicken wingettes and modulates expression of critical survival genes. *Poult. Sci.*, 98, 1461–1471, 2019.
80. Kang, H.J., Jo, C., Kwon, J.H., Kim, J.H., Chung, H.J., Byun, M.W., Effect of a pectin-based edible coating containing green tea powder on the quality of irradiated pork patty. *Food Control*, 18, 5, 430–435, 2007.
81. Xiong, Y., Li, S., Warner, R.D., Fang, Z., Effect of oregano essential oil and resveratrol nanoemulsion loaded pectin edible coating on the preservation of pork loin in modified atmosphere packaging. *Food Control*, 114, 107226, 2020.
82. Gedikoğlu, A., The effect of Thymus vulgaris and Thymbra spicata essential oils and/or extracts in pectin edible coating on the preservation of sliced bolognas. *Meat Sci.*, 184, 108697, 2022.
83. Nisar, T., Yang, X.I., Alim, A., Iqbal, M., Wang, Z.C., Guo, Y., Physicochemical responses and microbiological changes of bream (Megalobrama ambycephala) to pectin based coatings enriched with clove essential oil during refrigeration. *Int. J. Biol. Macromol.*, 124, 1156–1166, 2019.
84. Esmaeili, M. and Khodanazary, A., Effects of pectin/chitosan composite and bi-layer coatings combined with Artemisia dracunculus essential oil on the mackerel's shelf life. *J. Food Meas. Charact.*, 15, 3367–3375, 2021.

85. Nie, X., Gao, Z., Ren, X., Jiang, Q., Li, S., Jiang, C., He, F., Effect of pectin coating infused with gallic acid on the quality and shelf life of Japanese sea bass (Lateolabrax japonicas) fillets. *Food Bioprocess Technol.*, 13, 300–307, 2020.
86. Lan, W.Q., Lang, A., Chen, M.L., Xie, J., Combined effects of pectin–plant essential oil coating with vacuum packaging on the quality of large yellow croaker (Pseudosciaena crocea) during iced storage. *J. Food Saf.*, 42, 2, e12960, 2022.
87. Pellissery, A.J., Vinayamohan, P.G., Xue, J., Wang, X., Viju, L.S., Joseph, D., Venkitanarayanan, K., Efficacy of pectin-based caproic acid, caprylic acid, linalool, and cuminaldehyde coatings in reducing Salmonella Heidelberg on chicken eggs. *Front. Sustain. Food Syst.*, 6, 874219, 2022.
88. Ntzimani, A., Kalamaras, A., Tsironi, T., Taoukis, P., Shelf life extension of chicken cuts packed under modified atmospheres and edible antimicrobial coatings. *Appl. Sci.*, 13, 6, 4025, 2023.
89. Oliveira, G.D.S., Dos Santos, V.M., Rodrigues, J.C., Santana, Â. P., Conservation of the internal quality of eggs using a biodegradable coating. *Poult. Sci.*, 99, 12, 7207–7213, 2020.

13

Cellulosic Nanomaterials and Its Derivatives from Agro-Waste for Food Packaging Applications

Rahul Ranjan, Rohit Rai, Vikash Kumar, Smruti B. Bhatt and Prodyut Dhar*

School of Biochemical Engineering, Indian Institute of Technology (BHU), Varanasi, Uttar Pradesh, India

Abstract

Over the recent few decades, renewable and sustainable technology has emerged as a major innovation, particularly in the food packaging and processing sectors. The development of packaging materials mainly derived from renewable, sustainable, and biodegradable resources has become critical and a need of the hour. It helps in addressing some of the global issues with synthetic plastic due to their nonbiodegradability and nonrenewability. The conversion of agro-waste into biodegradable polymeric products has contributed to advancement in the food packaging industry. Packaging is essential for preserving the flavor, shelf life, aroma, odor, nutrients, and quality of food. Thus, natural, biobased, recyclable, biodegradable, and sustainable polymeric biomaterials with physical, chemical, and structural properties similar to conventional polymers are gaining popularity. Cellulose and its derivatives in the nano and micro ranges are among the most prevalent and extensively used polymeric materials in the packaging sector. Cellulose is abundantly available and can be derived from several renewable resources such as biomasses and agro-waste, which help in producing cost-effective biodegradable, nontoxic, and biocompatible materials for food-grade packaging. This chapter covers the fundamental structural insights of lignocellulose and its derivatives, and the compositional analysis of cellulose agro-waste sources, highlighting the modern pretreatment methodologies. Furthermore, the chapter highlights the different kinds of cellulose-derived polymers and composites, including cellulose

**Corresponding author*: prodyut.bce@iitbhu.ac.in

Santosh Kumar, Avik Mukherjee and Vimal Katiyar (eds.) Agro-Waste Derived Biopolymers and Biocomposites: Innovations and Sustainability in Food Packaging, (349–388) © 2024 Scrivener Publishing LLC

acetate, carboxymethyl cellulose, and cellulose nanofibers, discussing their applications and evaluating their properties in food packaging.

Keywords: Cellulose, biodegradable, biomass, shelf life, sustainable

13.1 Introduction

Food safety and health are becoming more and more significant to consumers, and concerns about food standards and food preservation composition have drawn a lot of attention. Degradation of food quality is one of the most serious issues, which is primarily caused by oxidation, and microbial metabolism, which can be influenced by physical factors such as temperature, pH, moisture, physical damage, odor, heat shock, UV radiation, and microbial attacks [1]. Packaging is a food preservation and storage method that maintains the quality of the food, improves shelf life, and prevents toxins and biological degradation [1, 2]. It is also essential for the transportation of foods such as fruits, vegetables, and meats, which protects from perishing and from abrasion, light, and shock. Different types of food necessitate distinct preservation, storage, and transportation conditions. For example, the preservation of perishable food items such as vegetables and fruits requires reduced respiration and transpiration rates, which can be accomplished by controlling the physical factors such as moisture, pH, temperature, and gas levels (oxygen, carbon dioxide, and ethylene). Petroleum-based synthetic plastics are widely utilized for food packaging applications; however, the use of plastic as a packaging material raises serious concerns about its degradability and recycling ability. The utilization of massive amounts of plastic leads to its accumulation in the environment and the degradation of soil quality. Therefore, alternative materials capable of replacing synthetic plastics as a packaging material are required. Cellulose and cellulose-derived polymers or their composites are potential candidates to meet the abovementioned requirements and the development of bio-bags/compostable bags. Cellulose is a naturally occurring biopolymer present in the cell walls of plants. It can be extracted from a wide range of agro-wastes including rice husk, wheat straw, maize stubs, grasses, bagasse, flax, giant trees, and cotton by using different physical, chemical, and enzymatic methods. Cellulose has good mechanical strength, thermal stability, low oxygen and vapor permeability, excellent film-forming ability, easily modified through chemical or physical approaches, good optical properties (transparency), and flexibility. It can be easily produced through industrially scalable approaches [3]. Biodegradability, biocompatibility, recyclability, sustainability, and renewability are some of the special features of cellulose. Different approaches have already been developed for the production of films from cellulose and its derivatives by casting, coating, electrospinning, layer-by-layer assembly, and extrusion methods [4]. The casting method is one of the simplest, most economical, and most widely used technology to produce cellulose-based food packaging material at the laboratory scale. In the casting method for films, two or more solutions are simply mixed and homogenized in a suitable solvent to produce a homogeneous suspension followed by casting and drying on a Teflon/glass mold. Recently, Moradian *et al.* (2021) developed a cellulose film from kraft pulp through carboxymethylation using isopropanol, monochloroacetic acid, and sodium hydroxide. The developed films with 1.5 wt % of pulp concentration had excellent tensile strength (103 MPa), tensile modulus (5.2 GPa), and tensile strain (8.7%) with optical transmittance ($\geq$ 90%), which makes it potentially suitable for food packaging [5]. Priyadarshi *et al.* (2021) developed CMC-based food packaging films for the preservation of meat by incorporating grapes extract and zinc oxide nanoparticles *via* solvent casting methods. The developed film showed good antibacterial properties against *E. coli* and *Listeria monocytogenes* and it reduced the lipid oxidation in meat by 88% after 15 days of refrigeration incubation [6].

Extrusion is an economical and commercial technique for the production of nanofibrillated cellulose to meet large-scale demand. A twin-screw extruder is the most used extrusion machine, which consists of a clamshell barrel, recirculation channels, and a valve. This machine can be operated in both batch and continuous modes for the production of films. First, cellulose suspension is fed into the continuous barrel followed by recirculation, and a die is used to get the desired shape of the films. Fourati *et al.* (2020) developed a square-shaped film using the twin-screw extrusion method for the first time. In this study, TEMPO-oxidized, kraft softwood was mixed with starch, glycerol, and water, and then transferred into the clamshell barrel and extruded in continuous mode at 110°C to remove all the water content from the mixture. Last, the obtained mixture was used to produce a film using a die [7]. Chen *et al.* (2020) also developed cellulose-based film using microcrystalline cellulose (MCC) and starch *via* screw extrusion method followed by hot pressing, and the produced film had a maximum tensile strength of 12.4 MPa and water vapor permeation of ~3.6 g/cm^2·hour·Pa [8]. Recently, Trigui *et al.* (2020) produced TEMPO-oxidized and carboxymethylated cellulose nanofiber-based film *via* screw extrusion techniques, and the developed film had a maximum tensile strength of ~ 75 MPa [9].

The continuous film-forming method is one of the potential techniques to produce films or packaging material with desired dimensions (length and width) at a commercial scale. In this method, gels or suspension is continuously passed over a continuously moving substrate or roller or conveyor belt equipped with doctor blades. The doctor blades are generally used to control the layer of thickness of wet monolayered films. The developed monolayer is then passed through the drying chamber followed by rolling. Leite *et al.* (2020) in their study developed a continuous film using cellulose nanocrystals (CNCs) and gelatin. In the preparation method, the suspension was passed over a conveyor belt followed by three steps of drying (one IR drying and two steps of oven drying at 80°C). A maximum of 12 m of the film with a thickness of 0.065 mm was produced in an hour. The produced film had the highest tensile strength of ~20 MPa with good UV barrier and thermal properties [10]. Recently, Droguet *et al.* (2022) developed CNC-based photonic film using (roll to roll) R2R techniques equipped with slot and die. In this work, a syringe pump was used to pass the suspension into the slot die, and a thickness filler gauge was used to maintain the uniform distance between the slot lip and the conveyor belt or substrate. Web holders were also used for adjusting the distance of 30 cm after each support. The maximum length of film produced was found to be 3.2 m, which depends on the casting length of the web holder, and finally, the wet film was passed through a blown air heating chamber [11].

The blending of nanoparticles (silver, zinc, and copper) [12], curcumin [13], anthocyanin [14], carmellose sodium [15], and cinnamaldehyde into cellulose or nanocellulose imparts antibacterial properties, which improve the shelf life of the foods. Even though cellulose-based films have advanced to industrial production, there are still several challenges to be overcome. Some of them are the following: (i) fibrillation is a highly energy-intensive process that turns biomass into cellulose fiber; (ii) numerous well-recognized pretreatment techniques have been developed, but the majority of them are time-consuming and chemically demanding, expensive, and utilizing extremely harmful chemicals; (iii) the selection of suitable, easily available lignocellulose biomass preferably with limited seasonal variation generated from noncommercial cellulosic feedstocks; and (iv) the hydrophilic nature of the cellulose and poor wet strength of films also limits their commercial application. These challenges provide an opportunity for scientists and researchers to develop more sustainable, economically viable, and scalable methodologies for the development of cellulose-based film from agro-waste. This chapter covered the basic understanding of cellulose, hemicellulose, holocellulose, bacterial cellulose, the derivatives of cellulose, compositional analysis of cellulose, agro-waste sources, pretreatment, and extraction methodologies particularly using trifluoroacetic acid (TFA) and ionic liquids (ILs), various types of cellulose-derived polymers, and composites such as cellulose acetate (CA), carboxymethyl cellulos (CMC), and cellulose nanofibers, and their potential application in food packaging sector.

13.2 Cellulose Structure and Its Nano-Derivatives

Cellulose is one of the most abundantly available biomaterials, which can be derived from plants, agricultural, algal, and bacterial sources. About 1.5×10^{12} tons of cellulose are produced annually through the plant photosynthesis process [16]. Rapid advances in cellulose research in the past few decades have enabled industries to make its extensive use for different applications such as paper making, textile, wastewater remediation, optical film, biodegradable film for food packaging, bioelectronic devices, piezoelectric devices, and biomedical and pharmaceutical applications. Cellulose, a class of innumerable natural polymers, is considered to have unique properties such as biodegradability, biocompatibility, nontoxicity, and renewability. It also offers a special feature of easy chemical and physical modification, which helps in the synthesis of different kinds of cellulose derivatives such as cellulose phosphate, CA, cellulose nitrate, sulfated cellulose, and TEMPO-oxidized cellulose. Cellulose is present in various fractions in different biomass sources such as bamboo (40 to 55%), barley straw (31 to 45%), switchgrass (45%), hemp (70 to 75%), peanut (49%), jute (60 to 65%), poplar (42 to 49%), sugarcane bagasse (40 to 50%), kapok (70 to 75%), and spruce (48%). Cotton is considered a pure form of cellulose, which contains almost 90% cellulose [17]. The physicochemical and structural properties of cellulose influence its potential application in different sectors and their subsequent industrial-scale utilization. It is critical to understand the structural makeup and evaluate its properties to choose the most appropriate delignification, pretreatment, and functionalization methods for the desired application.

Cellulose is a naturally derived biopolymer consisting of repeating monomeric units of D-glucopyranose joined together by β-1,4 glycosidic bonds. The number of monomeric units is highly dependent on the source from which cellulose has been extracted and it may range from 500 to 15,000 units. The backbone of cellulosic chains consists of reducing and nonreducing ends. The nonreducing end of cellulose at the C1 position of D-glucopyranose units (anomeric carbon) contributes to the bond formation (β-1,4 glycosidic linkage), while the aldehyde group is present at the reducing end of cellulose [17]. The hydrogen bonds that exist in between and among the neighboring cellulose chains have a great impact on the physiochemical properties such as reactivity, hydrophilicity, and crystallinity [18]. In the cellulose chain, intra- and intermolecular hydrogen bonds are present, which arise from the interaction of hydroxyl groups (-OH) with oxygen present in the glucose structure as well as the glycosidic linkage. In cellulose crystals, the chains are tightly and closely aligned combined with H-bonds, while the amorphous regions are less ordered and bonded through isotropic intermolecular hydrogen bonding. The amorphous segments, on the other hand, are randomly orientated with the formation of turns around their adjacent chains resulting in lowered density [19]. Six different polymorphs of cellulose exist, which include cellulose I, II, III_I, III_{II}, IV_I, and IV_{II} (Table 13.1). The cellulose I polymorph occurs naturally in plants, while the other forms are produced through chemical synthesis. Cellulose I polymorph is further subdivided into two distinct allomorphic forms I_α and I_β but the proportion of each allomorph varies from plant to plant species. The I_β allomorph of cellulose is the most common and abundantly present in lignocellulosic biomass derived from plants. Cellulose I_β belongs to the monoclinic $P2_1$ (space group) unit cell and has two chains of cellulose per unit cell. The cellulose I_α on the other hand is a member of the triclinic P_1 (space group) unit cell and has one chain per unit cell [20].

Generally, cellulose II is produced by a mild NaOH treatment process from cellulose I, which leads to the generation of the $P2_1$ subspace group with monoclinic crystal structures. Each unit cell has two antiparallel cellulose chains, which are more stable than the parallel chain present

Table 13.1 The dimensions and angles of a different polymorph of cellulose.

Unit cell	a (Å)	b (Å)	c (Å)	α (°)	β (°)	γ (°)
Cellulose Iβ	7.78	8.20	10.38	90	90	95.5
Cellulose Iα	6.72	5.96	10.40	118.8	114.8	80.38
Cellulose II	0.81	9.03	10.31	90	90	117.1
IIII	4.45	7.85	10.31	90	90	105.1
IIIII	4.45	7.64	10.36	90	90	106.96
IVI	8.03	8.13	10.34	-	-	-
IVII	7.99	8.1	10.34	-	-	-

in cellulose I. Cellulose III_1 and III_2 are synthesized from cellulose I and cellulose II, respectively, using liquid ammonia as a chemical reagent at −80°C. Cellulose III_1 has a monoclinic crystal structure with one parallel cellulose chain per unit cell, similar to cellulose I but in a different conformation [20–22]. The crystal structure of cellulose III_2 is still unidentified or unexplained. Both cellulose III_1 and III_2 regain their original structure when exposed to boiling water. Furthermore, cellulose IV_1 and IV_2 were prepared from cellulose III_1 and III_2, respectively. A thermochemical treatment method was used in the synthesis process of cellulose IV with the aid of hot glycerol. Gardiner *et al.* (1985) proposed that cellulose IV_1 had an orthorhombic crystal structure, with two cellulosic chains forming each unit cell. The cellulose IV_{II} structures were reported to have a similar size to cellulose IV_I but they had different conformations [23]. The cell wall of a plant contains fibrils of cellulose chains with a diameter in the nanometer (nm) range. These fibers bundle together to form microfibrils and their dimensions increase from nanometers (nm) to micrometers (μm) [24]. All the chemically synthesized polymorphs have less elastic modulus than cellulose I. Cellulose I has a maximum of 137-GPa tensile modulus, while cellulose II, cellulose III, and cellulose IV have a maximum modulus of 88, 87, and 75 GPa, respectively, which is mainly due to the huge variation in the inter- and intramolecular hydrogen bonding. The F-value (amount of force required to stretch the molecules by one degree) for each of the polymorph also differs, for example, cellulose I require 4.4 dynes of force, while others need a force of 2.87, 2.95, and 2.45 dyne for cellulose II, cellulose III, and cellulose IV, respectively [25]. Based on the abovementioned properties of different polymorphs of cellulose, it can be used for different and desired applications.

13.2.1 Hemicellulose

Hemicellulose is the second most prevalent polysaccharide and heteropolysaccharide present on the planet. Hemicellulose exhibits a significantly high degree of polymerization (DP; 100 to 300) and molecular weight; however, it has a lower DP than cellulose [26]. The forms of sugar and their corresponding amounts in hemicellulose vary between plant species. Even within the same plant species, hemicellulose can form different microstructures and their proportion differs in the plant's parts such as root, core, and branches. Hemicellulose typically consists of either a single pentose (xylose and arabinose) or a group of pentoses and hexoses (D-galactose, D-glucose, D-mannose, etc) [17, 27]. The hemicelluloses also contain the sugar rhamnose and fucose, as well as uronic acid at very low concentrations [28]. The main source of uronic acids including D-glucuronic ($C_6H_{10}O_7$), D-galacturonic ($C_6H_{10}O_7$), and methyl galacturonic acids ($C_7H_{12}O_7$), which are present in branched chains of hemicellulose. The β-(1,4)–linked xylopyranose segments found in different parts of the plant are thought to be representative structures of the hemicellulose. Sugar branching at the C_2 and C_3 positions of the hemicellulose adds complexity to the hemicellulose [29]. Classification of hemicellulose was done based on sugar involved in the synthesis of its main chain or backbone structures. Examples of the four most prevalent sugars that are involved in the synthesis of hemicellulose include (i) xylan, (ii) β-glucans, (iii) mannoglycans, and (iv) xyloglucans.

13.2.2 Xylan

Xylan is one of the most common and widely available types of sugar found in hemicellulose Figure 13.1(a). It is constituted of xylose subunits and is joined by β-1,4 bonds to units of feruloyl residues, 4-O-methyl-D-glucuronic acid residues, units of p-coumaroyl, and acetyl residues. Wheat and rice straws contain xylan as well as L-arabinofuranose in their side groups. Acidic xylans are also present in plants and are made up of 4-O-methyl-D-glucuronic xylopyranose, D-glucuronic acid, galactopyranose,

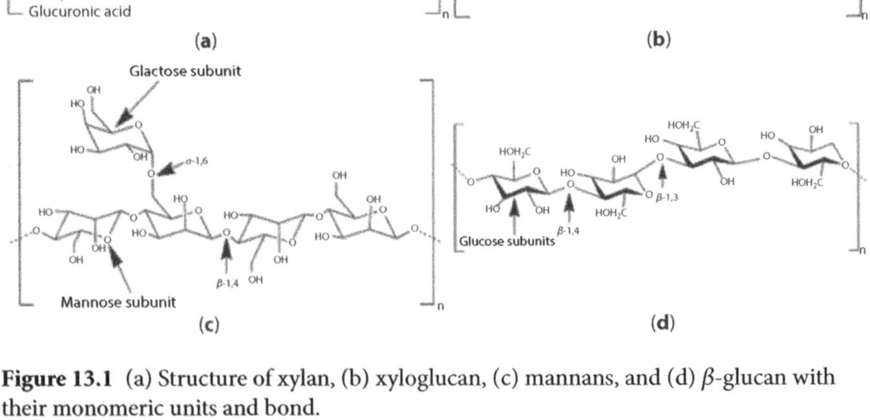

Figure 13.1 (a) Structure of xylan, (b) xyloglucan, (c) mannans, and (d) β-glucan with their monomeric units and bond.

methylgalacturonic acids, and furanose which are found as side chains in the complex xylan structures. Ferulic acids may interact with each other by constituting ester bonds (diesters) which connect two xylan chains or form the linkage between hemicelluloses and lignin [30]. Xylans are further divided into homoxylan and heteroxylan based on their branching and side chains. Homoxylans are xylans that comprise chains of xylose subunits joined with β-(1,3), β-(1,4) linkage, or both, whereas heteroxylans contain xylose in its backbone and are joined together by a β-(1,4). However, heteroxylan also contains some other functional groups, such as arabinose, 4-O-methyl-glucuronic acid, glucose, glucuronic acid, and galactose. Heteroxylans are further subdivided into glucuronoxylans, arabinoxylans, and glucuronoarabinoxylans based on their structures and abundance of functional groups in their branching chains. Glucuronoxylan is composed of xylose subunits with a major side group of 4-O-methyl-α-D-glucopyranosyl uronic acid (15 to 30 wt % present in hardwood lignocellulosic materials) and a minor side group of glucuronic acid covalently attached to the C2 in the backbone structures. Glucuronoxylan forms ester bonds with lignin residues in the presence of uronic acid groups. The amount of acetylation largely depends on the plant sources and the processing method used. To create the glucuronoxylan backbone, acetyl groups interact with the xylose residues at the C_2 or C_3 position [26, 31, 32]. The DP of glucuronoxylan mainly depends on the source of wood, hardwood has a DP of 200, while softwood has a DP of 100 [26, 33]. Arabinoxylan is another heteroxylan type that is frequently extracted from herbs. The xylose subunits that make up its backbone structures have arabinose attached to the C_3 position as the primary side chain. Acetylation also takes place in arabinoxylan, and acetyl groups are attached to the C_2 and C_3 positions of the xylose backbone. However, DP is much less than glucuronoxylans, with a range of 50–180 [34, 35]. Arabinoxylan also forms a covalent linkage with lignin due to the existence of ferulic acid at position C_5 in the arabinose residues [36, 37]. A heteroxylan consisting of both O-methyl-glucuronic acid (and/or glucuronic acid) groups and arabinose as their branch chain in a backbone of xylose is termed glucuronoarabinoxylans.

Xylans are hydrophilic biopolymers that contain a significant amount of free hydroxyl groups, which facilitate the functionalization process and lead to the formation of new materials with enhanced properties like hydrophobicity, thermal stability, and film-forming capacity. Due to its improved film properties such as water vapor, oxygen, and odor barrier properties after functionalization, xylan can be used to make films. Additionally, it can be utilized as a drug delivery system, emulsifier, thickener, and other food additives.

13.2.3 Mannans (Mannoglycans)

Mannans, a type of lignocellulose that accounts for 20-25% of the total dry weight of biomass, is predominantly found in softwood. It is classified into four major groups based on the branch chains and backbone structures: (i) homomannans, (ii) galactoglucomannans, (iii) glucomannans, and (iv) galactomannans. Galactose is commonly present in the branch chain of mannans. Homomannans are the most uncommon type of mannans lignocellulose, which is made up of mannopyranose subunits connected by β-1,4 linkages. Glucomannans are usually found in hardwood consisting of 3 to 5 wt % of total biomass. It is composed of glucose and mannose residues joined covalently with β-1,4 linkages. DP of glucomannans ranges from 70 to 100 and it has relatively low viscosity and is soluble in water due to the presence of side chains or branching in glucomannans structures. The glucomannans have branched chains in their structure at the C_3 position of it constituting monomers D-glucopyranose and D-mannopyranose. However, branching is more prevalent in mannose subunits [38]. The galactomannans mainly consist of two monosaccharide units, i.e., mannose and galactose. Mannoses form a β-1,4-linked main chain, whereas galactose forms a β-1,6–linked side chain that is randomly attached to the main backbone. The presence of galactose in the branched chain of galactomannans affects the solubility, interaction, and viscosity of other polymers. The ratio of mannose to galactose (galactosylated ratio) determines the solubility of the galactomannans. Galactomannans become insoluble in water when the ratio of mannose to galactose is less than 5%, but they become soluble when it is greater than 30%.

Galactoglucomannans have been largely studied in mannans form which accounts for 14 to 25% of the wood (mainly softwood). It is widely used in paper and pulp, cosmetics, pharmaceutical, food, and textile industries [39]. The backbone of galactoglucomannans is made up of D-glucose and D-mannose subunits that are covalently linked together by β-1,4 linkages and the α-1,6 linkages connect D-galactopyranose to mannose residues as well. Structures of mannans are shown in Figure 13.1(c). Mannans are naturally nontoxic, highly viscous, water-stable, and biodegradable substances, and they can be used as stiffening and emulsifying agents in the food industry because of their strong water stability. The excellent oxygen barrier capacity of galactoglucomannans makes it suitable for the production of film for food packaging [40].

13.2.4 Xyloglucan

Xyloglucans are a type of hemicellulose that consists of D-glucose units (Glu) joined together by β-1,4 linkages, which help in the formation of the xyloglucan backbone as shown in Figure 13.1(b). Seventy-five percent of D-glucopyranose in the xyloglucans structure can be replaced by α-1-6 linked xylose subunits (X) in the side chains. It is the key constituent of plant cell walls, typically extracted from biomasses such as tamarind [41], cranberry [42], chia [43], and most angiosperms and grasses [44]. Xyloglucans are grouped as (i) XXXGlu and (ii) XXGluGlu, depending on the frequency of their side chain distribution. The first group of xyloglucans, XXXGlu, had three of the four D-glucose subunits replaced by α-1,6 xylose subunits in the backbone [45]. XXGluGlu is the second type of xyloglucan polysaccharide, which is formed by the substitution of two xylose units in the regular repetitive backbone containing D-glucose units. XXGluGlu is produced by substituting two glucose units out of four with two xylose subunits in the regular repetitive backbone [46]. Because of its excellent water stability and ability to form viscous solutions, xyloglucan is frequently used as a stabilizer and thickener in the food industry. It also has good gelling, binding, and film-forming properties, which are useful in the fabrication of films and coatings for food packaging applications [47].

13.2.5 β-Glucan

It is a linear form of hemicellulose composed of D-glucose that is commonly found in grass, yeast, bacteria fungi, and cereals. It is similar to the cellulose structure, which only contains β-1,4 linkages, but the β-glucan has both β-1,4 and β-1,3 linkages. Cellotriosyl and cellotetraosyl are the two primary components of β-glucan. The cellotriosyl residues contain two continuous β-(1,4) and β-(1,3) bonds which hold them together. In contrast, the cellotetraosyl components have one β-(1,3) bond after three continuous repetitive β-(1,4) bonds. Understanding the structural characteristics of hemicellulose is critical because it has a considerable influence on the properties of polysaccharides, which in turn influences the properties of lignocellulosic biomass [46, 48]. The structure of β-glucan is shown in Figure 13.1(d). β-Glucan possesses strong emulsifying properties, foam-forming ability, thickening property, viscosity, and whip ability, which are highly suitable for food and bakery industries. It can be used for the production of prebiotic sausage, and gluten-free bread, as well as additives for dairy products [49].

13.2.6 Holocellulose

Holocellulose, a lignin-free fiber, contains only cellulose and hemicellulose. It can be extracted from various biomasses such as wheat straw, cocksfoot grass, apple pomace, passion fruit peels, and manau rattan. To produce holocellulose, various lignocellulosic biomasses have been processed using a variety of physical (grinding, homogenization, and ultrasonication) and chemical (TEMPO oxidation, citric acid, and peracetic acid treatment) methods [50–52]. Holocellulose has extremely good mechanical properties, colloidal stability, optical properties, vapor, oxygen barrier, biodegradability, and thermal stability [51, 52]. It has been used for various applications such as the preparation of aerogels, hydrogels, films [51], electrodes and nano papers [53], and electromagnetic shielding [54]. Several studies have been performed to produce food packaging film, for example, Long *et al.* (2023) in their study utilized *Camellia oleifera* to produce holocellulose using $NaClO_2$ and CH_3COOH for food packaging applications. The holocellulose-based synthesized film had a tensile strength of 123.48 MPa, which biodegraded within 75 days [51]. Wu *et al.* (2022) extracted holocellulose from wheat straw *via* ultrasonic and homogenization processes, and the produced holocellulose was further blended with graphene oxide to develop film for food packaging applications. The produced film had a tensile strength of 88.7 MPa and water vapor permeability of 0.75×10^{-10} g mm^{-1} hour^{-1} Pa^{-1} with improved hydrophobic and UV blocking properties [55].

13.2.7 Bacterial Cellulose

Brown (1886) first reported that bacterial cellulose can be produced by the bacterium *Acetobacter xylinum.* Bacterial cellulose can be found as a thick layer at the culture medium's liquid-air interface. Bacterial cellulose is free from other components such as hemicellulose, pectin, and lignin residues present in lignocellulosic biomass thus making it a very pure form of cellulose. A variety of bacteria from the genus *Gluconacetobacter* [56], *Aerobacter senengalensis* [57], *Alcaligenes faecalis-X3* [58], *Agrobacterium tumefaciens* [59], *Azotobacter* to *Achromobacter* make bacterial cellulose under certain culture conditions. Bacterial cellulose produced by *Gluconacetobacter xylinum* is of high purity and resembles the plant cellulose. Bacterial cellulose can be produced by agitated and static culture methods. In static culture methods, bacterial cellulose is produced in the form of a thick membrane, and in agitated culture methods, it is produced in the form of pellets, irregular shapes, or fibrous forms [60]. The properties of bacterial cellulose include biodegradability, high porosity, flexibility, transparency, biocompatibility, and higher mechanical strength. Bacterial cellulose has a very high Young's modulus of 15 GPa among the 2D organic materials, it has very high stiffness and high tensile strength due to its crystalline structure. Bacterial cellulose has a very high water-holding capacity and a large surface area per unit mass. The properties of bacterial cellulose can be improved by *ex situ* and *in situ* modifications [61]. The main materials that are currently being used for food packaging are mainly made up of polyethylene terephthalate, low- and high-density polyethylene, and other synthetic polymers. These synthetic materials are non-biodegradable and cannot be recycled, and thus, as an alternative bacterial cellulose has been used in food packaging [62]. A study was conducted to develop chitosan-bacterial cellulose-curcumin composite films using different molecular weights of chitosan, and the developed film showed high antioxidant and mechanical properties. The water vapor transmission rate of the film was found to be in the range of 45 to 61 $g/m^2 \cdot 24$ hours, and the oxygen permeability was in the range of 0.05 to 0.175 $g/cm^2 \cdot hour \cdot Pa$. It was found that the chitosan-bacterial cellulose-curcumin film was a good packaging material for meat-based food products [63]. A food packaging film was developed using TEMPO-oxidized bacterial cellulose that contained anthocyanin-rich purple potato extract and thymol. This film could block UV light and had very high antioxidant and antibacterial activity. This film had good water vapor permeability of 4.14×10^{-11} $g/m \cdot s \cdot Pa$, which helps to prevent the food from losing its texture, and had a good ability to preserve shrimps [64]. Edible food packaging film was developed with bacterial cellulose-glycerol-CMC olive oil/ginger oil for fruit packaging and enhanced the shelf life of oranges by 9 weeks in a refrigerator, 4 weeks at 30°C, and 2 weeks at 40°C [65].

13.3 Agro-Waste as Source of Cellulose

Agricultural waste is leftover part from crop cultivation or the waste coming out from the food processing and fodder industries that could be utilized to produce valuable compounds or materials. Historically, agricultural organic wastes from agro-industries have already been reused as fertilizer and animal fodder. However, the increase in agro-industrial production leads to waste accumulation. Landfilling and incineration are the two most frequent and common methodologies to dispose of agro-waste biomass and are hazardous to the ecosystem. Therefore, it is critical to handle agro-industrial waste legitimately and economically to protect the natural environment and possibly to produce products with added value. In this way, transforming waste into more valuable and profitable products has gained international attention and is consistent with the new strategy of the sustainable development goals (SDG). Agricultural wastes such as bagasse, rice straw, pea hulls, wheat straw, corn stalk, hemp, flake, jute, peanut shell, dried beet pulp, rice husk, and kenaf are rich sources of cellulose, hemicellulose lignin, and other bioactive compounds. Agricultural wastes can be potentially exploited as an alternative and cheap raw material for the production of cellulose to meet the global industrial demands for commercial applications as mentioned in Table 13.2. There are huge demands to produce cellulose-based products using less energy-intensive and easily scalable processes.

13.4 Extraction of Cellulose from Agro-Waste

A lot of research has been carried out on cellulose extraction from agricultural waste, which may pave the way for its industrial applications such as the production of biofuels (bioethanol, biomethane, biobutanol, and jet fuels), biomaterials (cellulose, holocellulose, and hemicellulose), and other value-added products. Usually, raw fibers are processed through several unit operations such as cutting, milling, or grounding into powdered form to increase surface area. Pretreatment is generally applied to disintegrate the intact structure and remove the two major components of lignocellulosic biomass (i.e., hemicellulose and lignin). While developing the best pretreatment strategy, it is crucial to take the complexity and variability of the lignocellulosic biomass into account. Different kinds of physical and chemical treatments, as well as physiochemical (mixture of both physical and chemical treatment together), have been established to extract cellulose from agricultural waste biomasses as shown in Figure 13.2. Physical pretreatment is commonly referred to as the size reduction process, which includes processes like grinding, sizing, and milling. Grinders and ball millers are the most used tools. This process incorporates various techniques such as grinding, crushing, milling, peeling, shredding, and chipping. Milling is the most used physical pretreatment technique for reducing biomass particle size. However, the final particle size depends on the technique used; for example, grinding or milling lignocellulosic biomass generates fragments with sizes ranging from 200 to 2000 μs, whereas chipping produces particles with sizes ranging from 1 to 3 cm [78]. Grinding and chipping significantly decrease the particle size and crystallinity of the biomass. It enhances the surface area of the regenerated biomass where the chemical agents can more easily act to produce cellulose as well as hydrolyze the biomass for glucose production [79].

Microwave irradiation is a green approach for the pretreatment of biomass. The explosions caused by microwave heating disintegrate the

Table 13.2 Recent studies on the production of cellulose from different agro-waste using different treatments and yields of cellulose, hemicellulose, and lignin.

Agro-waste	Treatment used	Cellulose (%)	Hemicellulose (%)	Lignin (%)	Reference
Ananus cosomus	NaOH (2 wt %) treatment and sodium chlorite (1.7%)	36.3	74.5	22.9	[66]
Oil palm empty fruit bunch	NaOH (4 wt %) treatment and sodium chlorite (1.7 wt %)	67–79	19–21	3.7–10	[67]
Red banana peduncle	NaOH treatment and sodium chlorite (0.7 wt %)	62	20	14	[68]
Corn waste residue	Sodium chlorite and sodium pyrosulfite	41	-	19	[69]
Banana rachis	NaOH (2.5M)	42	14	17	[70]
Garlic and onion	NaOH	41–42	16–20	34.5–39	[71]
Corn husk	NaOH (4 wt %) treatment and sodium chlorite (1.7 wt %)	29.3	40	11	[72]
Bagasse	Steam explosion (205°C) and high-pressure homogenization	58.08	3.3	8.5	[73]
Walnut shell	Microwave (525W), 5% v/v H_2O_2	79	0.3	6.2	[74]
Sugarcane bagasse	Microwave (525W), 5% v/v H_2O_2	85.6	0.9	4.8	[74]
Corn comb	Microwave (525W), 5% v/v H_2O_2	83	1.6	3.4	[74]
Cashew shell	NaOH (17.5% w/v) treatment and sodium hypochlorite (0.7 wt %)	28	17	-	[75]
Sugarcane bagasse	Methyltrioctylammonium chloride	45–65	-	-	[76]
Acer platanoides	Caustic soda treatment	45	9.3	8.7	[77]
Acer platanoides	H_2O_2	65	4	2.5	[77]

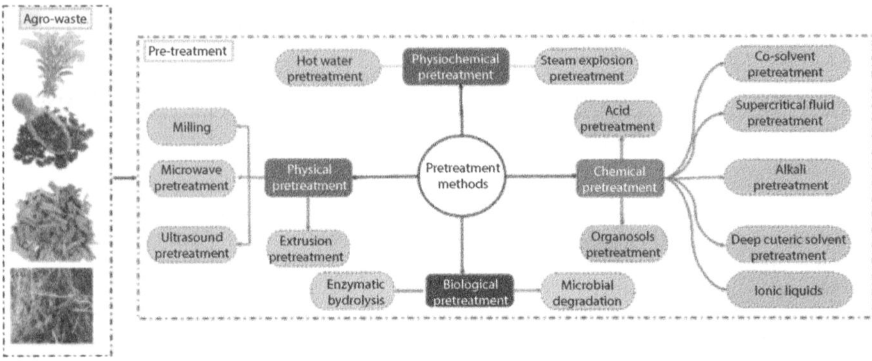

Figure 13.2 The source of agro-waste and pretreatment methods for cellulose extraction.

recalcitrant structures of the biomasses. Microwave along with alkali or acids improves the delignification rate. For example, Tsegaye *et al.* (2019) found that microwave pretreatment of wheat straw for 15 min at 160°C with 1.5% NaOH resulted in delignification of 69.48% [80]. Some of the advantages associated with the microwave-assisted pretreatment method included reduced time and energy consumption, fewer byproducts, uniform heating, cost-effectiveness, high yield, and environmentally friendly [81]. However, some of the disadvantages associated are (i) high energy requirement, (ii) high working capital, (iii) high operational cost, and (iv) availability of moisture in the biomasses [79]. Chemical treatment methods include the use of alkalis (NaOH, KOH, and $Ca(OH)_2$) and acids (H_2SO_4, CH_3COOH, HCl, and H_3PO_4). Several other approaches such as liquid ammonia, supercritical CO_2, and H_2O are also employed for pretreating biomasses. There are two ways to perform acid pretreatment: (i) with diluted acid at high temperatures and (ii) with concentrated acid at low temperatures [82]. Both methods have benefits and drawbacks. Although the dilute acid–assisted pretreatment process necessitates high temperatures making the overall process energy-intensive, the use of less acid is beneficial to the environment. Concentrated acids are hazardous to both humans and the environment, but the lower temperature makes them advantageous in terms of low energy requirements. Alkali pretreatment methods are used to break the bond between lignin-carbohydrates complexes through saponification and salvation reaction. Sodium hydroxide is a widely used alkali pretreatment agent, which breaks the ether or ester linkage between the lignin-carbohydrates complex [83]. Organosolvents such as acetone, C_2H_5OH, CH_3OH, and ethylene glycerol were used for the pretreatment of lignocellulosic biomass [78]. During this process, some organic acids had also been used as a catalyst such as salicylic acid, and oxalic acid. In this pretreatment method, linkages between hemicelluloses and lignin are disrupted by a solubilization reaction. Tang *et al.* (2017), for example, used ethanol and propylamine to pretreat corn stover at optimum temperatures and pressures of 140°C and 1 bar, respectively. They concluded that the maximum delignification of 81.7% was achieved in 40 min [84]. Zhang *et al.* (2013) utilized glycerol and hydrochloric acid for the pretreatment of sugarcane bagasse at 130°C to 210°C under ambient atmospheric pressure for 24 hours. The maximum recovery of the solid content during this pretreatment method was 77.2%. This pretreatment method had advantages such as simple recovery and low viscosity of the used solvent, which improves the solvent's ability to penetrate biomass, increasing lignin removal [85]. When organic solvents along with acid are used for the pretreatment, they corrode vessels in which the reaction was performed [84]. Another issue with this method is the high-temperature requirement and the production of inhibitory products during the reaction [86].

Supercritical fluids are considered an environmentally friendly, thermodynamically stable, greener, nontoxic, cheap, and sustainable method for the valorization of agro-wastes [87]. Supercritical fluids exist in a homogeneous phase with the features of both liquids and gasses. It has a low viscosity, which facilitates the permeation or diffusion of the solvent into the biomasses. Pasquini *et al.* (2005) proposed that bonds present in the lignocellulosic biomass are disrupted as a consequence of increased interactions between the supercritical liquids and biomass under applied high pressure [88]. Serna *et al.* (2017) pretreated rice straw with supercritical CO_2 at ideal temperatures and pressures of 80°C and 27 MPa and the greatest percentage of lignin removed was 90.6% [89]. Shaikh *et al.* (2021) used supercritical CO_2 and 20 wt % NaOH to extract cellulose from date palm waste at 340-bar pressure and 90°C temperature, resulting in 90% of the α-nanocellulose production [90].

According to the fundamental principles of green chemistry, the use of efficient and environmentally friendly solvents and chemical reagents is a critical factor in determining the renewability of any process. The conventional solvents that are currently used in industries are hazardous to the environment. Life cycle assessments were also studied to determine the best chemical agents suitable for the pretreatment of biomass and their environmental effects. For example, Smullen *et al.* (2019) discovered that pretreatment with sulfuric acid and sodium hydroxide had a greater impact on terrestrial acidification, eutrophication, and global warming. Ammonia- and methanol-assisted pretreatments had the lowest global warming impact suggesting that both reagents could be used for sustainable pretreatments [91]. As a result, the search for new environmentally friendly solvents has become more intense. Deep eutectic solvents (DESs) have received much attention in this regard over the past 20 years as an eco-friendly solvent for the pretreatment of biomasses.

13.5 Cellulose-Derived Biopolymers

Cellulose is a form of a structural polysaccharide made up of D-glucose units that are commonly obtained from plants, bacterial, algal, and animal sources. Cellulose, the most accessible biopolymeric material on the planet, is considered to have the ability to be used in the production of films or coating material for foods including fruits and vegetables. In 1838, Anselme Payen, a French scientist, extracted cellulose for the very first time from plant cells [92]. In the late 1800s, Hyatt Manufacturing Company developed the world's first cellulose-based insoluble bioplastic (celluloid). Celluloids are extremely crystalline by nature and have strong intermolecular and intramolecular H-bonding making them water-insoluble. C_2, C_3, and C_6 positions of cellulose monomeric units have free hydroxyl groups, which allows cellulose to interact with a wide range of reactants, and produce a variety of cellulose derivatives. It is possible to change its solubility, water uptake properties, crystalline nature, and capacity to form films by adopting several different approaches. For example, CA, a most common and widely known cellulose derivative, has been produced by complete or partial substitution of free hydroxyl groups from the native cellulose by acetyl groups [93]. CMC is another derivative of cellulose commonly formed by the replacement of free hydroxyl groups with carboxymethyl groups. Methyl cellulose (MC) is produced by substituting free hydroxyl groups of cellulose with methyl groups, while in the case of hydroxypropyl MC (HPMC), free hydroxyl groups are replaced with hydroxypropyl methyl [94]. These cellulose derivatives are sometimes also referred to as etherified cellulose. These water-soluble derivatives of cellulose are used to produce edible polymer films with desirable properties such as transparency, odorlessness, toughness, and flexibility. These derivatives also offer resistance to oils, lipids, and fats but are merely resistant to water.

13.5.1 Carboxymethyl Cellulose (CMC)

CMC is one of the most commonly used anionic polysaccharides (polyelectrolytes) produced by the etherification process. CMC is produced through the reaction of native cellulose in an aqueous sodium hydroxide solution.

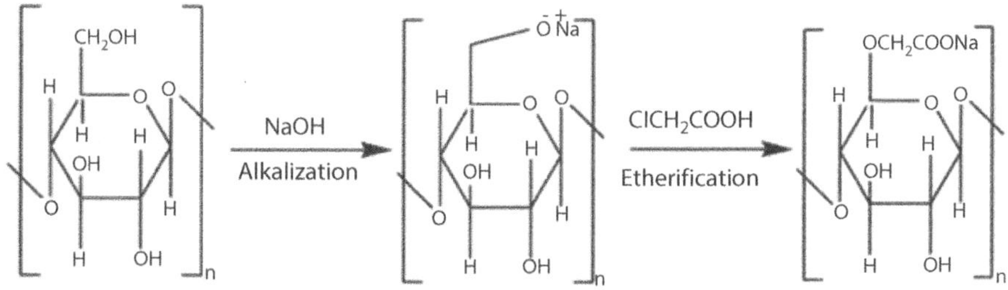

Figure 13.3 The mechanism of CMC production.

The reaction of the generated alkaline cellulose with monochloroacetic acid or sodium monochloroacetate ($C_2H_2ClNaO_2$) leads to the synthesis of CMC, and the mechanism of CMC production is discussed in Figure 13.3 [95].

It is produced by replacing the free hydroxyl group in cellulose fibers with carboxymethyl groups. The number of carboxymethyl groups attached to the cellulose after the etherification reaction will determine the degree of substitutions (DSs). The solubility of CMC depends on its DS. Although CMC with a DS less than 0.4 promotes swelling, CMC with a DS greater than 0.4 is water-soluble. CMC also has amphiphilic characteristics due to its water-insoluble polysaccharide backbone and abundance of water-soluble carboxyl groups. It can dissolve in an aqueous solvent or water due to the presence of carboxymethyl groups [94]. It has a brilliant film-forming capacity as well as it is benign, biodegradable, hydrophilic, and nontoxic. It is frequently used as a thickener, stabilizer, and flocculating and gelling agent as well as an emulsifier [96]. Furthermore, it has been designated as "Generally Regarded as Safe" by the Food and Drug Administration (USA) in 1973 [94]. It is a high molecular weight structural polysaccharide, which is ideal for the fabrication of film [97]. Depending on the extent of purity, CMC can be categorized as (i) technical, (ii) semi-purified, and (iii) purified. For example, cellulose gum is grouped under high-purity grade CMC and it is commercially used as a food additive [4]. CMC can be produced from various agricultural waste for the development of packaging material as listed in Table 13.3. Yaradoddi *et al.* (2020) extracted cellulose from sugarcane bagasse using acetic acid and sodium chlorite as a bleaching agent. The produced cellulose slurry was treated with sodium hydroxide and isopropanol, followed by etherification using sodium monochloroacetate. The fabrication of films was carried out by casting method, and the tensile strength of the developed films was in the range of 13.27 to 19.8 MPa. The maximum biodegradability of the films after 1 week was found to be 96%.

Table 13.3 Production of CMC from different agro-sources by etherification, degree of substitution (DS), and CMC yield.

Sources	Chemical used	DS	CMC yield	References
Tea leaves	Propan-2-ol, NaOH, and $C_2H_2ClNaO_2$	0.64–0.77	3.22–4.68	[100]
Palm fiber		0.54–0.6	5.91–8.13	
Bagasse		0.56–0.72	3.8–6.67	
Duckweed		0.53–0.65	5.98–6.14	
Coconut fiber		0.61–0.69	5.81–6.51	
Palm kernel cake		0.34–0.37	6.02–6.19	
Corn cobs	Propan-2-ol, NaOH, and $C_2H_2ClNaO_2$	1.18		[101]
Wheat straw	Propan-2-ol, NaOH, and $C_2H_2ClNaO_2$ (heat)	1.5	1.06	[102]
Rice hull	Propan-2-ol, NaOH, and $C_2H_2ClNaO_2$ (microwave)	0.7	0.94	[102]
Barley straw	Propan-2-ol, NaOH, and $C_2H_2ClNaO_2$ (microwave)	0.8	0.94	[102]
Palm bunch	Propan-2-ol, NaOH, $C_2H_2ClNaO_2$, C_2H_5OH, and CH_3OH	0.17–0.53	12–13.5	[103]
Elephant grass	Propan-2-ol, NaOH, and $C_2H_2ClNaO_2$	0.75	-	[104]

These properties of the films well suit their application for food packaging [98]. Gupta *et al.* (2022) utilized coconut coir and groundnut waste shells to produce cellulose followed by etherification using propan-2-ol, NaOH, and $C_2H_2ClNaO_2$ to synthesize CMC. The films were first blended with starch, olive oil, and glycerol, and the prepared solution was cast to generate films. The synthesized films from coir waste have a tensile strength of 15% and an opacity of 14%. The synthesized films from groundnut waste have a tensile strength of 18.25% and an opacity of 17.64%. According to the observation of the cited literature, CMC has the potential to replace nonbiodegradable and nonrecyclable plastic-based food packaging films due to its good mechanical and optical properties, biodegradability, and biocompatibility [99].

13.5.2 Cellulose Acetate (CA)

In 1865, Paul Schutzenberger used the wooden pulp to prepare CA through a chemical synthesis process involving acetic anhydride for the first time. CA can also be synthesized using cotton and acetic acid. During the acetylation of cellulose, acetic acid is employed as a clarifying agent. This is followed by dehydration and a reaction with acetic anhydride in the presence of a suitable chemical reagent (sulfuric acid and anhydrous acetic acid), which results in the formation of cellulose triacetate. Furthermore, it can also be hydrolyzed with acid to produce cellulose diacetate and CA, as shown in Figure 13.4.

DS has a greater bearing on the affinity of solubility and biodegradability of etherified cellulose. CA is water-insoluble, possesses good mechanical

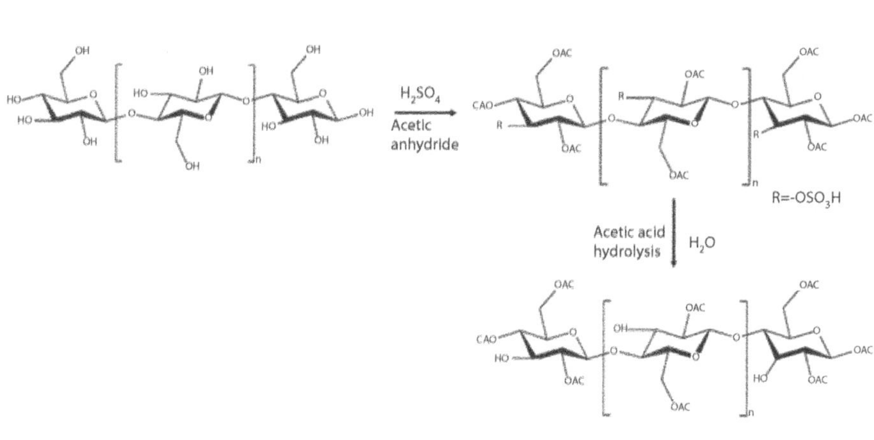

Figure 13.4 Mechanism of CMC production.

properties, and can be easily hydrolyzed. CA has a hydrophilic character and a significantly lower swelling and water retention ability due to the presence of hydroxyl groups in the cellulose backbones [4]. Several studies have been done to successfully utilize agro-waste to produce CA for film productions (Table 13.4). For example, Shaikh *et al.* (2009) developed CA-based film for food packaging from bagasse *via* acetic acid hydrolysis. The films were produced by a simple casting method using CA suspension with 14 MPa tensile strength after 4-6 h of drying [105]. Pereira *et al.* (2022) also developed CA film for bio-packaging from banana stems. Cellulose was first extracted by KOH hydrolysis followed by hydrogen peroxide bleaching. Acetylation was performed using acetic anhydride, acetic acid, and sulfuric acid, and the produced suspension was mixed with *Butia*

Table 13.4 Production of CA from different agro-waste using glacial acetic acid (GAA) and acetic anhydride (AA) with their yield and types of CA generated.

Source	Chemicals used	CA yield (%)	CA type	References
Oil palm branch	Glacial acetic acid (GAA) and acetic anhydride (AA)	52	Trihydrate	[107]
		41	Dihydrate	
Bamboo pulp	GAA and Glacial acetic acid (GA)	49	Trihydrate	[107]
		20	Dihydrate	
Raffia hookeri	GAA and GA	46	Trihydrate	[107]
		20	Dihydrate	
Musa paradisiaca	GAA and GA	38	Trihydrate	[107]
		39	Dihydrate	
Oil palm fruit fiber	GAA and GA	51	Trihydrate	[107]
		18	Dihydrate	
Sawdust	GAA and GA	57	Trihydrate	[107]
		21	Dihydrate	
Maize stem	GAA and GA	40	Trihydrate	[107]
		44	Dihydrate	

catarinensis seeds, and then cast to produce a thin film. The film also has antibacterial activities against *Staphylococcus aureus* and *E. coli*. The tensile strength and contact angle of films were in the range of 25 to 30 MPa and 96° to 101°, respectively [106]. The conversion of agricultural waste into more valuable products like CA can help in minimizing environmental pollution. It also has the potential to replace plastic-based packaging film due to its excellent mechanical properties, transparency, and simple fabrication methods.

13.5.3 Cellulose Nanofiber (CNF)

Microfibrilliated cellulose was created for the first time in the late 1970s and was patented by Turbak *et al.* in 1983 [108]. In 1995, researchers from the Centre de Recherches sur les Macromolécules Végétales in France produced cellulose from Tunicata (highly crystalline) at the nanoscale for the first time. They utilized it as a reinforcing component in latex. Zimmermann *et al.* (2004) isolated and published the first nanosized cellulose from wood pulp [109]. For the past 18 years, cellulose nanofibers have been a popular biodegradable, sustainable, and renewable biomaterial for research and development in different fields such as biomedical applications, food packaging, reinforcing materials, composite formulation, biosensors, electronics, papermaking, and adsorbents. A variety of agricultural wastes, including bamboo, rice straw, bagasse, flax, palm fruit waste, palm frond rachis, cotton, corn stalk, jute, soybean, and hemp, have also been used in the production of CNFs (Table 13.5). The CNFs that are typically produced mechanically from the natural fibers have already been treated through the physiochemical pretreatment method as mentioned in Figure 13.5.

In the mechanical treatment procedure, greater shear stress is produced, which causes CNFs to be extracted from the pretreatment fibers in a longitudinal orientation. However, this manufacturing process is highly energy-intensive, resulting in fiber delamination. The chemical pretreatment procedure can cut energy use by up to 98% [110]. High-pressure homogenization, micro-fluidization, ball milling, grinding, cryo-crushing, steam explosion, and ultrasonic treatments are some of the most common techniques used to produce CNFs [111]. Chemical processes like treatment with NaOH, H_2SO_4, TEMPO oxidation, phosphorylation, and ILs were also used to produce CNFs. Pennells *et al.* (2021) produced CNFs (nano paper, 6 nm to 5.4 m) from sorghum waste using a physiochemical process that included treating the material with NaOH and then homogenizing it under high pressure [112]. Amoroso *et al.* (2022) recently developed carrot-based CNF for

Table 13.5 Production of CNF from a different source of agricultural biomass using different treatment methods, their yields, and dimensions.

Source	Method	Dimension (nm)	Yield (%)	References
Sugarcane bagasse	Steam explosion and high-pressure homogenization	3–7	-	[73]
Coconut shell	Ball milling, ultrasonication, and NaOH	27	41.67	[114]
Pine apple leaves	Micro-fluidization and TEMPO oxidation	16–48	45.3	[115]
Sugarcane bagasse	Milling, bleaching, and phosphorylation	3.84	84.3	[116]
Alfa plant	NaOH and $NaClO_2$ treatment and TEMPO oxidation	1.4–4.6	40	[117]
Water hyacinth	KOH and $NaClO_2$ treatment	10–30	~43	[118]
Water hyacinth	No treatment	10–30	~83	[118]
Water hyacinth	Mild KOH	10–30	~50	[118]
Sugar palm	High-pressure homogenization	5.5	92	[119]
Sago pith	Homogenization, ultrasonication, and NaOH	15–30	-	[120]

food packaging through alkali treatment, high shear mixing, and micro-fluidization processes. The resulting film had an excellent oxygen transfer rate (44 cm^3 $m^{-2}day^{-1}$) as well as transmittance (86%) and tensile strength (125 MPa). The produced film was also used to package bananas, causing enzymatic browning to be delayed by a week [113]. Arun *et al.* (2022)

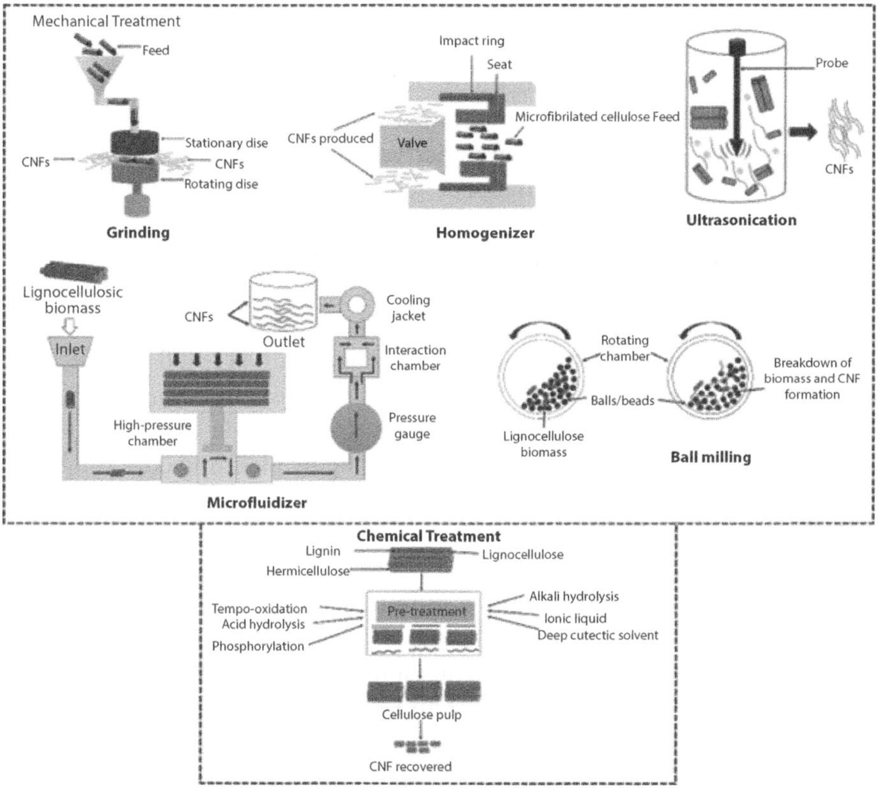

Figure 13.5 The different physical and chemical treatment processes for CNF production.

recently developed cellulose nanofiber from the waste of coconut shells, which were reinforced with polyvinyl alcohol to produce a thin and transparent film for food packaging. The CNFs were produced by the NaOH pretreatment method followed by bleaching with hydrogen peroxide, and the yield was 41.67%. The film had 43.4% transmittance, 6.35-MPa tensile strength, 31.5% antioxidant activities, and antibacterial properties against *Salmonella typhi* and *E. coli*. It was easily biodegradable within 45 days [114].

13.6 Food Packaging Applications of Cellulose and Its Derivatives

Food packaging is one of the most popular and widely used techniques of food preservation. In the world of food packaging, natural, environmentally benign, and biologically available polymers are becoming more and more popular. One of the most prevalent and extensively used natural biopolymers in the packaging sector is cellulose and its derived products. Some more recent studies on the development of food packaging material from cellulose have been mentioned in Table 13.6. Figure 13.6 depicts the production of cellulose and its derivatives through the amalgamation of various biomaterials and nanoparticles, as well as the film production method and its properties. Nanoparticles such as silver, copper, and zinc on blending with cellulose and its derivatives exhibit antibacterial properties. For example, when silver nanoparticles come into contact with water, they generate Ag^{2+}, which kills bacteria [121]. Cellulose and its derivatives on blending with curcumin [122], anthocyanin [123], β-cyclodextrin [124], sodium lignosulfonate [125], carvacrol [126], and cinnamaldehyde [15] showed great antibacterial activities. Recently, Ma *et al.* (2021) created a cellulose-based film for meat packaging by blending curcumin. The created film had an excellent elastic modulus (219.7 MPa), tensile strength (19 MPa), and transparency (85 to 90%). Additionally, the film demonstrated O_2 permeability of up to 90,029 cm^3 day^{-1} and photoinduced antibacterial properties (95% against *S. aureus*) [122].

You *et al.* (2022) created a CMC-based composite for fish packaging by combining blackcurrant anthocyanins and konjac glucomannan (tilapia). The developed film had good tensile and elongation properties, with values of 38.4 MPa and 3.67%, respectively. The films also had excellent water permeability 2.53 $g{\cdot}mm/m^2{\cdot}day{\cdot}kPa$, UV blocking abilities, and can extend fish shelf life by up to 5 days [123]. Zhang *et al.* (2021) created a CA-based film for beef packaging by coating it with cinnamaldehyde, and the developed film had an excellent tensile strength of 88.2 MPa as well as antioxidant properties. The developed film's water vapor permeability and oxygen transfer rate were 1.2×10^{-12} to 1.9×10^{-12} $g{\cdot}m/m^2{\cdot}s{\cdot}Pa$ and 2.6 to 3 $cm^3/m^2.day.0.1MPa$, respectively, and extended the shelf life of beef by 4 to 5 days [15]. Wen *et al.* (2021) developed TEMPO-oxidized bacterial cellulose-based film as well as composites with the incorporation of anthocyanin and thymol for the packaging of shrimp. The film was tested for water vapor permeability, in which TEMPO-oxidized bacterial cellulose film (6.10×10^{-11} $g{\cdot}m{\cdot}s^{-1}$ 1 Pa^{-1}) exhibited superior barrier properties than the composite one (5.20×10^{-11} g m^{-1} s^{-1} Pa^{-1}). The shelf life of the shrimps was also extended up to 48 hours using TEMPO-oxidized bacterial cellulose/anthocyanin/thymol film [127]. Karkhanis *et al.* (2021) synthesized a CNC composite film with poly(lactic acid) for the packaging of saltine crackers. When the shelf life of the cracker was tested using both poly(lactic acid) film and CNC/poly(lactic acid) film, the CNC/poly(lactic acid) film outperformed (86 hours) the poly(lactic acid) film (63.4 hours). The water vapor permeabilities of the CNC poly(lactic acid)

Table 13.6 Application of cellulose and its derivatives for packaging of different food items and their effect on the shelf life of food.

Composition of the cellulosic-based packaging materials	Food items packed	Effects on shelf life	References
Probiotic carboxymethyl cellulose-sodium caseinate films	Trout fillets	Shelf life up to 2 weeks	[131]
Cellulose nanocrystals and chitosan nanoparticles	Mango fruits	Shelf life of 20 days	[132]
Hydroxyethyl cellulose	Cheese	Shelf life extended from 28 to 56 days	[133]
Polyvinyl pyrrolidone-carboxymethyl cellulose-bacterial cellulose-guar gum	Fresh berries	Shelf life extended up to 15 days	[134]
zinc oxide-chitosan-cellulose acetate phthalate	Black grapefruits	Shelf life increased by 9 days	[135]
Carboxymethyl cellulose	Chicken fillets	Increase in shelf life	[136]
Hydroxy propyl methyl cellulose-gallic acid	Walnut	Active packaging material decreased the oxidation of walnuts and increased the shelf life	[137]
Polylactic acid-cellulose nanoparticle-propolis ethanolic extract-essential oil	Beef	Shelf life extended by 11 days	[138]
Chitosan-carboxymethyl cellulose	camel meat	Shelf life increased and the growth of *E. coli* inhibited	[139]
Cellulose nanocrystals-carboxymethyl cellulose	Red chillies	Shelf life increased and weight loss reduced	[140]
Nisin-dialdehyde cellulose	Pork	Shelf life of fresh pork extended by 6 days	[141]

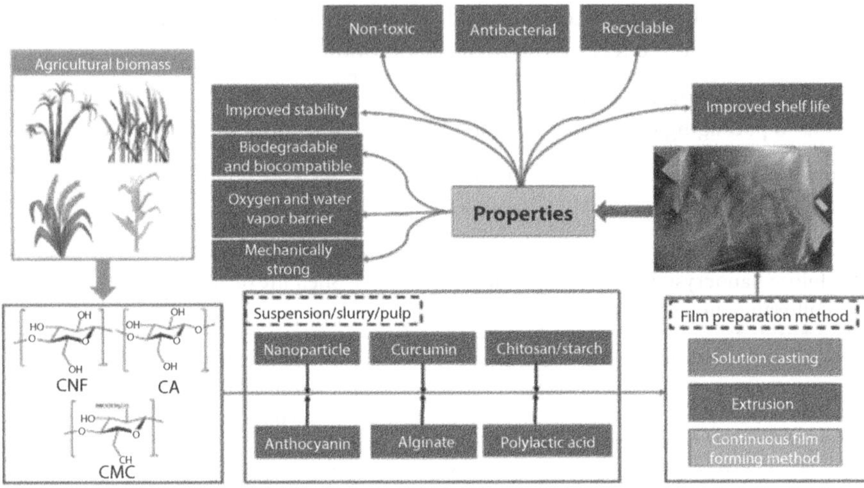

Figure 13.6 The production of cellulose and its derivatives through the blending of various biomaterials and nanoparticles, as well as the film production method and its properties.

and poly(lactic acid) films were 341 × 10^{-16} and 200 × 10^{-16} $kg/m^2/s/Pa$, respectively [128]. Francisco *et al.* (2020) developed a hydroxyethyl cellulose–based composite film by mixing acetylated cassava starch. The film (100% hydroxyethyl cellulose or 75% hydroxyethyl cellulose + 25% cassava starch) was used for coating guava, which increased the shelf life of the fruit by 13 days. The water vapor transmission rate also increased on increasing the concentration of hydroxyethyl cellulose and ranges between 12.84 and 18.94 $g\ h^{-1}\ m^{-2}$ [129]. Ruan *et al.* (2019) developed a composite film for pork packaging that included alginate, CMC, and an epigallocatechin gallate coating. The newly developed film increased the shelf life of pork by 7 days [130].

The studies mentioned above lead to the conclusion that cellulose and its derivatives have enormous potential for food packaging and can expand their global market in the years to come. It can also be used as an alternative to nonbiodegradable, nonrenewable, and nonrecyclable petroleum-based packaging materials.

13.7 Conclusion and Perspective

Synthetic plastic–based packaging materials are of great concern and threaten the environment and human lives. Synthetic plastic films are nonbiodegradable and nonrecyclable, which are simply dumped, landfilled, or burned, contributing to air, water, and soil pollution. Synthetic plastic-wrapped foods are the major source of micro and macro plastics inside humans and animals responsible for many diseases. Biomaterials like cellulose, starch, pectin, and their composites can be the best suitable alternatives to synthetic plastics for the development of food packaging materials. Especially cellulose and derived cellulose and their composites extracted from low cost, renewable, biodegradable, biocompatible, recyclable, and abundantly available agro-waste are the best choice for food packaging applications. However, some bottlenecks should be minimized, including the hydrophilic nature of the cellulose, the method of film production, the selection of proper raw material, and the use of chemicals and energy in the delignification and pretreatment process. Cellulose has excellent film-forming properties and can be easily blended with other biopolymers to generate a composite film with improved functional properties. However, the quality and utility of the packaging materials need further improvement for their commercial applications, and as a result, more research and development are required to meet the industrial targets.

References

1. Wang, J., Euring, M., Ostendorf, K., Zhang, K., Biobased materials for food packaging. *J. Bioresour. Bioprod.*, 7, 1, 1–13, 2022 Feb 1.
2. Gaikwad, K.K., Singh, S., Ajji, A., Moisture absorbers for food packaging applications. *Environ. Chem. Lett.*, 17, 2, 609–28, 2019 Jun 1.
3. Li, S. and Lee, P.S., Development and applications of transparent conductive nano-cellulose paper. *Sci. Technol. Adv. Mater.*, 18, 1, 620–33, 2017 Dec 31.
4. Liu, Y., Ahmed, S., Sameen, D.E., Wang, Y., Lu, R., Dai, J. *et al.*, A review of cellulose and its derivatives in biopolymer-based for food packaging application. *Trends Food Sci. Technol.*, 112, 532–46, 2021 Jun 1.
5. Moradian, M., Islam, Md.S., van de Ven, T.G.M., Insoluble regenerated cellulose films made from mildly carboxylated dissolving and Kraft pulps. *Ind. Eng. Chem. Res.*, 60, 15, 5385–93, 2021 Apr 21.
6. Priyadarshi, R., Kim, S.M., Rhim, J.W., Carboxymethyl cellulose-based multifunctional film combined with zinc oxide nanoparticles and grape seed extract for the preservation of high-fat meat products. *Sustain. Mater. Technol.*, 29, e00325, 2021 Sep 1.
7. Fourati, Y., Magnin, A., Putaux, J.L., Boufi, S., One-step processing of plasticized starch/cellulose nanofibrils nanocomposites via twin-screw extrusion of starch and cellulose fibers. *Carbohydr. Polym.*, 229, 115554, 2020 Feb 1.
8. Chen, J., Wang, X., Long, Z., Wang, S., Zhang, J., Wang, L., Preparation and performance of thermoplastic starch and microcrystalline cellulose for

packaging composites: Extrusion and hot pressing. *Int. J. Biol. Macromol.*, 165, 2295–302, 2020 Dec.

9. Trigui, K., De Loubens, C., Magnin, A., Putaux, J.L., Boufi, S., Cellulose nanofibrils prepared by twin-screw extrusion: Effect of the fiber pretreatment on the fibrillation efficiency. *Carbohydr. Polym.*, 240, 116342, 2020 Jul 15.
10. Leite, L.S.F., Ferreira, C.M., Corrêa, A.C., Moreira, F.K.V., Mattoso, L.H.C., Scaled-up production of gelatin-cellulose nanocrystal bionanocomposite films by continuous casting. *Carbohydr. Polym.*, 238, 116198, 2020 Jun 15.
11. Droguet, B.E., Liang, H.L., Frka-Petesic, B., Parker, R.M., De Volder, M.F.L., Baumberg, J.J. *et al.*, Large-scale fabrication of structurally coloured cellulose nanocrystal films and effect pigments. *Nat. Mater.*, 21, 3, 352–8, 2022 Mar.
12. Saravanakumar, K., Sathiyaseelan, A., Mariadoss, A.V.A., Xiaowen, H., Wang, M.H., Physical and bioactivities of biopolymeric films incorporated with cellulose, sodium alginate and copper oxide nanoparticles for food packaging application. *Int. J. Biol. Macromol.*, 153, 207–14, 2020 Jun 15.
13. Gupta, A., Briffa, S.M., Swingler, S., Gibson, H., Kannappan, V., Adamus, G. *et al.*, Synthesis of silver nanoparticles using curcumin-cyclodextrins loaded into bacterial cellulose-based hydrogels for wound dressing applications. *Biomacromolecules*, 21, 5, 1802–11, 2020 May 11.
14. Meng, Y., Luo, H., Dong, C., Zhang, C., He, Z., Long, Z. *et al.*, Hydroxypropyl guar/cellulose nanocrystal film with ionic liquid and anthocyanin for real-time and visual detection of NH3. *ACS Sustain. Chem. Eng.*, 8, 26, 9731–41, 2020 Jul 6.
15. Zhang, J., Guo, Z., Chen, S., Dong, H., Zhang, X., Qin, Y. *et al.*, High-barrier, strong, and antibacterial paper fabricated by coating acetylated cellulose and cinnamaldehyde for food packaging. *Cellulose*, 28, 7, 4371–84, 2021 May 1.
16. Klemm, D., Heublein, B., Fink, H.P., Bohn, A., Cellulose: Fascinating biopolymer and sustainable raw material. *Angew. Chem. Int. Ed.*, 44, 22, 3358–93, 2005.
17. Houfani, A.A., Anders, N., Spiess, A.C., Baldrian, P., Benallaoua, S., Insights from enzymatic degradation of cellulose and hemicellulose to fermentable sugars–a review. *Biomass Bioenergy*, 134, 105481, 2020 Mar 1.
18. Kondo, T., The relationship between intramolecular hydrogen bonds and certain physical properties of regioselectively substituted cellulose derivatives. *J. Polym. Sci. Part B Polym. Phys.*, 35, 4, 717–23, 1997.
19. Verma, P., *Thermochemical and Catalytic Conversion Technologies for Future Biorefineries*, vol. 1, p. 384, Springer Nature, Springer Singapore, 2022.
20. Srivastava, D., Ahopelto, J., Karttunen, A.J., Thermodynamic properties of crystalline cellulose allomorphs studied with dispersion-corrected density functional methods. *Molecules*, 27, 19, 6240, 2022 Jan.
21. Wada, M., Nishiyama, Y., Chanzy, H., Forsyth, T., Langan, P., The structure of celluloses. *Powder Diffr.*, 23, 2, 92–5, 2008 Jun.
22. Wada, M., Chanzy, H., Nishiyama, Y., Langan, P., Cellulose IIII crystal structure and hydrogen bonding by synchrotron x-ray and neutron fiber diffraction. *Macromolecules*, 37, 23, 8548–55, 2004 Nov 1.

23. Gardiner, E.S. and Sarko, A., Packing analysis of carbohydrates and polysaccharides. 16. The crystal structures of celluloses IVI and IVII. *Can. J. Chem.*, 63, 1, 173–80, 1985 Jan.
24. Endo, T., Aung, E.M., Fujii, S., Hosomi, S., Kimizu, M., Ninomiya, K. *et al.*, Investigation of accessibility and reactivity of cellulose pretreated by ionic liquid at high loading. *Carbohydr. Polym.*, 176, 365–73, 2017 Nov 15.
25. Dhar, P., Tarafder, D., Kumar, A., Katiyar, V., Effect of cellulose nanocrystal polymorphs on mechanical, barrier and thermal properties of poly(lactic acid) based bionanocomposites. *RSC Adv.*, 5, 74, 60426–40, 2015.
26. Timell, T.E., Recent progress in the chemistry of wood hemicelluloses. *Wood Sci. Technol.*, 1, 1, 45–70, 1967 Mar 1.
27. Kumar, R., Singh, S., Singh, O.V., Bioconversion of lignocellulosic biomass: Biochemical and molecular perspectives. *J. Ind. Microbiol. Biotechnol.*, 35, 5, 377–91, 2008 May.
28. Fares, K., Renard, C.M.G.C., Crepeau, M.J., Thibault, J.F., Characterization of hemicelluloses of sugar beet roots grown in Morocco. *Int. J. Food. Sci. Technol.*, 39, 3, 303–9, 2004.
29. Zhou, X., Li, W., Mabon, R., Broadbelt, L.J., A critical review on hemicellulose *pyrolysis. Energy Technol.*, 5, 1, 52–79, 2017.
30. Rakotoarivonina, H., Hermant, B., Aubry, N., Rabenoelina, F., Baillieul, F., Rémond, C., Dynamic study of how the bacterial breakdown of plant cell walls allows the reconstitution of efficient hemicellulasic cocktails. *Bioresour. Technol.*, 170, 331–41, 2014 Oct.
31. Verbelen, J.P. and Vissenberg, K., *The expanding cell*, 302 p., Springer Science & Business Media, Harvard Springer-Verlag Berlin Heidelberg, 2006.
32. Shen, D.K., Gu, S., Bridgwater, A.V., Study on the pyrolytic behaviour of xylan-based hemicellulose using TG–FTIR and Py–GC–FTIR. *J. Anal. Appl. Pyrolysis*, 87, 2, 199–206, 2010 Mar 1.
33. Carnachan, S.M. and Harris, P.J., Polysaccharide compositions of primary cell walls of the palms Phoenix canariensis and Rhopalostylis sapida. *Plant Physiol. Biochem.*, 38, 9, 699–708, 2000 Sep 1.
34. Pauly, M., Gille, S., Liu, L., Mansoori, N., de Souza, A., Schultink, A. *et al.*, Hemicellulose biosynthesis. *Planta*, 238, 4, 627–42, 2013 Oct 1.
35. Collard, F.X. and Blin, J., A review on pyrolysis of biomass constituents:-Mechanisms and composition of the products obtained from the conversion of cellulose, hemicelluloses and lignin. *Renew. Sustain. Energy Rev.*, 38, 594–608, 2014.
36. de O. Buanafina, M.M., Feruloylation in grasses: Current and future perspectives. *Mol. Plant*, 2, 5, 861–72, 2009 Sep 1.
37. Kumar, S., Integrated Waste Management: Volume II, in: *BoD – Books on Demand*, 486 p., 2011.
38. Du, Q., Liu, J., Ding, Y., Recent progress in biological activities and health benefits of konjac glucomannan and its derivatives. *Bioact. Carbohydr. Diet. Fibre*, 26, 100270, 2021 Nov 1.

39. Encina, L., Elgueta E, L., Rivas, B., Pereira, M., Sanhueza, F., Hydrogels derived from galactoglucomannan hemicellulose with inorganic contaminant removal properties. *RSC Adv.*, 11, 57, 35960–72, 2021.
40. Singh, S., Singh, G., Arya, S.K., Mannans: An overview of properties and application in food products. *Int. J. Biol. Macromol.*, 119, 79–95, 2018 Nov 1.
41. Adair, P., Sriprom, P., Narkrugsa, W., Phumjan, L., Manamoongmongkol, K., Permana, L. *et al.*, Preparation, characterization, and antimicrobial activity of xyloglucan-chitosan film from tamarind (tamarind indica L.) seed kernel. *Prog. Org. Coat.*, 179, 107486, 2023 Jun 1.
42. Hotchkiss, A., Nunez, A., Strahan, G., Chau, H., White, A., Marais, J. *et al.*, Cranberry xyloglucan structure and inhibition of escherichia coli adhesion to epithelial cells. *J. Agric. Food Chem.*, 63, 5622–5633, 2015 May 14.
43. Renoldi, N., Melchior, S., Calligaris, S., Peressini, D., Application of high-pressure homogenization to steer the technological functionalities of chia fibre-protein concentrate. *Food Hydrocoll.*, 139, 108505, 2023 May 1.
44. Rai, R., Kumar, V., Dhar, P., Recalcitrance of lignocellulosic biomass and pretreatment technologies: A comprehensive insight, in: *Thermochemical and Catalytic Conversion Technologies for Future Biorefineries*, vol. 1, P. Verma (Ed.), Springer Nature, Singapore, 2022, [cited 2022 Dec 9]. p. 13–52. (Clean Energy Production Technologies). Available from: https://doi.org/10.1007/978-981-19-4312-6_2.
45. Vincken, J.P., York, W.S., Beldman, G., Voragen, A.G., Two general branching patterns of xyloglucan, XXXG and XXGG. *Plant Physiol.*, 114, 1, 9–13, 1997 May.
46. Cui, S.W. and Wang, Q., Cell wall polysaccharides in cereals: Chemical structures and functional properties. *Struct. Chem.*, 20, 2, 291–7, 2009 Apr 1.
47. Mishra, A. and Vij Malhotra, A., Tamarind xyloglucan: A polysaccharide with versatile application potential. *J. Mater. Chem.*, 19, 45, 8528–36, 2009.
48. Izydorczyk, M.S., Macri, L.J., MacGregor, A.W., Structure and physicochemical properties of barley non-starch polysaccharides — I. Water-extractable β-glucans and arabinoxylans. *Carbohydr. Polym.*, 35, 3, 249–58, 1998 Mar 1.
49. Amini Sarteshnizi, R., Hosseini, H., Bondarianzadeh, D., Colmenero, F.J., Khaksar, R., Optimization of prebiotic sausage formulation: Effect of using β-glucan and resistant starch by D-optimal mixture design approach. *LWT - Food Sci. Technol.*, 62, 1, Part 2, 704–10, 2015 Jun 1.
50. Abebaw, N. and Baye, B., Extraction and characterization of Hibiscus Macrantus bast fibers as a reinforcement material for composite application. *Polym. Bull.* 80.7, 7407–7427, [Internet], 2022 Aug 1, [cited 2022 Dec 31]; Available from: https://doi.org/10.1007/s00289-022-04404-x.
51. Long, H., Gu, J., Jiang, J., Guan, L., Lin, X., Zhang, W. *et al.*, Mechanically strong and biodegradable holocellulose films prepared from Camellia oleifera shells. *Carbohydr. Polym.*, 299, 120189, 2023 Jan 1.

52. Wang, T., Jung, J., Zhao, Y., Isolation, characterization, and applications of holo-cellulose nanofibers from apple and rhubarb pomace using eco-friendly approach. *Food Bioprod. Process.*, 136, 166–75, 2022 Nov 1.
53. Rao, J., Lv, Z., Ding, Q., Chen, G., Hao, X., Bian, J. *et al.*, Rapid processing of holocellulose-based nanopaper toward an electrode material. *ACS Sustain. Chem. Eng.*, 9, 8, 3337–46, 2021 Mar 1.
54. Li, Y., Chen, Y., Liu, Y., Zhang, C., Qi, H., Holocellulose nanofibrils assisted exfoliation to prepare MXene-based composite film with excellent electromagnetic interference shielding performance. *Carbohydr. Polym.*, 274, 118652, 2021 Nov 15.
55. Wu, H., Xu, C., Li, J., Guan, Y., Mao, J., Kong, F. *et al.*, Development and properties of wheat straw nano-holocellulose and reduced graphene oxide composite films for active packaging materials. *Ind. Crops Prod.*, 189, 115816, 2022 Dec 1.
56. Souza, E.F., Furtado, M.R., Carvalho, C.W.P., Freitas-Silva, O., Gottschalk, L.M.F., Production and characterization of Gluconacetobacter xylinus bacterial cellulose using cashew apple juice and soybean molasses. *Int. J. Biol. Macromol.*, 146, 285–9, 2020 Mar 1.
57. Aswini, K., Gopal, N.O., Uthandi, S., Optimized culture conditions for bacterial cellulose production by Acetobacter senegalensis MA1. *BMC Biotechnol.*, 20, 1, 46, 2020 Aug 26.
58. Chen, S., Sun, S., Zhong, C., Wang, T., Zhang, Y., Zhou, J., Bioconversion of ligno-cellulose and simultaneous production of cellulase, ligninase and bioflocculants by Alcaligenes faecalis-X3. *Process Biochem.*, 90, 58–65, 2020 Mar 1.
59. Nicolas, W.J., Ghosal, D., Tocheva, E.I., Meyerowitz, E.M., Jensen, G.J., Structure of the bacterial cellulose ribbon and its assembly-guiding cytoskeleton by electron cryotomography. *J. Bacteriol.*, 203, 3, e00371–20, 2021 Jan 11.
60. Ahmed, J., Gultekinoglu, M., Edirisinghe, M., Bacterial cellulose micro-nano fibres for wound healing applications. *Biotechnol. Adv.*, 41, 107549, 2020 Jul 1.
61. Blanco Parte, F.G., Santoso, S.P., Chou, C.C., Verma, V., Wang, H.T., Ismadji, S. *et al.*, Current progress on the production, modification, and applications of bacterial cellulose. *Crit. Rev. Biotechnol.*, 40, 3, 397–414, 2020 Apr 2.
62. Cazón, P. and Vázquez, M., Bacterial cellulose as a biodegradable food packaging material: A review. *Food Hydrocoll.*, 113, 106530, 2021 Apr 1.
63. Xu, Y., Liu, X., Jiang, Q., Yu, D., Xu, Y., Wang, B. *et al.*, Development and properties of bacterial cellulose, curcumin, and chitosan composite biodegradable films for active packaging materials. *Carbohydr. Polym.*, 260, 117778, 2021 May 15.
64. Wen, Y., Liu, J., Jiang, L., Zhu, Z., He, S., He, S. *et al.*, Development of intelligent/active food packaging film based on TEMPO-oxidized bacterial

cellulose containing thymol and anthocyanin-rich purple potato extract for shelf life extension of shrimp. *Food Packag. Shelf Life*, 29, 100709, 2021 Sep 1.
65. Atta, O.M., Manan, S., Ul-Islam, M., Ahmed, A.A.Q., Ullah, M.W., Yang, G., Development and characterization of plant oil-incorporated carboxymethyl cellulose/bacterial cellulose/glycerol-based antimicrobial edible films for food packaging applications. *Adv. Compos. Hybrid Mater.*, 5, 2, 973–90, 2022 Jun 1.
66. dos Santos, R.M., Flauzino Neto, W.P., Silvério, H.A., Martins, D.F., Dantas, N.O., Pasquini, D., Cellulose nanocrystals from pineapple leaf, a new approach for the reuse of this agro-waste. *Ind. Crops Prod.*, 50, 707–14, 2013 Oct 1.
67. Shanmugarajah, B., Chew, I.M., Mubarak, N.M., Choong, T.S., Yoo, C., Tan, K., Valorization of palm oil agro-waste into cellulose biosorbents for highly effective textile effluent remediation. *J. Clean. Prod.*, 210, 697–709, 2019 Feb 10.
68. N, K.J., S, M.R., SB, K., R, G.R., AK, R., S, S. *et al.*, Extraction of cellulose nano-crystals from red banana peduncle agro-waste and application in environmentally friendly biocomposite film. *Polym. Compos.*, 43, 8, 4942–58, 2022.
69. Rahman MS, H., Mondal, M.I., Yeasmin, M.S., Sayeed, M.A., Hossain, M.A., Ahmed, M.B., Conversion of lignocellulosic corn agro-waste into cellulose derivative and its potential application as pharmaceutical excipient. *Processes*, 8, 6, 711, 2020 Jun.
70. Alzate-Arbeláez, A.F., Dorta, E., López-Alarcón, C., Cortés, F.B., Rojano, B.A., Immobilization of Andean berry (Vaccinium meridionale) polyphenols on nanocellulose isolated from banana residues: A natural food additive with antioxidant properties. *Food Chem.*, 294, 503–17, 2019 Oct 1.
71. Reddy, J.P. and Rhim, J.W., Extraction and characterization of cellulose microfibers from agricultural wastes of onion and garlic. *J. Nat. Fibers*, 15, 4, 465–73, 2018 Jul 4.
72. Xiao, S., Gao, R., Gao, L., Li, J., Poly(vinyl alcohol) films reinforced with nano-fibrillated cellulose (NFC) isolated from corn husk by high intensity ultra-sonication. *Carbohydr. Polym.*, 136, 1027–34, 2016 Jan 20.
73. Hongrattanavichit, I. and Aht-Ong, D., Nanofibrillation and characterization of sugarcane bagasse agro-waste using water-based steam explosion and high-pressure homogenization. *J. Clean. Prod.*, 277, 123471, 2020 Dec 20.
74. Harini, K. and Chandra Mohan, C., Isolation and characterization of micro and nanocrystalline cellulose fibers from the walnut shell, corncob and sugarcane bagasse. *Int. J. Biol. Macromol.*, 163, 1375–83, 2020 Nov 15.
75. Bamgbola, A.A., Adeyemi, O.O., Olubomehin, O.O., Akinlabi, A.K., Sojinu, O.S., Iwuchukwu, P.O., Isolation and characterization of cellulose from cashew (Anacardium occidentale L.) nut shells. *Curr. Res. Green Sustain. Chem.*, 3, 100032, 2020 Jun 1.

76. Ejaz, U., Muhammad, S., Ali, F.I., Hashmi, I.A., Sohail, M., Cellulose extraction from methyltrioctylammonium chloride pretreated sugarcane bagasse and its application. *Int. J. Biol. Macromol.*, 165, 11–7, 2020 Dec 15.
77. Rizwan, M., Gilani, S.R., Durrani, A.I., Naseem, S., Low temperature green extraction of Acer platanoides cellulose using nitrogen protected microwave assisted extraction (NPMAE) technique. *Carbohydr. Polym.*, 272, 118465, 2021 Nov 15.
78. Mankar, A.R., Pandey, A., Modak, A., Pant, K.K., Pretreatment of lignocellulosic biomass: A review on recent advances. *Bioresour. Technol.*, 334, 125235, 2021 Aug 1.
79. Rezania, S., Oryani, B., Cho, J., Talaiekhozani, A., Sabbagh, F., Hashemi, B. *et al.*, Different pretreatment technologies of lignocellulosic biomass for bioethanol production: An overview. *Energy*, 199, 117457, 2020 May 15.
80. Tsegaye, B., Balomajumder, C., Roy, P., Optimization of microwave and NaOH pretreatments of wheat straw for enhancing biofuel yield. *Energy Convers. Manage.*, 186, 82–92, 2019 Apr 15.
81. Mateo, S., Peinado, S., Morillas-Gutiérrez, F., La Rubia, M.D., Moya, A.J., Nanocellulose from agricultural wastes: Products and applications—A review. *Processes*, 9, 9, 1594, 2021 Sep.
82. Su, T., Zhao, D., Khodadadi, M., Len, C., Lignocellulosic biomass for bioethanol:Recent advances, technology trends, and barriers to industrial development. *Curr. Opin. Green Sustain. Chem.*, 24, 56–60, 2020 Aug 1.
83. Kumar, B., Bhardwaj, N., Agrawal, K., Chaturvedi, V., Verma, P., Current perspective on pretreatment technologies using lignocellulosic biomass: An emerging biorefinery concept. *Fuel Process. Technol.*, 199, 106244, 2020 Mar 1.
84. Tang, C., Shan, J., Chen, Y., Zhong, L., Shen, T., Zhu, C. *et al.*, Organic amine catalytic organosolv pretreatment of corn stover for enzymatic saccharification and high-quality lignin. *Bioresour. Technol.*, 232, 222–8, 2017 May 1.
85. Zhang, Z., Wong, H.H., Albertson, P.L., Doherty, W.O.S., O'Hara, I.M., Laboratory and pilot scale pretreatment of sugarcane bagasse by acidified aqueous glycerol solutions. *Bioresour. Technol.*, 138, 14–21, 2013 Jun.
86. Wei Kit Chin, D., Lim, S., Pang, Y.L., Lam, M.K., Fundamental review of organo-solv pretreatment and its challenges in emerging consolidated bioprocessing. *Biofuels Bioprod. Biorefin.*, 14, 4, 808–29, 2020.
87. Putrino, F.M., Tedesco, M., Bodini, R.B., de Oliveira, A.L., Study of supercritical carbon dioxide pretreatment processes on green coconut fiber to enhance enzymatic hydrolysis of cellulose. *Bioresour. Technol.*, 309, 123387, 2020 Aug 1.
88. Pasquini, D., Pimenta, M.T.B., Ferreira, L.H., Curvelo, A.A.S., Sugar cane bagasse pulping using supercritical CO2 associated with co-solvent 1-butanol/water. *J. Supercrit. Fluids*, 34, 2, 125–31, 2005 Jun 1.
89. Daza Serna, L.V., Orrego Alzate, C.E., Cardona Alzate, C.A., Supercritical fluids as a green technology for the pretreatment of lignocellulosic biomass. *Bioresour. Technol.*, 199, 113–20, 2016 Jan 1.

90. Shaikh, H.M., Anis, A., Poulose, A.M., Al-Zahrani, S.M., Madhar, N.A., Alhamidi, A. *et al.*, Isolation and characterization of alpha and nanocrystalline cellulose from date palm (Phoenix dactylifera L.) *Trunk Mesh. Polymers*, 13, 11, 1893, 2021 Jan.
91. Smullen, E., Finnan, J., Dowling, D., Mulcahy, P., The environmental performance of pretreatment technologies for the bioconversion of lignocellulosic biomass to ethanol. *Renewable Energy*, 142, 527–34, 2019 Nov 1.
92. Muqeet, M., Mahar, R.B., Gadhi, T.A., Ben Halima, N., Insight into cellulose-based-nanomaterials - A pursuit of environmental remedies. *Int. J. Biol. Macromol.*, 163, 1480–6, 2020 Nov 15.
93. He, X., Lu, W., Sun, C., Khalesi, H., Mata, A., Andaleeb, R. *et al.*, Cellulose and cellulose derivatives: Different colloidal states and food-related applications. *Carbohydr. Polym.*, 255, 117334, 2020 Nov 1.
94. Yildirim-Yalcin, M., Tornuk, F., Toker, O.S., Recent advances in the improvement of carboxymethyl cellulose-based edible films. *Trends Food Sci. Technol.*, 129, 179–93, 2022 Nov 1.
95. Wang, Y., Pillai, P.K.S., Nickerson, M.T., Effect of molecular mass and degree of substitution of carboxymethyl cellulose on the formation electrostatic complexes with lentil protein isolate. *Food Res. Int.*, 126, 108652, 2019 Dec 1.
96. Arancibia, C., Navarro-Lisboa, R., Zúñiga, R.N., Matiacevich, S., Application of CMC as thickener on nanoemulsions based on olive oil: Physical properties and stability. *Int. J. Polym. Sci.*, 2016, e6280581, 2016 Oct 18.
97. Almasi, H., Ghanbarzadeh, B., Entezami, A., Physicochemical properties of starch-CMC-nanoclay biodegradable films. *Int. J. Biol. Macromol.*, 46, 1–5, 2009 Oct 1.
98. Yaradoddi, J.S., Banapurmath, N.R., Ganachari, S.V., Soudagar, M.E.M., Mubarak, N.M., Hallad, S. *et al.*, Biodegradable carboxymethyl cellulose based material for sustainable packaging application. *Sci. Rep.*, 10, 1, 21960, 2020 Dec 15.
99. Gupta, H., Kumar, H., Gehlaut, A.K., Singh, S.K., Gaur, A., Sachan, S. *et al.*, Preparation and characterization of bio-composite films obtained from coconut coir and groundnut shell for food packaging. *J. Mater. Cycles Waste Manage.*, 24, 2, 569–81, 2022 Mar 1.
100. Huang, C.M., Chia, P.X., Lim, C.S., Nai, J.Q., Ding, D.Y., Seow, P.B. *et al.*, Synthesis and characterisation of carboxymethyl cellulose from various agricultural wastes. *Cellul. Chem. Technol.*, 51, 7–8, 665–72, 2017.
101. Singh, R.K. and Singh, A.K., Optimization of reaction conditions for preparing carboxymethyl cellulose from corn cobic agricultural waste. *Waste Biomass Valorization*, 4, 1, 129–37, 2013 Mar 1.
102. Biswas, A., Kim, S., Selling, G.W., Cheng, H.N., Conversion of agricultural residues to carboxymethylcellulose and carboxymethylcellulose acetate. *Ind. Crops Prod.*, 60, 259–65, 2014 Sep 1.
103. Suriyatem, R., Noikang, N., Kankam, T., Jantanasakulwong, K., Leksawasdi, N., Phimolsiripol, Y. *et al.*, Physical properties of carboxymethyl cellulose

from palm bunch and bagasse agricultural wastes: Effect of delignification with hydrogen peroxide. *Polymers*, 12, 7, 1505, 2020 Jul.

104. Wu, X.Q., Liu, P.D., Liu, Q., Xu, S.Y., Zhang, Y.C., Xu, W.R. *et al.*, Production of cellulose nanofibrils and films from elephant grass using deep eutectic solvents and a solid acid catalyst. *RSC Adv.*, 11, 23, 14071–8, 2021 Apr 13.
105. Shaikh, H.M., Pandare, K.V., Nair, G., Varma, A.J., Utilization of sugarcane bagasse cellulose for producing cellulose acetates: Novel use of residual hemicellulose as plasticizer. *Carbohydr. Polym.*, 76, 1, 23–9, 2009 Mar 2.
106. Pereira, N.R.L., Lopes, B., Fagundes, I.V., de Moraes, F.M., Morisso, F.D.P., Parma, G.O.C. *et al.*, Bio-packaging based on cellulose acetate from banana pseudostem and containing Butia catarinensis extracts. *Int. J. Biol. Macromol.*, 194, 32–41, 2022 Jan 1.
107. Israel, A.U., Obot, I.B., Umoren, S.A., Mkpenie, V., Asuquo, J.E., Production of cellulosic polymers from agricultural wastes. *J. Chem., NaN/NaN/NaN*, 5, 81–5, 2008.
108. Turbak, A.F., Snyder, F.W., Sandberg, K.R., Microfibrillated cellulose, a new cellulose product: Properties, uses, and commercial potential. *J. Appl. Polym. Sci. Appl. Polym. Symp.*, 37, 815–27, 1983.
109. Zimmermann, T., Pöhler, E., Geiger, T., Cellulose Fibrils for Polymer Reinforcement. *Adv. Eng. Mater.*, 6, 9, 754–61, 2004.
110. Nagarajan, K.J., Ramanujam, N.R., Sanjay, M.R., Suchart, S., Surya Rajan, B., Sathick Basha, K. *et al.*, A comprehensive review on cellulose nanocrystals and cellulose nanofibers: Pretreatment, preparation, and characterization. *Polym. Compos.*, 42, 4, 1588–630, 2021.
111. Rai, R. and Dhar, P., Biomedical engineering aspects of nanocellulose: A review. *Nanotechnology*, 33, 36, 362001, 2022 Jun 15.
112. Pennells, J., Cruickshank, A., Chaleat, C., Godwin, I., Martin, D., Sorghum as a novel biomass for the sustainable production of cellulose nanofibers. *Ind. Crops Prod.*, 171, 113917, 2021, https://doi.org/10.1016/j.indcrop.2021.113917.
113. Amoroso, L., De France, K.J., Milz, C.I., Siqueira, G., Zimmermann, T., Nyström, G., Sustainable cellulose nanofiber films from carrot pomace as sprayable coatings for food packaging applications. *ACS Sustain. Chem. Eng.*, 10, 1, 342–52, 2022 Jan 10.
114. Arun, R., Shruthy, R., Preetha, R., Sreejit, V., Biodegradable nano composite reinforced with cellulose nano fiber from coconut industry waste for replacing synthetic plastic food packaging. *Chemosphere*, 291, 132786, 2022 Mar 1.
115. Araya-Chavarría, K., Rojas, R., Ramírez-Amador, K., Sulbarán-Rangel, B., Rojas, O., Esquivel-Alfaro, M., Cellulose nanofibers as functional biomaterial from pineapple stubbles via TEMPO oxidation and mechanical process. *Waste Biomass Valorization*, 13, 3, 1749–58, 2022 Mar 1.
116. Messa, L.L., Faez, R., Hsieh, Y.L., Phosphorylated cellulose nanofibrils from sugarcane bagasse with pH tunable gelation. *Carbohydr. Polym. Technol. Appl.*, 2, 100085, 2021 Dec 25.

117. Kassab, Z., Boujemaoui, A., Ben Youcef, H., Hajlane, A., Hannache, H., El Achaby, M., Production of cellulose nanofibrils from alfa fibers and its nanoreinforcement potential in polymer nanocomposites. *Cellulose*, 26, 18, 9567–81, 2019 Dec 1.
118. Tanpichai, S., Biswas, S.K., Witayakran, S., Yano, H., Water hyacinth: A sustainable lignin-poor cellulose source for the production of cellulose nanofibers. *ACS Sustain. Chem. Eng.*, 7, 23, 18884–93, 2019 Dec 2.
119. Ilyas, R.A., Sapuan, S.M., Ibrahim, R., Abral, H., Ishak, M.R., Zainudin, E.S. *et al.*, Effect of sugar palm nanofibrillated cellulose concentrations on morphological, mechanical and physical properties of biodegradable films based on agro-waste sugar palm (Arenga pinnata (Wurmb.) Merr) starch. *J. Mater. Res. Technol.*, 8, 5, 4819–30, 2019 Sep 1.
120. Beh, J.H., Lim, T.H., Lew, J.H., Lai, J.C., Cellulose nanofibril-based aerogel derived from sago pith waste and its application on methylene blue removal. *Int. J. Biol. Macromol.*, 160, 836–45, 2020 Oct.
121. Ahmad, H., Celluloses as support materials for antibacterial agents: A review. *Cellulose*, 28, 5, 2715–61, 2021 Mar 1.
122. Ma, T., Chen, Y., Zhi, X., Du, B., Cellulose laurate films containing curcumin as photoinduced antibacterial agent for meat preservation. *Int. J. Biol. Macromol.*, 193, 1986–95, 2021 Dec 15.
123. You, P., Wang, L., Zhou, N., Yang, Y., Pang, J., A pH-intelligent response fish packaging film: Konjac glucomannan/carboxymethyl cellulose/blackcurrant anthocyanin antibacterial composite film. *Int. J. Biol. Macromol.*, 204, 386–96, 2022 Apr 15.
124. Goñi-Ciaurriz, L. and Vélaz, I., Antibacterial and degradable properties of β-cyclodextrin-TiO2 cellulose acetate and polylactic acid bionanocomposites for food packaging. *Int. J. Biol. Macromol.*, 216, 347–60, 2022 Sep 1.
125. Lu, X., Que, H., Guo, H., Ding, C., Liu, X., Qin, Y. *et al.*, α-Cellulose-based films:Effect of sodium lignosulfonate (SLS) incorporation on physicochemical and antibacterial performance. *Cellulose*, 28, 11, 7243–56, 2021 Jul 1.
126. Laroque, D.A., de Aragão, G.M.F., de Araújo, P.H.H., Carciofi, B.A.M., Active cellulose acetate-carvacrol films: Antibacterial, physical and thermal properties. *Packag. Technol. Sci.*, 34, 8, 463–74, 2021.
127. Wen, Y., Liu, J., Jiang, L., Zhu, Z., He, S., He, S. *et al.*, Development of intelligent/active food packaging film based on TEMPO-oxidized bacterial cellulose containing thymol and anthocyanin-rich purple potato extract for shelf life extension of shrimp. *Food Packag. Shelf Life*, 29, 100709, 2021 Sep 1.
128. Karkhanis, S.S., Stark, N.M., Sabo, R.C., Matuana, L.M., Potential of extrusion-blown poly(lactic acid)/cellulose nanocrystals nanocomposite films for improving the shelf-life of a dry food product. *Food Packag. Shelf Life*, 29, 100689, 2021 Sep 1.
129. Francisco, C.B., Pellá, M.G., Silva, O.A., Raimundo, K.F., Caetano, J., Linde, G.A. *et al.*, Shelf-life of guavas coated with biodegradable starch and cellulose-based films. *Int. J. Biol. Macromol.*, 152, 272–9, 2020 Jun 1.

130. Ruan, C., Zhang, Y., Sun, Y., Gao, X., Xiong, G., Liang, J., Effect of sodium alginate and carboxymethyl cellulose edible coating with epigallocatechin gallate on quality and shelf life of fresh pork. *Int. J. Biol. Macromol.*, 141, 178–84, 2019 Dec 1.
131. Mozaffarzogh, M., Misaghi, A., Shahbazi, Y., Kamkar, A., Evaluation of probiotic carboxymethyl cellulose-sodium caseinate films and their application in extending shelf life quality of fresh trout fillets. *LWT*, 126, 109305, 2020 May 1.
132. Dey, D., Dharini, V., Periyar Selvam, S., Rotimi Sadiku, E., Mahesh Kumar, M., Jayaramudu, J. *et al.*, Physical, antifungal, and biodegradable properties of cellulose nanocrystals and chitosan nanoparticles for food packaging application. *Mater. Today Proc.*, 38, 860–9, 2021 Jan 1.
133. Al-Moghazy, M., Mahmoud, M., Nada, A.A., Fabrication of cellulose-based adhesive composite as an active packaging material to extend the shelf life of cheese. *Int. J. Biol. Macromol.*, 160, 264–75, 2020 Oct 1.
134. Bandyopadhyay, S., Saha, N., Brodnjak, U.V., Sáha, P., Bacterial cellulose and guar gum based modified PVP-CMC hydrogel films: Characterized for packaging fresh berries. *Food Packag. Shelf Life*, 22, 100402, 2019 Dec 1.
135. Indumathi, M.P., Saral Sarojini, K., Rajarajeswari, G.R., Antimicrobial and biodegradable chitosan/cellulose acetate phthalate/ZnO nano composite films with optimal oxygen permeability and hydrophobicity for extending the shelf life of black grape fruits. *Int. J. Biol. Macromol.*, 132, 1112–20, 2019 Jul 1.
136. Nouri Ala, M.A. and Shahbazi, Y., The effects of novel bioactive carboxymethyl cellulose coatings on food-borne pathogenic bacteria and shelf life extension of fresh and sauced chicken breast fillets. *LWT*, 111, 602–11, 2019 Aug 1.
137. Aydogdu, A., Sumnu, G., Sahin, S., Fabrication of gallic acid loaded Hydroxypropyl methylcellulose nanofibers by electrospinning technique as active packaging material. *Carbohydr. Polym.*, 208, 241–50, 2019 Mar 15.
138. Shavisi, N., Khanjari, A., Basti, A.A., Misaghi, A., Shahbazi, Y., Effect of PLA films containing propolis ethanolic extract, cellulose nanoparticle and Ziziphora clinopodioides essential oil on chemical, microbial and sensory properties of minced beef. *Meat Sci.*, 124, 95–104, 2017 Feb 1.
139. Khezrian, A. and Shahbazi, Y., Application of nanocompostie chitosan and carboxymethyl cellulose films containing natural preservative compounds in minced camel's meat. *Int. J. Biol. Macromol.*, 106, 1146–58, 2018 Jan 1.
140. Li, H., Shi, H., He, Y., Fei, X., Peng, L., Preparation and characterization of carboxymethyl cellulose-based composite films reinforced by cellulose nano-crystals derived from pea hull waste for food packaging applications. *Int. J. Biol. Macromol.*, 164, 4104–12, 2020 Dec 1.
141. Wu, Y., Li, Q., Zhang, X., Li, Y., Li, B., Liu, S., Cellulose-based peptidopolysaccharides as cationic antimicrobial package films. *Int. J. Biol. Macromol.*, 128, 673–80, 2019 May 1.

14

Biodegradability of Biopolymers

Anamika Nayak and Debjani Dutta*

Department of Biotechnology, National Institute of Technology Durgapur, Durgapur, West Bengal, India

Abstract

Synthetic plastics are ubiquitous material that has transformed the food packaging industry. However, nonbiodegradable characteristics of conventional petroleum-based synthetic plastics are creating severe environmental problems, such as endangering aquatic life and degrading air quality. Therefore, enormous amounts of fossil-based plastics could be replaced by biodegradable plastics produced from biopolymers. The aim of this chapter is to highlight the biodegradability of different biopolymers, with a brief description of different stages of biodegradation and its possible mechanism, with particular emphasis on the factors affecting the biodegradation process. It is imperative to study the microbial as well as enzymatic degradation of biopolymers in order to revolutionize degradable biological materials without having any adverse environmental effects.

Keywords: Biodegradability, biopolymer, food packaging, life cycle

14.1 Introduction

Biopolymers are the material obtained from living organisms, whereas bioplastics are synthesized utilizing biodegradable polymers. The biodegradability of the polymers and their source are the two basic criteria, based on which biopolymers or bioplastics are classified. Globally, approximately 140 million tonnes of petroleum-derived synthetic polymers are manufactured each year with a notable proportion of these polymers being disposed into the environment as waste/refuse [1]. The majority

**Corresponding author*: debs_2000in@yahoo.com

Santosh Kumar, Avik Mukherjee and Vimal Katiyar (eds.) Agro-Waste Derived Biopolymers and Biocomposites: Innovations and Sustainability in Food Packaging, (389–420) © 2024 Scrivener Publishing LLC

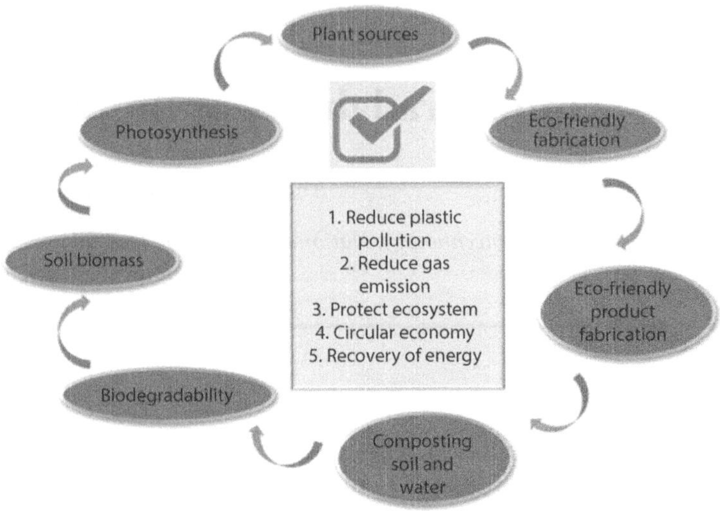

Figure 14.1 Advantages of biopolymers and their life cycle.

of common synthetic plastics, including polyethylene, polystyrene, polypropylene, and polyethylene terephthalate, are unable to microbial degradation, and they persist in nature for a longer duration [2]. Worldwide, plastics are utilized for various purposes and generate about 57 million tonnes of waste yearly [3]. Biodegradable polymers are also known to be "photo-degradable", "bio-erodible", or "hydro biodegradable" and fall under "environmentally degradable polymers". The "biodegradability" of a polymer depends not only on the raw material used for its production but also on the constitution of the end product and its chemical structure. The life cycle of biodegradable polymer is shown in Figure 14.1. Currently, two strategies are being investigated to reduce the environmental impact of synthetic polymers, i.e., firstly, the design of polymeric materials for long-term use which must combine inalterably and be made primarily from renewable resources [4–7], and secondly, technological advancement designed to produce polymers for a short duration which must be fast biodegradable [8–11].

14.2 Biodegradability of Traditional Food Packaging

The design of packaging material has a significant impact on the quality of the food product, and the traditionally used food packaging materials are mostly made up of paper, glass, metals (such as foils, aluminum, steel, and tinplate), and synthetic polymers. Nowadays, advancement in the field of food packaging research combines different materials to get the functional or aesthetic features of each material in the composite packaging. Commonly used thermoplastic food packaging includes polypropylene (PP) and polyethylene terephthalate (PET). PET is highly resistant to degradation in the environment and decomposes in tiny fragments that can last for up to 450 years [12, 13]. However, in a study, *Ideonella sakaiensis* 201-F6 bacterium degraded PET into non-hazard monomers utilizing two enzymes: mono(2-hydroxyethyl) terephthalic acid hydrolase (MHETase) and PET hydrolase (PETase). These two enzymes could be exploited in future technical applications for PET biodegradation systems [14]. By contrast, PET is the only plastic that can be recycled at a high rate without contaminating the environment [15].

Glass is nonbiodegradable; however, it can be recycled by mixing the small pieces of the glass or cullet into raw materials such as silica during manufacturing of a new glass [16]. The recycling of glass has been found to produce 0.529 kg CO_2 eq./kg glass based on the evaluated system boundaries, as reported by Otto *et al.* in 2016. In comparison to the virgin raw approach for glass manufacturing, this method yields a reduction of 40% in carbon dioxide emissions. According to a study, the reusability of a glass bottle can extend up to 50 times, leading to a significant reduction of 85% in the global warming potential [17]. Because glass does not decompose, a composting method is not possible [18]. Tinplate and aluminum are mostly used in food packaging applications. Aluminum is one of the safest packaging materials, although it is affected by the acidity of the food, and therefore, it is coated with plastic and alloyed with steel [19]. Production of one tin bottle results in emissions with a global warming potential of 0.245 kg CO_2 equivalent/0.093 kg [20]. Metal corrodes by microbial biofilm-induced electrochemical process, but it requires 200 years to degrade one aluminum can [19]. Paper and paperboard are comprised of plants and recycled cardboard and paper waste. The carbon footprint associated with the production of 0.03 kg of corrugated board utilized for butter packaging is 0.14 kg CO_2 eq. [21], 0.032 kg CO_2 eq. [22], or 0.38 kg CO_2 eq. [23]. A paper carton degrades over a period of 3 months, whereas a cardboard box degrades over a period of 2 months, which is equal to the degeneration of an apple core [13]. The extensive conversion of C to CO_2 in impregnated paper products demonstrated that they are totally biodegradable. The net CO_2 evolution also revealed that the plasticized cardboard components had undergone substantial biodegradation.

14.3 Biodegradability of Biopolymers

14.3.1 Cellulose-Based Biopolymers

Cellulose generally exists in crystalline form; the majority of cellulolytic microorganisms are fungi and eubacteria [24, 25]. These organisms can coexist harmoniously with non-cellulolytic species and together can decompose cellulose completely [25]. The hydrolytic systems of microorganisms produce cellulases, which belong to the glycoside hydrolases, and are the most important enzyme for extracellular cellulose degradation *via* the breaking down of the glycosidic bonds [26]. These two are the different types of cellulases, i.e., endoglucanases (EGs) and cellobiohydrolases (CBHs). EGs hydrolyze internal cellulose bonds in amorphous cellulose and form new end groups, whereas CBHs exclusively react with the end groups [27]. Two well-reported fungi that degrade cellulose are *Phanerochaete chrysosporium* and *Trichoderma reesei* [25, 26]. The fungi *Laetisaria arvalis* produces both hydrolytic enzymes and LPMOS and is thus more effective in cellulose degradation [28]. Additionally, Endoglucanases of thermophilic fungi have optimal activity in the pH range between 5 and 55 and temperature of 55 and 80°C. Other cellulose-degrading bacteria are *Streptomyces*, *Pseudomonas*, and *Cellulomonas* [25].

Cellulose is a biopolymer that undergoes rapid biodegradation [29]. The mean cellulose residence time was reported to be 81 to 495 days in temperate soil, and 31 to 61 days in tropical forest soil [30]. According to a study conducted under standard composting conditions, approximately 97±7% of cellulose was mineralized in laboratory settings after 47 days of incubation [31]. According to the ASTMD-5338-92 protocol, there is a cellulose mineralization rate of over 70% [32]. A study reported that the cellulose degradation in spruce needle litter occurred after a span of several years, as evident by the increase in cellulase activity within 1 to 2 years [33].

14.3.2 Starch-Based Biopolymers

Starch is a readily available energy source for microbes, and it is made up of linear amylose chains and branched amylopectin, both of which are composed of glucose molecules, linked by glycosidic linkages [24, 34]. Starch has a granular form and alternates between amorphous and semicrystalline layers at a higher structural level [35]. Glycoside hydrolases such as α-amylase facilitate the degradation of starch through the process of hydrolyzing glycosidic bonds, producing shorter fragments. Subsequently, a cascade of enzymes, including β-amylase, glucoamylase, and α-glucosidase, act upon these fragments by breaking down the α-glycosidic bonds through hydrolysis [36]. Several types of starch-degrading enzymes are produced by microorganisms including *Bacillus circulans, Klebsiella pneumoniae, Bacillus stearothermophillus,* and the fungus *Aspergillus oryzae* [26]. The copper enzyme LPMOS can catalyze oxidative cleavage of glucose linkages. The gelatinization process renders starch more amenable to biodegradation by disrupting intermolecular hydrogen bonding; therefore, the granular morphology of the material is compromised, leading to an augmentation in the number of amorphous regions that are susceptible to degradation.

The degradation rates of starch frequently conform to an asymptotic pattern, with a considerable proportion of starch undergoing degradation within a span of 3 days after its introduction into the soil. Cheshire *et al.* (1969) studied that the starch underwent complete degradation within a period of 35 days under laboratory conditions when subjected to soil incubation [37]. The study was implemented by the ASTM D5338-92 protocol to examine typical composting conditions and the results indicated that the starch underwent complete mineralization after an incubation period of 44 days at 58°C. Shahbaz (2017) observed a correlation between starch concentration and its breakdown, noting a 50% reduction when the starch content is less than 10% of the soil organic matter (SOM) [38]. The decrease in the decomposition rate is attributed to the reduction in enzymatic activity because when there is a limited amount of starch present the production of extracellular enzymes by microbes becomes too difficult compared to the energy yield obtained through metabolic processes.

14.3.3 Protein-Based Biopolymers

Protein-based biopolymers are made up of zein, gluten, whey protein, soy protein, maize protein, and gelatin and have attracted considerable interest because of their superior film-forming potential, cost-effectiveness, and biodegradability [39, 40]. Because of the unique structure of proteins, particularly their strong intermolecular binding *via* covalent bonds, protein-based films offer a wide variety of functional capabilities that typically surpass the mechanical qualities of polysaccharide and lipid-based films [41]. Wheat gluten is a type of agricultural biopolymer comprising water-soluble monomeric 'gliadins' with molecular weights ranging between 30 and 50 kDa and a water-insoluble 'glutenin' with molecular weight up to 300 kDa. Additionally, low-molecular-weight proteins such as albumins and globulins are also present in wheat. β-lactoglobulin constitutes the majority of whey protein, contributing to 57% of the total protein, while α-lactalbumin is the second most abundant whey protein.

Soy protein comprises two predominant globular proteins, i.e., glycinin and β-conglycinin. According to Uranga *et al.* (2019), gelatin is a complex mixture of protein fragments with diverse molecular weights ranging from 100 to 300 kDa, in addition to high-molecular-weight aggregates and peptide fractions (100 kDa), which are derived from collagen present in animal skin, bones, byproducts (e.g., bovine and pork), and fish [42]. The molecular weight of gelatin is influenced by both its origin and extraction process. The biodegradation of gelatin biopolymers can be facilitated by indigenous microorganisms, releasing nontoxic organic byproducts such as carbon dioxide and water into the environment. The findings indicate that single gelatin films exhibited a notable biodegradation rate ranging from 18 to 25% within 3 days [43–45].

14.3.4 Polylactide (PLA)

PLA is an aliphatic polyester derived from renewable sources such as biomass, rice bran, corn, potato, and starch. There are three stereo forms of PLA, namely, poly(L-lactide) (L-PLA), poly(D-lactide) (D-PLA), and poly(DL-lactide) (DL-PLA). L-PLA and D-PLA are semicrystalline, while DL-PLA is relatively more amorphous. Several reports suggest that enzymes play a significant role in the degradation of PLA, while the crystallinity of PLA exhibits an inverse correlation with its degradation [46–48]. Li *et al.* (1999) reported that the rate of degradation of polylactic acid (PLA) exhibits a direct correlation with the increased concentration of L-lactate units [49]. The hydrolysis of the ester group results in the release of lactic acid molecules, which subsequently reduce its degree of polymerization. Under specific composting conditions, PLA is expected to undergo degradation into carbon dioxide and water within a period of 90 days [50]. Ghorpade *et al.* (2001) have reported that the efficient composting of PLA can be achieved by incorporating it into pre-composted yard waste in a proportion of approximately 30% by weight [50]. Ghorpade *et al.* (2001) studied the degradation patterns of polylactic acid (PLA) products utilized in agricultural practices evaluated immediately following the removal of the test specimens from the reactor [51]. As per the findings of the authors, the PLA soft film samples underwent degradation within a span of 3 weeks, whereas the PLA band and PLA rope samples underwent decay within a period of 6 weeks under composting conditions. PLA has been verified to undergo natural decomposition in soil or compost, despite its relatively lower susceptibility to breakdown in the natural environment when compared to other biodegradable aliphatic polymers like PCL [50]. According to reports, the byproducts of PLA hydrolysis can be fully consumed by bacteria or fungi [52–54]. Tokiwa *et al.* (1998) reported that the proteinase K derived from the fungus *Tritirachium album* exhibited the ability to degrade PLA [55]. Protein A and Savinase, obtained from *Bacillus subtilis* and *Bacillus lentis*, respectively, are commercially available proteases that exhibit superior capability in degrading PLA [56]. The PLA-degrading enzyme of *Amycolatopsis* strain 41 has been proven to be effective in degrading both silk and PLA plate materials [57]. *Pseudonocardiaceae* and *Amycolatopsis*, with other related genera of *Streptoalloteichus*, *Lentzea*, *Saccharothrix*, and *Kibdelosporangium* actinomycetes, belong to the phylogenetic family responsible for the degradation of PLA [58]. According to the findings of Pranamuda and Tokiwa (2001), the sole fungal strain, i.e., *T. album*, is capable of degrading PLA. Low-molecular-weight oligomer degradation was observed in both *Pseudomonas putida* and *Fusarium moniliforme* regardless of their stereoisomeric properties. However, the high-molecular polymeric form was found to be resistant to degradation by these microorganisms [59]. The incorporation of gelatin at a concentration of 0.1% (w/v) into the basal medium resulted in an increase in the rate of PLA degradation attributed to the induction of enzymes responsible for the degradation of PLA. The study revealed that proteinase K was the primary catalyst responsible for PLA degradation, as opposed to subtilisin BPN and Carlsberg, as determined by a newly developed fluorescent assay [60].

14.3.5 Polyhydroxybutyrate (PHB)

PHB is a naturally occurring polymer that is synthesized by various bacterial species as a carbon and energy storage reservoir. PHB has garnered the interest of scholars owing to its derivation from economically viable and renewable raw materials, as well as its polymerization process that necessitates moderate conditions, thereby minimizing its ecological footprint. Furthermore, degradation of PHB can result in the production of nontoxic substances under both aerobic and anaerobic environmental conditions. In 1963, Chowdhury *et al.* (1963) made the initial observation that microorganisms belonging to the *Pseudomonas*, *Bacillus*, and *Streptomyces* species are capable of degrading PHB [61]. Several aerobic and anaerobic microorganisms including *Comamonas sp.*, *Pseudomonas lemoigne*, *Variovorax paradoxus*, *Aspergillus fumigatus*, *Alcaligenes faecalis*, *Illyobacter delafieldi*, and *Acidovorax faecalis* are capable of degrading PHB [62]. Furthermore, *Pseudomonas stutzeri* and *Comamonas testosteroni* have been identified as PHB-degrading organisms in lake water and seawater. A thermotolerant strain of *Aspergillus* is also capable of decomposing 90% of PHB film following a five-day incubation period at 50°C [70]. The hydrolysis of PHB yields R-3-hydroxybutyric acid [63].

14.3.6 Polyhydroxybutyrate-Co-Valerate (PHBV)

PHB is made from renewable resources and is thus completely biodegradable. Compared to polypropylene, it possesses better physical properties for food packaging applications and is fully nontoxic. However, the most disadvantageous facts of PHB are its low break extension (6%) compared to polypropylene, low impact strength, and its decomposition at 10°C above its melting point. Therefore, PHBV has been developed by incorporating the monomers of hydroxy valerate into the polymer to compensate for PHB's poor low-impact strength. Under aerobic conditions, PHBV is completely broken down into water molecules and carbon dioxide. Both 3-hydroxybutyrate and 3-hydroxy valerate are produced *via* the extracellular degradation of PHBV [64]. When a PHBV contains 41% 3-hydroxy valerate, its degradation rate in an aqueous solution of PHA depolymerase reaches its maximum. After 14 days in aerobic sewage, a PHBV film comprised of 18% 3-hydroxy valerate demonstrates 100% weight loss, while the homopolymer (PHB) exhibits just 36% weight loss [65]. The rate of PHBV degradation is higher in anaerobic sewage than in aerobic sewage, estuarine sediment, soil, and seawater [64]. Hankermeyer *et al.* (1999) reported a 6% weight loss of PHBV after 50 weeks in a simulated landfill environment made up of shredded municipal waste [66]. To study the degradation of Biopol bottles containing 8% 3-hydroxy valerate and 92% of 3-hydroxybutyrate in Switzerland's Lake Lugano under extreme conditions (no sunshine exposure, and seasonal change in dissolved oxygen concentration) of the freshwater environment [67], steady degradation was observed with a rate of 10 to 20 mg/d. Conversely, bottles comprised of polyethylene-starch blend (13 and 9% starch) of similar shape and size exhibited no degradation even after 1 year.

14.4 Mechanisms and Pathways of Biopolymer Degradation

The term biodegradation pertains to the biochemical process of microbial decomposition of substances that leads to carbon recycling, organic substance mineralization (including CO_2, H_2O, and salts), and the generation of fresh biomass [68]. Biodeterioration refers to the degradation of biodegradable substances into minuscule fragments through the collaborative influence of microorganisms, other decomposer organisms, and/or abiotic factors [69]. The depolymerization process involves the enzymatic and free radical-mediated breakdown of polymers resulting in the formation of monomers, dimers, and oligomers produced through the action of microorganisms that leads to the gradual reduction of the molecular weight of polymeric molecules through cleavage. During assimilation, certain molecules can traverse the plasma membrane and subsequently bind to receptors located on a microbial cell, while the remaining molecule persists within the extracellular milieu and undergoes diverse modifications. The process of incorporating transported molecules into the microbial metabolism within the cytoplasm results in the generation of energy, production of fresh biomass, formation of storage vesicles, and a wide range of primary and secondary metabolites. During the mineralization stage, a variety of simple and complex metabolites, including organic acids, aldehydes, terpenes, and antibiotics, may be secreted and subsequently released into the extracellular environment. Upon complete oxidation of intracellular metabolites, various simple molecules such as carbon dioxide (CO_2), nitrogen (N_2), methane (CH_4), water (H_2O), and various salts are liberated into the surrounding environment.

14.4.1 Abiotic Degradation

Polymeric materials can undergo substantial alterations under mechanical, sunlight, thermal, and chemical stresses when subjected to environmental factors such as weather, aging, and burial. Abiotic factors primarily contribute to the degradation of the polymeric framework, thereby promoting unfavorable modifications. The abiotic parameters have the potential to act as a beneficial synergistic factor or as an initiator of the biodegradation process. Mechanical degradation is a phenomenon that arises from the application of compression, tension, and shear forces, encompassing a variety of constraints such as air and water turbulences, load-induced aging, bird-induced damage, and snow pressure. The interaction between mechanical stresses and other abiotic factors exhibits a synergistic effect in field conditions. Thermoplastic films, including low tunnel films and mulches, undergo various forms of mechanical degradation when exposed to field conditions [70].

The influence of light radiation is a critical aspect of the process of abiotic degradation. In photosensitive material, photon energy can generate unstable states. Luminescence, photoionization, thermal radiation, and fluorescence are the processes that accomplish energy transfer. Sometimes impurities present in the manufactured product affect the material's resistance involuntarily. In certain cases, intentional incorporation of photosensitive molecular structures, such as prooxidants, into the polymer framework is carried out to induce light degradation of macromolecules, which can be triggered by the light intensity and duration of exposure [71]. Polyolefin manufacturers employ this approach to enhance the degradability of plastic bags, packaging materials, agricultural films, and similar products [72]. The Norrish reactions, namely, Norrish I and Norrish II, pertain to the photodegradation of polymers *via* photoionization and chain scission, respectively. Photodegradation has the potential to induce oxidative mechanisms, crosslinking phenomena, or Norrish-type reactions [73]. Recent studies have demonstrated that the Norrish II reaction takes place during the process of photodegradation of polycaprolactone (PCL) and PLA [74].

The thermoplastic polymers undergo a phase transition from a solid to a liquid state upon reaching their melting temperature; for example, the degradation of L-PLA takes place within the temperature range of 159 to 178°C, which is contingent upon its molecular weight. The melting temperature of PHB-co-hydroxyvalerate (PHBV) varies between 137 and 169°C, depending on the proportion of hydroxy valerate present. Similarly, PHB exhibits a melting temperature of 175°C [75]. Temperature can have an impact on the organization of macromolecular frameworks. Semicrystalline polymers that are biodegradable, such as polybutylene adipate (PBA), PCL, L-PLA, and cellulose, exhibit a combination of amorphous and crystalline regions [76]. The volume and mobility of polymeric chains are influenced by the glass transition temperature (Tg). For instance, PBT has a Tg of 25°C, L-PLA has a Tg of 50°C, PBS has a Tg ranging from 10 to 45°C, and PHB has a Tg of 5°C.

Chemical transformation is another significant aspect contributing to abiotic degradation. The characteristics of macromolecules may be altered by the interaction of agrochemicals and atmospheric contaminants with polymers [77]. Oxygen is the most potent chemical agent that triggers the degradation of materials. The atmospheric forms of oxygen, namely O_2 and O_3, can cleave covalent bonds and generate free radicals. The oxidative degradation rate is influenced by the presence of unsaturated links and branched chains within the polymeric structure [78]. The oxidation can lead to the production of free radicals, which may increase photodegradation. The oxidative degradation process generates peroxyl radicals that lead to chain scissions, akin to the byproducts of Norrish reactions. Hydrolysis is another form of chemical degradation [79–81]. Polymers that possess hydrolyzable covalent bond groups, such as ester, carbamide (urea), ether, amide, urethane, and anhydride, can be cleaved by H_2O through hydrolysis.

The process of hydrolysis is influenced by various factors such as temperature, water activity, duration, and pH. The production of copolymers with regulated longevity necessitates the careful choice of monomers possessing the requisite hydrophilic characteristics [82, 83]. The chemical degradation is restricted by the crystalline domain due to its ability to hinder the diffusion of O_2 and H_2O. On the other hand, the amorphous domain facilitates hydrolytic and oxidative degradation more readily.

14.4.2 Biodeterioration and Depolymerization

Microbial growth on biopolymeric material is the root cause of biodeterioration. Superficial degradation refers to the alteration of a material's physical, mechanical, and chemical properties. The growth and development of microorganisms are impacted by the composition and properties of polymer materials. Furthermore, the distinct environmental variables, including but not limited to climate, humidity, and atmospheric pollutants, are crucial determinants [84]. Microorganisms can form consortia through a structured organization known as a biofilm, which operates synergistically to cause significant damage to various materials and generate uncomplicated molecules [85–92]. Each of these substances functions as a reservoir of growth factors that facilitate the proliferation of microorganisms [93]. Recent studies have indicated that certain bacterial species have the capacity to derive sustenance from atmospheric pollutants [94, 95]. According to Zanardini *et al.* (2000), a multitude of polymer materials has been subjected to the deposition of aliphatic, aromatic, and sulfur dioxide from urban air [95]. The presence of these adsorbed contaminants may facilitate the colonization of additional microbial species on the material. *Physical way:* The attachment of microbial species to material surfaces can be attributed to the excretion of a type of adhesive substance, which is a multifaceted matrix consisting of polysaccharides and proteins [97]. This slimy substance permeates porous materials, altering the pore size and distribution as well as the moisture and thermal transfer levels. The mucus matrix acts as a barrier between microorganisms and hazardous elements such as ultraviolet radiation and dehydration. Inside the material, filamentous microorganisms produce a mycelia structure. The size of pores increases as a result of the mechanical action of organisms that penetrate materials and cause fissures. This causes the material's durability and resistance to degrade [98].

Chemical Way: Microorganisms' extracellular polymers can function as surfactants that enhance interactions between hydrophilic and hydrophobic phases, thereby facilitating the penetration of microorganisms.

In addition, the development of microorganisms, which is encouraged by the deposition of atmospheric contaminants in the presence of slime, can accelerate biodeterioration [95, 99]. Many different forms of microbial flora that grow into materials impact chemical biodeterioration. Chemolithotrophic bacteria utilize inorganic compounds such as hydrogen sulfide, ammonia, nitrites, thiosulphates, and elemental sulfur as sources of energy and electrons. Active chemicals include *Nitrobacter* spp., *Thiobacillus* spp., and *Nitrosomonas* spp. [93, 99–101]. Chemoautotrophic species rely only on organic substrates for carbon, electrons, and energy [102]. It generates organic acids such as fumaric, oxaloacetic, gluconic, glyoxylic, glutaric, and oxalic [103]. Succinic acid, lactic acid, adipic acid, and butanediol are released during the biotic or abiotic hydrolysis of certain polymers, including PBA, PBS, and PLA [104–106]. Polymer hydrolysis is started by water infiltration and results in the formation of monomers and oligomers. Progressive deterioration forms pores and thus changes the microstructure of the polymer matrix pores, which liberates oligomers and monomers. Concomitantly, the degradation products with some acid-base characteristics alter the pH inside the pores [106]. These acids function in a variety of ways. Some may react with material constituents and accelerate surface degradation [84]. Others are capable of sequestering the matrix's cations, such as Al^{3+}, Ca^{2+}, Mn^{2+}, and Si^{4+}, to produce a stable compound. The primary contributor to biodegradation is organic acids, which have been found to be more efficient in fixing cations compared to mineral acids [99]. Furthermore, certain microorganisms, including filamentous bacteria and fungi, have the capability to elongate their mycelial structure by utilizing organic acids as a means of carbon nourishment [107]. Oxidation processes may potentially lead to chemical biodeterioration. Some fungi and numerous chemolithotrophic organisms can absorb iron from the matrix *via* oxidation. To retrieve iron atoms, they employ certain proteins embedded in cell membranes that capture siderophores or iron-chelating substances released by other bacteria [102]. In photosynthetic structures, siderophores can undergo redox reactions in the presence of oxygen. Several extracellular enzymes, particularly peroxidases, can couple the catalytic breakdown of hydrocarbons with the oxidation of cations [108–111].

Enzymatic Way: Some materials such as polyurethane, polyvinylchloride, and polyamide are considered recalcitrant polymers that are nevertheless susceptible to microbial biodeterioration [112–114]. The biosynthesis of esterase, urease, lipase, and protease are responsible for the microbial susceptibility of these polymers [84]. The hydrolysis of specific bonds by enzymes necessitates the presence of cofactors, including coenzymes synthesized by microorganisms and cations present in the material matrix [102]. Two distinct mechanisms, namely, bulk and surface erosion, can lead to the biodegradation of thermoplastic polymers [115, 116]. In bulk erosion, fragments of the entire polymer mass are lost, and bond cleavage causes a change in molecular weight. Chemicals (such as H_2O, transition metals, acids, bases, and radicals) and radiations provoke this lysis. Due to their large size, it is difficult to fit inside the matrix structure. In contrast, surface erosion causes the loss of matter while leaving the molecular weight of polymers unchanged. Bulk erosion of the polymer occurs when chemical diffusion through the material occurs more quickly than polymer bond cleavage. Bond cleavage takes place primarily at the matrix surface if it occurs faster than chemical diffusion [115, 116]. Several authors have described several polymer erosion mechanisms [117], surface erosion for aromatic-aliphatic copolyesters [81], polyanhydrides [118], and PHB [80].

14.4.3 Biofragmentation

Fragmentation is a process of lysis that is essential for the subsequent phenomenon of assimilation. Polymers with elevated molecular weight are incapable of traversing the cellular wall or cytoplasmic membrane. Hence, the amalgamation of oligomers and monomers necessitates the cleavage of multiple bonds. Scission energy can be derived from diverse origins, such as radiation, heat, physical force, living organisms, and chemical reactions. Microorganisms employ various mechanisms such as enzymatic secretion or free radical generation to break down polymers. Enzymes function as catalytic proteins through the reduction of activation energy for chemical reactions to occur. Although these proteins have remarkable specificity and wide diversity, they are prone to degradation by radiation, heat, surfactants, and so forth. Exoenzymes mostly catalyze processes at the polymer's edges, whereas endopeptidase and endoesterases carry out their catalytic action along the polymer chain. Constitutive enzymes are produced throughout cellular life without any substrates. Cells that recognize substrate and generate molecular signals produce inducible enzymes. In this instance, the establishment of the cell machinery must take place over a latent period before enzymes are produced instantaneously. The induced enzyme ceases to function when the substrate supply becomes depleted and increases with time. When enzymes are released into the extracellular environment, they can be obtained as free catalysts (i.e., soluble in aqueous or lipophilic fluids) or as enzymes attached to particulate (such as clays, SOM, and sediment). Enzymes that have been fixed frequently increase their catalytic activity. In addition, they are resistant to autocatalytic denaturation, specifically from proteases [119]. Enzymes that have been secreted can continue their function long after the producing cells have died. Following oxidoreductases, the first number specifies the type of enzymes catalyzing the specific chemical reaction: hydrolases, transferases, isomerases, lyases, and ligases.

Enzymatic Hydrolysis: Most enzymes that are involved in fragmentation are hydrolases and oxidoreductases. Amylases, cutinase, and cellulases are hydrolases that soil microorganisms easily produce to hydrolyze biopolymers (such as starch, cellulose, and cutin). To promote biodegradability, these polymers are sometimes co-extruded with polyesters in industrial composites [120, 121]. It has been discovered that several enzymes can depolymerize (co)polyesters [122–125]. Endopeptidases specifically target amide bonds, whereas lipases and esterases specifically target carboxylic bonds.

Enzymatic Oxidation: The scission reactions by specific enzymes are challenging due to the presence of crystalline regions, steric hindrances, and hydrophobic zones. Monooxygenases and di-oxygenases (also known as oxidoreductases) combine oxygen atoms to generate easily fragmentable alcohol or peroxyl groups. Peroxidases catalyze other reactions that lead to the formation of smaller molecules. Hemoproteins are enzymes containing an iron atom in a prosthetic group that can donate or absorb electrons (in their reduced or oxidized forms). Peroxidases are enzymes that catalyze reactions between peroxyl molecules and electron acceptor groups. Oxidases, the third class of oxidoreductases, are metalloproteins containing copper. Typically, ligninolytic microorganisms are responsible for their production. According to the literature, one oxidase catalyzes hydroxylation reactions, and the other catalyzes oxidation reactions [126, 127]. Lignins are natural polymers with three dimensions associated with hemicellulose and cellulose. The association provides lignin a major role in the formation of novel materials derived from lignocellulosic sources. It is because of the macromolecular framework of lignin, which is challenging for microbes to break down and only ligninolytic microorganisms can do so.

Radical Oxidation: The elevation of the polarity of the molecule is influenced by introducing hydroxyl moiety and the formation of carboxyl or carbonyl functionalities. The increased hygroscopic nature of the compound renders it more susceptible to biological degradation. Furthermore, certain oxidation occurrences facilitated by diverse enzymes yield free radicals that stimulate chain reactions, thereby expediting polymer degradation. Nonetheless, enzymatic degradation is impeded by the presence of crystalline structures and molecular networks that exhibit a high degree of organization due to the limited accessibility of their interiors [128]. The oxidative molecule H_2O_2, which exhibits high reactivity, is generated by various soil decomposers such as brown-rot fungi. It can catalyze the enzymatic biodegradation of cellulose molecules [129].

14.4.4 Assimilation and Mineralization

In assimilation, atoms of polymeric fragments integrate inside the microbial cells. These special events provide microorganisms with essential elements (i.e., oxygen, nitrogen, carbon, sulfur, and phosphorus) along with energy and electrons to form their cell structure. In assimilation, microorganisms ingest polymeric materials from the environment as their nutritional substrate and can grow and reproduce. Molecules that have been assimilated may have previously undergone biodegeneration or biofragmentation. For the purpose of assimilation, the monomers in the vicinity of microbial cells exhibit permeability across the cellular membranes. Due to certain membrane carriers, certain monomers can be easily carried into the cell. Certain molecules that are impermeable to membranes cannot be absorbed and may undergo biotransformation to generate molecules that are either assimilable or non-assimilable. Molecules that undergo transportation are subjected to oxidation processes within the cellular environment *via* catabolic pathways, which produce ATP and requisite structural constituents that are essential for cellular operations. The production of energy necessary for cellular activity, structure, and reproduction is facilitated by three fundamental catabolic mechanisms: aerobic respiration, anaerobic respiration, and fermentation. The pathways are contingent upon the microorganisms' capacity to thrive under either oxygen-rich or oxygen-deprived circumstances.

Aerobic Respiration: A wide range of microorganisms can utilize oxygen as the ultimate electron acceptor in the process of aerobic respiration. The microorganisms necessitate cell substrates that are oxygenated. The energy yield resulting from basic catabolic pathways, including glycolysis, amino acid catabolism, β-oxidation, and purine and pyrimidine catabolism, is constrained. Furthermore, the electron transport mechanisms that facilitate the conversion of oxygen to water through oxidative phosphorylation yield a greater amount of energy [130].

Anaerobic Respiration: Anaerobic respiration necessitates the involvement of multiple microorganisms as oxygen is not a viable terminal electron acceptor. Total oxidation can be achieved by modifying electron transport in membrane systems. They employ alternative electron acceptors aside from oxygen, such as NO_3^-, SO_2^{-4}, S, CO_2, Fe^{3+}, and fumarate [131]. Furthermore, this metabolic pathway yields a greater quantity of ATP molecules in comparison to incomplete oxidation.

Fermentation: Fermentation is caused in specific microorganisms due to the lack of electron transport mechanisms. The exclusive means of energy production for this organism is through fermentation, which constitutes an incomplete oxidation cycle. The cellular mechanism produces ultimate electron acceptors by utilizing internal organic compounds. Regnault's research indicates that fermentation generates organic and inorganic compounds, including CO2, acetate, lactate, ethanol, and butanediol, which are discharged into the surrounding environment [132]. Other organisms commonly utilize these compounds as sources of carbon due to their reduced energy potential. The ecotoxicological implications of mineral compounds produced by bacteria are typically negligible as they conform to biogeochemical cycles. On the other hand, microbial organic compounds that are expelled or altered may provide ecotoxic risks under certain circumstances and at certain degrees.

14.5 Biodegradation of Biopolymer-Based Food Packaging

14.5.1 Microbial Degradation

Polymers have the ability to undergo anaerobic biodegradation in sediments and landfills, aerobic biodegradation in natural environments, and a combination of aerobic and anaerobic biodegradation in soil and compost. Aerobic biodegradation yields carbon dioxide and water, while anaerobic biodegradation yields methane and water. The mineralization process necessitates the involvement of microorganisms that facilitate the degradation of polymers into their constituent monomers, which are subsequently utilized for the excretion of byproducts. Microorganisms that commence the process of polymer degradation through macromolecular breakdown exhibit heightened activity within the soil. The primary phase of polymer degradation encompasses the establishment of soil microorganisms on the polymer surface, which subsequently results in the excretion of extracellular depolymerase. The acceleration of polymer biodegradation was observed to be directly proportional to the increase in the total contact area between microbial degraders and polymers. The communities associated with biopolymers displayed a significant increase in certain fungi, primarily classified under the phylum *Ascomycota*, and exhibited a decreased level of fungal diversity when compared to the adjacent bulk soils [133–137]. These studies indicate that polymer-degrading fungi preferentially colonize and multiply on polymers [138–141]. Polymer-degrading bacterial populations are also common in bulk soil [142]. Unicellular bacteria colonized by cell division from soil polymer contact sites. As a colonization mechanism, smooth and nonpolar film surfaces may hinder the production of aqueous films of adequate thickness and continuity [143–146]. However, motile bacteria may colonize the polymer surface *via* the hyphal network (as described in the "fungal highway" notion) [147, 148].

14.5.2 Enzymatic Degradation

The process of enzymatic hydrolysis of food packaging made from biopolymers is heterogeneous in nature and is characterized by the interaction mode between the enzymes and the polymeric chain. The microorganisms present in each environment are known to synthesize depolymerase enzymes that facilitate the hydrolysis of the bonds that exist within the polymer structure. This procedure yields monomers and oligomers possessing a diminished molecular mass (roughly 600 Da), which can be harnessed by other microorganisms [149]. Under certain soil temperature and pH conditions, enzymatic hydrolysis exhibits a higher rate compared to abiotic hydrolysis [76, 150, 151]. In the biodegradation of polymers in soil, enzymatic depolymerization is widely considered to be the rate-limiting step, as evidenced by the fact that oligomers and monomers are metabolized by soil microorganisms much faster than polymers [152, 153]. Several polymer properties affect its enzymatic activity. For example, crystallinity is the most significant factor in hydrolyzability [122, 142, 154, 155]. Semicrystalline polyesters utilized in biodegradable food packaging films contain nanometer-scale amorphous and crystalline phases [156, 157]. The disordered polyester backbone has greater mobility in the amorphous phases, encouraging esterase binding and hydrolytic cleavage. Contrastingly, crystalline phases in the polyester chains impair esterase binding and hydrolysis due to strong chain-chain interactions arranged in a highly ordered manner. Several bacterial and fungal esterases have been found to hydrolyze aliphatic and aliphatic-aromatic copolyesters [55, 122, 143, 158]. Cutinase and lipase are examples of such esterases. In this context, a schematic representation of polymer degradation has been depicted in Figure 14.2.

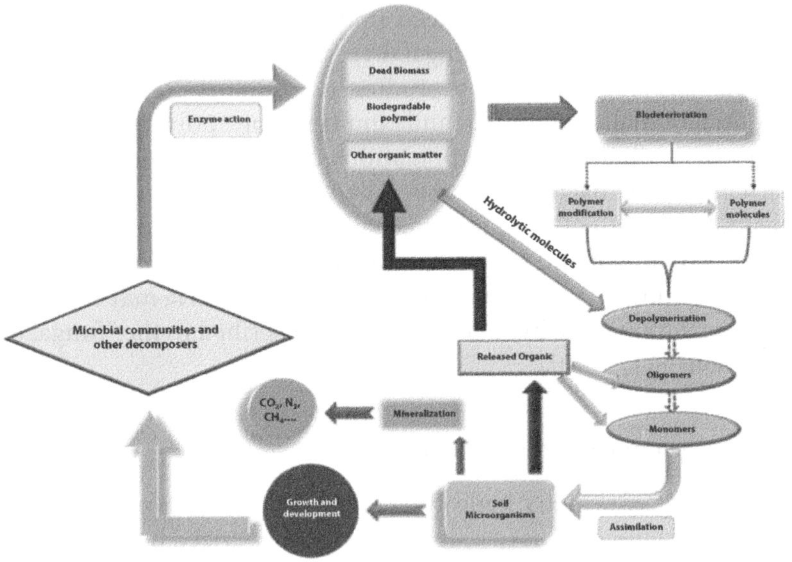

Figure 14.2 Schematic representation of polymer degradation.

14.6 Conclusion and Perspective

Biodegradable polymers reduce the impact of plastic production and refining on the environment. Because biodegradable polymers are derived from renewable feedstocks such as agricultural and food waste, there is a tremendous opportunity for research to capitalize on this economic sustainability. Currently, biodegradable polymers replace only about 1% of all synthetic plastics. The development of biodegradable packaging is influenced by several factors, including policy and legislative changes and the global demand for food and energy resources. The extent of biodegradation of bioplastics is significantly influenced by various environmental factors, including but not limited to the pH level of the medium, moisture content, oxygen concentration, and temperature. Furthermore, it is important to note that the structure and composition of biopolymers or biocomposites have a significant impact on the biodegradation process, and this should be taken into consideration during their production phases. The biodegradability of bioplastics can be improved by altering the biopolymer composition, which may involve incorporating materials that have a high concentration of soluble sugars. Despite the enhanced mechanical properties, such as high tensile strength, associated with the production of biocomposites from bioplastics, the biodegradation process may not always be favorable and can become interrupted at certain stages. Factors such as the end of life of biopolymers and processing costs need positive consideration to avoid the adverse impact on biodiversity and the environment. Biobased polymers are increasingly used for food packaging applications to extend the shelf life of a variety of high-quality food products including perishable foods. Biodegradable packaging can be utilized in active packaging systems, modified environment packaging, and edible packaging. However, proper studies need to be done on the interaction between biopolymers and food components before applying them for food packaging purposes.

References

1. Shah, A.A., Hasan, F., Hameed, A., Ahmed, S., Biological degradation of plastics: A comprehensive review. *Biotechnol. Adv.*, 26, 246–265, 2008, https://doi. org/10.1016/j.biotechadv.2007.12.005.
2. Schmidt, T.M., DeLong, E.F. and Pace, N.R., Analysis of a marine picoplankton community by 16S rRNA gene cloning and sequencing. *J. Bacteriol.*, 173, 14, 4371–4378, 1991.
3. Kumar, R., Verma, A., Shome, A., Sinha, R., Sinha, S., Jha, P.K., Kumar, R., Kumar, P., Shubham, Das, S., Sharma, P., Prasad, P.V.V., Impacts of plastic pollution on ecosystem services, sustainable development goals, and need to focus on circular economy and policy interventions. *Sustainability (Switzerland)*, 13, 9963, 2021, https://doi.org/10.3390/su13179963.
4. Ashori, A., Wood-plastic composites as promising green-composites for automotive industries! *Bioresour. Technol.*, 99, 4661–4667, 2008, https://doi.org/10.1016/j.biortech.2007.09.043.
5. Sudin, R. and Swamy, N., Bamboo and wood fibre cement composites for sustainable infrastructure regeneration. *J. Mater. Sci.*, 41, 6917–6924, 2006, https://doi.org/10.1007/s10853-006-0224-3.
6. A short review: Manufacturing problems and solutions of natural fiber reinforced composites. *J. Appl. Emerg. Sci.*, 9, 2, 68, 2019, https://doi.org/10.36785/buitems.jaes.275.
7. Wambua, P., Ivens, J., Verpoest, I., Natural fibres: Can they replace glass in fibre reinforced plastics? *Compos. Sci. Technol.*, 63, 1259–1264, 2003, https://doi.org/10.1016/S0266-3538(03)00096-4.
8. Avérous, L. and le Digabel, F., Properties of biocomposites based on lignocellulosic fillers. *Carbohydr. Polym.*, 66, 480–493, 2006, https://doi.org/10.1016/j.carbpol.2006.04.004.
9. J.L.-P., Large-scale production, properties and commercial applications of polylactic acid polymers. *Polym. Degrad. Stab.*, 59, 1–3, 145–152, 1998. Elsevier. (n.d.), https://www.sciencedirect.com/science/article/pii/S0141391097001481 (accessed November 26, 2022).

10. Lörcks, J., Properties and applications of compostable starch-based plastic material. *Polym. Degrad. Stab.*, 59, 245–249, 1998, https://doi.org/10.1016/s0141-3910(97)00168-7.
11. Bastioli, C., Properties and applications of mater-Bi starch-based materials. *Polym. Degrad. Stab.*, 59, 263–272, 1998, https://doi.org/10.1016/S0141-3910(97)00156-0.
12. Orset, C., Barret, N., Lemaire, A., How consumers of plastic water bottles are responding to environmental policies? *Waste Manage.*, 61, 13–27, 2017, https://doi.org/10.1016/j.wasman.2016.12.034.
13. Bertling, J., Bertling, R., Hamann, L., *Kunststoffe in der Umwelt: Mikro- und makroplastik*, 2018.
14. Yoshida, S., Hiraga, K., Takehana, T., Taniguchi, I., Yamaji, H., Maeda, Y., Toyohara, K., Miyamoto, K., Kimura, Y., Oda, K., A bacterium that degrades and assimilates poly(ethylene terephthalate). *Science* (1979), 351, 1196–1199, 2016, https://doi.org/10.1126/science.aad6359.
15. Petcore Europe, PET collection and recycling rates in Europe significantly increased in 2016, Press Release, 2017.
16. Geueke, B., Groh, K., Muncke, J., Food packaging in the circular economy: Overview of chemical safety aspects for commonly used materials. *J. Clean. Prod.*, 193, 491–505, 2018, https://doi.org/10.1016/j.jclepro.2018.05.005.
17. Otto, S., Strenger, M., Maier-Nöth, A., S., M., Food packaging and sustainability–Consumer perception vs. correlated scientific facts: A review, *Elsevier. J. Clean.*, 298, 126733, 2021, https://www.sciencedirect.com/science/article/pii/S0959652621009537 (accessed November 25, 2022).
18. Bajpai, P., Biobased polymers: Properties and applications in packaging. Elsevier, 2019.
19. Manzardo, A., Loss, A., Mazzi, A., Scipioni, A, Organization life-cycle assessment (OLCA): Methodological issues and case studies in the beverage-packaging sector. Environmental footprints of packaging, 47–73, 2016.
20. Navarro, A., Puig, R., Martí, E., Bala, A., Fullana-i-Palmer, P., Tackling the relevance of packaging in life cycle assessment of virgin olive oil and the environmental consequences of regulation. *Environ. Manage.*, 62, 277–294, 2018, https://doi.org/10.1007/s00267-018-1021-x.
21. Sheane, R., Lewis, K., Hall, P., Holmes-Ling, P., Kerr, A., Stewart, K., Webb, D., Identifying opportunities to reduce the carbon footprint associated with the Scottish dairy supply chain-main report. Edinburgh: Scottish Government, 2011.
22. Finnegan, W., Goggins, J., Clifford, E., Zhan, X., Environmental impacts of milk powder and butter manufactured in the Republic of Ireland. *Sci. Total Environ.*, 579, 159–168, 2017, https://doi.org/10.1016/j.scitotenv.2016.10.237.
23. Flysjö, A., Greenhouse gas emissions in milk and dairy product chains: Improving the carbon footprint of dairy products, 2012.

24. Kögel-Knabner, I., The macromolecular organic composition of plant and microbial residues as inputs to soil organic matter: Fourteen years on. *Soil Biol. Biochem.*, 105, A3–A8, 2017, https://doi.org/10.1016/j.soilbio.2016.08.011.
25. Pérez, J., Muñoz-Dorado, J., de La Rubia, T., Martínez, J., Biodegradation and biological treatments of cellulose, hemicellulose and lignin: An overview. *Int. Microbiol.*, 5, 53–6,3 2002, https://doi.org/10.1007/s10123-002-0062-3.
26. W., R., Microbial hydrolysis of polysaccharides. *Annu. Rev. Microbiol.*, 50, 1, 183–212, 1996. Go.Gale.Com. (n.d.), https://go.gale.com/ps/i.do?id=GALE%-7CA189635514&sid=googleScholar&v=2.1&it=r&linkaccess=abs&issn=00664227&p=HRCA&sw=w (accessed November 24, 2022).
27. Johansen, K.S., Lytic polysaccharide monooxygenases: The microbial power tool for lignocellulose degradation. *Trends Plant Sci.*, 21, 926–936, 2016, https://doi.org/10.1016/j.tplants.2016.07.012.
28. Navarro, D., Rosso, M.N., Haon, M., Olivé, C., Bonnin, E., Lesage-Meessen, L., Chevret, D., Coutinho, P.M., Henrissat, B., Berrin, J.G., Fast solubilization of recalcitrant cellulosic biomass by the basidiomycete fungus Laetisaria arvalis involves successive secretion of oxidative and hydrolytic enzymes. *Biotechnol. Biofuels*, 7, 1–14, 2014, https://doi.org/10.1186/s13068-014-0143-5.
29. Gómez, E.F. and Michel, F.C., Biodegradability of conventional and bio-based plastics and natural fiber composites during composting, anaerobic digestion and long-term soil incubation. *Polym. Degrad. Stab.*, 98, 2583–2591, 2013, https://doi.org/10.1016/j.polymdegradstab.2013.09.018.
30. Hayakawa, C., Funakawa, S., Fujii, K., Kadono, A., Kosaki, T., Effects of climatic and soil properties on cellulose decomposition rates in temperate and tropical forests. *Biol. Fertil. Soils*, 50, 633–643, 2014, https://doi.org/10.1007/s00374-013-0885-4.
31. Degli-Innocenti, F., Tosin, M., Bastioli, C., Evaluation of the biodegradation of starch and cellulose under controlled composting conditions. *J. Environ. Polym. Degrad.*, 6, 197–202, 1998.
32. Jayasekara, R., Harding, I., Bowater, I., Lonergan, G., Biodegradability of a selected range of polymers and polymer blends and standard methods for assessment of biodegradation. *J. Polym. Environ.*, 13, 231–251, 2005, https://doi.org/10.1007/s10924-005-4758-2.
33. Lähdesmäki, P. and Piispanen, R., Degradation products and the hydrolytic enzyme activities in the soil humification processes. *Soil Biol. Biochem.*, 20, 287–292, 1988, https://doi.org/10.1016/0038-0717(88)90005-3.
34. Polman, E.M.N., Gruter, G.J.M., Parsons, J.R., Tietema, A., Comparison of the aerobic biodegradation of biopolymers and the corresponding bioplastics: A review. *Sci. Total Environ.*, 753, 141953, 2021, https://doi.org/10.1016/J.SCITOTENV.2020.141953.
35. Haslam, M., The decomposition of starch grains in soils: Implications for archaeological residue analyses. *J. Archaeol. Sci.*, 31, 1715–1734, 2004, https://doi.org/10.1016/j.jas.2004.05.006.

36. German, D.P., Chacon, S.S., Allison, S.D., Substrate concentration and enzyme allocation can affect rates of microbial decomposition. *Ecology*, 92, 1471–1480, 2011, https://doi.org/10.1890/10-2028.1.
37. Cheshire, M.v., Mundie, C.M., Shepherd, H., Transformation of 14C glucose and starch in soil. *Soil Biol. Biochem.*, 1, 117–130, 1969, https://doi.org/10.1016/0038-0717(69)90002-9.
38. Shahbaz, M., Kuzyakov, Y., Heitkamp, F., Decrease of soil organic matter stabilization with increasing inputs: Mechanisms and controls. *Geoderma*, 304, 76–82, 2017, https://doi.org/10.1016/j.geoderma.2016.05.019.
39. Zubair, M. and Ullah, A., Recent advances in protein derived bionanocomposites for food packaging applications. *Crit. Rev. Food Sci. Nutr.*, 60, 406–434, 2020, https://doi.org/10.1080/10408398.2018.1534800.
40. Hassan, B., Chatha, S.A.S., Hussain, A.I., Zia, K.M., Akhtar, N., Recent advances on polysaccharides, lipids and protein based edible films and coatings: A review. *Int. J. Biol. Macromol.*, 109, 1095–1107, 2018, https://doi.org/10.1016/j.ijbiomac.2017.11.097.
41. Coltelli, M.B., Wild, F., Bugnicourt, E., Cinelli, P., Lindner, M., Schmid, M., Weckel, V., Müller, K., Rodriguez, P., Staebler, A., Rodríguez-Turienzo, L., Lazzeri, A., State of the art in the development and properties of protein-based films and coatings and their applicability to cellulose based products: An extensive review. *Coatings*, 6, 1, 2016, https://doi.org/10.3390/coatings6010001.
42. Uranga, J., Puertas, A.I., Etxabide, A., Dueñas, M.T., Guerrero, P., de la Caba, K., Citric acid-incorporated fish gelatin/chitosan composite films. *Food Hydrocoll.*, 86, 95–103, 2019, https://doi.org/10.1016/j.foodhyd.2018.02.018.
43. Martucci, J.F. and Ruseckaite, R.A., Biodegradation behavior of three-layer sheets based on gelatin and poly (lactic acid) buried under indoor soil conditions. *Polym. Degrad. Stab.*, 116, 36–44, 2015, https://doi.org/10.1016/j.polymdegradstab.2015.03.005.
44. Martucci, J.F. and Ruseckaite, R.A., Three-layer sheets based on gelatin and poly(lactic acid), part 1: Preparation and properties. *J. Appl. Polym. Sci.*, 118, 3102–3110, 2010, https://doi.org/10.1002/app.32751.
45. Martucci, J.F. and Ruseckaite, R.A., Biodegradation of three-layer laminate films based on gelatin under indoor soil conditions. *Polym. Degrad. Stab.*, 94, 1307–1313, 2009, https://doi.org/10.1016/j.polymdegradstab.2009.03.018.
46. Nakamura, K., Tomita, T., Abe, N., Kamio, Y., Purification and characterization of an extracellular poly(L-lactic acid) depolymerase from a soil isolate, amycolatopsis sp. strain K104-1. *Appl. Environ. Microbiol.*, 67, 345–353, 2001, https://doi.org/10.1128/AEM.67.1.345-353.2001.
47. Pranamuda, H., Chollakup, R., Tokiwa, Y., Degradation of polycarbonate by a polyester-degrading strain, Amycolatopsis sp. strain HT-6. *Appl. Environ. Microbiol.*, 65, 4220–4222, 1999, https://doi.org/10.1128/aem.65.9.4220-4222.1999.

48. Lee, S.H. and Song, W.S., Enzymatic hydrolysis of polylactic acid fiber. *Appl. Biochem. Biotechnol.*, 164, 89–102, 2011, https://doi.org/10.1007/s12010-010-9117-7.
49. Li, S. and McCarthy, S., Influence of crystallinity and stereochemistry on the enzymatic degradation of poly(lactide)s. *Macromolecules*, 32, 4454–4456, 1999, https://doi.org/10.1021/ma990117b.
50. Ghorpade, V.M., Gennadios, A., Hanna, M.A., Laboratory composting of extruded poly(lactic acid) sheets. *Bioresour. Technol.*, 76, 57–61, 2001, https://doi.org/10.1016/S0960-8524(00)00077-8.
51. Kimura T, S.Y., Ishida, Y., Ihara, N., Degradability of biodegradable plastic polylactic acid products. *JSAM*, 64, 115–120, 2020.
52. Gnanavel, G., Valli, V.P.M.J., Thirumarimurugan, M., Kannadasan, T., Degradation of plastics waste using microbes. *Int. J. Pharm. Chem. Sci.*, 54, 12212–4, 2013.
53. Li, S., Tenon, M., Garreau, H., Braud, C., Vert, M., Enzymatic degradation of stereocopolymers derived from L-, DL- and meso-lactides. *Polym. Degrad. Stab.*, 67, 85–90, 2000, https://doi.org/10.1016/S0141-3910(99)00091-9.
54. Tsuji, H., Mizuno, A., Ikada, Y., Blends of aliphatic polyesters. III. Biodegradation of solution-cast blends from poly(L-lactide) and poly(ε-caprolactone). *J. Appl. Polym. Sci.*, 70, 2259–2268, 1998, https://doi.org/10.1002/(SICI) 1097-4628(19981212)70:11<2259::AID-APP20>3.0.CO;2-J.
55. Tokiwa, Y. and Suzuki, T., Hydrolysis of polyesters by lipases. *Nature*, 270, 76–78, 1977, https://doi.org/10.1038/270076a0.
56. Oda, Y., Yonetsu, A., Urakami, T., Tonomura, K., Degradation of polylactide by commercial proteases. *J. Polym. Environ.*, 8, 29–32, 2000, https://doi.org/10.1023/A:1010120128048.
57. Tokiwa, Y., Konno, M., Nishida, H., Isolation of silk degrading microorganisms and its poly(L-lactide) degradability. *Chem. Lett.*, 28, 4, 355–356 1999, https://doi.org/10.1246/cl.1999.355.
58. Pranamuda, H. and Tokiwa, Y., Degradation of poly(L-lactide) by strains belonging to genus Amycolatopsis. *Biotechnol. Lett.*, 21, 901–905, 1999, https://doi.org/10.1023/A:1005547326434.
59. Torres, A., Li, S.M., Roussos, S., Vert, M., Screening of microorganisms for biodegradation of poly(lactic acid) and lactic acid-containing polymers. *Appl. Environ. Microbiol.*, 62, 2393–2397, 1996, https://doi.org/10.1128/aem.62.7.2393-2397.1996.
60. Vichaibun, V. and Chulavatnatol, M., A new assay for the enzymatic degradation of polylactic acid. *Science Asia*, 29, 297–300, 2003.
61. Chowdhury, F.H. and Neale, S.M., Acid behavior of carboxylic derivatives of cellulose. Part II. Oxycellulose. *J. Polym. Sci. A*, 1, 2893–2904, 1963, https://doi.org/10.1002/pol.1963.100010912.
62. R.S.-I.J., Biodegradable polymers, Ripublication.Com. *Int. J. Appl. Chem.*, 13, 179–196, 2017, https://www.ripublication.com/ijac17/ijacv13n2_02.pdf (accessed November 26, 2022).

63. Doi, Y., Rheological properties of ring polymers and their derivatives. Nihon Reoroji Gakkaishi, 50, 1, 57–62, 2022.
64. Luzier, W.D., Materials derived from biomass/biodegradable materials. *Proc. Natl. Acad. Sci. U.S.A.*, 89, 839–842, 1992, https://doi.org/10.1073/PNAS.89.3.839.
65. Guillet, J.E., Huber, H.X., Scott, J., *Studies of the biodegradation of synthetic plastics*, Special Publication - Royal Society of Chemistry, United Kingdom, 1992, https://doi.org/10.3/JQUERY-UI.JS.
66. Hankermeyer, C.R. and Tjeerdema, R.S., Polyhydroxybutyrate: Plastic made and degraded by microorganisms. *Rev. Environ. Contam. Toxicol.*, 159, 1–24, 1999, https://doi.org/10.1007/978-1-4612-1496-0_1.
67. Brandl, H., Bachofen, R., Mayer, J., Wintermantel, E., Degradation and applications of polyhydroxyalkanoates. *Can. J. Microbiol.*, 41, 13, pp.143–153, 1995, https://doi.org/10.1139/m95-181.
68. Alarie, A., Ecologie microbienne du sol. *Soil Sci.*, 114, 160, 1972, https://doi.org/10.1097/00010694-197208000-00014.
69. Kurtböke, I., Ivshina, I., Blackall, L.L., Microbial biodeterioration and biodegradation. *Microbiol. Aust.*, 39, 115–116, 2018, https://doi.org/10.1071/MA18036.
70. Briassoulis, D., Analysis of the mechanical and degradation performances of optimised agricultural biodegradable films. *Polym. Degrad. Stab.*, 92, 1115–1132, 2007, https://doi.org/10.1016/j.polymdegradstab.2007.01.024.
71. Koutny, M., Lemaire, J., Delort, A.M., Biodegradation of polyethylene films with prooxidant additives. *Chemosphere*, 64, 1243–1252, 2006, https://doi.org/10.1016/j.chemosphere.2005.12.060.
72. Wiles, D.M. and Scott, G., Polyolefins with controlled environmental degradability. *Polym. Degrad. Stab.*, 91, 1581–1592, 2006, https://doi.org/10.1016/j.polymdegradstab.2005.09.010.
73. Nakamura, H., Nakamura, T., Noguchi, T., Imagawa, K., Photodegradation of PEEK sheets under tensile stress. *Polym. Degrad. Stab.*, 91, 740–746, 2006, https://doi.org/10.1016/j.polymdegradstab.2005.06.003.
74. Tsuji, H., Echizen, Y., Nishimura, Y., Photodegradation of biodegradable polyesters: A comprehensive study on poly(l-lactide) and poly(ε-caprolactone). *Polym. Degrad. Stab.*, 91, 1128–1137, 2006, https://doi.org/10.1016/j.polymdegradstab.2005.07.007.
75. Ojumu, T.v., Yu, J., Solomon, B.O., Production of polyhydroxyalkanoates, a bacterial biodegradable polymer. *Afr. J. Biotechnol.*, 3, 18–24, 2004, https://doi.org/10.5897/AJB2004.000-2004.
76. Lucas, N., Bienaime, C., Belloy, C., Queneudec, M., Silvestre, F., Nava-Saucedo, J.E., Polymer biodegradation: Mechanisms and estimation techniques - A review. *Chemosphere*, 73, 429–442, 2008, https://doi.org/10.1016/j.chemosphere.2008.06.064.
77. Briassoulis, D., The effects of tensile stress and the agrochemical Vapam on the ageing of low density polyethylene (LDPE) agricultural films. Part I.

Mechanical behaviour. *Polym. Degrad. Stab.*, 88, 489–503, 2005, https://doi.org/10.1016/j.polymdegradstab.2004.11.021.
78. Bewa, H., *Matériaux polymères biodégradables et applications: Note de Synthèse II*, Ademe, 2006.
79. He, Y., Qian, Z., Zhang, H., Liu, X., Alkaline degradation behavior of polyesteramide fibers: Surface erosion. *Colloid Polym. Sci.*, 282, 972–978, 2004, https://doi.org/10.1007/s00396-003-1011-1.
80. Tsuji, H. and Ikada, Y., Properties and morphology of poly(L-lactide) 4. Effects of structural parameters on long-term hydrolysis of poly(L-lactide) in phosphate-buffered solution. *Polym. Degrad. Stab.*, 67, 179–189, 2000, https://doi.org/10.1016/S0141-3910(99)00111-1.
81. Müller, R.J., Witt, U., Rantze, E., Deckwer, W.D., Architecture of biodegradable copolyesters containing aromatic constituents. *Polym. Degrad. Stab.*, 59, 203–208, 1998, https://doi.org/10.1016/s0141-3910(97)00186-9.
82. Yew, G.H., Mohd Yusof, A.M., Mohd Ishak, Z.A., Ishiaku, U.S., Water absorption and enzymatic degradation of poly(lactic acid)/rice starch composites. *Polym. Degrad. Stab.*, 90, 488–500, 2005, https://doi.org/10.1016/j.polymdegradstab.2005.04.006.
83. le Digabel, F. and Avérous, L., Effects of lignin content on the properties of lignocellulose-based biocomposites. *Carbohydr. Polym.*, 66, 537–545, 2006, https://doi.org/10.1016/j.carbpol.2006.04.023.
84. Lugauskas, A., Levinskaite, L., Pečiulyte, D., Micromycetes as deterioration agents of polymeric materials. *Int. Biodeterior. Biodegrad.*, 52, 233–242, 2003, https://doi.org/10.1016/S0964-8305(03)00110-0.
85. Flemming, H.C., Relevance of biofilms for the biodeterioration of surfaces of polymeric materials. *Polym. Degrad. Stab.*, 59, 309–315, 1998, https://doi.org/10.1016/s0141-3910(97)00189-4.
86. Gu, J.D., Microbial colonization of polymeric materials for space applications and mechanisms of biodeterioration: A review. *Int. Biodeterior. Biodegrad.*, 59, 170–179, 2007, https://doi.org/10.1016/j.ibiod.2006.08.010.
87. Gu, J.D., Microbiological deterioration and degradation of synthetic polymeric materials: Recent research advances. *Int. Biodeterior. Biodegrad.*, 52, 69–91, 2003, https://doi.org/10.1016/S0964-8305(02)00177-4.
88. Gu, J.D., Mitton, D.B., Ford, T.E., Mitchell, R., Microbial degradation of polymeric coatings measured by electrochemical impedance spectroscopy. *Biodegrad.*, 9, 39–45, 1998, https://doi.org/10.1023/A:1008252301377.
89. Gu, J.-D., Roman, M., Esselman, T., Mitchell, R., The role of microbial biofilms in deterioration of space station candidate materials. *Int. Biodeterior. Biodegrad.*, 41, 25–33, 1998, https://doi.org/10.1016/s0964-8305(98)80005-x.
90. Gu, J.D., Lu, C., Thorp, K., Crasto, A., Mitchell, R., Fiber-reinforced polymeric composites are susceptible to microbial degradation. *J. Ind. Microbiol. Biotechnol.*, 18, 364–369, 1997, https://doi.org/10.1038/sj.jim.2900401.

91. Gu, J.D., Ford, T.E., Mitchell, R., Susceptibility of electronic insulating polyimides to microbial degradation. *J. Appl. Polym. Sci.*, 62, 1029–1034, 1996, https://doi.org/10.1002/(sici)1097-4628(19961114)62:7<1029::aid-app8>3.3.co;2-p.
92. Gu, J.D., Ford, T., Thorp, K., Mitchell, R., Microbial growth on fiber reinforced composite materials. *Int. Biodeterior. Biodegrad.*, 37, 197–204, 1996, https://doi. org/10.1016/s0964-8305(96)00035-2.
93. Crispim, C.A. and Gaylarde, C.C., Cyanobacteria and biodeterioration of cultural heritage: A review. *Microb. Ecol.*, 49, 1–9, 2005, https://doi.org/10.1007/s00248-003-1052-5.
94. Nuhoglu, Y., Oguz, E., Uslu, H., Ozbek, A., Ipekoglu, B., Ocak, I., Hasenekoglu, I., The accelerating effects of the microorganisms on biodeterioration of stone monuments under air pollution and continental-cold climatic conditions in Erzurum, Turkey. *Sci. Total Environ.*, 364, 272–283, 2006, https://doi.org/10.1016/j.scitotenv.2005.06.034.
95. Zanardini, E., Abbruscato, P., Ghedini, N., Realini, M., Sorlini, C., Influence of atmospheric pollutants on the biodeterioration of stone. *Int. Biodeterior. Biodegrad.*, 45, 35–42, 2000, https://doi.org/10.1016/S0964-8305(00)00043-3.
96. Mitchell, R. and Gu, J.D., Changes in the biofilm microflora of limestone caused by atmospheric pollutants. *Int. Biodeterior. Biodegrad.*, 46, 299–303, 2000, https://doi. org/10.1016/S0964-8305(00)00105-0.
97. Cappitelli, F., Principi, P., Sorlini, C., Biodeterioration of modern materials in contemporary collections: Can biotechnology help? *Trends Biotechnol.*, 24, 350–354, 2006, https://doi.org/10.1016/j.tibtech.2006.06.001.
98. Bonhomme, S., Cuer, A., Delort, A.M., Lemaire, J., Sancelme, M., Scott, G., Environmental biodegradation of polyethylene. *Polym. Degrad. Stab.*, 81, 441–452, 2003, https://doi.org/10.1016/S0141-3910(03)00129-0.
99. Warscheid, T. and Braams, J., Biodeterioration of stone: A review. *Int. Biodeterior. Biodegrad.*, 46, 343–368, 2000, https://doi.org/10.1016/S0964-8305(00)00109-8.
100. Rubio, C., Ott, C., Amiel, C., Dupont-Moral, I., Travert, J., Mariey, L., Sulfato/thiosulfato reducing bacteria characterization by FT-IR spectroscopy: A new approach to biocorrosion control. *J. Microbiol. Methods*, 64, 287–296, 2006, https://doi.org/10.1016/j.mimet.2005.05.013.
101. Roberts, D.J., Nica, D., Zuo, G., Davis, J.L., Quantifying microbially induced deterioration of concrete: Initial studies. *Int. Biodeterior. Biodegrad.*, 49, 4, 227–234, 2002, https://doi.org/10.1016/S0964-8305(02)00049-5.
102. Pelmont, J., Mangin, P., Kahn, R., Depondt, P., Kéroack, D., Stafford, L., ... & Demailly, J. P. *Glossaire de biochimie environnementale*. EDP sciences, 2005.
103. Jennings, D.H. and Lysek, G., *Fungal biology: Understanding the fungal lifestyle*, 1997.
104. Trinh Tan, F., Cooper, D.G., Marić, M., Nicell, J.A., Biodegradation of a synthetic co-polyester by aerobic mesophilic microorganisms. *Polym. Degrad. Stab.*, 93, 1479–1485 2008, https://doi.org/10.1016/j.polymdegradstab.2008.05.005.

105. Lindström, A., Albertsson, A., H., M., Development of a solid-phase extraction method for simultaneous extraction of adipic acid, succinic acid and 1, 4-butanediol formed during hydrolysis of poly, *Elsevier.* (n.d.). *J. Chromatogr. A*, 1022, 1–2, 171–177, 2004, https://www.sciencedirect.com/science/article/pii/S0021967303017679 (accessed November 24, 2022).
106. Göpferich, A., Mechanisms of polymer degradation and erosion. *Biomaterials*, 17, 117–128, 1996, https://doi.org/10.1016/0142-9612(96)85755-3.
107. Hakkarainen, M., Karlsson, S., Albertsson, A.C., Rapid (bio)degradation of polylactide by mixed culture of compost microorganisms - low molecular weight products and matrix changes. *Polym. (Guildf)*, 41, 2331–2338, 2000, https://doi.org/10.1016/S0032-3861(99)00393-6.
108. Otsuka, Y., Sonoki, T., Ikeda, S., Kajita, S., Nakamura, M., Katayama, Y., Detection and characterization of a novel extracellular fungal enzyme that catalyzes the specific and hydrolytic cleavage of lignin guaiacylglycerol β-aryl ether linkages. *Eur. J. Biochem.*, 270, 2353–2362, 2003, https://doi.org/10.1046/J.1432-1033.2003.03545.X.
109. Hofrichter, M., Review: Lignin conversion by manganese peroxidase (MnP). *Enzyme Microb. Technol.*, 30, 454–466, 2002, https://doi.org/10.1016/S0141-0229(01)00528-2.
110. Zapanta, L.S. and Tien, M., The roles of veratryl alcohol and oxalate in fungal lignin degradation. *J. Biotechnol.*, 53, 93–102, 1997, https://doi.org/10.1016/S0168-1656(96)01678-1.
111. Enoki, A., Itakura, S., T., H., The involvement of extracelluar substances for reducing molecular oxygen to hydroxyl radical and ferric iron to ferrous iron in wood degradation by wood. *J. Biotechnol.*, 53, 2–3, 265–272, 1997. Elsevier. (n.d.), https://www.sciencedirect.com/science/article/pii/S0168165697016829 (accessed November 24, 2022).
112. Szostak-Kotowa, J., Biodeterioration of textiles. *Int. Biodeterior. Biodegrad.*, 53, 3, 165–170, 2004, https://doi.org/10.1016/S0964-8305(03)00090-8.
113. Howard, G.T., Biodegradation of polyurethane: A review. *Int. Biodeterior. Biodegrad.*, 49, 4, 245–252, 2002, https://doi.org/10.1016/S0964-8305(02)00051-3.
114. Wackett, L.P., Biodegradation of plastics. *Environ. Microbiol. Rep.*, 13, 73–74, 2021, https://doi.org/10.1111/1758-2229.12924.
115. Pepic, D., Zagar, E., Zigon, M., Krzan, A., Kunaver, M., Djonlagic, J., Synthesis and characterization of biodegradable aliphatic copolyesters with poly(ethylene oxide) soft segments. *Eur. Polym. J.*, 44, 904–917, 2008, https://doi.org/10.1016/j.eurpolymj.2007.11.035.
116. von Burkersroda, F., Schedl, L., Göpferich, A., Why degradable polymers undergo surface erosion or bulk erosion. *Biomaterials*, 23, 4221–4231, 2002, https://doi.org/10.1016/S0142-9612(02)00170-9.
117. Siepmann, J. and Göpferich, A., Mathematical modeling of bioerodible, polymeric drug delivery systems. *Adv. Drug Deliv. Rev.*, 48, 229–247, 2001, https://doi. org/10.1016/S0169-409X(01)00116-8.

118. Göpferich, A. and Tessmar, J., Polyanhydride degradation and erosion. *Adv. Drug Deliv. Rev.*, 54, 911–931, 2002, https://doi.org/10.1016/S0169-409X(02)00051-0.
119. Mateo, C., Palomo, J.M., Fernandez-Lorente, G., Guisan, J.M., Fernandez-Lafuente, R., Improvement of enzyme activity, stability and selectivity via immobilization techniques. *Enzyme Microb. Technol.*, 40, 1451–1463, 2007, https://doi.org/10.1016/j.enzmictec.2007.01.018.
120. Ratto, J.A., Stenhouse, P.J., Auerbach, M., Mitchell, J., Farrell, R., Processing, performance and biodegradability of a thermoplastic aliphatic polyester/starch system. *Polym. (Guildf)*, 40, 6777–6788, 1999, https://doi.org/10.1016/S0032-3861(99)00014-2.
121. Vroman, I. and Tighzert, L., Biodegradable polymers. *Materials*, 2, 307–344, 2009, https://doi.org/10.3390/ma2020307.
122. Mueller, R.J., Biological degradation of synthetic polyesters-Enzymes as potential catalysts for polyester recycling. *Process Biochem.*, 41, 2124–2128, 2006, https://doi.org/10.1016/j.procbio.2006.05.018.
123. Gebauer, B. and Jendrossek, D., Assay of poly(3-hydroxybutyrate) depolymerase activity and product determination. *Appl. Environ. Microbiol.*, 72, 6094–6100, 2006, https://doi.org/10.1128/AEM.01184-06.
124. Marten, E., Müller, R.J., Deckwer, W.D., Studies on the enzymatic hydrolysis of polyesters - I. Low molecular mass model esters and aliphatic polyesters. *Polym. Degrad. Stab.*, 80, 485–501, 2003, https://doi.org/10.1016/S0141-3910(03)00032-6.
125. Walter, T., Augusta, J., Müller, R.J., Widdecke, H., Klein, J., Enzymatic degradation of a model polyester by lipase from Rhizopus delemar. *Enzyme Microb. Technol.*, 17, 218–224, 1995, https://doi.org/10.1016/0141-0229(94)00007-E.
126. Chiellini, E., Corti, A., del Sarto, G., D'Antone, S., Oxo-biodegradable polymers - Effect of hydrolysis degree on biodegradation behaviour of poly(vinyl alcohol). *Polym. Degrad. Stab.*, 91, 3397–3406, 2006, https://doi.org/10.1016/j.polymdegradstab.2006.05.021.
127. Chiellini, E., Corti, A., D'Antone, S., Solaro, R., Biodegradation of poly (vinyl alcohol) based materials. *Prog. Polym. Sci. (Oxford)*, 28, 963–1014, 2003, https://doi.org/10.1016/S0079-6700(02)00149-1.
128. Müller, R.J., Kleeberg, I., Deckwer, W.D., Biodegradation of polyesters containing aromatic constituents. *J. Biotechnol.*, 86, 87–95, 2001, https://doi.org/10.1016/S0168-1656(00)00407-7.
129. Green, F. and Highley, T.L., Mechanism of brown-rot decay: Paradigm or paradox. *Int. Biodeterior. Biodegrad.*, 39, 113–124, 1997, https://doi.org/10.1016/S0964-8305(96)00063-7.
130. Moussard, C. *Biochimie structurale et métabolique*. De Boeck Supérieur, 2006.
131. Madigan, M. T., Clark, D. P., Stahl, D., & Martinko, J. M. *Brock biology of microorganisms* 13th edition. Benjamin Cummings, 2010.
132. regnault_07_3, (n.d.).

133. Bailes, G., Lind, M., Ely, A., Powell, M., Moore-Kucera, J., Miles, C., Inglis, D., Brodhagen, M., Isolation of native soil microorganisms with potential for breaking down biodegradable plastic mulch films used in agriculture. *J. Visualized Exp.*, 75, e50373, 2013, https://doi.org/10.3791/50373.
134. Mergaert, J., Webb, A., Anderson, C., Wouters, A., Swings, J., Microbial degradation of poly(3-hydroxybutyrate) and poly(3- hydroxybutyrate-co-3-hydroxyvalerate) in soils. *Appl. Environ. Microbiol.*, 59, 3233–3238, 1993, https://doi.org/10.1128/aem.59.10.3233-3238.1993.
135. Moore-Kucera, J., Cox, S.B., Peyron, M., Bailes, G., Kinloch, K., Karich, K., Miles, C., Inglis, D.A., Brodhagen, M., Native soil fungi associated with compostable plastics in three contrasting agricultural settings. *Appl. Microbiol. Biotechnol.*, 98, 6467–6485, 2014, https://doi.org/10.1007/s00253-014-5711-x.
136. Muroi, F., Tachibana, Y., Kobayashi, Y., Sakurai, T., Kasuya, K.I., Influences of poly(butylene adipate-co-terephthalate) on soil microbiota and plant growth. *Polym. Degrad. Stab.*, 129, 338–346, 2016, https://doi.org/10.1016/j.polymdegradstab.2016.05.018.
137. Koitabashi, M., Noguchi, M.T., Sameshima-Yamashita, Y., Syuntaro, H., Suzuki, K., Yoshida, S., Watanabe, T., Shinozaki, Y., Tsushima, S., Kitamoto, H.K., Degradation of biodegradable plastic mulch films in soil environment by phylloplane fungi isolated from gramineous plants. *AMB Express*, 2, 1–10, 2012, https://doi.org/10.1186/2191-0855-2-40.
138. Kim, D.Y. and Rhee, Y.H., Biodegradation of microbial and synthetic polyesters by fungi. *Appl. Microbiol. Biotechnol.*, 61, 300–308, 2003, https://doi.org/10.1007/s00253-002-1205-3.
139. Kim, M.N., Lee, A.R., Yoon, J.S., Chin, I.J., Biodegradation of poly(3-hydroxybutyrate), Sky-Green and Mater-Bi by fungi isolated from soils. *Eur. Polym. J.*, 36, 1677–1685, 2000, https://doi.org/10.1016/S0014-3057(99)00219-0.
140. Sang, B.I., Hori, K., Tanji, Y., Unno, H., Fungal contribution to *in situ* biodegradation of poly(3-hydroxybutyrate-co-3-hydroxyvalerate) film in soil. *Appl. Microbiol. Biotechnol.*, 58, 241–247, 2002, https://doi.org/10.1007/s00253-001-0884-5.
141. Kamiya, M., Asakawa, S., Kimura, M., Molecular analysis of fungal communities of biodegradable plastics in two Japanese soils. *Soil Sci. Plant Nutr.*, 53, 568–574, 2007, https://doi.org/10.1111/j.1747-0765.2007.00169.x.
142. Suyama, T., Tokiwa, Y., Ouichanpagdee, P., Kanagawa, T., Kamagata, Y., Phylogenetic affiliation of soil bacteria that degrade aliphatic polyesters available commercially as biodegradable plastics. *Appl. Environ. Microbiol.*, 64, 5008–5011, 1998, https://doi.org/10.1128/aem.64.12.5008-5011.1998.
143. Witt, U., Müller, R.J., Deckwer, W.D., Biodegradation behavior and material properties of aliphatic/aromatic polyesters of commercial importance. *J. Environ. Polym. Degrad.*, 5, 81–89, 1997, https://doi.org/10.1007/BF02763591.

144. Wang, G. and Or, D., Aqueous films limit bacterial cell motility and colony expansion on partially saturated rough surfaces. *Environ. Microbiol.*, 12, 1363–1373, 2010, https://doi.org/10.1111/j.1462-2920.2010.02180.x.
145. Tecon, R. and Or, D., Bacterial flagellar motility on hydrated rough surfaces controlled by aqueous film thickness and connectedness. *Sci. Rep.*, 6, 19409, 2016, https://doi.org/10.1038/srep19409.
146. Zeng, J.B., Huang, C.L., Jiao, L., Lu, X., Wang, Y.Z., Wang, X.L., Synthesis and properties of biodegradable poly(butylene succinate-co-diethylene glycol succinate) copolymers. *Ind. Eng. Chem. Res.*, 51, 12258–12265, 2012, https://doi.org/10.1021/ie300133a.
147. Kohlmeier, S., Smits, T.H.M., Ford, R.M., Keel, C., Harms, H., Wick, L.Y., Taking the fungal highway: Mobilization of pollutant-degrading bacteria by fungi. *Environ. Sci. Technol.*, 39, 4640–4646, 2005, https://doi.org/10.1021/es047979z.
148. Warmink, J.A. and van Elsas, J.D., Migratory response of soil bacteria to Lyophyllum sp. strain Karsten In soil microcosms. *Appl. Environ. Microbiol.*, 75, 2820–2830, 2009, https://doi.org/10.1128/AEM.02110-08.
149. Weiss, M.S., Abele, U., Weckesser, J., Welte, W., Schiltz, E., Schulz, G.E., Molecular architecture and electrostatic properties of a bacterial porin. *Sci. (1979)*, 254, 1627–1630, 1991, https://doi.org/10.1126/science.1721242.
150. Haider, T.P., Völker, C., Kramm, J., Landfester, K., Wurm, F.R., Plastics of the future? The impact of biodegradable polymers on the environment and on society. *Angew. Chem. – Int. Ed.*, 58, 50–62, 2019, https://doi.org/10.1002/anie.201805766.
151. de Jong, S.J., Arias, E.R., Rijkers, D.T.S., van Nostrum, C.F., Kettenes-Van Den Bosch, J.J., Hennink, W.E., New insights into the hydrolytic degradation of poly(lactic acid): Participation of the alcohol terminus. *Polym. (Guildf)*, 42, 2795–2802, 2001, https://doi.org/10.1016/S0032-3861(00)00646-7.
152. Zumstein, M.T., Schintlmeister, A., Nelson, T.F., Baumgartner, R., Woebken, D., Wagner, M., Kohler, H.P.E., McNeill, K., Sander, M., Biodegradation of synthetic polymers in soils: Tracking carbon into CO2 and microbial biomass. *Sci. Adv.*, 4, eaas9024, 2018, https://doi.org/10.1126/sciadv.aas9024.
153. Witt, U., Evaluation of the biodegradability of copolyesters containing aromatic compounds by investigations of model oligomers. *J. Environ. Polym. Degrad.*, 4, 9–20, 1996, https://doi.org/10.1007/BF02083878.
154. Chen, D.R., Bei, J.Z., Wang, S.G., Polycaprolactone microparticles and their biodegradation. *Polym. Degrad. Stab.*, 67, 455–459, 2000, https://doi.org/10.1016/S0141-3910(99)00145-7.
155. Shirahama, H., Kanamori, N., Yasuda, H., Synthesis and biodegradability of monoalkylated caprolactone copolymers. *J. Oleo Sci.*, 50, 191–200, 2001, https://doi.org/10.5650/jos.50.191.
156. Gan, Z., Kuwabara, K., Yamamoto, M., Abe, H., Doi, Y., Solid-state structures and thermal properties of aliphatic-aromatic poly(butylene

adipate-co-butylene terephthalate) copolyesters. *Polym. Degrad. Stab.*, 83, 289–300, 2004, https://doi.org/10.1016/S0141-3910(03)00274-X.

157. Kuwabara, K., Gan, Z., Nakamura, T., Abe, H., Doi, Y., Crystalline/amorphous phase structure and molecular mobility of biodegradable poly(butylene adipate-co-butylene terephthalate) and related polyesters. *Biomacromolecules*, 3, 390–396, 2002, https://doi.org/10.1021/bm0156476.
158. Mukai, K., Doi, Y., Sema, Y., Tomita, K., Substrate specificities in hydrolysis of polyhydroxyalkanoates by microbial esterases. *Biotechnol. Lett.*, 15, 601–604, 1993, https://doi.org/10.1007/BF00138548.

15

Migration Concerns of Biopolymer-Based Food Packaging

Vikramsingh Thakur and Bhabani K. Satapathy*

Department of Materials Science and Engineering, Indian Institute of Technology, Delhi, New Delhi, India

Abstract
The need for sustainable food packaging materials has led to the development of bioplastics such as starch, poly(lactic acid) (PLA), and polyhydroxyalkanoate (PHA). These materials have similar properties to traditional polymers but are more environmentally friendly. In food packaging, biopolymers are often combined to improve their performance. Various additives such as metal ions, metal oxides, clays, antimicrobial agents, and antioxidants have been investigated to improve the performance of biopolymers. Nanocoatings of inorganic fillers are also used to improve the aroma and gas barrier properties of packaging materials. However, the migration of additives from nanocomposites to food can cause health problems, and the release of such chemicals and nanofillers is beyond acceptable limits. Mathematical diffusion models are used to understand the release behavior of additives. This chapter of the book examines the migration of food packaging materials, particularly starch, PHA, and PLA, and the potential risks posed by migrating substances. Interactions between food and packaging materials were also analyzed to better understand migration processes.

Keywords: Migration concerns, food packaging, biopolymers, diffusion models, nanocoatings, nanocomposites

**Corresponding author*: bhabaniks@gmail.com

Santosh Kumar, Avik Mukherjee and Vimal Katiyar (eds.) Agro-Waste Derived Biopolymers and Biocomposites: Innovations and Sustainability in Food Packaging, (421–450) © 2024 Scrivener Publishing LLC

15.1 Introduction

Considerable efforts have been made to construct packaging from more sustainable materials to conserve and uphold the food because the impact of conventional synthetic polymers from petroleum sources is a concern for the environment. As alternatives, the fascinating bioplastics such as poly(lactic acid) (PLA), starch, and polyhydroxyalkanoates (PHAs) are in the market since they can be processed using standard conversion machinery and have a great combination of inherent qualities along with industrial scalability at reasonable costs. In carbohydrate polymers, starch is one of the most abundant polysaccharides. The primary sources of starch are corn, wheat, potato, and cassava. PLA is used in disposable dinnerware and for packaging of perishable foods, including fruit juices, bottled water, and curd, due to its exhibiting properties similar to polystyrene. PHB is a semicrystalline, isotactic, stereoregular polymer that is highly biodegradable and has characteristics comparable to polypropylene in terms of material composition. The packaging film remaining in contact with food demands gas barrier resistance and excellent mechanical strength. As a result, combining the two biopolymers with or without modification by either incorporating the other functional components [antioxidant (AO), plasticizer, and fillers] or by chemical modification is usually necessary to increase their performance. These aforementioned methodologies have been discussed thoroughly in the chapter.

Moreover, in food packaging, the various systems based on PHAs, PLA, and starch have been widely evaluated and show benefits in terms of numerous features of interest. The common additives in biopolymers that have been studied vastly are metal ions (silver, gold, and copper), metal oxide (ZnO and TiO_2), clays [montmorillonite (MMT), mica, halloysite, and sepiolite], and antimicrobial and AO agents. The nanocoating of inorganic filler as a barrier layer is also used to develop a new strategy for the improvement in aroma and gas (oxygen) barrier properties. The migration of such additives from nanocomposites to food leading to consumer ingestion can potentially cause various health-related issues. The release of such chemicals and nanofillers must be below the maximum acceptable limits in food. The process of migration of typical additives may take place through Fickian diffusion. However, it may also happen through swelling, mechanical abrasion, and hydrolysis. The release of additives from the nanocomposites to food can change with the temperature, polymer-additives interactions, incorporation of hybrid filler, and presence of food simulants. Theoretically, mathematical diffusion models are used to understand the release behavior of such migrants (reagents, additives, etc.). In this book chapter, we include the migration concern pertaining to food packaging applications that are of potential commercial interest for certain food-specific biopolymers such as starch, PHAs, and PLA. The interaction between the food and packaging material and its associated potential risk due to their migration into food are also discussed.

15.2 Migration Concerns of Biopolymer-Based Nanocomposite

Biopolymers generally have lower barrier properties (WVTR, 5 to 30 g/m^2/day) than conventional polymers (1 to 15 g/m^2/day), and additives are most commonly added to biodegradable plastics to improve their properties. Currently, accessible biopolymers, including starch, PHAs, and PLA, seem most appealing commercially, as they have been processed with the standard processing equipment for bulk production. Given the advantages associated with these biopolymers, such as eco-friendliness, their commercial uses to replace traditional polymers as packaging materials on a broad scale have remained limited, primarily because they have some disadvantages in their physical and functional qualities (barrier, thermomechanical, dimensional stability, etc.). Notable issues are mostly processing-related restrictions and diminished barrier qualities against water and gasses.

Consequently, some unfavorable interactions between packaging materials and food and resulting substance migration may be more or less relevant for certain specific ingredients than for others. Plastic materials and their desired product must not release the migrants of more than 10 mg/dm^2 of total component/dm^2 of food contact area when transferring their constituents to food simulants. Such specifications further limit to 60 mg of total migrant/kg of food simulant for young children and infants according to the commission directives 2006/141/EC [1]. Food simulants are used in place of actual food not only to study and estimate the migration of nanoparticles (NPs) to packaged but also to make their subsequent identification easier. Broadly, two categories of food simulants are used for the migration test. They are (i) aqueous in nature: distilled water (simulant A), 3% acetic acid (simulant B), 10 % ethanol (simulant C); and (ii) fatty food simulants: rectified olive oil (simulant D) [2]. Simulant D can be replaced by synthetic triglycerides, sunflower oil, and corn oil. For technical reasons related to the test method, "substitute tests" that use test media under the typical substitute test circumstances may be applicable if the migration test utilizing fatty food simulants tends to be more impractical. The substitute tests use all the test media, including modified polyphenylene oxide, iso-octane, and 95% ethanol in an aqueous solution, under test settings that are analogous to the test parameters for simulant D [3].

The essential step in developing of new food packaging materials is to evaluate the possible migration/release of a constituent from the packaging to the packaged food. As the migrants' release depends on the food and its interaction with the polymer matrix, the migration mainly occurs via the swelling and the diffusion process. The usage of food simulants for standardized migration tests does not replicate the real-time situation of food and film interaction. The chemical qualities of the food imitators may vary, making them representative of a variety of foods, e.g., foods that are hydrophilic (water-based), lipophilic (fatty), or amphiphilic (both watery and fatty). For example, in the immersion process of film into food simulants, one tends to overestimate the process of migration of NPs. In contrast to the actual food packaging application, only one side of the film is in contact with the packaged food products. Considering the above real-time scenario, Dias *et al.* (2018) proposed that when diffusion occurs on one surface of the materials, the actual release is four times slower as compared to that through two sides of the material [4, 5].

15.2.1 Starch-Based Nanocomposites

One of the most studied, easily accessible, and reasonably priced carbohydrates in food packaging is starch, which can be obtained from renewable sources. Attributing to the hydrophilic characteristics, starch-based films exhibit water uptake resulting in less water vapor barrier qualities, despite having excellent oxygen barrier capacity. Starch combined with nanomaterials leads to an increase in these properties by enhancing the water barrier properties and mechanical properties. Starch is the most promising natural source of thermoplastic polymers for food packaging because of its abundance and affordability. Starch contains two types of polysaccharides, amylose (linear), and amylopectin (branched), and their ratio mainly depends on the source of starch. Rice starch, for example, contains more amylose than potato starch, which contains more amylopectin. Furthermore, various tissues within a plant can synthesize starches with varying amylopectin-to-amylose ratios. Corn endosperm tissue, for example, produces a starch that is high in amylopectin, whereas waxy maize produces a starch that is entirely amylopectin [6, 7].

The processing of starch is difficult due to the presence of hydrogen bonds and intermolecular forces. Hence, the use of a plasticizer is essential to make the thermoplastic material processable. The different plasticizers used to process starch are urea, sorbitol, glycerin, and glycerol. The glass transition temperature (T_g) reduces plasticized starch, hence providing flexibility and ease of processing [8, 9]. Different fillers, like clay, silver, and zinc oxide, were added to starch to improve properties such as antibacterial, gas/water barrier, mechanical strength, and processability. In starch-based food packaging, migration concerns arise due to the presence of filler and plasticizers that reduce the processing temperature and melting temperature (T_m), improving the processability. Various compositions of amylose-content starch were studied by Zhu *et al.* (2022) to analyze the extent of migration of plasticizer films. It has been found that the weak interaction between the plasticizer and the esterified amylose in the plasticizer-rich phase resulted in the more substantial relaxation and larger inter-planer spacing (d) of the crystalline region. Due to this, amylose-containing starch films are attributed to higher diffusion coefficients (D) of plasticizer migration, as shown in Figure 15.1 [10].

Liu *et al.* (2021) analyzed propionylated waxy (PW-N) and normal starch–based (PN-N) nanocomposites with varying amylopectin content and the migration behavior of plasticizer (triacetin) in the nanocomposite of nanoclay (barrier additives). The release of triacetin improves the interaction between clay and starch, leading to exfoliated structure formation. Liu *et al.* (2021) also reported in the highly branched amylopectin

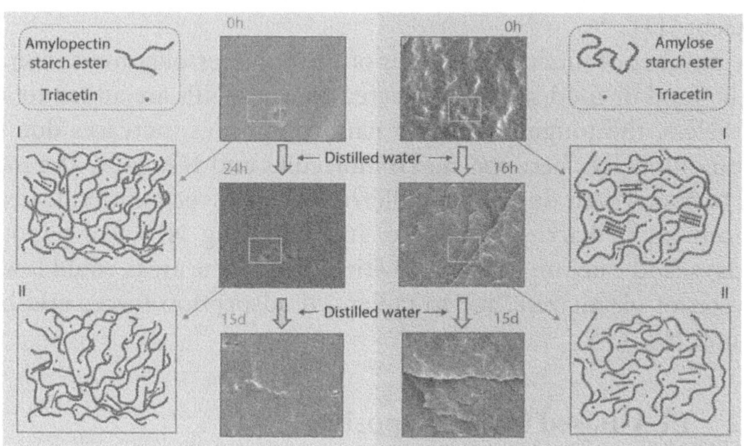

Figure 15.1 Schematic representation of plasticizer (triacetin) migration from hydrophobic starch–based films with amylopectin and amylose. Reproduced with permission from ref. [10]. Copyright 2022 Elsevier.

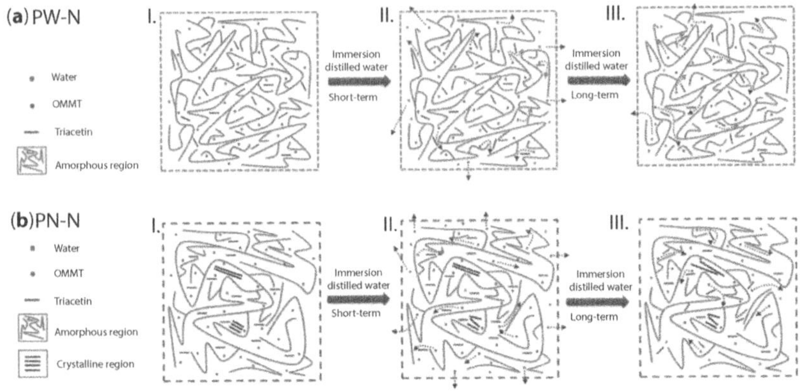

Figure 15.2 Migration path of plasticizer and structural change in different starch grades. Reproduced with permission from ref. [11]. Copyright 2021 Elsevier.

(PW-N) that there are no plasticizers and starch-rich amorphous regions in short-term migration, where homogeneous dispersion of clay promotes the migration of plasticizers to obey Fick's law. However, in the PN-N, the short-term migration is mainly due to the amorphous region of plasticizer-rich domains. The concentration differences of triacetin in the nanocomposites obey Fick's second law, where the propionylated starch and clay show improved interaction due to plasticizer migration, eventually leading to an ordered clay structure, as illustrated in Figure 15.2. The insufficient hindrance from propionylate, normal starch, shows a higher diffusion coefficient [11].

The number of functional groups of the polymer/oligomers interacting with the food or food simulants decreases when NPs are added to starch; nevertheless, the toughness of the nanocomposites increases due to the creation of linkages between starch molecules and NPs [10]. The increase in the tortuous path due to nanofiller/nanoclay incorporation leads to an enhancement in barrier resistance. Incorporating NPs like nano-silica, nanoclays, and calcium carbonate with high specific areas could lower the migration of plasticizers due to enhanced adsorption forces and barrier resistance [12, 13].

15.2.2 PLA-Based Nanocomposites

PLA, a member of the aliphatic polyester family, is a thermoplastic polymer that may be degraded by direct hydrolysis of the ester link. The stereochemical structure of this high-molecular-weight aliphatic polyester may be produced by the polymerization of the dimer of lactic acid (LA), resulting in the isomeric forms of LA. This class of polymer can readily be altered to produce a low- or high-molecular-weight PLA with tunable inherent characteristics, which can either be amorphous or crystalline. The primary applications of PLA are in the pharmaceutical and biomedical industries, as well as in the packaging of perishable foods [14]. In food packaging, PLA formulations that include a stabilizer, plasticizers, nanofillers, and other biopolymers that can enhance the qualities of the end product may potentially change the migration behavior of PLA. These low-molecular-weight additives potentially diffuse through polymeric matrix when used in food packaging that remains in contact with food. Among various oligomers of PLA, like lactose, LA, and the cyclic dimer of LA, only LA appears in the list of approved monomers according to the European Directive EC 10/2011.

Similarly, e-CGR 184.1061 of the U.S. Code of Federal Regulation of Food and Drug Administration (FDA) certified monomer LA as a safe food ingredient. However, the other oligomers of PLA and their migration-related concerns were discussed in section 15.3. To minimize the oxidation effect in PLA, synthetic AOs are primarily used because natural AOs are costly. Sirait *et al.* (2018) studied the migration of AOs in different food simulants from PLA films, and the substances that migrated at 29°C and 40°C were monomers, AOs, and solvents, as shown in Table 15.1. PLA stability was improved by using synthetic AOs such as butylated hydroxytoluene (BHT) and tert-butyl hydroxy quinone (TBHQ). However, BHT migrates more quickly than TBHQ, this is due to the fact that TBHQ contains more -OH groups compared to BHT leads to a better interaction with the PLA matrix. Reportedly, the 1-hydroxyl group of BHT has higher diffusivity during migration [15].

Various studies have found that incorporating nanoclay loading to improve water vapor and oxygen gas barrier properties tends to slow down the migration of LA [16]. The inclusion of nanocellulose in PLA/nanocellulose composites was also studied, where the improved interaction of nanofillers and polymer matrix was reiterated. Also, the highly polar solvent like isooctane (food simulant), which is absorbed by PLA, causes swelling and enhances the migration of nanocrystals and LA. The surface erosion of PLA-based nanocomposites was observed by Fortunati *et al.* (2017) at 40°C, which shows the appearance of holes, as PLA are prone to alcohol. The surfactant-modified cellulose nanocrystal shows a 10-μm-diameter hole on the surface. On the other hand, neat PLA and PLA-nanocrystal of cellulose have a small hole compared to surfactants containing nanocomposites, as shown in Figure 15.3. The appearance of big holes on the

Table 15.1 Total migrant of TBHQ and BHT antioxidants from PLA films. Reproduced with permission from ref. [13]. CC BY 4.0.

Antioxidant	Temperature	Food simulant	Total migration (mg/dm^2)
Tert-butyl hydroxy quinone (TBHQ)	29°C (room temperature)	DI water	33.6
		15% ethyl alcohol	55.2
		3% ethanoic acid	3.2
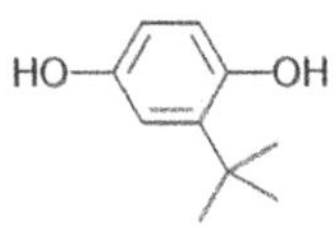	40°C	DI eater	231.2
		15% ethyl alcohol	763.2
		3% ethanoic acid	250.8
Butylated hydroxytoluene (BHT)	29°C (room temperature)	DI eater	110.0
		15% ethyl alcohol	38.0
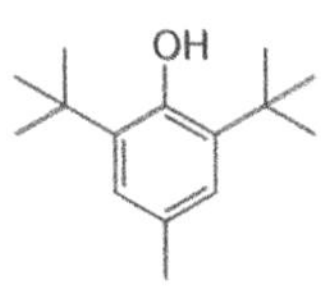		3% ethanoic acid	9.2
	40°C	DI eater	297.6
		15% ethyl alcohol	919.1
		3% ethanoic acid	467.2

surface can enhance the interaction with a polar solvent (food simulants) and reduce the barrier properties of nanocomposites [17].

The reinforcement of PLA with NPs in active food packaging applications enhances the shelf life of food. For example, PLA/ZnO nanocomposite films are antimicrobial with improved shelf life. Li *et al.* (2017) used nano-ZnO packaging film for fresh apples with PLA at 4°C for 14 days, and the results showed that the fruit when packaged with such materials, improved tissue firmness, sensory value, and total phenolic with inhibition of browning index and microbial growth [18]. PLA/ZnO nanocomposites film was also studied at a refrigerated temperature for 8 days for *Otolithes ruber* (fish) by Heydari-Majd *et al.* (2019), where the migration of Zn^{2+} ions showed 1.551 mg/100 g of fish at a refrigerated temperature [19]. In another study by Vasile *et al.* (2017) for PLA/ZnO nanocomposites containing bimetallic copper and silver, the films were subjected to food simulants (A, B, and C) for 2 hours at 70°C. The increase in migration of Zn and Cu was observed with an increasing ZnO:Cu ratio, with maximum migration of 10 mg/dm^2 of Zn [20].

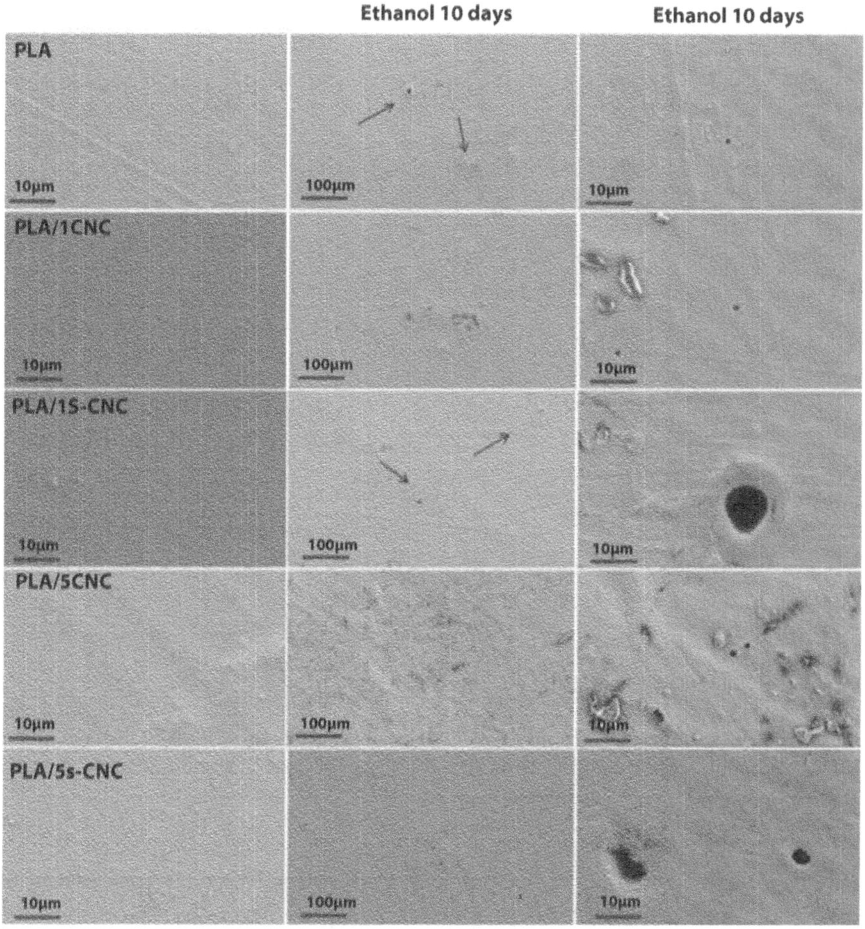

Figure 15.3 Microstructure of PLA and PLA-based nanocomposites in ethanol at 40°C. Reproduced from ref. [15]. Copyright 2012 Elsevier.

The TiO_2-Ag–containing PLA-based nanocomposites were studied for food simulants A, B, and C by Li *et al.* (2017). It was observed that the migration in acidic simulant (B) is more as compared to organic simulant (C). It is also observed that in hybrid nanocomposites, the interaction of various nanofillers and PLA plays a vital role, as nanofillers with lower interaction with PLA can increase migration behavior. The migration of silver showed a linear variation with the storage time. Comparatively, the silver-based NPs' migration ratio was found to be higher than that of titanium-based NPs and is attributed to the higher surface-to-volume ratio of Ag NPs, as shown in Figure 15.4 [21].

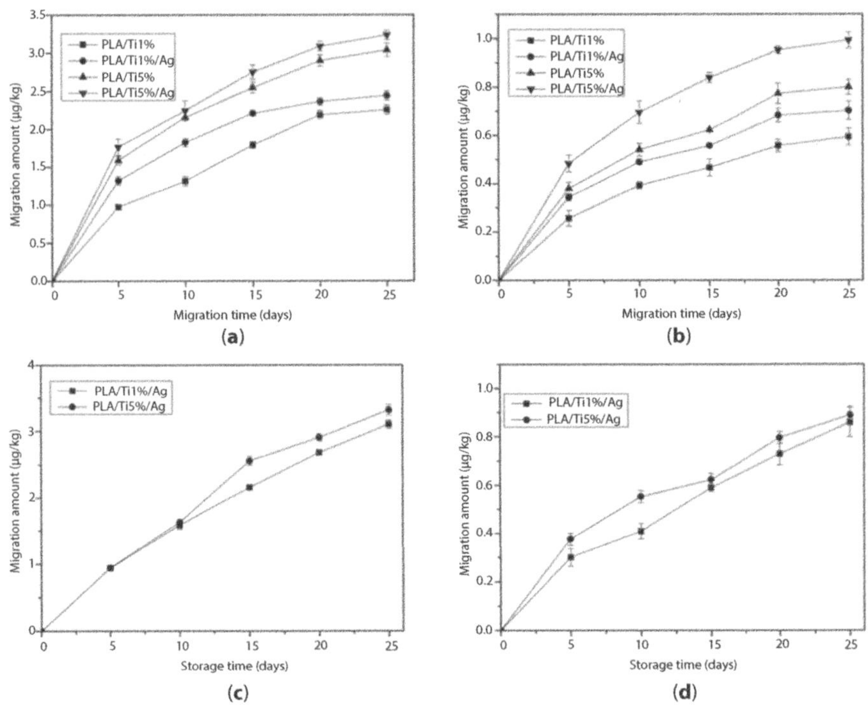

Figure 15.4 Migration of Ti from PLA/TiO_2 nanocomposites for (a) 3 % acetic acid and (b) 50 % ethanol. Migration of Ag from PLA/TiO_2/Ag for (c) 3 % acetic acid and (d) 50 % ethanol. Reproduced with permission from ref. [17]. CC BY 4.0

It is well known that PLA possesses properties like easy processing at lower cost and good optical properties compared to other biopolymers. The main drawback of fragility and moderate gas barrier of PLA can potentially be overcome via the incorporation of nanoclay, which is one of the most attractive and affordable nanofillers, which can functionally act as a reinforcing agent with excellent barrier properties due to its sheet-like structure. The migration issues related to clay-based nanocomposites are extensively discussed in section 15.4.2. Concisely, food packaging materials of PLA should have minimum migration of additives during storage and processing with controlled gas permeability.

15.2.3 Polyhydroxyalkanoates (PHA)-Based Nanocomposites

PHAs are aliphatic polyesters of different hydroxyalkanoates, which are produced using several microorganism-assisted biochemical reactions that are typically cultured on agricultural raw materials. Fermentation processes are used to produce PHAs on a large scale for commercial usage. The physical and chemical features of PHAs, which depend significantly on the monomer composition, are strong UV resistance, biocompatibility, and nontoxicity. Polyhydoxybutyrate (PHB) is prepared by microbial fermentation using *Bacillus megaterium*, *Ralstonia eutropha*, etc. Being highly brittle and having a low gas barrier, PHB generally blends with polyhydroxybutyrate-co-hydroxyvalerate (PHBV) or/and nanofillers are added to improve the properties of PHB. Dhar *et al.* (2015) have reported the use of crystalline cellulose (CNC) to improve the gas barrier performance of the PHB. The overall migration significantly increased with the addition of CNC. The migration level of such CNC-loaded PHAs drops even at a minimal level of 1 to 2 wt % of CNC. This may possibly be attributed to the restriction with polymer chain mobility and a good extent of nanofiller dispersion. The extent of migration increases with CNC addition of more than 3 wt %, which might be because of food simulant (isooctane) absorbed by PHB and causing the release of CNC and low-molecular-weight oligomers [22]. The migration concerns of PHBV/functionalized methyl ester CNC (CNC-me) nanocomposites film produced via solution casting were recently studied by Yu *et al.* (2014) and the result showed better interfacial adhesion with the PHBV matrix [23]. The incorporation of CNC-me increases the gas barrier properties and lowers overall migration by increasing the crystallinity. The extent of the overall migrations result for PBHV/CNC-me is shown in Figure 15.5.

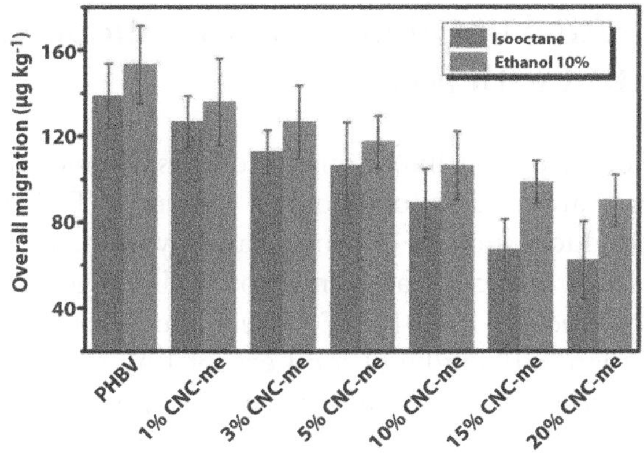

Figure 15.5 Overall migration of PHBV/CNC-me nanocomposites films [19].

The overall migration was reduced in PHBV-based nanocomposites with nanofillers that include ZnO, carbon nanofibers (CNFs), and multi-walled carbon nanotubes (MWCNTs) due to their relatively better adhesion with the matrix with enhancement in crystallinity. The heterogeneous nucleation effect of nanofiller is responsible for leading not only to specific crystalline structures but also to increase the extent of crystallinity. CNC/Ag nanofiller combination has also been studied with the PLA and PHBV as the matrices, and it was found that an increase in CNC may cause aggregation effects leading to a reduction in crystallinity [24–26]. It is reported that PHB-based nanocomposites containing ZnO NPs showed better antibacterial, barrier, and migration properties. The incorporation of 5 wt % of ZnO tends to reduce water vapor permeability by ~66 to ~38%. The enhancement in crystallinity tends to lower the migration in the existence of food simulants up to the extent of 5 wt % of their loading [27]. In the study by Salomez *et al.* (2019), poly(butylene succinate-co-butylene adipate) (PBSA) and PHBV nanocomposite films containing 5 wt % of Cloisite 30B were prepared by melt mixing. The overall migration was estimated gravimetrically in standard food simulants for long-term storage (EU 10/2011). The lower migration in nanocomposite film was not because of decreasing diffusivity of additives, but due to the barrier effect. The UV irradiation effect may lead to an enhancement of biopolymer crystallinity, as observed by differential scanning calorimetry (DSC), that may be attributed to the reduction in overall migration (Figure 15.6). The UV treatment reportedly caused a significant change in PHBV when in contact with ethanol [28].

15.3 Migration of Oligomers From Biopolymers in Contact with Food

The primary source of non-intentionally added substances (NIAS) in biopolymers, which are used as food contact materials (FCM), is caused by degradation products such as oligomers and byproducts. The oligomers of PLA include cyclic lactide and linear or cyclic oligomers that are not on the list of Directive 10/2011/EC, where the permissible value must remain lower than 0.01 mg/kg of food except LA. Mutsuga *et al.* (2008) studied PLA oligomers migration in water, 20 % ethanol, and 4% acetic acid for short- and long-term contact time at a temperature range of 20°C to 95°C. Near the glass transition temperature (T_g) of PLA, LA migration was 0.008 to 0.040 mg/dm^2 in food simulants. The migration of PLA has

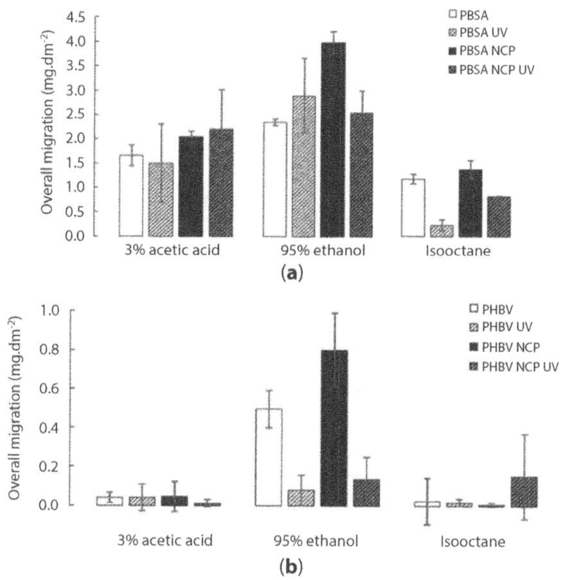

Figure 15.6 Overall migration in standard food simulants for (a) PBSA, PBSA UV, PBSA NCP, and PBSA NCP UV aged materials and (b) PHBV, PHBV UV, PHBV NCP, and PHBV NCP UV aged film [28].

been intensively studied, where migrants may include LA and its linear (lactoyl LA) and cyclic dimer along with oligomers of PLA mentioned in Table 15.2 (nos. 3 to 13) and lactide cyclic dimer of LA. The lactide level and LA migration increased with acidity and temperature. The PLA with high D-LA levels shows an increase in migration of LA over a half-year time at 40°C, as the initiation of decomposition occurs at a temperature above the T_g [29]. The oligomers of PLA, including linear and cyclic, are shown in Table 15.2. Reportedly, a total of 15 oligomers have been obtained in the chemical reaction between food simulants and PLA [30, 31].

The PLA oligomers are of two kinds, and they are (a) HO–$[LA]_n$–H and (b) CH_3–CH_2–O–$[LA]_n$–H. According to Ubeda *et al.* (2019), cyclic oligomers of PLA were not found in any food simulants. This is because the reaction with food simulants breaks down the cyclic structure, leading to the formation of new linear oligomers [30]. However, these oligomers can migrate from the polymer and remain dissolved in food as microplastics, potentially causing toxicity. Migration tests have confirmed the formation of new oligomers due to the reaction between PLA and food simulants. In PHA/CNC films, the interaction of ethanol has been observed to result in the appearance of a white, brittle substance after storage time, possibly

Table 15.2 Oligomer of PLA and their migration in different three food simulants with the number of PLA monomers repeated (* L for linear, C for cyclic, O1 for HO–$[LA]_n$–H and O_2 for CH_3–CH_2–O–$[LA]_n$–H).

Sr. no.	PLA Oligomers	n	Migration in 95 % ethanol	Migration in 10% ethanol	Migration 3% acetic acid
1	L: O1	3		×	×
2	L: O2	2	×	×	
3	L: O1	4	×	×	×
4	L: O2	3	×	×	
5	L: O1	5	×	×	×
6	L: O2	4	×	×	
7	L: O1	6	×	×	×
8	C-LA	5			
9	L: O1	7	×	×	×
10	L: O2	5	×	×	
11	C-LA	6			
12	L: O1	8	×	×	×
13	L: O2	6	×	×	
14	L: O1	9	×	×	×
15	C-LA	7			
16	L: O1	10	×	×	×
17	C-LA	8			
18	L: O2	7	×	×	
19	L: O1	11	×	×	×
20	L: O2	8	×	×	
21	C-LA	9			
22	L: O1	12	×	×	×

(*Continued*)

Table 15.2 Oligomer of PLA and their migration in different three food simulants with the number of PLA monomers repeated (* L for linear, C for cyclic, O1 for HO–$[LA]_n$–H and O_2 for CH_3–CH_2–O–$[LA]_n$–H). (*Continued*)

Sr. no.	PLA Oligomers	n	Migration in 95 % ethanol	Migration in 10% ethanol	Migration 3% acetic acid
23	L: O2	9	×	×	
24	L: O1	13	×	×	×
25	L: O2	10	×		
26	C-LA	10			
27	L: O1	14			
28	L: O2	11	×		
29	L: O1	15			
30	L: O2	12	×		
31	C-LA	11			
32	L: O2	13	×		
33	C-LA	12			
34	L: O2	14	×		
35	L: O2	15	×		
36	C-LA	13			
37	C-LA	14			
38	C-LA	15			
39	C-LA	16			

due to the presence of an oligomer unit. PHB and PHBV are nontoxic, but further research is needed to determine the nature of degradation product release in the presence of food simulants.

A polymer blend of PLA derived from fossil fuels was used to examine the nonvolatile content in the biodegradable films used for food packaging. The nonvolatile components of a biodegradable PLA blend mix were effectively analyzed using an extractive UPLC-MS (QTOF) method, which was preceded by sample dissolution/precipitation treatment. The analysis

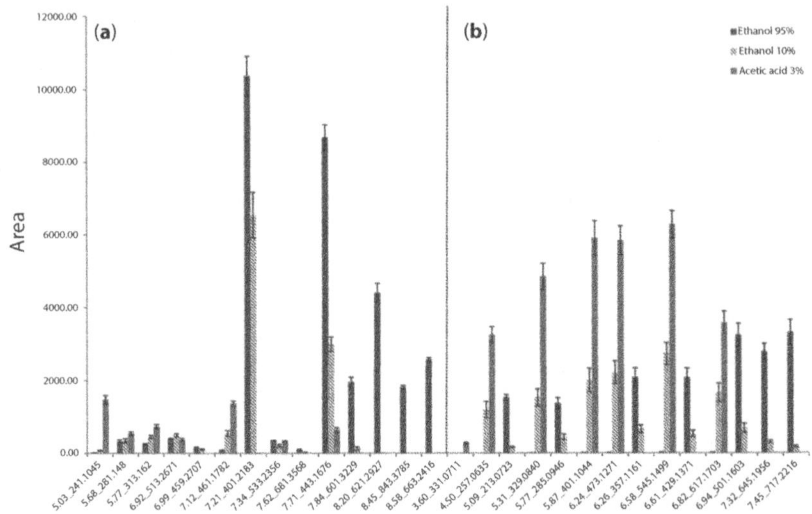

Figure 15.7 Area intensities of nonvolatile components for PLA-based films for three different food simulants (a) for film and (b) after migration. Reproduced from ref. [27]. Copyright 2021 Elsevier.

revealed that there was no formation of new compounds during the extrusion process, and the composition of PLA granules and films was found to be relatively similar [32]. The highest-intensity chemicals in PLA blend samples and migration assays proved to have originated from the polyester component of the mixture. They remain the cyclic oligomers of butanediol, adipic acid, and phthalic acid, although numerous oligomers (cyclic) were found in the pellets, membrane, and film samples with medium intensity and could not be found in the migration solution.

On the other hand, it was found that ring opening led to the formation of new linear oligomers. The samples and the drift both had a few plasticizers and other chemicals. It is crucial to remember to gather comprehensive information about migrants in food models, as it is necessary to examine migrant time in solutions and project the film samples. The area intensities of nonvolatile component for PLA-based films for three different food simulants is shown in Figure 15.7 [32].

15.4 Migration of Nanomaterials

The usage of nanomaterials in food packaging applications for better performance has lately raised the concern that migration leads to toxicity effects [33]. Commercially, food packaging solutions with nano-inclusions/fillers, etc., are either coated or composited with nanofillers like nanosilver, nanoclay, CNC, and TiO_2. The risk assessment of migration is debatable among researchers regarding the extent of migration and its safety.

15.4.1 Silver Nanoparticle–Incorporated Biopolymer Composite Films

Silver NP (AgNP) migration in food packaging is regulated by the European Union (EU): 10/2011/EU. The toxicological properties of AgNPs depend on their size, shape, and environmental conditions. The well-known method for analyzing silver NP migration in films was mainly analyzed by immersing the nanocomposite film in different food simulants like (a) water, (b) 15% v/v ethanol, and (c) 3% v/v acetic acid. The defined area of the film sample is dipped into the food simulant kept on an orbital shaker, and the simulant sample is collected at different time intervals. The collected simulant samples are analyzed using the inductively coupled plasma optical spectroscopy system (ICP-OES) [34]. The migration of AgNPs mainly depends on (a) the interaction between the polymer matrix and AgNPs and (b) the interaction of migrant and food simulants. Ortega *et al.* (2022) prepared starch film containing AgNPs prepared by active compounds of lemon juice (AgNP L) showed increased release of NPs for a shorter time (0 to 8 hours) compared to the AgNPs synthesized *in situ* in native corn starch with 25% amylose. The release mechanism observed for the *in situ* synthesized AgNP remained in disagreement with the Fickian process due to the second sigmoid of the release [35]. Such an observation of a second release curve may be starch-induced swelling attributed to the kinetic-constrained relaxation phenomena. When it comes to the osmotic response of starch, food simulants can penetrate into starch macromolecules, leading to an increase in the swelling and rearrangement of the structure. The release of AgNPs can be due to both Fickian and non-Fickian diffusion, in which the relaxation of the polymer matrix and diffusion process may occur. Simultaneously, the migration test setup is shown in Figure 15.8a An indicative representation of the migration percentage of Ag release after 7 days of immersion in food simulants is shown in Figure 15.8b and 15.8c [11]. The AgNPs are being used as antimicrobial additives in food packaging with a maximum permissible limit of ~0.01 mg/m^3 for all forms of silvers, according to the United States National Institute for Occupational Safety and Health (US-NIOSH). In this regard, the concept of the threshold of toxicological concern (TTC) is a helpful feature for NIAS risk measurement.

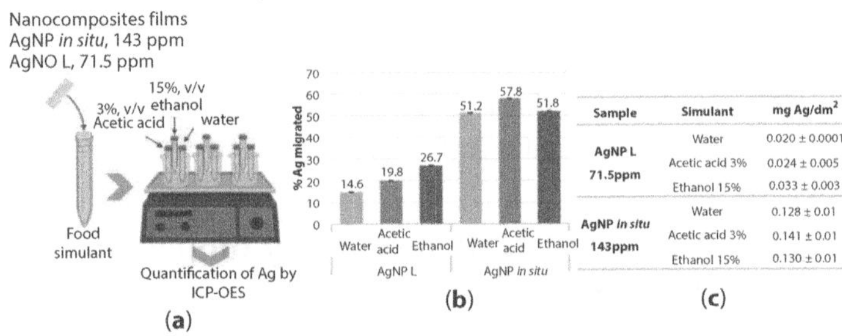

Figure 15.8 (a) Migration testing for silver-based biopolymer films (b) migration percentage of Ag released (c) after 1 week of migration [35].

15.4.2 Clay-Based Biopolymer Films

The usage of nanoclays has increased in food packaging films because of low prices, abundant availability, reinforcing effects, and barrier properties. In food packaging, PLA- and nanoclay-based nanocomposites are widely explored, as nanoclay provides excellent gas barrier properties. PLA-clay–reinforced polymer films do not provide a prominent rise in Fe, Ca, Mg, or Al, thus making it a safer option to be used in food packaging, according to N.10/2011 commission regulations [36]. Various Al release possibilities in the presence of food simulants were studied, and migration was found to be ~51.65 and ~24.13 ng/cm^2 for two commercially explored food-grade clays [37, 38]. Farhoodi *et al.* (2014) investigated the release of Si and Al from clay for 90 days and found that Si release was 23 % higher than Al [38]. According to the EU regulation, for nanoclay, the overall migration value is ~10 mg/dm^2. Nanoclay release depends on factors like polymer/clay interaction, temperature, and food simulants. Homogeneously dispersed nanoclay provides lower migration, but the nanoclay size of less than 5 nm has a diffusion coefficient, and hence, the possibility of migration needs to be investigated [39]. Nanocomposite films made from soybean polysaccharide were filled with Cloisite 30B nanoclay. The release of clay NPs into the food was examined, and such films were employed in food packaging applications like in bread. The findings showed that Cloisite 30B releases and migrates from deionized water (Food Simulant A), but NPs remain undetected in bread. This observation was supported by the study of Salarbashi *et al.* (2018), which found that Cloisite 30B enhances the mechanical and thermal properties of nanocomposite films in addition to preventing the growth of *S. typhi 1690*, *L. monocytogenes*, and *S. epidermis* [40]. Cytotoxic assay and antibacterial test for commercial clays like Cloisite 20A and Cloisite 30B showed no cytotoxicity.

MMT, on the other hand, upon long-time exposure, can cause not only cytotoxic effects but also eventually be absorbed in the body within 2 hours [41]. Souza *et al.* (2019) developed chitosan-based nanocomposites with MMT-clay containing *Zingiber officinale* Rosc (GEO) and *Rosmarinus officinalis* L (REO) essential oils. They found that lipophilic simulants release phenolic moieties faster in GEO and slower in hydrophilic simulants as compared to REO. Thus, clay improves the shelf life of poultry meat by restricting the release of phenolic compounds [42]. It should be emphasized that each clay mineral has a different toxicity profile, particularly the organo-modified nanoclays, and that each situation must be evaluated individually. To investigate the cytotoxicity of clay minerals, human cell lines were used. Experiments revealed that sodium MMT had no cytotoxic or mutagenic effects. Due to the presence of quaternary ammonium in MMT, which leads to generating oxidative stress in the biological system, Cloisite 30B nanoclays may have a hazardous effect. Clay NP stability in the solution is crucial because unstable clay results in clumping (agglomerate) together and falling apart, leaving deposits in the sediments [43]. The organo-modification of Na-bentonite has also shown similar effects. Its toxicity was examined using modifiers, for example, octadecyl trimethyl ammonium chloride (TMSA) and hexadecyl trimethyl ammonium bromide (HDTA). Cell models created *in vitro* were utilized to study the cytotoxicity. The cell viability and toxicity were both triggered and increased by the modifiers. To further assess their toxicity, Wistar rats were exposed orally to nanoclay for approximately 90 days at 40 mg/kg/day, and the morphological investigation of primary organ tissues did not indicate any harmful or notable toxic consequences [44].

15.4.3 Other Nanomaterials as Fillers in Biopolymer Films

The usage of inorganic nanomaterials in biopolymer films has gained interest due to the improvement in toughness and antimicrobial properties [45]. For example, various metals and their oxides, most commonly ZnO and titanium dioxide (TiO_2), were studied with regard to migration aspects in food packaging materials [46]. The regulation for using such metal oxides is limited to regulation regarding food contact packaging films. The FDA-approved inorganic NPs have been mentioned in Table 15.3 with their specific migration limit (SML). TiO_2 is mainly used in the food industry as an E171 for texture enhancer and whitening agent. TiO_2 is an FDA-approved coloring additive for up to 1% in food and is primarily used in food products such as curd, candies, cream, sugar toppings, and cheese. Besides TiO_2, zinc and its oxides, chlorides, sulfates, stearates, gluconates, and sulfate

Table 15.3 FDA-approved inorganic nanoparticles according to FDA [42–44].

NPs/regulation	Details	SML
TiO_2/FDA	FDA approves TiO_2 as a coating agent and preservative at 2.2 PPM dry coatings, regardless of size.	-
TiO_2 EU2020/1245	Surface treatment of TiO_2 with fluoride modifies alumina as an additive (FCM No. 1077)	Up to 25 % by weight, including nanoforms
Titanium nitride, nanoparticles EU 10-2011	Used as gas barrier additives for enhancement in shelf life	No migration of titanium nitride in packaging film
Zinc oxide	For the unplasticized polymers, up to 2% by weight is approved. Used mainly as transparent UV light absorbers in food packaging (FCM No. 1050)	ZnO nanoparticles do not migrate in nano-size when used with unplasticized bio-polymers. For migration soluble ionic zinc, safety evaluation should be considered.
Zinc EU 10/2011	As an additive in contact with food expressed as zinc For the films in contact with food and used as additives with food contact materials (FCM No. 402)	25 mg/kg of simulant or food
Aluminum EU 10-2011/EU 2020/1245	Packaging film should not release aluminum exceeding its permissible values.	1 mg/kg of simulant or food
Copper EU 10-2011/EU 2020/1245	Packaging film should not release copper exceeding its permissible values.	5 mg/kg of simulant or food

(*Continued*)

Table 15.3 FDA-approved inorganic nanoparticles according to FDA [42–44]. (*Continued*)

NPs/regulation	Details	SML
Zinc EU 10-2011/EU 2020/1245	Packaging film should not release zinc exceeding its permissible values.	5 mg/kg of simulant or food
Iron EU10-2011/EU 2020/1245	Packaging film should not release iron exceeding its permissible limits.	48 mg/kg of simulant or food
Graphite EU 10/2011	Used as additives for thermal stability and water vapor barrier resistance	Approved additive without specific SML limit list
Silver EU 86437	Packaging film should not release silver exceeding its permissible values.	0.05 mg/kg of simulant or food

are on the list of "Generally Recognized as Safe" (GRAS) materials (FDA 2011). Some studies show that ZnO NPs are carcinogenic to eukaryotic cells upon migration from food packaging materials [47, 48]. Because of an increase in oxidative stress, which results in oxidative DNA damage by ZnO NPs, cytotoxicity reportedly showed activity against healthy cells. In contradiction, some articles suggest that the ZnO NPs cannot penetrate normal cells of animals and humans on migration from food packaging films [49, 50]. The lower MW NPs are absorbed faster in the body and may cause damage to the cells. In a study by Maisanaba *et al.* (2016) on mice, carbon nanotubes (CNT) were injected into the mice, and it was found that they had an asbestos-like toxic effect [44].

15.5 Biopolymer-Based Nanocomposite Films and Coatings

The use of NPs as nanofillers in nanocomposite films and coatings is an effective way to improve their functional properties and can alter the migration process [51, 52]. Huang *et al.* (2018) studied the migration of AOs, where the inorganic silicon oxide (SiOx) layer was coated over the PLA composite films containing up to 1% AOs. SiOx layers were deposited over the PLA film using plasma-enhanced chemical vapor deposition with

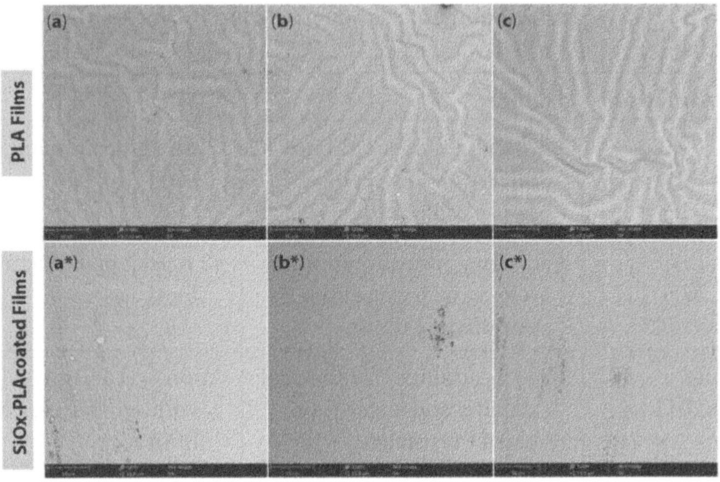

Figure 15.9 SEM images of PLA and SiOx/PLA-coated films for 1% antioxidants (a) TBHQ, (b) BHA, and c) BHT [45].

hexamethyldisiloxane as a reactive monomer in the presence of oxygen (oxidizing gas) and argon (ionizing gas). SiOx is used as the barrier layer to reduce the migration of AOs from PLA composite films by mechanical interlocking, clogging of the submicroscopic leakage holes, and filling of the crevices of the PLA surface, thus providing the additional inorganic barrier/protective coating, as shown in Figure 15.9. The AOs, TBHQ, butylated hydroxyanisole (BHA), and BHT migration were reduced by ~46, ~47, and ~46%, respectively, compared to uncoated films [53]. In food packaging, various inorganic layers are deposited via vacuum coating to enhance the aroma and oxygen barrier properties of the packaging material. Metallization of aluminum material is most frequently used with polyester-based materials as the substrate. The inorganic layer of AlOx or SiOx of ~10- to 100-nm thickness as nanocoating should potentially be evaluated according to the migration standard as the path forward to understand the efficacy of such approaches in enhancing the functionality of such packaging materials.

15.6 Effect of Polymer and Migrant Geometry on Migration

The heavy molecules as migrant greater than voids in the biopolymer matrix can be oligomers, fillers, and monomer residues. The diffusion of the migrant in the biopolymer is higher than the diffusion of the biopolymer itself. The flexible migrant of high molecular weight can enhance the transportation of a low-molecular-weight rigid migrant; however, a large molecule of migrant requires a free volume of the size of the rigid block to migrate. The polymer also affects the diffusion of migrants by relaxation of its segments, affecting the free volume and interaction of rigid blocks (nanofiller) and polymer. Considering that hole-free volume (HFV) remains unaffected by temperature-induced redistribution, the migration consequentially remains only limited to the local reorganization of a particular free volume. With the thermal expansion of the polymer, the free volume for diffusion (HFV) of the migrant is proportional to $T\text{-}T_g + K_\beta$, where T_g and T are the glass transition temperature and reference temperature, respectively, of the polymer and K_β is constant for the shape of translating block and polymer. Thus, the diffusion of the migrant mainly depends on the T_g of the polymer [54, 55].

15.7 Diffusion Modeling of Migration in Food Packaging

'Migration' refers to releasing compounds from packaging material to food, where the migrants can interact with the food [56]. The migration analyzing methods, such as swelling, surface desorption, and ion dissolution, are considered the most relevant, though apart from these, the diffusion mechanism also remains quite relevant. The diffusion of components in a polymer/biopolymer mainly follows Fick's second law as given in Equation 15.1 [57].

$$\frac{\partial C^p{}_A}{\partial t} = D^p{}_A \frac{\partial^2 C^p{}_A}{\partial x^2} \tag{15.1}$$

where $C^p{}_A$ is the migrant concentration of component A at the time (t), $D^p{}_A$ is the diffusion coefficient of A, and x is the lineal direction of mass transport.

Some reported diffusion models were proposed in the literature with limited or no migration test data. The diffusion models of Barner and Uhlman (2018) and Gavriil *et al.* (2018) are examples based on the estimation of diffusion coefficient, nature of migrants, and physical properties of polymers. The critical assumption/limitations for using these diffusion models may be (a) the distribution of the migrant in the matrix is considered isotropic and (b) aspects of migrations, like mass transfer, morphology, partitioning, shape, and size of migrants and its polarity, and plasticizer types are not considered in the models [58, 59].

An experimental correlation between the diffusion coefficient, temperature, and molecular weight (MW) of migrants was reported by Brandsch *et al.* (2002) for a wide range of polymers [60]. The diffusion coefficient of migrant (D_p) in the polymer at zero solvent concentration for the migrants of MW of approximately 150 to 1900 g/mol with the Arrhenius-like relationship given below may be applicable (Equation 15.2).

$$D_p = 10^4 \exp\left[A_p - 0.1351 M_r^{\frac{2}{3}} + 0.003 M_r - \frac{10454R}{RT} \right] \mathrm{cm}^2 \big/ \mathrm{s} \quad (15.2)$$

where $A_p = A'_p - \tau/T$, R is the gas constant (8.3145 J/mol K) at temperature T, A'_p is the matrix's particular characteristics, and M_r is the migrant's relative molecular mass.

Due to the amount of absorbed solvent, the polymer's swelling might accelerate the solvent's diffusion rate and the migrants, deviating from Fick's second law. Equation 15.3 (related to the diffusion coefficient depending on the solvent concentration DF) describes the diffusion of chemicals dispersed at the interface across the solvent and the polymers.

$$D_F = D_P \exp(\gamma C_F) \quad (15.3)$$

where C_F is food simulant concentration, and γ is an *adjustable parameter.*

Similarly, Equation 15.4, given by Crank, was used to calculate the total amount of diffusing substance (M_t) that left the film at a time 't' [60].

$$M_t = 2M_0 (D_A t / \pi)^{\frac{1}{2}} \mathrm{mg/cm}^3 \quad (15.4)$$

where M_0 is the original concentration of migrants.

The concentration of migrant at a boundary condition, where *x*=*L*, *L* is the thickness at the 'polymer film surface', which may be calculated by using Equation 15.5, as given below.

$$M_t = 4 \frac{M_o}{L} \mathrm{erf}(D_F t / \pi)^{1/2} \quad (15.5)$$

Mathematical models were proposed to obey Fick's second law of diffusion for the flat plate type of geometry of material and boundary conditions below (Equation 15.6).

$$\frac{M_t}{M_\infty} = 4\left(\frac{D_{As}t}{\pi L^2}\right)^{1/2} \tag{15.6}$$

To investigate the long-term migration, in which $M_t/M_\infty > 0.67$, where the suffixes s and l stand for short- and long-term migration, respectively, Equation 15.7 may be applicable.

$$\frac{M_t}{M_\infty} = 1 - \frac{8}{\pi^2}\exp\left(\frac{-\pi^2 D_{Al}t}{L^2}\right) \tag{15.7}$$

where M_∞ is the migration at equilibrium with diffusion exponent n, the value of n is based on the release and transport mechanisms, where $n < 0.5$ for Fickian diffusion and $0.5 < n < 1$ is for non-Fickian transport (Equation 15.8) [61].

$$\frac{M_t}{M_\infty} = 1 - \sum_{x=0}^{\infty} \frac{8}{(2n+1)^2\pi^2} e\left[\frac{(2n+1)^2}{L^2} D_{Al}t\right] \tag{15.8}$$

15.8 Conclusion and Future Perspectives

This chapter discussed an overview of the migration concern of biopolymer-based composites and nanocomposites used in food packaging. Specifically, discussions were pertaining to the starch, PLA, and PHA polymeric materials that are mainly used in food contact packaging applications. The various additives as an inclusion in the composite and nanocomposites and their profilometric and mechanistic aspect of release were discussed while correlating their functional performances. The chapter also summarized various strategies to lower migration, such as functionalization and coating. The AO in biodegradable nanocomposites leads to enhanced overall migration, which can be lowered by the addition of various nanofillers based on their selectivity and specificity. Thermodynamic properties like polarity and solubility are understood as essential factors in determining the role and extent of migration. Migrants with low solubility in food simulants will retain in the polymer matrix and vice versa, leading to their likely absorption by the matrix. Swelling of the polymer matrix also plays a vital role in the presence of food simulants. Such outcomes are attributed to the high migration rates due to increased spatial gaps between additives and molecules. The migration risks of different nanofillers, such as nanosilver, nanoclay, CNT, ZnO, and TiO_2, were discussed with their considered SML. The diffusion modeling for short-term and long-term migration may help to understand the release of migrants. The migration testing of biobased materials at the glass transition temperature and under refrigerated storage conditions could affect stability attributes to alter the migration limits.

References

1. European Commission, Commission Regulation (EU) No 10/2011 of 14 January 2011 on plastic materials and articles intended to come into contact with food. *Off. J. Eur. Union*, 12, 1–89, 2011.
2. Basumatary, I.B., Mukherjee, A., Kumar, S., Chitosan-based composite films containing eugenol nanoemulsion, ZnO nanoparticles and Aloe vera gel for active food packaging. *Int. J. Biol. Macromol.*, 242, 124826, 2023.
3. Battum, D.V. and Lierop, J.V., *Materials and Articles in Contact with Foodstuffs, Guide For Examination of Plastic Food Contact Materials*, p. 194, CEN TC, Brussels, 1997.
4. Lee, P. and Meisel, D., Adsorption and surface-enhanced Raman of dyes on silver and gold sols. *J. Phys. Chem.*, 86, 17, 3391–3395, 1982.
5. Dias, M.V. *et al.*, Thermal and morphological properties and kinetics of diffusion of antimicrobial films on food and a simulant. *Food Packag. Shelf Life*, 16, 15–22, 2018.
6. Khan, B. *et al.*, Thermoplastic starch: A possible biodegradable food packaging material—a review. *J. Food Process Eng.*, 40, 3, e12447, 2017.
7. Romainor, A.N., Chin, S.F., Lihan, S., Antimicrobial starch-based film for food packaging application. *Starch-Stärke*, 74, 3-4, 2100207, 2022.
8. Nouri, L. and Nafchi, A.M., Antibacterial, mechanical, and barrier properties of sago starch film incorporated with betel leaves extract. *Int. J. Biol. Macromol.*, 66, 254–259, 2014.
9. Zhong, K. *et al.*, Starch derivative-based superabsorbent with integration of water-retaining and controlled-release fertilizers. *Carbohydr. Polym.*, 92, 2, 1367–1376, 2013.
10. Zhu, J. *et al.*, Modelling and assessment of plasticizer migration and structure changes in hydrophobic starch-based films. *Int. J. Biol. Macromol.*, 195, 41–48, 2022.

11. Liu, Y. *et al.*, Controlling plasticizer migration based on crystal structure and micromorphology in propionylated starch-based food packaging nanocomposites. *Carbohydr. Polym.*, 273, 118621, 2021.
12. Zhu, J. *et al.*, Plasticization effect of triacetin on structure and properties of starch ester film. *Carbohydr. Polym.*, 94, 2, 874–881, 2013.
13. Ji, S. *et al.*, Application of a bio-based polyester plasticizer modified by hydro-silicon-hydrogenation reaction in soft PVC films. *Polym. Adv. Technol.*, 30, 4, 1126–1134, 2019.
14. Ahmed, J. and Varshney, S.K., Polylactides—chemistry, properties and green packaging technology: A review. *Int. J. Food Prop.*, 14, 1, 37–58, 2011.
15. Sirait, S.M., Widyahapsari, D.A.N., Irawan, C., Migration test of polylactic acid (PLA) packaging that modified with BHT (Butyl Hydroxy Toluene) and TBHQ (Tert Butyl Hydroxy Quinon) synthetic antioxidant in food simulant. *Orient. J. Chem.*, 35, 2, 552, 2019.
16. Di Maio, L. *et al.*, Bionanocomposite polylactic acid/organoclay films: Functional properties and measurement of total and lactic acid specific migration. *Packag. Technol. Sci.*, 27, 7, 535–547, 2014.
17. Fortunati, E. *et al.*, Effects of modified cellulose nanocrystals on the barrier and migration properties of PLA nano-biocomposites. *Carbohydr. Polym.*, 90, 2, 948–956, 2012.
18. Li, W. *et al.*, Effects of PLA film incorporated with ZnO nanoparticle on the quality attributes of fresh-cut apple. *Nanomaterials*, 7, 8, 207, 2017.
19. Heydari-Majd, M. *et al.*, A new active nanocomposite film based on PLA/ZnO nanoparticle/essential oils for the preservation of refrigerated Otolithes ruber fillets. *Food Packag. Shelf Life*, 19, 94–103, 2019.
20. Vasile, C. *et al.*, New PLA/ZnO: Cu/Ag bionanocomposites for food packaging. *Express Polym. Lett.*, 11, 7, 531–544, 2017.
21. Li, W. *et al.*, Development of antimicrobial packaging film made from poly (lactic acid) incorporating titanium dioxide and silver nanoparticles. *Molecules*, 22, 7, 1170, 2017.
22. Dhar, P. *et al.*, Poly (3-hydroxybutyrate)/cellulose nanocrystal films for food packaging applications: Barrier and migration studies. *Polym. Eng. Sci.*, 55, 10, 2388–2395, 2015.
23. Yu, H., Yan, C., Yao, J., Fully biodegradable food packaging materials based on functionalized cellulose nanocrystals/poly (3-hydroxybutyrate-co3-hydroxyvalerate) nanocomposites. *RSC Adv.*, 4, 104, 59792–59802, 2014.
24. Díez-Pascual, A.M. and Diez-Vicente, A.L., ZnO-reinforced poly (3-hydroxybutyrate-co-3-hydroxyvalerate) bionanocomposites with antimicrobial function for food packaging. *ACS Appl. Mater. Interfaces*, 6, 12, 9822–9834, 2014.
25. Yu, H.-Y. *et al.*, Reinforcement of transparent poly (3-hydroxybutyrate-co-3-hydroxyvalerate) by incorporation of functionalized carbon nanotubes as a novel bionanocomposite for food packaging. *Compos. Sci. Technol.*, 94, 96–104, 2014.

26. Sanchez-Garcia, M., Lagaron, J., Hoa, S., Effect of addition of carbon nanofibers and carbon nanotubes on properties of thermoplastic biopolymers. *Compos. Sci. Technol.*, 70, 7, 1095–1105, 2010.
27. Díez-Pascual, A.M. and Díez-Vicente, A.L., Poly (3-hydroxybutyrate)/ZnO bionanocomposites with improved mechanical, barrier and antibacterial properties. *Int. J. Mol. Sci.*, 15, 6, 10950–10973, 2014.
28. Salomez, M. *et al.*, A comparative study of degradation mechanisms of PHBV and PBSA under laboratory-scale composting conditions. *Polym. Degrad. Stab.*, 167, 102–113, 2019.
29. Mutsuga, M., Kawamura, Y., Tanamoto, K., Migration of lactic acid, lactide and oligomers from polylactide food-contact materials. *Food Addit. Contam.*, 25, 10, 1283–1290, 2008.
30. Ubeda, S. *et al.*, Migration of oligomers from a food contact biopolymer based on polylactic acid (PLA) and polyester. *Anal. Bioanal. Chem.*, 411, 16, 3521–3532, 2019.
31. Bradley, E., *Biobased materials used in food contact applications: An assessment of the migration potential-FSA PROJECT A03070*, The Food and Environment Research, London, 2010.
32. Aznar, M. *et al.*, Determination of non-volatile components of a biodegradable food packaging material based on polyester and polylactic acid (PLA) and its migration to food simulants. *J. Chromatogr. A*, 1583, 1–8, 2019.
33. Markarian, J., Automotive and packaging offer growth opportunities for nanocomposites. *Plast. Addit. Compd.*, 7, 6, 18–21, 2005.
34. Kim, M.H. *et al.*, Kinetic and thermodynamic studies of silver migration from nanocomposites. *J. Food Eng.*, 243, 1–8, 2019.
35. Ortega, F. *et al.*, Starch nanocomposite films: Migration studies of nanoparticles to food simulants and bio-disintegration in soil. *Polymers*, 14, 9, 1636, 2022.
36. Pereira, M. *et al.*, *Biodegradable nanocomposite for food packaging application*, 2014.
37. Echegoyen, Y., Rodríguez, S., Nerín, C., Nanoclay migration from food packaging materials. *Food Addit. Contam. Part A*, 33, 3, 530–539, 2016.
38. Farhoodi, M. *et al.*, Migration of aluminum and silicon from PET/clay nanocomposite bottles into acidic food simulant. *Packag. Technol. Sci.*, 27, 2, 161–168, 2014.
39. Franz, R., Bott, J., Störmer, A., Considerations for and guidance to testing and evaluating migration/release of nanoparticles from polymer based nanocomposites. *Nanomaterials*, 10, 6, 1113, 2020.
40. Salarbashi, D. *et al.*, Characterization of a green nanocomposite prepared from soluble soy bean polysaccharide/Cloisite 30B and evaluation of its toxicity. *Int. J. Biol. Macromol.*, 120, 109–118, 2018.
41. Wang, C. *et al.*, Fullerenol@ nano-montmorillonite nanocomposite as an efficient radioprotective agent for ameliorating radioactive duodenal injury. *Chem. Eng. J.*, 427, 131725, 2022.

42. Souza, V.G.L. *et al.*, *In vitro* bioactivity of novel chitosan bionanocomposites incorporated with different essential oils. *Ind. Crops Prod.*, 140, 111563, 2019.
43. Tullio, S. and Chalcraft, D., Converting natural nanoclay into modified nanoclay augments the toxic effect of natural nanoclay on aquatic invertebrates. *Ecotoxicol. Environ. Saf.*, 197, 110602, 2020.
44. Maisanaba, S. *et al.*, Effects of the subchronic exposure to an organomodified clay mineral for food packaging applications on wistar rats. *Appl. Clay Sci.*, 95, 37–40, 2014.
45. Kumar, S. *et al.*, Nanomaterials in food packaging, in: *Biopolymer-Based Food Packaging*, pp. 336–367, 2022.
46. Mulla, M.Z. *et al.*, Poly lactic acid (Pla) nanocomposites: Effect of inorganic nanoparticles reinforcement on its performance and food packaging applications. *Molecules*, 26, 7, 1967, 2021.
47. Sewalt, V. *et al.*, The generally recognized as safe (GRAS) process for industrial microbial enzymes. *Ind. Biotechnol.*, 12, 5, 295–302, 2016.
48. Espitia, P.J.P. *et al.*, Zinc oxide nanoparticles: synthesis, antimicrobial activity and food packaging applications. *Food Bioprocess Technol.*, 5, 5, 1447–1464, 2012.
49. Nohynek, G.J. *et al.*, Safety assessment of personal care products/cosmetics and their ingredients. *Toxicol. Appl. Pharmacol.*, 243, 2, 239–259, 2010.
50. Zhaohui, W. *et al.*, Understanding of Commission Regulation (EU) No 10/2011 on plastic materials and articles intended to come into contact with food. *China Plastics*, 25, 07, 83, 2011.
51. Majumder, R. *et al.*, Biopolymers, in: *Biopolymer-Based Food Packaging*, pp. 29–65, 2022.
52. Basumatary, I.B. *et al.*, Edible films and coatings, in: *Biopolymer-Based Food Packaging*, pp. 445–475, 2022.
53. Huang, C. *et al.*, Antioxidant migration resistance of SiOx layer in SiOx/PLA coated film. *Food Addit. Contam. Part A*, 35, 2, 366–376, 2018.
54. Fang, X. and Vitrac, O., Predicting diffusion coefficients of chemicals in and through packaging materials. *Crit. Rev. Food Sci. Nutr.*, 57, 2, 275–312, 2017.
55. Welle, F., A new method for the prediction of diffusion coefficients in poly (ethylene terephthalate). *J. Appl. Polym. Sci.*, 129, 4, 1845–1851, 2013.
56. Picó, Y., Chapter 9-Safety assessment and migration tests, in: *Nanomaterials for Food Packaging*, M.Â.P.R. Cerqueira (Eds.), pp. 249–275, Elsevier, Netherlands, 2018.
57. Crank, J., *The Mathematics of Diffusion*, 2nd Edition, p. 32, Oxford Science Publications, London, 1975.
58. Barner, W. and Uhlman, K., Contaminant transport mechanisms in karst terrains and implications on remediation, in: *Karst Geohazards*, pp. 207–212, Routledge, 2018.

59. Gavriil, G., Kanavouras, A., Coutelieris, F.A., Food-packaging migration models: A critical discussion. *Crit. Rev. Food Sci. Nutr.*, 58, 13, 2262–2272, 2018.
60. Brandsch, J. *et al.*, Migration modelling as a tool for quality assurance of food packaging. *Food Addit. Contam.*, 19, S1, 29–41, 2002.
61. Thakur, V. *et al.*, Biodegradable polymers for food packaging applications, in: *Biodegradable Polymers and Their Emerging Applications*, S. Saha and C. Sarkar (Eds.), pp. 169–189, Springer Nature Singapore, Singapore, 2023.

Index

β-1,4 glycosidic bond, 351
β-1,4 glycosidic linkage, 72

3D printing, 296, 308

Acetylation, 249
Acid chloride, 249
Active coatings,
  fruits and vegetables, 337–338
  meat, poultry, and aquatic animal products, 338
Active films, 192–195
  casting, 335–336
  extrusion, 336–337
  spraying, 336
Active packaging, 373
Agricultural, 244
Agro waste, 156, 209, 244
Algae, 243
Alkaline treatment, 125
Alkalization, 365
Amylopectin, 250
Anhydro-D-glucopyranose, 72
Animal fiber, 36, 46, 48–50
Animals, 243
Antibacterial, 349, 350, 359, 369, 371
Anticarcinogenic, 14
Anti-inflammatory, 14
Antimicrobial, 3, 5–6, 14–15, 18–19, 23, 30–31, 221–223
Antimicrobial agents, 33, 35
Antioxidant, 216–219
Apple peel, 216
Arabinoxylan, 84

Bacterial cellulose, 350, 351, 358, 359, 364, 372
Ball milling, 369–371
Bamboo fiber, 44, 45, 56
Banana fiber, 43
Banana peel, 214, 223
Biobased feedstock, 33
Biobased plastics, 296
Biocompatibility, 3
Biocompatible, 347, 374–375
Biodegradability, 297, 308, 389, 390, 392, 393, 406
Biodegradable, 91–92, 98–99, 101, 103, 106–109, 111–114, 155, 156, 164, 166, 169, 171–175, 258, 261, 266–269, 281, 285, 289, 293, 347, 348, 359, 365, 367, 369, 371, 374, 375
Biodeterioration, 396, 399, 400
Biological pretreatment, 129–130
Biomass, 296, 297, 347, 348, 350–352, 356–360, 362, 364
Bionanocomposite, 83, 84, 86
Bioplastic, 364
Biopolymer matrix, 130
Biopolymers, 1–3, 6–9, 21, 155, 156, 167, 267, 268, 270, 272, 274–276, 278, 280–282, 284–286, 288, 290, 292, 294, 295, 298, 299, 314, 389, 392–394, 402, 404–407, 422–423, 427, 430, 432, 437

Bleaching, 365, 368, 370–371
Blends, 254, 255, 266, 268
Bombyx mori, 187, 200
Butylated hydroxy toulene (BHT), 427–428, 442

Carbon nanotube (CNT), 432, 441, 446
Carboxymethyl, 351, 364, 365, 373
Carboxymethyl cellulose, 165, 166
Cassava, 243
Casting, 349–350, 365, 368, 374
Cellulose, 71–76, 117–122, 125–127, 153, 155–174, 295, 296, 299, 302, 303, 308, 309, 392, 398, 402, 403
Cellulose acetate, 165, 168, 351, 367, 373
Cellulose esters, 155, 167, 168
Cellulose ethers, 155, 164
Cellulose nanocrystals, 155, 156, 160
Cellulose nanofiber, 106, 348, 349, 351, 369, 371
Cellulose nitrate, 168
Cellulose sulfate, 169
Chemical composition, 117
Chemical indicator, 229
Chitin, 267, 269, 270, 272–275, 278, 286–288, 290
Chitosan, 76, 81–83, 267–285, 295, 299, 302–303
Citrus fruit, 210, 217, 222
Clay, 422, 425–427
Coating, 267–269, 274, 277–286, 289, 290, 292–294
Cocos nucifera, 91
Compatibilizers, 81, 83
Composite, 359, 369, 372, 374
Composite films, 278, 281, 283, 286, 289, 290, 292, 293
Composite packaging, 295, 308, 314
Corncob, 69–80, 215
Corona, 92
Crosslinking, 252
Crystallization, 302, 306–308

Degradation, 389–407
Degree of esterification, 324–325, 327, 334
Degree of substitution, 357, 364, 366
Degumming, 190–191
Depolymerization, 397, 399, 405
D-glucose, 352–356, 360, 364
Diffusion, 422, 424
Diffusion modelling, 443–445

Ecosystem, 3
Edible coating, 196–197
Edible, 247
Enzyme-assisted extraction, 333–334
Eri cocoon, 190
Essential oil, 106–107, 114
Etherification, 153, 161, 162, 365–367
Ethyl cellulose, 164
Extraction of chitosan, 270, 272, 274, 275, 287
Extrusion method, 349, 362, 374

Fibroin, 187–190, 192–200
Fillers, 130–133, 295, 305, 307
Films, 241, 243, 244, 248, 249, 251–254, 256, 257, 259, 260, 264–266, 269, 427–429, 431–433, 437–442
Fish and meat products, 281–283
Flexural modulus, 94, 99, 104–105
Flexural strength, 94, 98, 100, 104–105, 117–118
Food, 245
Food allergy, 197
Food containers, 135
Food packaging, 33–36, 39–44, 46–48, 50–52, 54, 58, 59, 295, 299, 302–303, 308–314, 389–391, 396, 405, 407
Fragmentation, 401–403
Free volume, 443

Fruits and vegetables, 283–285
Functional agents, 33, 208
Functional properties, 267, 269, 270, 275, 276, 286
Functionalization, 161–164

Glass transition temperature (T_g), 425, 432, 446
Glucose, 242
Grape pomace, 213, 218, 222
Green alternative, 33
Grinding, 358, 360, 369, 371

HDPE, 93–95, 103
Hemicellulose, 6, 8–11, 13, 73–78, 115–121, 125–126
Hemp, 92
Hemp fiber, 41, 43
High-methoxyl pectin, 324–325
Hole free volume (HFV), 443
Homogalacturonan, 322–324
Homogenization, 358, 361, 369–371
Horticultural produce, 196
Husk, 91–92, 103, 109, 112
Hybrid, 95–96, 98–100, 102–105, 108–113
Hydrodynamic cavitation, 331–332
Hydrophillic, 350, 355, 364, 368, 375
Hydrophobicity, 303
Hydropropylation, 251
Hydrothermal extraction, 328–329
Hydroxylethyl cellulose, 167
Hydroxypropyl cellulose, 165, 167, 168

Impact strength, 96–99, 103–105, 110
Industrialization, 2, 4, 92

Jackfruit, 246
Jute fiber, 41, 43

Kenaf, 92
Kenaf fibers, 40, 43

Lactic acid, 254
Lactide, 299–302, 305, 306
Leaf fibers, 35, 39
Legume, 245
Life cycle, 390
Lignocellulose, 70
Lignocellulosic biomass, 156, 157
Lipid, 245
Litchi, 246

Maize, 243
Mangifera, 245
Mango kernel, 213, 218
Man-made fibers, 39
Marine source, 273, 274, 285, 288
Medicine, 245
Methyl cellulose, 164, 165
Microstructure, 92–93
Microwave pretreatment, 123
Microwave-assisted extraction, 332
Migration
  concerns, 423, 425, 431
  geometry, 442, 445
  migrants, 423–424, 436, 443–444
  test, 423–424, 433, 438, 443, 446
Moisture or humidity indicator, 228
Mulberry silk, 189

Nanocellulose, 72–79, 427, 431
Nanocoating, 422, 441
Nanocomposite, 35, 36, 39, 44, 47, 52, 56, 254, 256, 257, 267, 268, 421–424, 426–432, 437–439, 441, 445
Nanocrystals, 295, 302
Nanofiller, 161, 168
Nanomaterial, 36, 51, 53
Nanoparticles (NPs), 164, 166, 168, 423–424, 426, 437, 440–441
Native, 247

Octanoyl, 249, 250
Oligomers, 426, 432–436, 442
Organosolv, 127–128

Oxidation, 252
Oxygen and carbon dioxide indicators, 225

Packaging, 241, 243, 246–248, 250, 251, 254–262, 264, 266–270
Packaging film, 268, 277, 279, 281, 283, 285
Packaging material, 33, 34, 36, 39–42, 44, 45, 47, 48, 50–52, 59
Palm sheath, 133
Perishable foods, 267, 269, 281
Permeability, 348, 368–359
Petrochemical, 1, 3
pH indicators, 228
Phosphorylation, 369, 370
Pineapple leaf fiber, 43
Pineapple peel, 212, 218
Plant fiber, 35, 37–40
Plants, 243
Plastic, 347–348, 367, 369, 374
Plasticizer, 33, 35, 42, 422, 425–427, 436, 444
Pollution, 2–4, 11, 21
Poly(lactic acid) (PLA), 84, 254, 259, 267, 269, 295–306, 308–314, 421–423, 426–430, 432–436, 438, 441–442, 445
Polyethylene (PE), 295, 296, 298, 309, 311
Polyhydroxy butyrate, 79
Polyhydroxyalkanoates (PHAs), 421–423, 430–432
Polyhydroxybutyrate (PHB), 395
Polyhydroxybutyrate-co-valerate (PHBV), 396
Polylactide (PLA), 394
Polymerization, 296, 299, 301, 305
Polymorph, 352–353
Polyphenols, 208–215
Polypropylene, 84
Polysaccharide, 242
Pomegranate peel, 214, 219, 221
Potato, 243
Potato peel, 212, 217
Prerequisite, 3
Pretreatment, 122, 124–127, 129, 347, 351, 360, 362–364, 369, 371, 375
Production of corn, 71
Protein, 245, 393–395, 399–402
Purification, 3

Ramie fiber, 41
Regulation, 427, 438–441
Rice husk, 45
Rice straw, 155, 159, 160, 165, 170, 172

Seeds, 242
Sericin, 187–188, 190
Sericulture wastes, 189
Shelf life, 268–269, 276, 277, 281–283, 348, 350, 359, 372–374
Silanization, 163
Silicon oxide (SiOx), 441–442
Silk protein, 190, 192, 196–197
Silk-based composite, 46
Silkworm, 187–190
Silver, 422, 425, 428–429, 437–438, 441, 446
Specific migration limit (SML), 439–441, 446
Stalk fiber, 35, 37
Starch, 241–269, 242, 392–394, 396, 402, 422–425, 427, 437, 445
Stereocomplexation, 304–306, 308
Subcritical water extraction, 333
Succinylation, 251, 265
Sugarcane bagasse, 115–118, 130
Sustainable, 153, 155, 164, 174
Synthetic polymer, 302

Tapioca starch, 131, 133–134
Temperature indicator, 229
TEMPO-oxidation, 349, 351, 358, 359, 370, 372

Tensile strength, 349–350, 353, 358, 365, 367–372
Tert-butyl quinone (TBHQ), 427–428, 442
Thermoplastic, 243, 244, 253, 254, 261
Thermosetting, 91, 93, 96
Titanium dioxide, 422, 429–430, 437, 439–440, 446
Toxicity, 197–200
Transmittance, 349, 370
Tuber, 244

Ultrasound pretreatment, 124
Ultrasound-assisted extraction, 329–331
Urbanization, 2

Valorization, 3, 5, 21–23, 25
Value-added, 1–2, 4–5, 9, 21, 23–24, 26
Van der Waals force, 72
Vanya silk, 189

Wheat, 243
Wheat straw, 215
White-rot fungi, 130
Wood fiber, 35, 37–40
Wool fiber, 38, 39, 46, 47

Xylan, 77
Xylo-oligosaccharides, 71

Zea mays, 70
Zinc oxide (ZnO), 425, 440
Zinc phenylphosphonate (PPZn), 308

Printed and bound by CPI Group (UK) Ltd, Croydon, CR0 4YY

17/09/2026

14964277-0006